S0-AUG-120

Synthesis and Properties of Advanced Catalytic Materials

MATERIALS RESEARCH SOCIETY SYMPOSIUM PROCEEDINGS VOLUME 368

Synthesis and Properties of Advanced Catalytic Materials

Symposium held November 29–December 1, 1994, Boston, Massachusetts, U.S.A.

EDITORS:

Enrique Iglesia

University of California-Berkeley
Berkeley, California, U.S.A.

Peter W. Lednor

Shell Research B.V.
Amsterdam, The Netherlands

Dick A. Nagaki

Union Carbide Corporation
South Charleston, West Virginia, U.S.A.

Levi T. Thompson

University of Michigan
Ann Arbor, Michigan, U.S.A.

MATERIALS RESEARCH SOCIETY
Pittsburgh, Pennsylvania

Single article reprints from this publication are available through
University Microfilms Inc., 300 North Zeeb Road, Ann Arbor, Michigan 48106

CODEN: MRSPDH

Published by:

Materials Research Society
9800 McKnight Road
Pittsburgh, Pennsylvania 15237
Telephone (412) 367-3003
Fax (412) 367-4373

Library of Congress Cataloging in Publication Data

Synthesis and properties of advanced catalytic materials : symposium held November 29-December 1, 1994, Boston, Massachusetts, U.S.A. / editors, Enrique Iglesia, Peter W. Lednor, Dick A. Nagaki, Levi T. Thompson

p. cm.—(Materials Research Society symposium proceedings, ISSN 0272-9172 ; v. 368)

Includes bibliographical references and index.

ISBN 1-55899-270-7 (hardcover : alk. paper)

1. Catalysts—Congresses. I. Iglesia, Enrique II. Lednor, Peter W. III. Nagaki, Dick A. IV. Thompson, Levi T. V. Series: Materials Research Society symposium proceedings ; v. 368.

TP159.C3S96 1995 95-22117
660'.2995—dc20 CIP

Manufactured in the United States of America

Contents

*Invited Paper

PART III: SUPPORTED METAL CATALYSTS

*Invited Paper

PART IV: MOLECULAR AND CLUSTER DERIVED CATALYTIC MATERIALS

PART V: METAL OXIDES

*Invited Paper

PART VI: MONOLITHS AND FOAMS

PART VII: POROUS MATERIALS

*Invited Paper

*Invited Paper

Preface

This volume is dedicated to the papers presented at the symposium on Synthesis and Characterization of Novel Catalytic Materials at the 1994 MRS Fall Meeting, held in Boston, Massachusetts November 28–30, 1994. This symposium is a follow-up to a similar, very successful symposium held at the 1990 MRS Fall Meeting organized by E.W. Corcoran, Jr. and M.J. Ledoux. Progress in materials science and engineering has had a significant impact on the development of new catalysts, in particular through synthesis, characterization and shaping. Catalysts are of major importance in the petroleum and chemical industries, and are used increasingly to solve environmental problems.

As with the previous symposium, the objective of this symposium was to provide a forum for scientists and engineers from diverse backgrounds to discuss new materials, novel preparations, and new characterization methods for catalysis research. The appropriateness of this interdisciplinary approach to research is validated by the excellent attendance at the meeting. There were 77 presentations from across the world and 55 papers are included in this volume. The symposium covered a wide variety of topics that included high surface area metal nitrides and metal carbides, carbon, supported metal catalysts, molecular and cluster derived catalytic materials, metal oxides, monoliths and foams, and porous materials. The organizers tried to keep a distinct focus of the subject matter in each session and yet achieve a certain continuity between each session. We have also tried to maintain this clear focus and continuity in the publication of this volume.

Enrique Iglesia
Peter W. Lednor
Dick A. Nagaki
Levi T. Thompson

June 1995

Acknowledgments

We wish to thank the following sources of financial support for this symposium: Air Products & Chemicals, Catalytica, DSM Research, DYCAT International, Exxon Research and Engineering Company, Haldor Topsoe A/S, General Motors Corporation, MEL Chemicals, Mobil Research and Development, Shell Development Company, Union Carbide Corporation, and W.R. Grace & Company. Their support is especially appreciated in these times of tight budgets and growing expenses.

We wish to express our appreciation to all of the symposium contributors for their excellent presentations, to the session chairs for generating a stimulating atmosphere and discussion, yet keeping the meeting on schedule. And we wish to thank the other participants of the symposium for their lively contributions and enthusiastic interest.

We wish to acknowledge the help that we received from the many attendees who reviewed manuscripts. Also the help of the various coworkers of Dick A. Nagaki at Union Carbide in reviewing several manuscripts is gratefully acknowledged.

We are grateful to the staff of the MRS for their support and help in the organization of the symposium. As the MRS meeting continues to grow in size and scope, the MRS staff continue to put on an excellent meeting.

Finally, we would like to thank our own organizations for allowing us to devote our time to this symposium.

MATERIALS RESEARCH SOCIETY SYMPOSIUM PROCEEDINGS

Volume 336— Amorphous Silicon Technology-1994, E.A. Schiff, A. Matsuda, M. Hack, M.J. Powell, A. Madan, 1994, ISBN: 1-55899-236-7

Volume 337— Advanced Metallization for Devices and Circuits—Science, Technology, and Manufacturability III, S.P. Murarka, K.N. Tu, A. Katz, K. Maex, 1994, ISBN: 1-55899-237-5

Volume 338— Materials Reliability in Microelectronics IV, P. Børgesen, W. Filter, J.E. Sanchez, Jr., K.P. Rodbell, J.C. Coburn, 1994, ISBN: 1-55899-238-3

Volume 339— Diamond, SiC and Nitride-Wide-Bandgap Semiconductors, C.H. Carter, Jr., G. Gildenblat, S. Nakamura, R.J. Nemanich, 1994, ISBN: 1-55899-239-1

Volume 340— Compound Semiconductor Epitaxy, C.W. Tu, L.A. Kolodziejski, V.R. McCrary, 1994, ISBN: 1-55899-240-5

Volume 341— Epitaxial Oxide Thin Films and Heterostructures, D.K. Fork, J.M. Phillips, R. Ramesh, R.M. Wolf, 1994, ISBN: 1-55899-241-3

Volume 342— Rapid Thermal and Integrated Processing III, J.J. Wortman, J.C. Gelpey, M.L. Green, S.R.J. Brueck, F. Roozeboom, 1994, ISBN: 1-55899-242-1

Volume 343— Polycrystalline Thin Films—Structure, Texture, Properties and Applications, M. Parker, K. Barmak, R. Sinclair, D.A. Smith, J. Floro, 1994, ISBN: 1-55899-243-X

Volume 344— Materials and Processes for Environmental Protection, C. Adkins, P.N. Gadgil, L.M. Quick, K.E. Voss, 1994, ISBN: 1-55899-244-8

Volume 345— Flat Panel Display Materials, J. Batey, A. Chiang, P. Holloway, 1994, ISBN: 1-55899-245-6

Volume 346— Better Ceramics Through Chemistry VI, C. Sanchez, M.L. Mecartney, C.J. Brinker, A. Cheetham, 1994, ISBN: 1-55899-246-4

Volume 347— Microwave Processing of Materials IV, M.F. Iskander, R.J. Lauf, W.H. Sutton, 1994, ISBN: 1-55899-247-2

Volume 348— Scintillator and Phosphor Materials, M.J. Weber, P. Lecoq, R.C. Ruchti, C. Woody, W.M. Yen, R.-Y. Zhu, 1994, ISBN: 1-55899-248-0

Volume 349— Novel Forms of Carbon II, C.L. Renschler, D. Cox, J. Pouch, Y. Achiba, 1994, ISBN: 1-55899-249-9

Volume 350— Intermetallic Matrix Composites III, J.A. Graves, R.R. Bowman, J.J. Lewandowski, 1994, ISBN: 1-55899-250-2

Volume 351— Molecularly Designed Ultrafine/Nanostructured Materials, K.E. Gonsalves, G.-M. Chow, T.D. Xiao, R.C. Cammarata, 1994, ISBN: 1-55899-251-0

Volume 352— Materials Issues in Art and Archaeology IV, P.B. Vandiver, J.R. Druzik, J.L. Galvan Madrid, I.C. Freestone, G.S. Wheeler, 1995, ISBN: 1-55899-252-9

Volume 353— Scientific Basis for Nuclear Waste Management XVIII, T. Murakami, R.C. Ewing, 1995, ISBN: 1-55899-253-7

Volume 354— Beam-Solid Interactions for Materials Synthesis and Characterization, D.E. Luzzi, T.F. Heinz, M. Iwaki, D.C. Jacobson, 1995, ISBN: 1-55899-255-3

Volume 355— Evolution of Thin-Film and Surface Structure and Morphology, B.G. Demczyk, E.D. Williams, E. Garfunkel, B.M. Clemens, J.E. Cuomo, 1995, ISBN: 1-55899-256-1

Volume 356— Thin Films: Stresses and Mechanical Properties V, S.P. Baker, P. Børgesen, P.H. Townsend, C.A. Ross, C.A. Volkert, 1995, ISBN: 1-55899-257-X

Volume 357— Structure and Properties of Interfaces in Ceramics, D.A. Bonnell, U. Chowdhry, M. Rühle, 1995, ISBN: 1-55899-258-8

MATERIALS RESEARCH SOCIETY SYMPOSIUM PROCEEDINGS

Volume 358—Microcrystalline and Nanocrystalline Semiconductors, R.W. Collins, C.C. Tsai, M. Hirose, F. Koch, L. Brus, 1995, ISBN: 1-55899-259-6

Volume 359—Science and Technology of Fullerene Materials, P. Bernier, D.S. Bethune, L.Y. Chiang, T.W. Ebbesen, R.M. Metzger, J.W. Mintmire, 1995, ISBN: 1-55899-260-X

Volume 360—Materials for Smart Systems, E.P. George, S. Takahashi, S. Trolier-McKinstry, K. Uchino, M. Wun-Fogle, 1995, ISBN: 1-55899-261-8

Volume 361—Ferroelectric Thin Films IV, S.B. Desu, B.A. Tuttle, R. Ramesh, T. Shiosaki, 1995, ISBN: 1-55899-262-6

Volume 362—Grain-Size and Mechanical Properties—Fundamentals and Applications, N.J. Grant, R.W. Armstrong, M.A. Otooni, T.N. Baker, K. Ishizaki, 1995, ISBN: 1-55899-263-4

Volume 363—Chemical Vapor Deposition of Refractory Metals and Ceramics III, W.Y. Lee, B.M. Gallois, M.A. Pickering, 1995, ISBN: 1-55899-264-2

Volume 364—High-Temperature Ordered Intermetallic Alloys VI, J. Horton, I. Baker, S. Hanada, R.D. Noebe, D. Schwartz, 1995, ISBN: 1-55899-265-0

Volume 365—Ceramic Matrix Composites—Advanced High-Temperature Structural Materials, R.A. Lowden, J.R. Hellmann, M.K. Ferber, S.G. DiPietro, K.K. Chawla, 1995, ISBN: 1-55899-266-9

Volume 366—Dynamics in Small Confining Systems II, J.M. Drake, S.M. Troian, J. Klafter, R. Kopelman, 1995, ISBN: 1-55899-267-7

Volume 367—Fractal Aspects of Materials, F. Family, B. Sapoval, P. Meakin, R. Wool, 1995, ISBN: 1-55899-268-5

Volume 368—Synthesis and Properties of Advanced Catalytic Materials, E. Iglesia, P. Lednor, D. Nagaki, L. Thompson, 1995, ISBN: 1-55899-270-7

Volume 369—Solid State Ionics IV, G-A. Nazri, J-M. Tarascon, M. Schreiber, 1995, ISBN: 1-55899-271-5

Volume 370—Microstructure of Cement Based Systems/Bonding and Interfaces in Cementitious Materials, S. Diamond, S. Mindess, F.P. Glasser, L.W. Roberts, J.P. Skalny, L.D. Wakeley, 1995, ISBN: 1-55899-272-3

Volume 371—Advances in Porous Materials, S. Komarneni, D.M. Smith, J.S. Beck, 1995, ISBN: 1-55899-273-1

Volume 372—Hollow and Solid Spheres and Microspheres—Science and Technology Associated with their Fabrication and Application, M. Berg, T. Bernat, D.L. Wilcox, Sr., J.K. Cochran, Jr., D. Kellerman, 1995, ISBN: 1-55899-274-X

Volume 373—Microstructure of Irradiated Materials, I.M. Robertson, L.E. Rehn, S.J. Zinkle, W.J. Phythian, 1995, ISBN: 1-55899-275-8

Volume 374—Materials for Optical Limiting, R. Crane, K. Lewis, E.V. Stryland, M. Khoshnevisan, 1995, ISBN: 1-55899-276-6

Volume 375—Applications of Synchrotron Radiation Techniques to Materials Science II, L.J. Terminello, N.D. Shinn, G.E. Ice, K.L. D'Amico, D.L. Perry, 1995, ISBN: 1-55899-277-4

Volume 376—Neutron Scattering in Materials Science II, D.A. Neumann, T.P. Russell, B.J. Wuensch, 1995, ISBN: 1-55899-278-2

Prior Materials Research Society Symposium Proceedings available by contacting Materials Research Society

PART I

High Surface Area Metal Nitrides

PREPARATION, CHARACTERIZATION AND CATALYTIC ACTIVITY OF NIOBIUM OXYNITRIDE AND OXYCARBIDE IN HYDROTREATMENT

H.S. KIM*, C. SAYAG*, G. BUGLI*, G. DJEGA-MARIADASSOU* AND M. BOUDART**
* Université P&M Curie, Laboratoire Réactivité de Surface, CNRS URA 1106, 4, Place Jussieu, Case 178, 75252 Paris Cedex 05, France
** Stanford University, Laboratory of Chemical Engineering, Stanford, CA 94305

ABSTRACT

Oxynitride and oxycarbide of niobium present metallic and acidic catalytic functions as evidenced by molecular probe reactions : isomerization of cyclohexane and hydrodenitrogenation of 1-4 tetrahydroquinoline. The proximity of the two functions leads to the concept of "dual site" by selective inhibition of the metallic and acid sites. Substitution of nitrogen of the oxynitride by carbon to form the oxycarbide produces a large enhancement of the metallic character of the surface. The oxynitride was shown to present a more acidic and a less hydrogenating activity as compared to the activity of the oxycarbide. Both compounds of niobium have been compared to molybdenum oxynitride.

INTRODUCTION

Insertion of nitrogen or carbon atoms into the lattice of non-noble transition metals such as Mo and W confers to them some of the catalytic properties of noble metals [1]. The addition of carbon or nitrogen lattice makes the cubic lattice parameter a_o of the non-noble metal to increase (Table I) and subsequently the width of its d band (proportional to $1/a_o^5$) to decrease. Consequently, the density of d electronic states rises at the Fermi level. So properties of nitrides or carbides of nonnoble transition metals (Groups 4 to 6) shift towards those of metals on the right hand side of the Periodic Table (Groups 8 to 10).

TABLE I. Lattice parameter of niobium compounds

Compounds	Crystal Structure	a_o (pm)
Nb	bcc	330.6
NbN	fcc	439.2
NbC	fcc	447.0

bcc : body-centred cubic lattice. fcc : faced-centred cubic lattice. a_o : parameter of the cubic lattice

Furthermore, oxygen can be incorporated into the lattice of the nitride or carbide, either during the passivation step following their preparation and before exposure to air, or during exposure to controlled pulses of pure oxygen [2-5]. As a consequence, acidic sites appear at the surface of these materials. Bifunctional behavior of "*oxynitrides* " or "*oxycarbides* " can then be expected, the balance between metallic and acidic properties depending upon the nature of the transition metal.

The present paper deals with niobium oxynitride (NbN_xO_y) or oxycarbide (NbC_xO_y).

The metallic behavior of these materials will be evidenced through specific catalytic reactions such as (de)hydrogenation and C-C bond hydrogenolysis. Their catalytic activity for

Mat. Res. Soc. Symp. Proc. Vol. 368

isomerization and cracking reactions will show their acid character. Bifunctional behavior will be evidenced in both isomerization and hydrodenitrogenation (HDN) reactions. From this dual catalytic behavior, a concept of *dual active site* as already previously assumed [2,5] will be considered, simultaneously with a possible non-stoichiometry of NbN_xO_y and NbC_xO_y.

EXPERIMENTAL

Preparation of oxynitrides and oxycarbides

Niobium oxide Nb_2O_5 (Prolabo, Aldrich or Janssen) was converted to niobium oxynitride by exposing it to pure ammonia, increasing the temperature from room temperature (RT) to 770 K at 150 K h^{-1} and remaining at 770 K for 12 h.

Niobium oxycarbide was prepared by direct carburization of Nb_2O_5 in presence of beads of a commercial $NiMo/Al_2O_3$ (Procatalyse) using an 80%CH_4 / 20%H_2 mixture as the sample was heated from 603 to 1103 K at 30 K h^{-1}. X-Ray diffraction patterns showed that beads of $NiMo/Al_2O_3$ are carburized at a lower temperature than Nb_2O_5 and catalyzes the carburization of niobium oxide permitting to lower its temperature of carburization from 1220 to 1100 K and to produce a higher specific surface area. Beads of carburized $NiMo/Al_2O_3$ were then mechanically separated from the powder of niobium oxycarbide by visual and magnetic screening (Scheme 1).

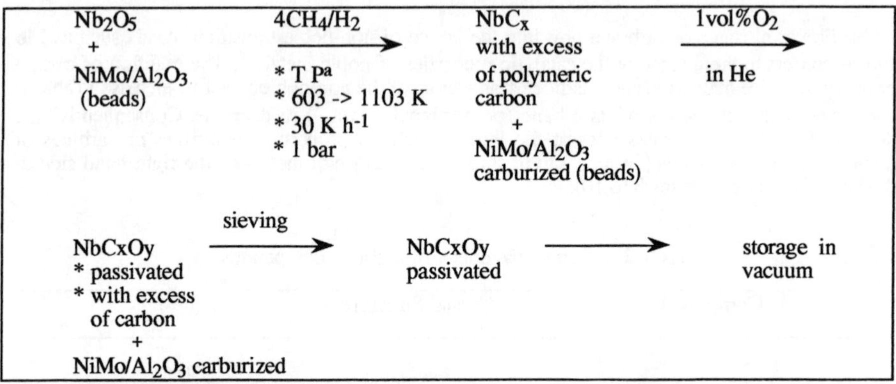

Scheme 1. Preparation of niobium oxycarbide [a] : temperature programmed

Both the fresh nitride and carbide were subsequently passivated to prevent exothermic and bulk conversion to niobium oxide, by flowing a 1vol% O_2 / He mixture (10 Lh^{-1}), at RT for 0.5 h. Samples were then exposed to air or stored in vacuum.

Pretreatments before catalytic run measurements

Oxynitrides of niobium were charged into the reactor and reduced in flowing hydrogen (1 bar) at 680 K for 2 h. Niobium oxycarbides, as in the case of tungsten [6] and molybdenum [2,7] carbides, were submitted to a pretreatment that removes excess polymeric carbon produced during the carburization process (Scheme 2) : NbC_xO_y samples were treated in flowing hydrogen (50 bar) at 720 K for 12 h before run measurements.

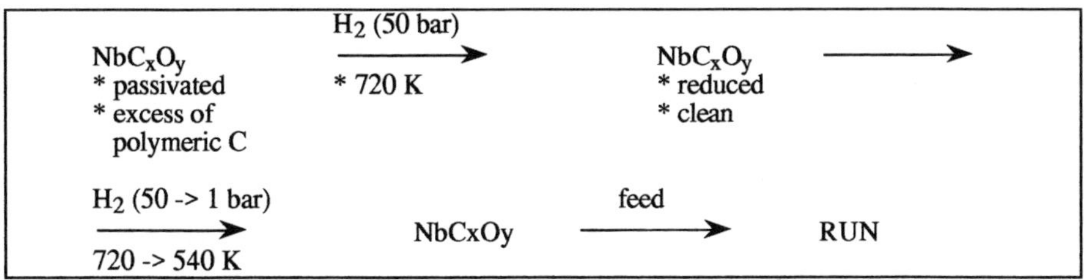

Scheme 2. Activation of passivated niobium oxycarbide before catalytic run

Standard catalytic run

Gases and reactants were used without prior purification : H_2 (Air Liquide, custom grade 99.95%), He (Air Liquide, 99.5%), O_2 (Air Liquide, 99.5%), NH_3 (Air Liquide, 99.5%), 1vol% CH_4/H_2 (Air Liquide, 99.5%), 1,2,3,4 tetrahydroquinoline, orthopropylaniline, propylbenzene (Fluka), cyclohexane (Janssen).

Reactions were performed as a gas-solid process and were carried out in a stainless steel flow reactor [8] at a total pressure of 50 bar (hydrogen at 41.4 bar, cyclohexane at 8.5 bar, used as solvent, reactant at 0.06 bar) between 543 and 673 K. The catalyst was pressed, ground and sieved to yield grains between180 and 355 μm. Hydrogen flow was controlled by a mass flow controller Brooks 5850 TR and the hydrogen pressure by a pressure controller Brooks 5866 located at the outlet of the reactor. The feed, a solution of 1wt% of reactant (1-4THQ, PB) in cyclohexane, was delivered as a liquid mixture to the flow manifold by a piston pump (Gilson model 302). The flow rate of H_2 was 25 cm^3 min^{-1} and liquid flow rate was 0.025 cm^3 min^{-1} with about 0.25 g of catalyst.

The products were analyzed on-line by a gas chromatograph with a 50 m long, 0.32 mm i.d. capillary column (Methyl Silicone Gum).

BET measurements

A Quantachrome-Quantasorb Jr. was used for the surface measurements in dynamic conditions. After a pretreatment of the solid in flowing nitrogen at 670 K the amount of nitrogen adsorbed and desorbed was determined by using a catharometer detector. The specific area was obtained by means of the BET method from the nitrogen desorbed at different partial pressures.

Chemical analysis

Elemental analysis of the solids, before and after catalysis, was obtained from the Service Central d'Analyse du Centre National de la Recherche Scientifique à Vernaison (France). Oxygen, nitrogen, hydrogen and carbon amounts were determined by a combustion method and analysis of the effluent gases. Niobium analysis was performed by plasma emission spectroscopy.

Powder X-Ray Diffraction (XRD)

A Siemens D500 automatic diffractometer with a CuKα monochromatized radiation source was used for the XRD powder patterns of the various solid phases. The identification of the different phases was made with data from the JCPDS library.

RESULTS AND DISCUSSION

Characterization of the catalysts

Table II reports the main physico-chemical characteristics of NbN_xO_y and NbC_xO_y.

TABLE II. Physico-chemical characteristics of NbN_xO_y and NbC_xO_y

	NbNxOy	NbCxOy
BET surface area m^2g^{-1}	60	50
Crystal structure	fcc	fcc
Lattice parameter a_o* (pm)	430.1	446.9
Chemical composition	$NbN_{1.14}O_{0.82}$	near NbC low O content

* Stoichiometric NbN : a_o = 439.2 pm   stoichiometric NbC : a_o = 447.0 pm

Reactions as molecular probes of surface reactivity

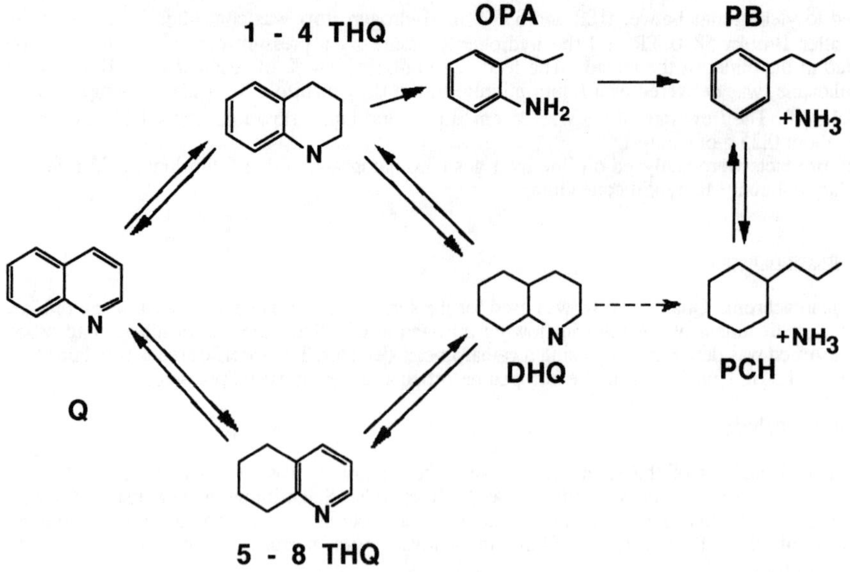

Figure 1. Simplified reaction network for HDN of 1-4 THQ, according to Satterfield [9].

Hydrodenitrogenation (HDN) of 1-4 THQ in cyclohexane ($c-C_6$) is a test reaction for :

- *the metallic function* of the materials : dehydrogenation of c-C_6 into benzene, hydrogenation of propylbenzene(PB) into propylcyclohexane, hydrogenation of 1-4 THQ into decahydroquinoline (DHQ), hydrogenolysis of C-C (formation of ethyllbenzene or ethyl-cyclohexane) or C-N (HDN) bonds.

- *the acidic function* of the materials : isomerization of c-C_6 into methyl-cyclopentane, cracking.

Figure 1 presents the reaction network for the HDN of 1-4 THQ, according to Satterfield [9].

Model reaction for testing the bifunctionality of catalysts :

The isomerization of c-C_6 over a bifunctional catalyst can be considered to occur according to the reaction network shown in Figure 2, where a metallic function favors the dehydrogenation of the saturated molecule (c-C_6) and an acidic function favors the formation of carbonium-ion intermediates in skeletal rearrangements (and cracking) of hydrocarbons.

The two catalytic functions have to co-exist side-by-side, and the cooperation between them and their close proximity are essential to their operation [10]. The cyclo-alkane is first dehydrogenated on the metallic site with formation of cyclohexene which migrates to an acidic site where it is rearranged. The isocycloalkene then desorbs and moves back to the metallic site for hydrogenation to the final isocycloalkane.

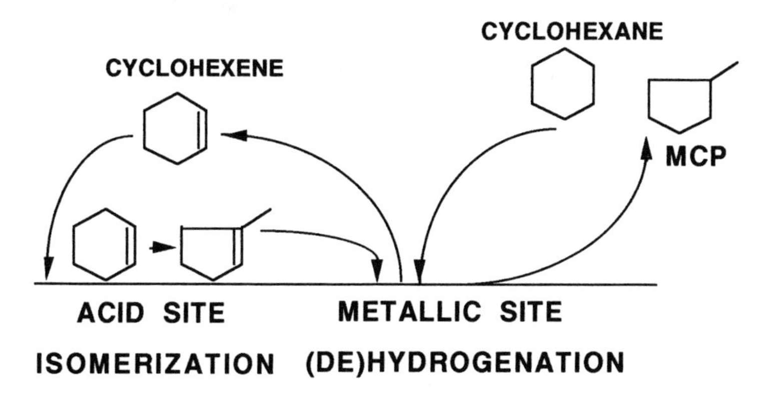

Figure 2. Reaction network of cyclohexane, a molecular probe of isomerization mechanism on oxygen-exposed niobium nitride or carbide

Since the intermediate must move from one function to the other, it can be expected that the present materials work as ideal catalysts where both functions are in close proximity, leading to the concept of "dual active site".

Model for dual bifunctional site on oxynitride or oxycarbide of non-noble transition metal

Gouin and Gouin et al. [11, 12] have carried out density measurements on molybdenum oxynitrides and found that γ-Mo_2N presents a non-stoichiometric character. They also found that oxygen can be present in the lattice. Figure 3 can be considered as a model [11] for dual a bifunctional site. This scheme shows vacancies in both the metal and non-metal sublattices of γ-Mo_2N as well as oxygen in the sub-lattice of the p-element. From this model, the metallic function can be considered as being linked to the transition metal, whereas the acidic function can be attributed to surface OH groups linked to the oxygen atoms.

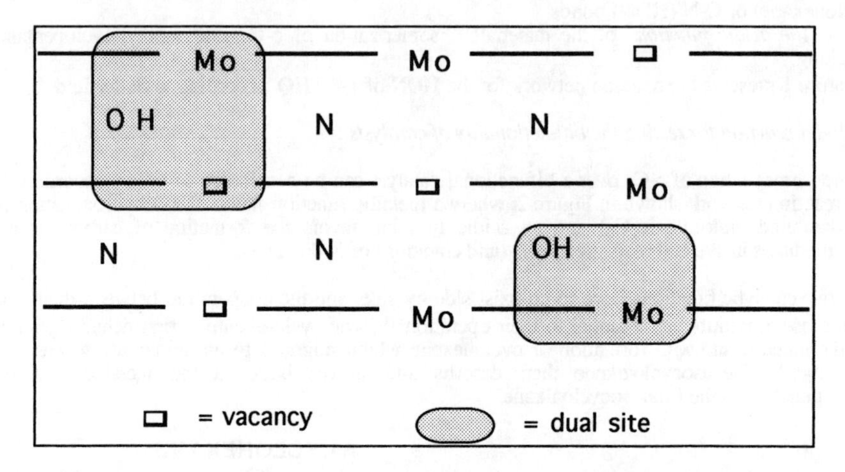

Figure 3. Model for non-stoichiometric molybdenum oxynitride according to Gouin [11].

Metallic function of oxynitrides

The (de)hydrogenation activity of materials is, in the present context, a probe reaction for their metallic function.

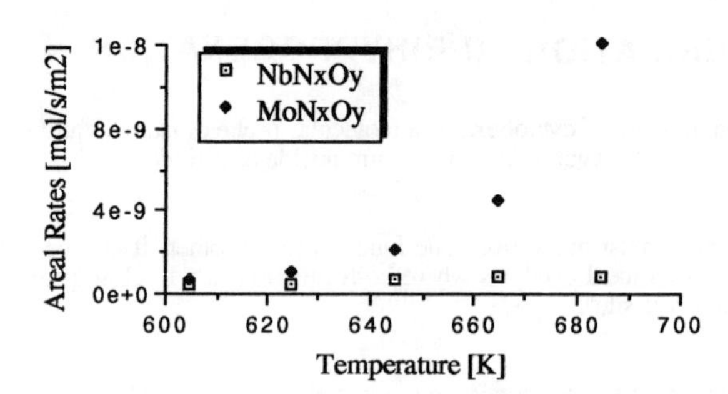

Figure 4. Metallic function : Dehydrogenation of cyclohexane on NbC_xO_y (Standard conditions). Comparison with MoN_xO_y [8].

Dehydrogenation of cyclohexane and toluene hydrogenation. As can be seen in Figure 4, the dehydrogenation of cyclohexane to benzene is quite weak over NbN_xO_y related to that of MoN_xC_y taken as a reference[8]. This tendency is confirmed by the very low activity of NbN_xO_y in toluene hydrogenation, as shown in Figure 5 [8], compared to MoN_xO_y. Accordingly, propylbenzene hydrogenation to propylcyclohexane was not observed quantitatively (less than 1%) on NbN_xO_y even under 50 bar of hydrogen.

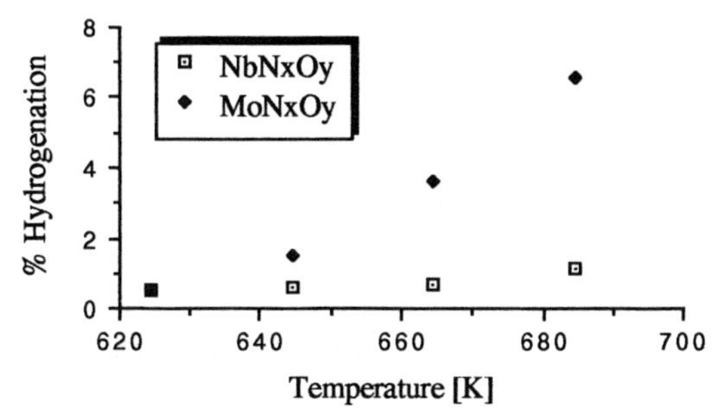

Figure 5. Metallic function : Toluene hydrogenation over NbN_xO_y (Standard conditions) Comparison with MoN_xO_y[8].

Acid function of oxynitrides.

The isomerization and cracking activity of materials is a probe reaction for their acid function. Figure 6 shows that the rate of *isomerization of cyclohexane to methylcyclopentane* is higher over NbN_xO_y compared to MoN_xO_y. This result can be considered as characteristic for surface acidity, as proposed in Figure 2 for bifunctional catalysts.The activity of the niobium oxynitride begins at 623 K with less than 0.1 % conversion to reach 8.1 % at 683 K.

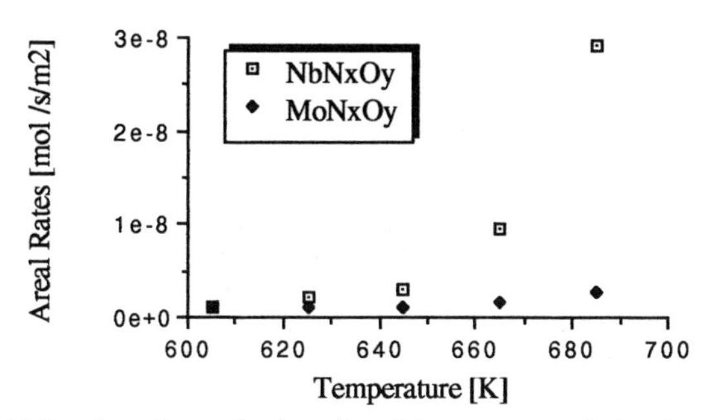

Figure 6. Acid function : Isomerization of cyclohexame to methylcyclopentane over NbN_xO_y (Standard conditions). Comparison with MoN_xO_y [8].

The intermediate carbo-cation, linked to the surface acid site of the material in the isomerization reaction, can also act as the intermediate for the cracking reaction at high temperature : Figure 7 shows that NbN_xO_y also cracks the cyclohexane to lighter hydrocarbons at 683 K quite better than does MoN_xO_y [8].
In conclusion, the comparison between oxynitrides of niobium and molybdenum clearly shows that NbN_xO_y presents an important surface activity in isomerization and cracking, and a weak (de)hydrogenation function. This dual activity can now be checked in terms of proximity of the two kinds of sites, leading to the concept of "dual surface active sites" as described in Figure 3.

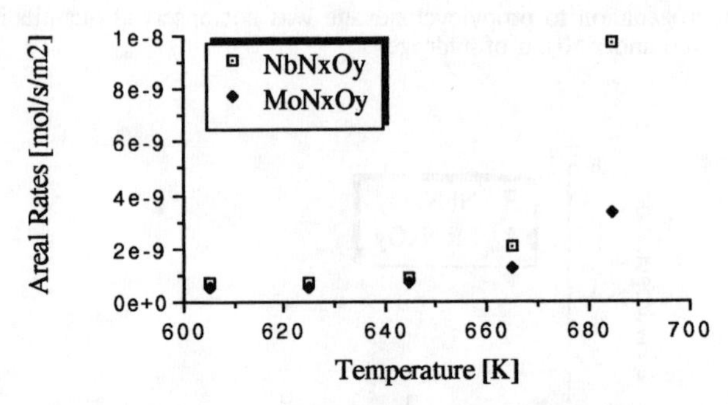

Figure 7. Acid function : Cracking of cyclohexane into light hydrocarbons on NbN_xO_y (standard conditions). Comparison with MoN_xO_y [8].

<u>Bifunctionality of oxynitrides : dual sites evidenced by steric inhibition</u>

Assumption will be made hereafter that alkenes or alkanes will chemisorbed on a metallic site for (de)hydrogenation. An alkene will be considered to chemisorb on an acidic site to isomerize. Similarly, basic compounds such as nitrogen containing molecules, will be assumed to adsorb preferentially on the acidic part of the "dual site".

If the two kinds of sites, - metallic and acidic -, are side-by-side as expected from the model surface described by Gouin [11,12] for the oxynitride of molybdenum, a reaction demanding for a metallic site should be inhibited by a basic reactant adsorbed on the adjacent acid site. On the contrary, a reaction demanding for an acidic site should be inhibited by the competitive adsorption of a stronger basic reactant on the same acidic site.

Dual site and Inhibition of dehydrogenation

Figure 8 shows the inhibiting effect of 1-4 THQ adsorption on the dehydrogenation of cyclohexane to benzene. Above 650 K, 1-4 THQ begins to react (HDN) and dehydrogenation of cyclohexane occurs again. The same tendency is observed in presence of decahydroquinoline or orthopropylaniline in the feed.

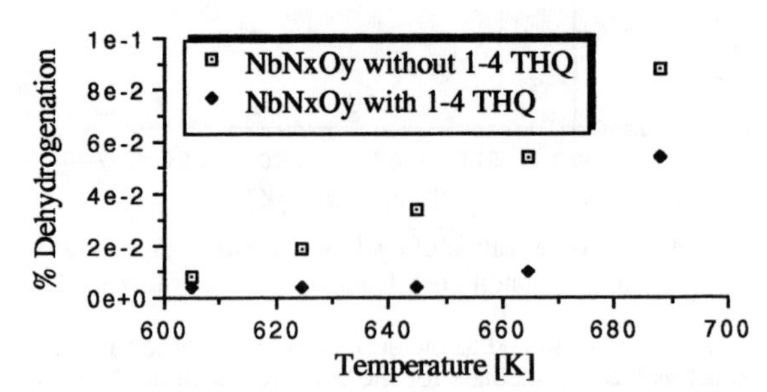

Figure 8. Dual Surface Sites of NbN_xO_y : Inhibition of cyclohexane dehydrogenation (%conversion) by chemisorption of 1-4 THQ (standard conditions) vs Temperature

Dual site and Inhibition of isomerization and cracking

Figure 9 reports the effect of the 1-4 THQ competitive adsorption on acid sites, on the transformation of cyclohexane to methylcyclopentane : the conversion lowers from 8.1 % to 0.3 % at 683 K. Again the same tendency was observed in presence of DHQ or OPA.

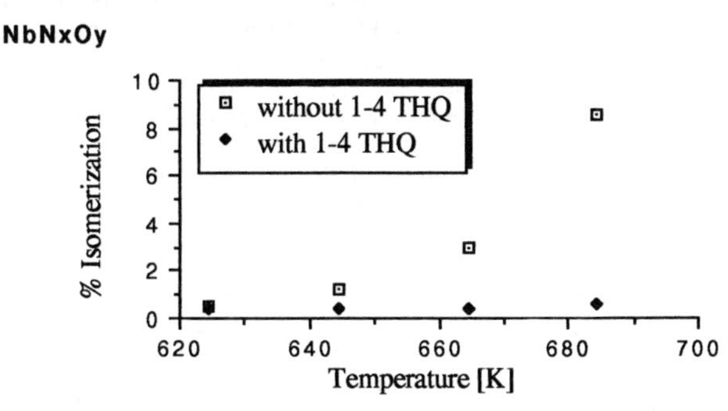

Figure 9. Dual Surface Sites of NbN_xO_y : Inhibition of the isomerization of cyclohexane (% conversion) into methylcyclopentane, by chemisorption of 1-4 THQ, vs temperature

Figure 10 shows that the cracking activity of NbN_xO_y simultaneous to lowers in presence of 1-4 THQ as expected. The loss in cracking is about 90 % (cracking conversion varying from 2.7 % to 0.3 % at 683 K).

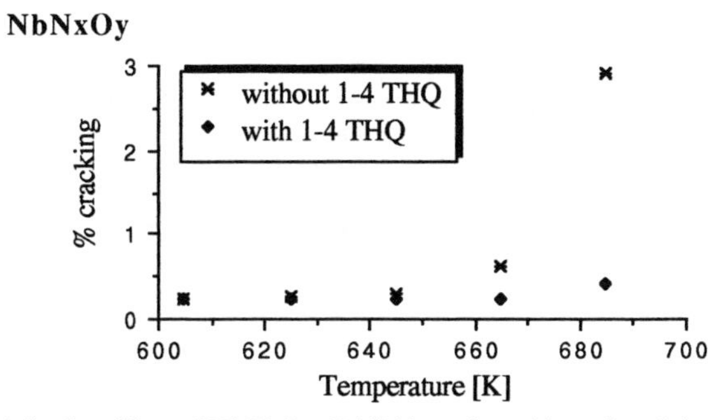

Figure 10. Dual Surface Sites of NbN_xO_y : Inhibition of cracking of cyclohexane into light hydrocarbons by chemisorption of 1-4 THQ

Dual behavior of oxynitrides in a complex reaction : Hydrodenitrogenation of 1-4 THQ

The reactions of HDN of 1-4 THQ can be considered to lead to PCH or PB and NH_3 according to the network in Figure 2. As can be seen on this figure, HDN necessitates several kinds of functions : a metallic function for 1-4 THQ hydrogenation to decahydroquinoline and hydrogenation of propylbenzene to propylcyclohexane. But this HDN reaction also needs for hydrogenolysis of the C-N bond a metallic/acid function. Secondary reactions can also occur, such as hydrogenolysis of C-C bonds (for instance propylbenzene ->ethylbenzene) on metallic sites, and cracking/isomerization on acid ones, as observed in this work. The general tendency observed here is that the higher is the metallic function, the higher is *(i)* the total HDN conversion and *(ii)* the selectivity in propylcyclohexane. So it will be found that NbN_xO_y is less active than MoN_xO_y in the main products of the reaction.

HDN of 1-4 THQ on niobium oxynitride ; comparison with molybdenum oxynitride

Table III shows that the total percentage in HDN is low for NbN_xO_y with, correlatively, a low percentage of PCH and even a lower percentage of benzene, whereas MoN_xO_y is 100% active in HDN in the same conditions, leading mainly to PCH. These results confirm that the higher is the hydrogenating function, the higher is the total conversion and the selectivity in PCH. Figure 12 shows the product distribution over NbN_xO_y versus temperature. At high temperature, the oxynitride also cracks the products, leading to lighter compounds.

TABLE III. Total HDN conversion and amounts formed in propylbenzene (PB) and propylcyclohexane (PCH)

	Total Conversion in HDN at 643 K	PCH %	PB %
NbN_xO_y	9.7	1.3	0.7
MoN_xO_y	100	78	8

Effect of "C" substituted to "N" : Evolution of the catalytic activity from the oxynitride to the oxycarbide

The effect of carbon in the oxycarbide can be seen in two reactions : the hydrogenation of PB into PCH, one of the step of the HDN pathway (Fig. 1) and the global HDN reaction.

Hydrogenation of PB to PCH over NbC_xO_y

NbN_xO_y presents a very low activity for PB hydrogenation in the standard conditions of reaction. On the contrary, Figure 11 shows that substituting N by C to give NbC_xO_y, the conversion at 543 K is already about 75 % for the oxycarbide. The conversion goes through a maximum of 100 % at 583 K, then lowers according to thermodynamic data. At 663 K, 50 % conversion is observed whereas a small fraction of ethylbenzene appears, a probe for the hydrogenolysis of the C-C bond of the propyl group of PB. As already shown with MoC_xO_y[8] the

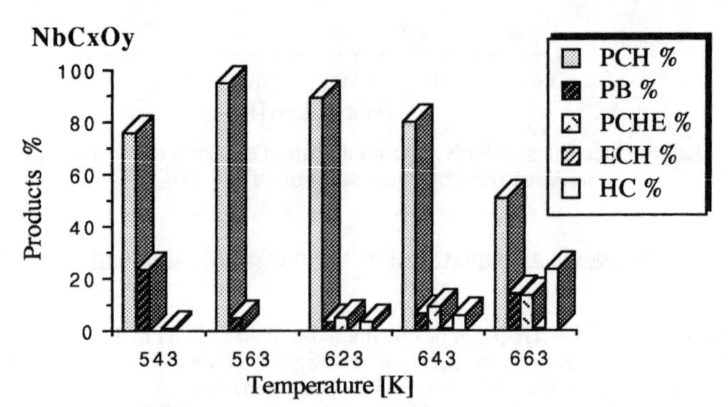

Figure 11. Effect of addition of Carbon into Niobium lattice : Product distributions for the hydrogenation of propylbenzene over NbC_xO_y (Standard Conditions) vs Temperature. Note that NbN_xO_y does not hydrogenate PB.

substitution of N by C dramatically modifies the sorptive properties of the resulting material (lowering of the adsorption strength of the reactant), favoring here the hydrogenation process.

HDN of 1-4 THQ over NbC_xO_y.

NbC_xO_y also appears to be quite better than NbN_xO_y in HDN of 1-4 THQ, in the standard conditions defined in the experimental section. This result confirms the tendency observed between the niobium and the molybdenum oxynitrides : the higher the hydrogenation function, the higher the HDN activity. Figure 12 compares the product distributions and the total conversions in HDN for both catalysts. Both data demonstrate that the carbide leads to high yields in propylcyclohexane and high total HDN conversion : 98% at 643 K compared to 9.7 % for the oxynitride. A deactivation, not reported here, exists on the oxycarbide ; initial conversion leads to 100 % conversion on the oxycarbide at 563 K for only 16 % in the same conditions for the MoN_xC_y [8]. Note that the molybdenum oxycarbonitride presented a higher specific surface area and did not suffer a so drastic deactivation. Another characteristic for both oxynitride and oxycarbide of niobium is their activity in cracking.

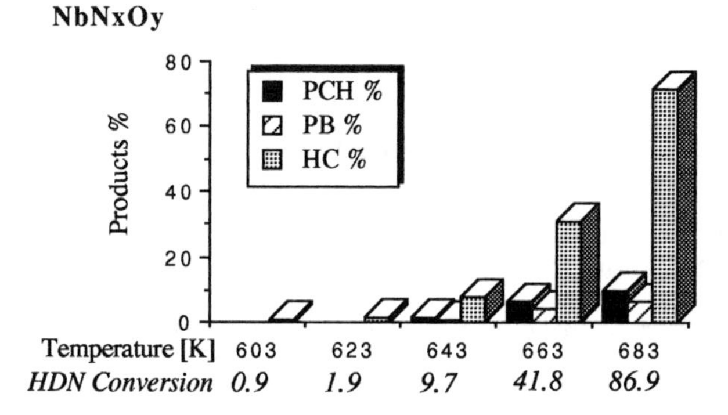

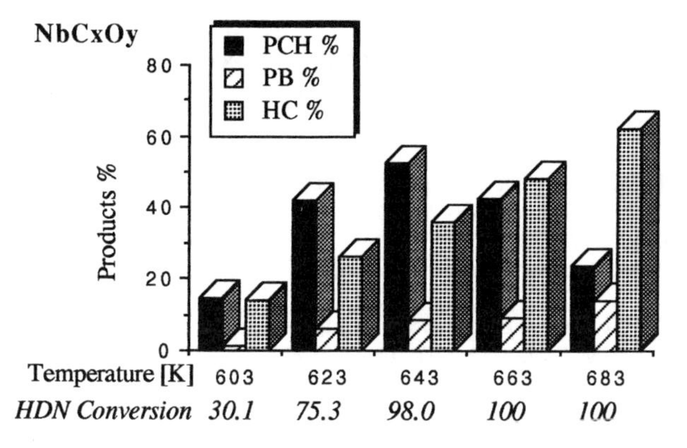

Figure 12. Effect of substitution of N by C in the lattice of Nb : comparison between NbNxOy and NbCxOy in the total conversion of 1-4 THQ hydrodenitrogenation and product distribution .

CONCLUSION

NbN_xO_y presents a strong acid character and a low hydrogenating function.On the contrary, the oxycarbide of niobium has a strong metallic function, near from that of a molybdenum nitride after carburization by cracking of the hydrocarbon feed at high temperature (typically 680 K) [8]. Thus insertion of carbon into the niobium lattice is shown to bring a higher metallic character.

NbN_xO_y is not able to easily exchange N by C in its structure and NbC_xO_y had to be prepared by direct carburization by a CH_4/H_2 mixture.

Ribeiro et al. [2] have shown that oxygen provokes an inhibition of hydrogenolysis of n-hexane and leads to its isomerization. In the case of NbC_xO_y, the HDN reaction (C-N hydrogenolysis) can be strongly inhibited by the surface oxygen. As NbC_xO_y has a lower oxygen content (3wt% for 10 in the case of NbN_xO_y) it can more easily hydrogenolyze 1-4 THQ.

The bifunctionality observed in HDN is confirmed by the isomerization of cyclohexane. A "dual site", whose metallic and acidic functions can be selectively inhibited by selected adsorbates, resembles an ideal bifunctional site where both function are adjacent.

Analysis of the products of the HDN of 1-4 THQ confirms the results of Lee et al [13] : even at high hydrogen pressure, the main route for HDN goes through the OPA rather through the DHQ route [14].

Bibliography

1 R.B. Levy and M. Boudart, Science, **181**, 547 (1973).

2 F.H. Ribeiro, R. Dalla Betta, M. Boudart, J. Baumgartner and E. Iglesia, J. Catal. **130**, 86 (1991).

3 A. Frennet, G. Leclercq, G. Maire and F. Bouillon, in Proceed. 10th Intern. Congr. Catalysis, (Budapest, 1992), p 144 .

4 M.J. Ledoux, C.P. Huu, H. Dunlop, J. Guille, in Proceed. 10th Intern. Congr. Catalysis, (Budapest, 1992), p 149.

5 V. Keller, Strasbourg University Thesis (1993).

6 L. Volpe and M. Boudart, J. Solid State Chem. **59**, 332 (1985).

7 J.S. Lee, S.T. Oyama and M. Boudart, J. Catal. **106**, 125 (1987).

8 C. Sayag, P&M Curie (Paris VI) University Thesis (1993).

9 J.C. Schlatter, S.T. Oyama, J.E. Metcalfe and J.M. Lambert Jr., Ind. Eng. Chem. Res. **27**, 1648 (1988).

10 M. Boudart and G. Djéga-Mariadassou, Kinetics of Heterogeneous Catalytic Reactions, Princeton University Press, Princeton, N.J. (1984).

11 X. Gouin, Rennes University Thesis (1993).

12 X. Gouin, R. Marchand, P. L'Haridon and Y. Laurent, J. Solid State Chem. **109**, 175 (1994).

13 K.S. Lee, H. Abe, J.A. Reimer and A.T. Bell, J. Catal. **139**, 34 (1993).

14 H.S. Kim, University P&M Curie (Paris VI) Thesis, 1993.

PREPARATION AND CHARACTERIZATION OF NEW MOLYBDENUM NITRIDE OR OXYNITRIDE PHASES

ROGER J. MARCHAND, X. GOUIN, F. TESSIER AND Y. LAURENT
URA 1496 CNRS "Verres et Céramiques", University of Rennes I, Campus de Beaulieu, F-35042 RENNES Cedex, France.

ABSTRACT

Several methods of synthesizing molybdenum nitride or oxynitride fine powders are presented.

We have prepared a γ-Mo_2N type oxynitride phase by reacting ammonia with MoO_3. The surface area and morphology of the oxynitride powders depend on the synthesis conditions. Characterization of the solids by elemental analysis, X-ray and neutron diffraction, and thermogravimetric analysis shows dramatic modification of the stoichiometry of conventional Mo_2N nitride. Aging at room temperature under air results in decreasing the material surface area. The initial surface area can be recovered be fine tuning of experimental conditions.

$MoCl_5$ and Ca_3N_2 are reacted in a molten $CaCl_2$ medium leading to a new Mo_2N structure type.

The reaction between molybdenum sulfide and NH_3 produces two different phases depending on the reaction conditions. They are structurally related to δ-MoN.

INTRODUCTION

In this paper treating of possibilities to access to molybdenum-nitrogen compounds, three types of reactions are considered :

- the molybdenum oxide + ammonia reaction, which forms a cubic γ-Mo_2N type oxynitride phase.
- the molybdenum pentachloride + calcium nitride reaction, which forms, in molten calcium chloride medium, a new α-Mo_2C type molybdenum nitride modification.
- the molybdenum sulfide + ammonia reaction, which forms hexagonal δ-MoN type nitride phases.

RESULTS

Cubic γ-Mo_2N type oxynitride phase

Synthesis conditions

In the well-known Boudart's temperature-programmed method of fcc γ phase synthesis from MoO_3 and NH_3, an appropriate heating rate is chosen in order to favor a topotactic transformation process [1-3]. In Table I are reported the reaction conditions we used to prepare two different Mo_2N-A and Mo_2N-B nitrided powders [4].

While low surface area Mo_2N-B powders consist of spherical grains, high surface area Mo_2N-A powders consist of platelets. A homogeneous and controlled platelet size can be obtained by using a MoO_3 precursor resulting from oxidation of Mo_2N-A powder :

$$\text{Commercial } MoO_3 \xrightarrow[700°C]{NH_3} Mo_2N\text{-}A_1 \xrightarrow[340\text{-}700°C]{air} \boxed{MoO_3 \text{ precursor}} \xrightarrow[700°C]{NH_3} Mo_2N\text{-}A_2.$$

The characteristics of the powders, which depend on the chosen oxidation temperature, are given in Table II and illustrated by Fig. 1.

Mat. Res. Soc. Symp. Proc. Vol. 368 © 1995 Materials Research Society

TABLE I

Experimental preparation conditions of γ-Mo_2N type phases and resulting surface area values.

	Mo_2N-A (*)	Mo_2N-B (*)	Mo_2N Volpe and Boudart [1]
Amount of commercial MoO_3 precursor (g) ($S = 2\ m^2.g^{-1}$)	2-3	2-3	1
Rate of temperature increase (°C.min^{-1})			
- from 20 to 360°C	10	20	6
- from 360 to 450°C	1	20	0.6
- from 450°C to final T	1	20	2
Final temperature (°C)	700	780	710
Step time (h)	10-12	0.5	1
Ammonia flow rate (l.h^{-1})(**)	35	35	-
Surface area ($m^2.g^{-1}$)	115-120	15-20	170-220

(*) without oxygen passivation (**) cooling under ammonia

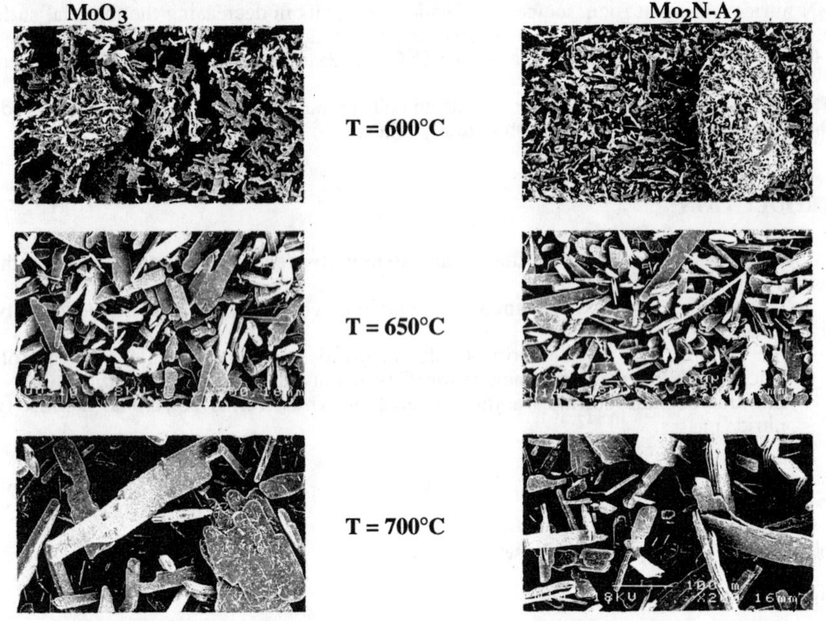

FIG. 1 - *Platelet size of MoO_3 and corresponding Mo_2N-A_2 powders in function of MoO_3 preparation temperature (x 200).*

TABLE II

MoO_3 powders prepared by oxidation of Mo_2N-A (115-120 $m^2.g^{-1}$) at different temperatures and corresponding Mo_2N-A_2 powders.

Oxidation temperature (°C)	340	440	600	650	700
MoO_3					
Surface area ($m^2.g^{-1}$)	4	3	1	-	0.2
Average length of platelets (µm)	-	-	10	50	100
Mo_2N-A_2					
Surface area ($m^2.g^{-1}$)	-	-	115	103	70
Average length of platelets (µm)	-	-	10	50	100

Composition of the powders

Chemical analysis results of the Mo_2N-A and Mo_2N-B powders, reported in Table III, point out the presence of oxygen in the anionic network and higher nitrogen contents than in Mo_2N. On the basis of these data and according to X-ray peak intensity calculations and to neutron diffraction results, the experimental formulation is near to $Mo_2O_{0.4}N_{1.2}$, which is also in good agreement with thermogravimetric measurements.

TABLE III

Chemical analysis of Mo_2N-A and Mo_2N-B powders (*).

	Mo_2N-A	Mo_2N-B	Mo_2N (calc.)
N(wt. %)	8.2	7.9	6.8
O (wt. %)	2.0	3.4	0
N/Mo	0.63	0.61	0.5
O/Mo	0.13	0.23	0
(N+O)/Mo	0.76	0.84	0.5
a (Å)	-	4.20	4.163 JCPDS file no 25-1366

(*) after 400°C H_2 surface treatment.

It is possible to totally remove oxygen by long time nitriding, but the nonmetal atoms/metal atoms ratio in the nitride composition remains close to 0.8, which is much higher than 0.5 in Mo_2N ($Mo_2N_{1.6}$: N calc. % = 10.46 ; N obs. % = 10.1).

Powder aging and regeneration

Exposure to ambient air causes aging of the Mo_2N-A powders which results in a progressive decrease in surface area without changing either the X-ray diagram or the N/Mo atomic ratio. The initial surface area can be recovered by a regeneration treatment at 400°C in a H_2 atmosphere (heating rate : 1°C.min^{-1}).

Orthorhombic distortion of the cubic phase

Heating the powders in H_2 at temperatures higher than 400°C induces bulk modifications. A loss of cubic symmetry is observed when a $Mo_2N_{1.6}$ nitride sample is heated at 550°C in a H_2 atmosphere, revealed as shown in Fig. 2 by a splitting of the hhk peaks into three peaks whereas the 111 and 222 hhh peaks remain single peaks. The orthorhombic parameters are :

a = 4.181 (5) Å b = 4.125 (5) Å c = 4.050 (5) Å.

The corresponding experimental formulation is $Mo_2N_{1.06}$.

New α-Mo_2C type molybdenum nitride modification

The preparation of this new α-Mo_2N nitride is performed thanks to a reaction in molten salt medium [5] by using $MoCl_5$ as molybdenum source and Ca_3N_2 as nitriding agent, according to :

$$2\ MoCl_5 + 5/3\ Ca_3N_2 \xrightarrow[850\text{-}900°C]{CaCl_2} Mo_2N + 5\ CaCl_2 + 7/6\ N_2\uparrow .$$

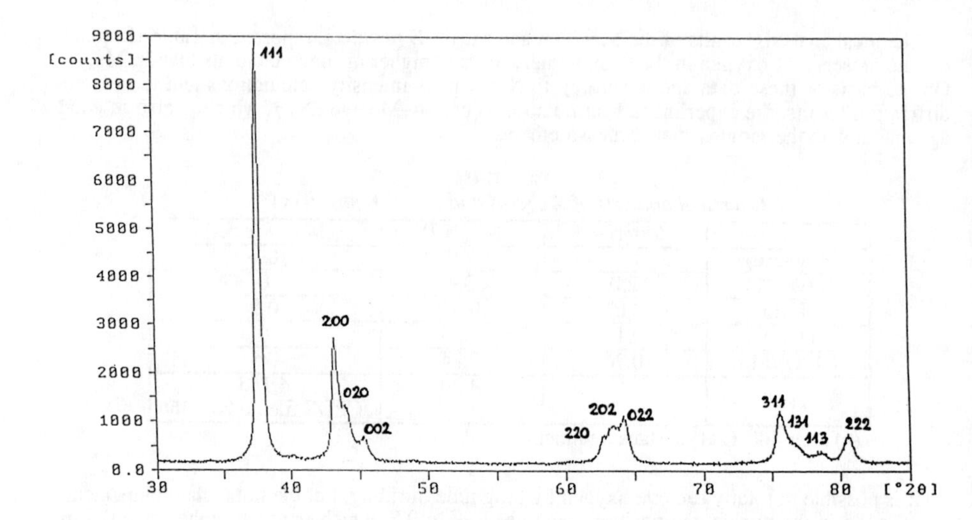

FIG. 2 - *X-ray diffraction powder pattern of the orthorhombic γ-Mo_2N distorted phase (Cu Kα).*

After dissolving at 850-900°C under inert atmosphere calcium metal in anhydrous molten calcium chloride (melting point : 782°C), bubbling of pure nitrogen results in Ca_3N_2 formation.

The reaction with molybdenum chloride leads to a black nitrogen containing molybdenum phase which is isotypic to α-Mo_2C molybdenum carbide.

The XRD powder pattern can be indexed with the hexagonal parameters :

$$a = 3.012\ (1)\ \text{Å} \qquad c = 4.735\ (2)\ \text{Å}.$$

These values are very close to the hexagonal parameter values of α-Mo_2C [6] :

$$a_{Mo_2C} = 3.0124\ (4)\ \text{Å} \qquad c_{Mo_2C} = 4.7352\ (7)\ \text{Å}.$$

In Table IV are collected the X-ray diffraction data of the new molybdenum-nitrogen phase.

The calcium chloride molten bath influences the nature of the reaction product. If the amount of $CaCl_2$ is not large enough, the following reaction can be observed :

$$MoCl_5 + 5/2\ Ca_3N_2 \rightarrow Ca_5MoN_5 + 5/2\ CaCl_2.$$

TABLE IV

Observed and calculated interplanar d spacings and relative intensities of α-Mo_2N.

h k l	$d_{obs.}$ (Å)	$d_{calc.}$ (Å)	I/Io
1 0 0	2.613	2.608	20
0 0 2	2.372	2.367	25
1 0 1	2.287	2.286	100
1 0 2	1.752	1.753	20
1 1 0	1.505	1.506	17
1 0 3	1.350	1.350	15
1 1 2	1.270	1.271	13
2 0 1	1.257	1.258	10

Hexagonal δ-MoN type nitride phases

A new route to access to nitrides consists in using sulfides as starting products. The reaction of ammonia with a sulfide can form a nitride with H_2S gas formation :

$$MS_x \xrightarrow[T]{NH_3} MN_y + x\,H_2S\uparrow.$$

Starting from MoS_2, molybdenum nitride phases related to hexagonal δ-MoN are prepared according to the reaction :

$$MoS_2 + 4/3\,NH_3 \xrightarrow{700\text{-}900°C} \text{"MoN"} + 2\,H_2S\uparrow + 1/6\,N_2\uparrow.$$

Nitridation of commercial MoS_2

Depending on the reaction conditions, two different nitride phases closely related to the δ-MoN JCPDS file no. 25-1367 are obtained. "L-MoN" is mostly prepared at lower temperatures (750°C), "H-MoN" at higher temperatures (850°C), but mixtures of L-MoN and H-MoN can also be observed at these two temperatures depending on other sensitive parameters such as reaction time, ammonia space velocity or thickness of MoS_2 powder in contact with flowing ammonia.

The results in Table V show that the two phases L-MoN and H-MoN differ from MoN in their chemical composition. The difference in nitrogen content is the most significant in the case of L-MoN which has also the lowest density.

TABLE V

Chemical analysis and density values of L-MoN and H-MoN(*).

	L-MoN (obs.)		H-MoN (obs.)		δ-MoN (calc.)
N (wt. %)	17.0	16.3	14.5	13.6	12.74
S (wt. %)	-	0.3	0.1	-	0
N/Mo	1.44	1.37	1.18	1.12	1
d ($g.cm^{-3}$)	7.1 - 7.2		7.8 -7.9		9.18

(*) after 300 C H_2 treatment

According to chemical analysis and density measurements, the experimental formulations are :

L-MoN : $Mo_{0.71\ \ 0.02}N$ H-MoN : $Mo_{0.87\ \ 0.02}N$

Comparison of the X-ray diffraction data with ≅ WC type related ≅ hexagonal δ-MoN points out a similarity in the case of H-MoN except the presence of a few extra weak peaks, and a significant difference in the case of L-MoN. In particular a 3.6 % relative volume decrease of the unit cell is observed. The XRD powder patterns can be indexed in orthorhombic unit cells with :

L-MoN		H-MoN	
$a \sim 3\,a_{hex.}$	= 16.994 (3) Å	$a = 2\,a_{hex.}$	= 11.452 (2) Å
$b \sim \rceil 3/2\,a_{hex.}$	= 4.906 (1) Å	$b = \rceil 3/2\,a_{hex.}$	= 4.959 (1) Å
$c \sim c_{hex.}$	= 5.520 (1) Å	$c = c_{hex.}$	= 5.607 (1) Å

High surface area MoS_2 precursor

Preparation and characterization

Fine reactive MoS_2 powders are prepared at 300-400 C from MoO_3 in a potassium thiocyanate melt (melting point : 173 C), according to first experiments [7,8]. After adjustment of the reaction parameters surface areas higher than 200 $m^2.g^{-1}$ are obtained. The corresponding MoS_2 morphology is illustrated by the SEM micrograph of Fig. 3.

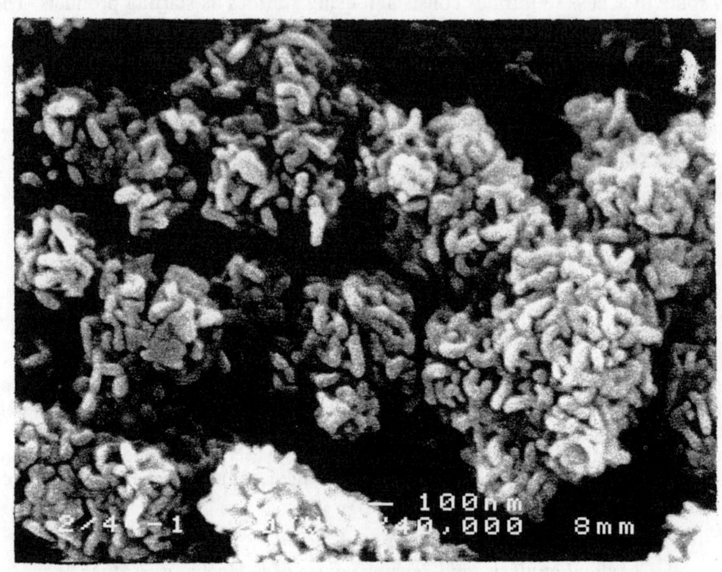

FIG. 3 - *High surface area MoS_2 SEM micrograph (x 40000).*

Nitridation

The results are summarized in Table VI. Pure L-MoN is prepared with relatively high surface area. High surface area MoS_2 + L-MoN mixtures could be attractive for catalytic applications. Similar results have been obtained with tungsten compounds [9].

TABLE VI

Nitridation of high surface area MoS_2 at different temperatures.

Sample no.	Temperature (°C)	XRD analysis	S ($m^2.g^{-1}$)
1	650	MoS_2 + L-MoN	140
2	690	L-MoN	50
3	750	L-MoN + H-MoN	20
4	800	H-MoN + L-MoN	10

References

1. L. Volpe and M. Boudart, J. Solid State Chem., **59**, 332 (1985).
2. L. Volpe and M. Boudart, Catal. Rev.-Sci. Eng., **27**, 515 (1985).
3. S.T. Oyama, Catal. Today, **15**, 179 (1992).
4. X. Gouin, R. Marchand, P. L'Haridon and Y. Laurent, J. Solid State Chem., **109**, 175 (1994).
5. Extramet S.A., French Patent No. 87 00097 FR 2 609 461 (4 January 1987).
6. Nat. Bur. Stand. (U.S.) Monogr., **25**, 21 (1984), JCPDS file no. 35-787.
7. J. Milbauer, Z. Anorg.Allgem. Chem., **42**, 433 (1904).
8. D.H. Kerridge and S.J. Walker, J. Inorg. Nucl. Chem., **39**, 1579 (1977).
9. F. Tessier, R. Marchand and Y. Laurent, to be published.

ACTIVITY AND STRUCTURE OF SUPPORTED MOLYBDENUM NITRIDE HYDROTREATING CATALYSTS

GREGORY M. DOLCE, CRAIG W. COLLING, AND LEVI T. THOMPSON
University of Michigan, Department of Chemical Engineering, Ann Arbor, MI, 48109-2136

ABSTRACT

Molybdenum nitrides have been shown to be active hydrotreating catalysts. We explored the hydrotreating activities of supported Mo nitrides. It was observed that the hydrodenitrogenation and hydrodesulfurization activities were inverse functions of the molybdenum loading. By using oxygen chemisorption and carbon monoxide temperature programmed desorption, we were able to characterize the particle morphology of the nitrides.

INTRODUCTION

Heavier crudes and coal derived liquids are now being used as fuel precursors. Since these liquids tend to be richer in sulfur, nitrogen, and oxygen than light petroleum crudes, new catalysts are being developed which are more efficient than typical commercial catalysts. Molybdenum nitrides have been shown to be effective for hydrodenitrogenation and hydrodesulfurization. Schlatter et al.[1] reported that molybdenum nitrides were as active as a Ni-Mo-S/Al_2O_3 catalyst for quinoline HDN. Others have shown that molybdenum nitrides are active for HDN, HDS, and hydrodeoxygenation (HDO)[2,3]. We recently reported that supported molybdenum nitrides are active for pyridine HDN[4] and found that the activity was a strong inverse function of molybdenum loading. The present work focuses on the hydrotreatment of more complex compounds, such as quinoline, benzothiophene, and benzofuran. These compounds are more representative of compounds typically found in coal derived liquids and heavy oils. The effects of synthesis conditions, molybdenum loading, and surface structure on the activity are explored.

EXPERIMENTAL

Catalyst Synthesis

The supported molybdenum nitrides were prepared from a series of γ-Al_2O_3 supported molybdena. Low molybdenum loadings (< 10 wt% Mo) were produced via an incipient wetness technique similar to that employed by Colling and Thompson[4]. Higher loadings (> 10 wt% Mo) were achieved using the equilibrium adsorption method which was used by Wang and Hall[5]. The supported molybdate was nitrided in a temperature programmed manner in NH_3 flowing at 150 cm^3/min. The sample was heated quickly from room temperature to 350° C. Two different heating ramps (β_1 and β_2) were used to heat the sample from 350° to 450° C and from 450° to 700° C, respectively. After remaining at 700° C for one hour, the sample was quickly cooled and passivated for 2 hours in 0.98% O_2/He to prevent bulk oxidation. Parameters such as the space velocity and heating rates were varied in order to study their effects on the catalytic properties of the nitrides (Table I).

Structural Characterization

X-ray diffraction (XRD) and high resolution transmission electron microscopy (HRTEM) were used to determine the structures of the supported Mo nitrides. The XRD patterns were collected using a Rigaku DMAX-B diffractometer and CuKα radiation (λ = 1.542 Å). Samples for HRTEM were prepared by grinding and ultrasonically dispersing the powders in isopropanol.

Mat. Res. Soc. Symp. Proc. Vol. 368 © 1995 Materials Research Society

A drop of the resulting suspension was then applied to holey carbon Cu grids. HRTEM was performed at 400 kV, using a JEOL 4000 EX transmission electron microscope, which has a point-to-point resolution of less than 1.75 Å.

Sorption Analysis

Oxygen uptakes were measured in a dynamic fashion at 195 K using a Quantasorb Sorption Analyzer (Model QS-17). The as-prepared materials were pretreated for 3 hours at 673 K in H_2 (99.99%, Air Products) flowing at 20 cm^3/min, then cooled to 195 K in He.

Table I: Supported Mo nitride synthesis parameters.

Catalyst Code	Loading (wt% Mo)	MHSV (hr^{-1})	ß1 (K/h)	ß2 (K/h)	N/Mo
MN04+++	4	17 (+)	100 (+)	200 (+)	0.97
MN04+-+	4	17 (+)	40 (-)	200 (+)	0.91
MN08+++	8	17 (+)	100 (+)	200 (+)	0.61
MN08+-+	8	17 (+)	40 (-)	200 (+)	1.03
MN14+++	14	17 (+)	100 (+)	200 (+)	0.62
MN14+-+	14	17 (+)	40 (-)	200 (+)	0.75

The CO temperature programmed desorption (TPD) experiments were performed on an Altamira AMI-M which was retrofitted with a Fisons Sensorlab 200D Mass Spectrometer. Between 50 and 250 mg of each sample was placed in a Pyrex U-tube on a quartz wool plug. Before adsorbing CO, the passivated materials were reduced at 673 K in H_2 flowing at 30 cm^3/min for 3 h, flushed overnight at 723 K with Ar (Air Products, 99.998%, with Matheson Gas Purifier Model 6406) flowing at 10 cm^3/min , and then cooled to room temperature in flowing Ar. The materials were purged overnight in Ar at 723 K to insure that the surface and bulk of the reduced catalyst were free of residual gases. The materials were dosed by passing a stream of CO (99.5%, Matheson) flowing at 10 cm^3/min over a packed bed of ascarite and 4 Å molecular sieves to remove carbon dioxide and water, respectively, and then over the catalyst for 30 min. The materials were then flushed with Ar for 45 min and heated in Ar flowing at 30 cm^3/min from room temperature at 15 K/min. The reactor effluent was monitored with a thermal conductivity detector and the mass spectrometer.

Activity Studies

The nitrides were evaluated using quinoline HDN, benzothiophene HDS, and benzofuran HDO as test reactions. The reactions were run in stainless steel batch micro-reactors for 1-4 hours at 1000 psi H_2 (cold). The reactors were loaded with 10 mg of catalyst and 100 mg of reactant, pressurized, and placed in a heated fluidized sandbath. HDN and HDO experiments were run at 390° C, and HDS experiments were run at 320° C. Product analysis was performed using a Hewlett Packard 5890 gas chromatograph. The activities of the supported nitrides were compared to those of typical commercial sulfide catalysts, Ni-Mo/Al_2O_3 and Co-Mo/Al_2O_3 (Crosfield 504 and 477).

RESULTS AND DISCUSSION

Structural Characterization

The only discernible x-ray diffraction peaks in the patterns of the 4 and 8% loaded oxides were due to γ-Al_2O_3. Monolayer coverage by MoO_3 on the 190 m^2/g support would correspond to a loading of ~10 wt% Mo. The absence of peaks other than those for γ-Al_2O_3 suggested that the Mo oxide domains on the surfaces of the low loaded materials were small, highly dispersed, and/or amorphous. Diffraction peaks due to MoO_3 were observed for the high loaded materials. There was no evidence of any other crystalline species in these materials.

Nitridation of the 4 and 8% materials did not significantly change their diffraction patterns. Domains in these materials remained highly dispersed and x-ray amorphous. The average crystallite sizes for these materials were up to 15 Å for the 4% materials and between 20 and 35 Å for the 8% materials, as measured by TEM. The diffraction patterns for the high loaded materials indicated the presence of γ-Mo_2N. There was no evidence of unreacted Mo oxides and no nitrides other than γ-Mo_2N were produced. Under conditions similar to those used in this work, γ-Mo_2N was the only nitride produced during the temperature programmed nitridation of MoO_3[2,8,9]. Average γ-Mo_2N crystallite sizes ranged between 50 and 90 Å.

Oxygen Chemisorption

Oxygen chemisorption was used to further probe the structure of the supported Mo nitrides. For unsupported nitrides, it has been shown that oxygen provides a measure of the available nitride surface area. Oxygen does not adsorb on the γ-Al_2O_3 support[6]. As shown in Figure 1, the amount of oxygen that adsorbed per Mo atom did not change significantly for different molybdenum loadings. This behavior implies that the nitride dispersion is constant with loading. We attempted to explain this behavior using two simple models of the particle structure: a two dimensional plate-like particle and a three dimensional spherical particle. Assuming that O_2 chemisorbs non-specifically on the nitride domains, the results suggest that the domains were two-dimensional (Figure 2). As the size of these particles increases with increasing loading, they maintain this highly dispersed morphology.

Activity Studies

The molybdenum nitrides demonstrated activities which were generally comparable to those of the commercial sulfide catalysts (Table II). These activities were reproducible to within

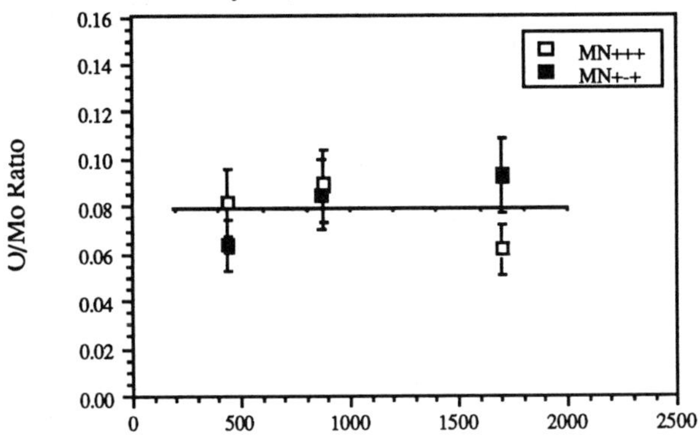

Figure 1: The amount of O_2 that sorbed per Mo atom as a function of catalyst loading.

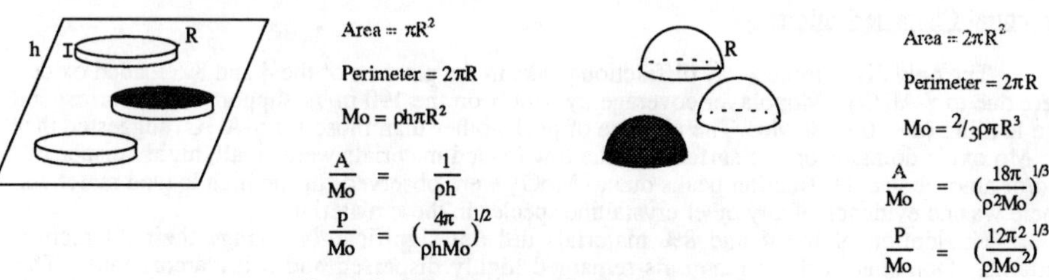

Figure 2: Models of particle morphology.

approximately 10%. The supported nitrides were most active for HDS, followed by HDO and HDN. One goal of this work was to investigate the effects of the synthesis parameters on catalytic activity. We observed that the most influential parameter was the molybdenum loading. Changes in nitriding conditions had little effect on the activities of the materials. For HDN and HDS, the nitride activities decreased with an increase in Mo loading, as shown in Figure 3. This trend was not observed for HDO. The decrease in activity with loading suggested that the most active sites were not "on top" of the particles. Instead we believe that the most active HDN and HDS sites were located at the perimeter of the nitride particles. Product distributions indicated that the nitrides had better hydrogen economies (moles H_2 consumed per mole reactant consumed) than the commercial catalysts. For HDN, the commercial catalysts formed more fully hydrogenated products, while the nitrides formed more unsaturated products including benzenes and cyclohexenes (Figure 4). Similar trends were observed for HDO. This behavior may indicate the existence of different reaction pathways for the nitrides and the commercial catalysts.

Table II: Supported Mo nitride Hydrotreating Activities

Catalyst	Quinoline HDN		Benzothiophene HDS		Benzofuran HDO	
	(nmol/g_{cat}/s)	(µmol/mol Mo/s)	(nmol/g_{cat}/s)	(µmol/mol Mo/s)	(nmol/g_{cat}/s)	(µmol/mol Mo/s)
MN04+++	124	296	1180	2830	68	162
MN04+-+	84	201	1060	2540	80	193
MN08+++	121	145	1670	2000	160	191
MN08+-+	100	120	1620	1940	206	247
MN14+++	110	75	2080	1420	398	272
MN14+-+	140	96	2080	1420	260	178
$Ni-Mo/Al_2O_3$	332	252	8960	6790	2270	1720
$Co-Mo/Al_2O_3$	1260	1300	10130	10410	5130	5280

Thermal Desorption Spectroscopy

We expected CO TPD to reveal differences in the character of the surface sites[10]. CO desorbed from all the materials between 300 and 400 K. No other desorption peaks were observed. While the desorption temperature was constant for all the materials, the amount of CO increased with decreasing Mo loading (Figure 5). For unsupported Mo nitrides, CO dissociated on adsorption and no CO desorbed. Carbon monoxide did not adsorb on γ-Al_2O_3. Associative adsorption of CO was unique to the supported materials, while dissociative adsorption occurred on both supported and unsupported nitrides. Since both the particle support interface and associative adsorption of CO were unique to the supported materials, we have tentatively

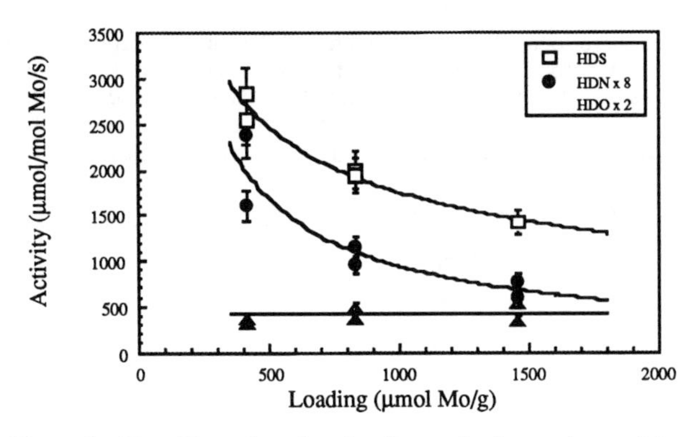

Figure 3: The effect of catalyst loading on hydrotreating activity.

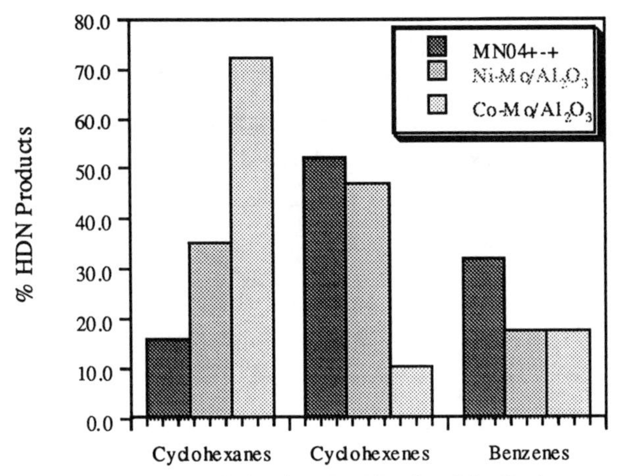

Figure 4: Quinoline HDN Product Distribution

assigned associative adsorption of CO to the perimeter regions. This hypothesis is supported by Figure 5, which shows that the CO/Mo ratio decreased with loading as would be expected if associative CO adsorption occurred at the perimeter. This trend is similar to that seen with respect to the HDN and HDS activities. A plot of pyridine HDN activity vs. CO/Mo yielded a straight line (Figure 6), with a slope corresponding to a turnover frequency of 0.0015 s^{-1} at 633 K.

SUMMARY

The hydrotreating activities and selectivities of a series of γ-Al_2O_3 supported molybdenum nitrides have been determined. Although nitride domains in the low loaded materials were too highly dispersed to be characterized by XRD, other information suggested that they were two dimensional plates. XRD showed that the higher loaded materials contained γ-Mo_2N crystallites. Activity studies showed that the Mo nitrides were active for HDN, HDS, and HDO. They were more hydrogen efficient than commercial sulfide catalysts for HDN and HDO. Both the HDN and HDS activity decreased with increasing loading which suggested that the most active sites were located at the perimeter of the nitride particles. Carbon monoxide TPD showed that CO adsorbed associatively to the supported nitrides. This behavior was not observed over the unsupported nitrides. The amount of adsorbed CO decreased with increasing loading suggesting again that adsorption took place primarily at the perimeter sites. A plot of pyridine HDN activity versus CO/Mo ratio yielded a turnover frequency of 0.0015 s^{-1}.

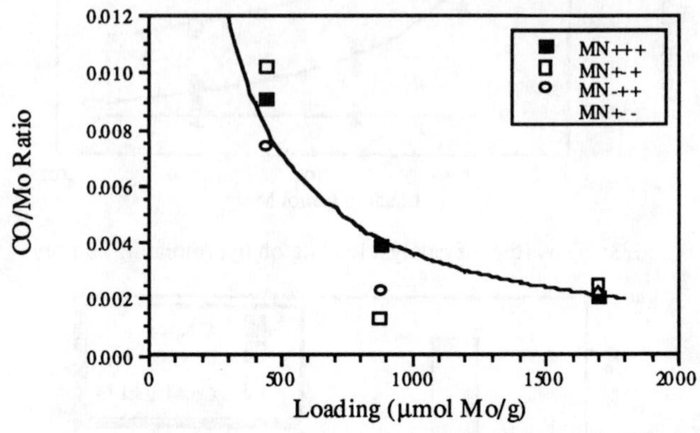

Figure 5: The amount of associative CO that desorbed as a function of catalyst loading.

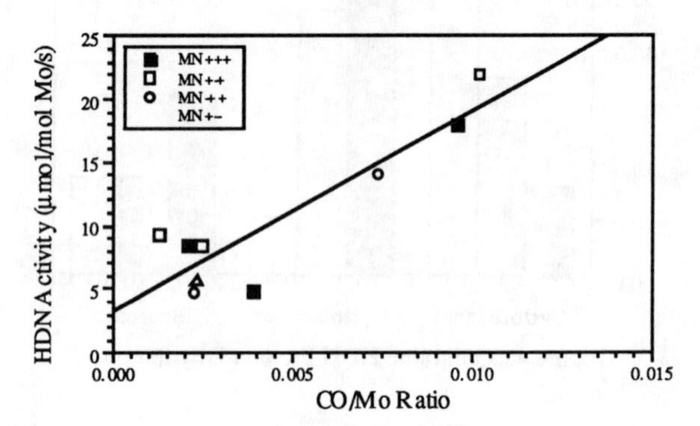

Figure 6: The pyridine HDN activity of the Mo nitrides as a function of the CO/Mo ratio.

References

1. J.C. Schlatter, S.T. Oyama, J.E. Metcalfe, and J.M. Lambert, Ind. Eng. Chem. Res. **27**, 1648 (1988).
2. K.S. Lee, H. Abe, J.A. Reimer, and A.T. Bell, J. Catal. **139**, 34 (1993).
3. H. Abe and A.T. Bell, Catal. Let. **18**, 1 (1993).
4. C.W. Colling and L.T. Thompson, J. Catal. **146**, 193 (1994).
5. L. Wang and W.K. Hall, J. Catal. **77**, 232 (1982).
6. B.S. Parekh and S.W. Weller, J. Catal. **47**, 100 (1977).
7. W. Zmierczak, G. MuraliDhar, and F.E. Massoth, J. Catal. **77**, 432 (1982).
8. E.J. Markel and J.W. Van Zee, J. Catal. **126**, 643 (1990).
9. L. Volpe and M. Boudart, J. Solid State Chem. **59**, 332 (1985).
10. P.C. Stair, J. Am. Chem. Soc. **104**, 4044 (1982).

RELATIVE ACTIVITY AND SELECTIVITY OF NANOSCALE Mo_2N, Mo_2C AND MoS_2 CATALYSTS SYNTHESIZED BY LASER PYROLYSIS

R. OCHOA*, G. T. HAGER*, W.T. LEE**, S. BANDOW***, E. GIVENS*
and P.C.EKLUND* · **
*Center for Applied Energy Research,**Department of Physics and Astronomy, University of Kentucky Lexington, KY 40506.*** Institute of Molecular Science, Myodaiji, Okazaki , 444 Japan.

INTRODUCTION

Nanoscale and well dispersed catalyst particles offer a large number of advantages: no diffusion resistance, easy accessibility to reactants, and a large number of active sites. In coal liquefaction, highly dispersed catalysts are especially needed because the catalyst particles are only able to influence reactions within their immediate vicinity .

Laser pyrolysis constitutes a new method for the preparation of ultrafine particle catalysts. This technique involves a gas phase pyrolysis reaction of two or more molecular species sustained by the heat generated through the absorption of CO_2 laser energy into vibrational-rotational excitations of at least one of the reactant gas species. This process allows the production of nanoscale particles due to the fast growth and rapid heating /cooling rates (100,000 °C/s) in the reaction zone defined by the intersection of a reactant gas stream and the high-power infrared laser beam. Average production rates vary between 5-20 g/min and the resulting catalysts have an average particle size on the order of 5-10 nm. The catalysts produced by this method offer the potential of high activity, especially in those reactions or processes in which the degree of catalyst dispersion is directly related to its activity, as in the case of coal liquefaction [1].

Using this technique, a variety of Fe-based catalysts such as α-Fe, Fe_3C, Fe_7C_3, and $Fe_{1-x}S$ [2] have been produced, and the catalytic activity of some of them has been evaluated for coal liquefaction and Fischer Tropsch synthesis [3-5]. The results show that these catalysts exhibit only moderate activity. However, due to the complexity of the coal structure and coal liquefaction process it is difficult to elucidate the significance that the particle size and dispersion have in the conversion of coal. Catalytic reactions using model molecules may shed some light in that respect.

High surface area Mo_2N and Mo_2C synthesized by Temperature Programmed Reduction [10] have been reported to exhibit high activity for heteroatom removal in the hydrotreatment of naphtha and upgrading of coal liquids . These materials were shown to be highly resistant to deactivation and transformation to the sulfide phase [6].

In this work results are presented on the catalytic activity of Mo_2N, Mo_2C and MoS_2 synthesized by laser pyrolysis and they are compared to that of a molybdenum promoted sulfated hematite (8%$Mo/Fe_2O_3/SO_4$) which has been reported to promote high conversion of coal [7,8]. The activity of these catalysts was evaluated for model hydrocarbon and heteroatom reactions that are representative of some functionalities present in coal and its residua. These reactions are important or relevant for the simultaneous conversion and upgrading of coal liquids and petroleum residuum into higher quality products. The model compounds selected were naphthalene to simulate hydrogenation of aromatics in coal, diphenyl ether to simulate the removal of etheric oxygen, benzothiophene and quinoline to simulate sulfur and nitrogen removal from aromatic rings in coal and petroleum residua.

EXPERIMENTAL

Nanoscale Mo_2N, Mo_2C and MoS_2 were synthesized by laser pyrolysis from the reaction of $Mo(CO)_6$ in the presence of ammonia, ethylene or a mixture ethylene/hydrogen sulfide. Following the synthesis, the catalysts were passivated in a flow of 5% O_2/He for several hours

Mat. Res. Soc. Symp. Proc. Vol. 368 © 1995 Materials Research Society

before removal from the reaction apparatus. Details of the experimental apparatus and conditions are given elsewhere [2,9]. The molybdenum-promoted, sulfated hematite was prepared by the aqueous precipitation of ferric ammonium sulfate and ammonium molybdate in the presence of urea, followed by calcination in air at 475°C [7].

The catalyst structure and morphology have been characterized by x-ray diffraction, high resolution TEM and Raman Spectroscopy [9]. The surface area was determined by the nitrogen BET method. The average composition of these particles was obtained by elemental analysis and the chemical state of the surface was characterized by XPS. Irreversible chemisorption of CO was used to measure the number of active sites on the catalyst surface.

Catalytic activity was evaluated using a stainless steel bomb microreactor, 22 cm^3 in volume, which was pressurized at 800 psi of hydrogen . A 5% catalyst loading with respect to a solution of the model compound was dissolved in hexadecane. The reactor was maintained at 380°C for reaction periods of 15, 30 and 60 min while agitating at 440 rpm in a fluidized sand bath. The effect of the presence of added sulfur in the reaction was studied by adding dimethyl disulfide (DMDS) in 20% excess of the stoichiometric amount required to convert Mo_2N and Fe_2O_3 to MoS_2 and FeS_2, respectively. The liquid products were analyzed by gas chromatography with a fused-silica 30 m capillary DB5 column with Flame Ionizing Detector detection. In the model compound reactions, the concentration of the liquid phase products is given as a mole percentage of the amount initially loaded in the reactor.

RESULTS AND DISCUSSION

Characterization

X-ray diffraction and High Resolution TEM have shown that the particles produced consist of fcc Mo_2N Mo_2C and hexagonal MoS_2 [9]. No evidence of bulk oxide formation has been found using these tests. Crystallite size was calculated from the full width at half maximun of the corresponding x-ray diffraction lines and the particle size from the value of the surface area of the particle using the formula D_p=6/ ρS_g, where S_g is the surface area of the particle and ρ its density. Table I summarizes the results of these calculations. The difference observed between particle size and crystallite size indicates that these particles are partially agglomerated.

TABLE I
STRUCTURAL CHARACTERIZATION OF CATALYST

Catalyst	Crystallite Size (XRD)	Surface Area (BET)	Particle Size (BET)	Composition (Elemental Analysis)	Site Density x10^{14}cm^{-2}
fcc Mo_2N	3 nm	63 m^2/g	10 nm	$Mo_2N_{0.77}C_{0.5}O_{1.88}$	0.92
fcc Mo_2C	2 nm	75 m^2/g	8.5 nm	$Mo_2C_5O_{0.9}$	0.24
hex MoS_2	2 nm	86 m^2/g	4.3 nm	$MoS_{2-\delta}$	
hex $Mo/Fe_2O_3/SO_4^{2-}$	20 nm	150 m^2/g		$0.0005Mo/Fe_2O_3/0.16SO_4$	

In the Mo-carbide and nitride, high concentration of oxygen observed in the results from elemental analysis can be accounted for as an oxide monolayer MoO_3 on the particle surface. Such monolayer would represent about 60 % of the volume of a 5 nm diameter particle. XPS analysis confirmed the existence of this oxide layer on the particle surface, together with the presence of substantial amounts of surface carbon [9]. Oxygen is most probably incorporated into the samples during the passivation treatment (flowing 5% O_2 in He) to which these particles are subjected prior to their removal from the pyrolysis reaction chamber.

Titration of the active sites of Mo_2N and Mo_2C UFP with CO yielded a surface density of active sites approximately half of that reported by Oyama et al [10] for Mo_2N and Mo_2C prepared by TPR. This low density of sites may be due to the obstruction of these sites by the carbon deposited during synthesis together with the incomplete removal of the MoO_3coating by H_2reaction prior to the CO absorption experiments. Notice also that the site density of Mo_2N is almost four times higher than that of Mo_2C This agrees with the fact that the synthesis of the nitride does not involve C_2H_4, which may decompose to surface carbon.

CATALYTIC ACTIVITY

Hydrogenation of Naphthalene

Figure 1 shows the conversion of naphthalene over Mo_2N and $Mo/Fe_2O_3/SO_4$.

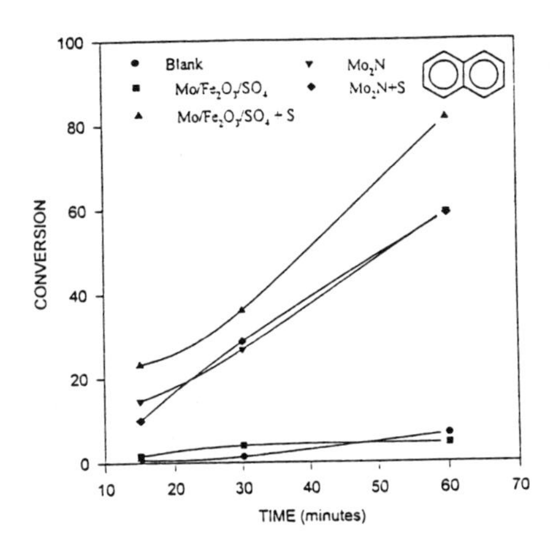

Figure 1
Naphthalene conversion over Mo_2N and $Mo/Fe_2O_3/SO_4$)

It is observed that under thermal conditions (blank), less than 10 % was converted to tetralin. Both catalysts Mo_2N and $Mo/Fe_2O_3/SO_4$) gave tetralin as the primary reaction product of the reaction, with little or no decalin detected. In the presence of sulfur the sulfated hematite gave higher naphthalene conversion (82%) than without additional sulfur (4%), and higher conversion than Mo_2N. This is consistent with the general finding that Fe_2O_3 is rapidly converted to pyrrhotite ($Fe_{1-x}S$) which is considered the active phase of the iron catalyst [1,7]. Addition of sulfur in the presence of Mo_2N did not affect the conversion values. The activity of these catalysts was also compared to that of Mo_2C and MoS_2 for a reaction period of 30 minutes. Mo_2C gave similar values of conversion of naphthalene to tetralin (26%) , whereas MoS_2 gave the highest conversion overall (46%).

Deoxygenation of Diphenyl Ether

Figure 2 shows the conversion and product distribution of diphenyl ether over Mo_2N as a function of time. Mo_2N exhibited higher activity than $Mo/Fe_2O_3/SO_4$ in the reaction of diphenyl ether. The main products of this reaction were benzene, phenol, cyclohexane and cyclohexanol. As seen in Figure 2, about 53% of diphenyl ether was converted over Mo_2N, compared to 35% over the sulfated hematite. The addition of sulfur reduced considerably the activity of Mo_2N to produce an overall conversion of less than 15%. It is observed in Figure 3 that Mo_2N exhibited higher selectivity toward the production of benzene than phenol (50% higher), and that only small amounts of cyclohexane or cyclohexanol were detected in the product mixture. In the case of the sulfated hematite, the proportion benzene/phenol is about 1:1, indicating that only bond cleavage has occurred, and that no oxygen removal was accomplished. Table II presents the results of the conversion (%Conv) of diphenyl ether, percent oxygen removal (%HDO) together with the product distribution observed for Mo_2C and MoS_2. The product distribution is expressed in mol %.

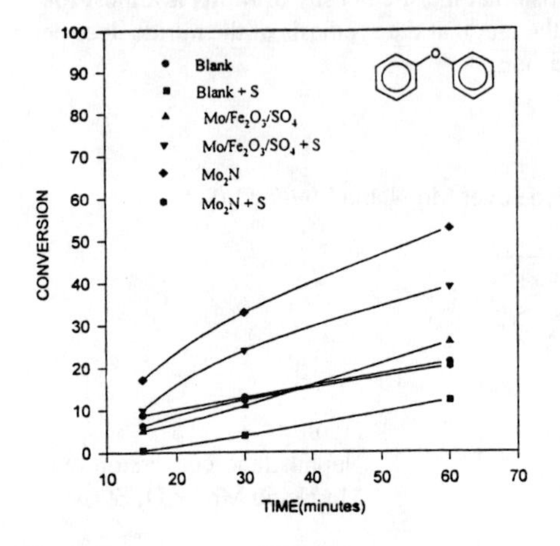

Figure 2
Conversion of Diphenyl Ether as a function of time over Mo_2N and $Mo/Fe_2O_3/SO_4$

Figure 3
Product distribution obtained from the reaction of diphenyl ether over Mo_2N

TABLE II
CONVERSION OF DIPHENYL ETHER

Catalyst	%Conv	% HDO	%DPE	%B	%POH	%CH	%COH
Mo_2N	33	10	50	30	16	3	1
Mo_2C	30	9	53	26	16	4	0
MoS_2	39	23	42	38	11	8	1
$Mo/Fe_2O_3/SO_4{}^{2-}$	24	0	85	5	7	0	1.5

DPE=diphenyl ether B=benzene POH= Phenol CH=cyclohexane COH=cyclohexanol

Similar values of overall conversion were obtained for the Mo_2C and Mo_2N. However, considering that Mo_2C contains about 1/4 of the measured number of active sites of Mo_2N, this catalyst results to be more active. As in the case of the naphthalene reaction, MoS_2 showed higher conversion of diphenyl ether than the other catalysts, and also gave a higher production of hydrogenated products.

Desulfurization of Benzothiophene

Mo_2N and $Mo/Fe_2O_3/SO_4$ exhibited higher activity toward the conversion of benzothiophene than the thermal blank. After 30 minutes of reaction, almost 100% of benzothiophene was converted. The main products detected were ethylbenzene and ethylcyclohexane, as well as dihydrobenzothiophene as an intermediate product. The product evolution with respect to time is shown in Figure 4. This figure shows that ethyl benzene constituted over 80% of the resulting products. The fact that dihydrobenzothiophene appears as an intermediate product suggests that the reaction pathway appears to go via the saturation of the five member ring in benzothiophene, followed by hydrogenolysis and desulfurization to produce ethylbenzene.

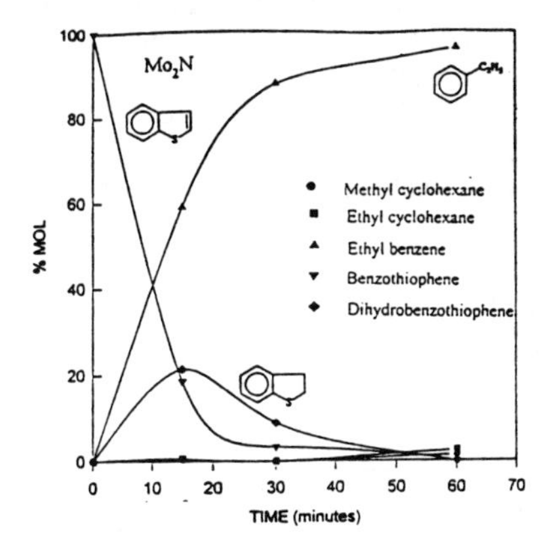

Figure 4
Product evolution of the conversion of benzothiophene over a) Mo_2N and b) $Mo/Fe_2O_3/SO_4$

Table III displays the conversion of benzothiophene (%Conv), percentage of sulfur removal (%HDS) and product distribution over Mo_2C and MoS_2 after 30 minutes of reaction time.

TABLE III
CONVERSION OF BENZOTHIOPHENE

CATALYST	%CONV	%HDS	BZT	HBZT	EB	ECH	OTHER
Mo_2N	96	88	3	9	88	0	0
Mo_2C	98	92	2	6	90	2	0
MoS_2	100	100	0	0	94	5	1
$Mo/Fe_2O_3/SO_4$	100	100	0	3	94	3	0

BZT= benzothiophene HBZT=dihydrobenzothiophene EB=ethylbenzene
ECH=ethylcyclohexane OTHER= cyclohexane, methylcyclohexane.

After 30 min, all catalysts gave comparable values of conversion and product distribution. Mo_2N gave the lowest % HDS (88%) while MoS_2 and $Mo/Fe_2O_3/SO_4$ gave 100%. As observed in the case of the diphenyl ether reaction, Mo_2N gave higher concentration of hydrogenated products in the form of ethylcyclohexane production (about 6% of the total).

Quinoline Denitrogenation

The activity of the present catalyst for denitrogenation of quinoline proved to be low. High conversions of quinoline to tetrahydroquinoline were obtained from all catalysts tested. However, poor selectivity for HDN was observed. Besides tetrahydroquinoline, the other main products of the reaction were o-propylaniline, cyclohexenopyridine and deca-hydroquinoline. Table IV lists the values of final conversions of quinoline, product distribution and % HDN expressed as a mol percentage of denitrogenated products over total products of the reaction. Note that Mo_2N alone was not capable of removing the nitrogen from the quinoline molecule. Longer reaction time periods (2 hours) and higher temperatures (400 C) gave HDN values of about 2.4 %. This low activity was unexpected, since there are reports of higher conversions and selectivity toward the production of propylbenzene [10]. When sulfur was added, % HDN increased from 0 to 3.1 % at 380°C, and the denitrogenated products as detected by GC-MS were mixtures of propylcylohexane and propylcyclohexene. As in the previous cases, Mo_2C gave a similar product distribution as Mo_2N. MoS_2 without sulfur gave a larger proportion of cyclohexenopyridine, cis- and trans-decahydroquinoline than the other catalysts. This catalyst yielded only 1%HDN and the main denitrogenated product was propylbenzene. In the presence of sulfur, the HDN activity of MoS_2 increased to 13% and the ratio of propylbenzene to propylcyclohexane+propylcyclohexene was 13:1.

TABLE IV
CONVERSION OF QUINOLINE

CATALYST	CONV	HDN	THQ	O-PA	CHPYD	DHQ	PB	PCH+ PCHE
Mo_2N	93	0	82	2	4	5	0	0
Mo_2N+S	93	3	71	7	7	5	0	3
MoS_2	94	1	80	3	4	6	1	0
MoS_2+S	95	13	56	13	6	6	1	13
$Mo/Fe_2O_3/SO_4^{2-}$	94	0	83	5	2	3	0	1

THQ=tetrahydroquinoline o-PA=o-propylaniline DHQ=decahydroquinoline(c,t)
CHP=cyclohexenopyridine PB=propylbenzene PCHE+PCHE=propylcyclohexane+ propylcyclohexene

CONCLUSIONS

Nanoscale Mo_2N, Mo_2C and MoS_2 produced by laser pyrolysis exhibited higher activity for heteroatom removal than $Mo/Fe_2O_3/SO_4$. MoS_2 appears to be the most active catalyst overall. Mo_2N gave similar conversion and product distribution as Mo_2C. On the basis of the number of active sites measured by CO chemisorption, Mo_2C was found to be more active than Mo_2N. A low number of active sites on the particles compared to values reported in the literature is attributed to coverage of carbon and oxide on the particle surface. More experiments are underway to study the benefits from removing or avoiding altogether this coating and reduce oxygen in the particles.

REFERENCES

1. F.J. Derbyshire Catalysis in Coal Liquefaction, IEA Coal Research, 1988, p. 16.
2 X.X. Bi, B. Ganguly, G.P. Huffman, F.E. Huggins, M. Endo, P.C. Eklund, J. Mater. Res. **8**(7), 1666 (1993).
3. G.T. Hager, E. Givens, F. Derbyshire, ACS Prep. Div. Fuel Chem. **38**(3), 1087 (1993).
4. R. O'Brien, L. Xu, X.X. Bi, P. Eklund, B. Davis, to be published, Appl. Catalysis, 1994.
5. J. Stencel, P.C. Eklund, X.X. Bi, F. Derbyshire, Catalysis Today, **15**, 285 (1992).
6. S. T. Oyama, R. Kapoor,C. Sudhakar, Prep of Am. Chem. Soc. Div. Fuel Chemistry, **37**(1) 1992.
7. G.T. Hager, X.X.Bi, P.C. Eklund, E. Givens, F. Derbyshire, Energy & Fuels, **8**, 88(1994).
8. V. Pradhan, J. Tierney, I. Wender. Energy & Fuels, **5**, 497 (1991).
9. X.X. Bi, Proc. Mat. Res. Soc. Symp. Proc. T. Boston, 1994.
10 S.T. Oyama, J. C. Schlatter, J. Metcalfe, J. Lambert, Ind. Eng. Chem. Res. **27**, 1639 (1988).

GAS PHASE SYNTHESIS OF MOLYBDENUM AND/OR IRON NITRIDES, CARBIDES AND SULFIDES

Michael R. Close and Jeffrey L. Petersen
Department of Chemistry, West Virginia University, Morgantown, WV 26506-6045

ABSTRACT

The thermolytic decomposition of $Mo(CO)_6$ with hydrogen sulfide or ammonia vapor (in a He carrier stream) at temperatures ranging from 300 to 1100 °C produces high surface area molybdenum sulfides (MoS_2 or Mo_2S_3) or molybdenum carbides (hexagonal Mo_2C) and carbonitrides, (hexagonal MoN(C) or cubic $Mo_2N(C)$), respectively. The MoS_2 surface areas range from 16.7 to 82.0 m^2/g, while the surface areas of molybdenum carbides and carbonitrides vary from 14.9 to 21.1 m^2/g. The maximum surface area for MoS_2 is achieved at 500 °C and decreases with increasing or decreasing temperature. The surface area of the carbonitrides formed from 300 to 800 °C increases with increasing temperature up to 950 °C, where lower surface area Mo_2C is formed. Crystallographically pure hexagonal MoN is prepared by decomposing $Mo(CO)_6$ in pure ammonia. $Fe(CO)_5$ decompositions in ammonia produce Fe_xZ (where $5.8 \geq x \geq 1.6$ and Z=C and N), and in some cases elemental Fe. Hexagonal $Fe_3N(C)$ forms when $Fe(CO)_5$ is thermolyzed in ammonia from 300 to 600 °C, with surface areas ranging from 9.5 to 13.7 m^2/g, whereas orthorhombic Fe_3C and cubic Fe are produced at 700, 800, 900 and 1000 °C with surface areas of 6.7, 7.6, 2.2 and 2.0 m^2/g, respectively. Within the same phase, the surface areas of the carbonitrides increase with increasing reaction temperature. These iron and molybdenum carbonitrides catalyze the conversion of CO/H_2 to alkanes and methanol. Based on preliminary catalytic studies, the highest rate of methane (2850 g/kg/hr at 374 °C) and methanol (440 g/kg/hr at 284 °C) formation was accomplished with an FeMo carbonitride prepared by decomposing $Mo(CO)_6$ and $Fe(CO)_5$ in ammonia at 800 °C.

INTRODUCTION

Heterogeneous catalysts are typically prepared on high surface area supports due in part to the difficulty of preparing high surface area bulk materials. Low surface areas and incomplete phase formation are characteristic of bulk catalysts prepared by conventional solid state syntheses. Producing bulk materials with high surface areas and purity at lower temperatures has motivated researchers to investigate molecular mixing methods, such as chemical vapor deposition, sol-gel processing[1] spray pyrolysis[2] and co-precipitation.[3] Alkali or transition metal modified MoS_2 catalysts have been shown to produce higher alcohols from syn-gas.[4] In addition, transition metal nitrides and carbides have proven to be effective hydrogenation and alcohol forming catalysts.[5] In this research effort, vapor phase pyrolysis has been used to produce finely divided molybdenum based sulfides, nitrides and carbides for use as potential catalysts for higher alcohol synthesis. The variables of temperature, gas composition and flow rate have been evaluated as they relate to phase formation, morphology, composition and material surface area.

EXPERIMENTAL

Chemicals

$Mo(CO)_6$ was obtained from Strem Chemical Co. and purified by sublimation prior to use. $Fe(CO)_5$ was obtained from Aldrich Chemical Co. and used as received. Helium was dried and deoxygenated before use. H_2S and ammonia were obtained from Matheson Gas Products and used as received.

Mat. Res. Soc. Symp. Proc. Vol. 368 © 1995 Materials Research Society

Gas Phase Reactor

The reaction tube used in this study was fabricated from 3 feet of 25 mm fused silica tubing. A liquid cooled jacket was added to the intake end of the tube to facilitate higher yields and maintain greater temperature control. Either end of the tube has Urry-type fused silica connectors. A Lindberg split tube furnace was used to heat the reaction tube. He, H_2S and ammonia gas flow rates were controlled by Matheson rotameters, which were calibrated at WVU.

$Mo(CO)_6$ Reactions

In a typical reaction, 2 grams of $Mo(CO)_6$ were added to a flow tube, which was then connected to the reaction tube at ambient temperature under flowing He. The reaction tube was purged with He and/or a reactive gas (NH_3 or H_2S) and heated to the target temperature. Then, the tube containing the $Mo(CO)_6$ was heated to 80 °C and the gas flows adjusted to appropriate rates. Upon completion of the reaction, the products were isolated in a dry box or in air depending on their reactivity. For the reactions involving H_2S, the He rates ranged from 550 to 720 ml/min and the H_2S rates were constant at 26 ml/min. For the two sets of ammonia reactions the He flow rate was 610 ml/min and the NH_3 rate was 50 ml/min for the first, while for the second set of reactions the NH_3 flow rate was 84 ml/min (no He was used). A typical yield for the H_2S reactions was 41%, based on MoS_2, while the ammonia reactions yielded an average of 51%, based on MoN.

$Fe(CO)_5$ Reactions

For $Fe(CO)_5$ decompositions, eight reactions were completed at temperatures ranging from 300 to 1000 °C in 100 °C increments. In each synthesis, the reaction apparatus was assembled to include a U tube for liquid $Fe(CO)_5$. The furnace temperature and gas flow rates were then adjusted to the appropriate values (He flow rate 100 ml/min through the U tube; NH_3 flow rate: 170 ml/min), 2.0 ml of $Fe(CO)_5$ was injected into the "U" tube at ambient temperature and the injection valve was resealed, thus allowing the He to transport the $Fe(CO)_5$ into the furnace region. The completion of the reaction was noted when all of the liquid $Fe(CO)_5$ had disappeared. All the materials were isolated in the dry box, due to the pyrophoric nature of the iron containing products.

Materials Characterization

X-ray powder diffraction studies were performed on a Philips PW 1800 diffractometer. X-ray photoelectron spectroscopy of the MoS_2 materials were performed at Union Carbide Corp. Analyses of the C/N amounts in the MoN/MoC materials were performed at METC using combustion analyses on a Perkin Elmer CNH analyzer. The BET method (using N_2 gas as the adsorbate) was used for the surface area analyses. Catalytic studies were performed on 500 mg of sample in a flow tube reactor (CO/H_2 = 1, SV=5000-9000 l/kg cat./hr, 750 psi).

RESULTS AND DISCUSSION

MoS_2 Preparations

MoS_2 has been prepared from a thermolytic reaction between $Mo(CO)_6$ and H_2S at 300, 500, 800, 900, 1000 and 1100 °C. X-ray powder diffraction patterns correspond to MoS_2 for 500-1100 °C materials. The pattern for the 300 °C material has extremely broad peaks, although the broad peak positions correspond with crystalline MoS_2. Elemental S and MoS_2 are the only two identifiable phases in each of these compounds. From ESCA data, the S:Mo ratio of selected MoS_2 samples is approximately 2. For samples prepared at 500, 800 and 1100, the Mo and S binding energies match those expected for MoS_2 (S 2p: 162.1 ev, 163.2 ev; Mo 3d: 229.2 ev, 232.4 ev), while the material prepared at 300 °C has electron binding energies of 161.2 ev (single) and 227.1 ev, 230.2 ev for 2p and 3d electrons, respectively. The surface areas of MoS_2 prepared at 300, 500, 900 and 1100 °C were 18.1, 82.0, 29.6 and 16.7 m^2/g, respectively.

The surface area decrease going from 500 to 300 °C is accompanied by a decrease in crystallinity and decrease in 2s and 3d binding energies. This binding energy shift would correspond to a lower oxidation state and/or a modified environment for Mo. When the Mo $d_{5/2}$ peak of 227.1 ev is compared to available reference materials, Mo_2C and Mo(0) carbonyl compounds yield the closest agreement. Thus, it is likely that the 300 °C material contains molybdenum in an oxidation state near zero. The coalescence of the sulfur 2p peaks indicates a distribution of sulfur environments are present. A variety of S environments would be expected for a material lacking any long range order. Therefore, it is probable that a material with short range order corresponding to that in MoS_2 is forming at 300 °C, causing the significant peak broadening in the X-ray powder pattern and a relatively lower surface area.

$Mo(CO)_6$ Decompositions in He/NH_3 and NH_3

The reactions of $Mo(CO)_6$ with He (610 ml/min) and NH_3 (50 ml/min) was studied at 200, 300, 400, 500, 600, 700, 800, 950 and 1100 °C . Four distinct carbonitride or carbide phases were observed in materials produced over this temperature range. A mixture of hexagonal MoN and cubic Mo_2N (or $Mo_{16}N_7$) are present at 700 and 800 °C. Hexagonal Mo_2C and a mixture of hexagonal Mo_2C and Mo are present at 950 and 1100 °C, respectively. X-ray powder diffraction peaks for materials produced at 200 - 600 °C were very broad and corresponded to cubic Mo_2N. C and N percentages for these materials are related inversely and follow the crystallographic trend. The surface areas of these nitrides/carbides range from 14.9 to 21.2 m^2/g (see Fig. 1).

In the second series of $Mo(CO)_6$ decompositions in NH_3 (84 ml/min), the products obtained at 700, 800, 900 and 1000 °C correspond to hex. MoN, hex. MoN, a mixture of MoN and hexagonal Mo_2C and Mo_2C, respectively. The material produced at 800 °C was crystallographically pure MoN. In these two studies, two significant variations were made in the experimental conditions. In addition to using only NH_3, the flow rate was lowered by a factor of 8, which resulted in longer heating times. Transition metal nitrides are unstable thermodynamically with respect to N_2 formation. In the case of MoN, thermal decomposition occurs between 850 and 950 °C.

In contrast to the inverse surface area/temperature correlation observed in the MoS_2 studies, the surface areas of the molybdenum carbonitrides increase with increasing temperature for a given phase. Factors affecting material particle size include the metal carbonyl flux through the system (a function of metal carbonyl vapor pressure and gas flow rates), the furnace temperature, the reactive gas composition, and the resident time of metal in the hot zone. Because the ammonia/$Mo(CO)_6$ reactions were conducted using the same $Mo(CO)_6$ flux rates (i.e., the same gas flow rates, molybdenum carbonyl vapor pressure and thus the same resident heating time), the only variable in these reactions is the furnace temperature. The variation in furnace temperature effects the rate of decomposition (induced by the thermal gradient at the furnace/water jacket interface) and annealing temperature. These two factors have opposing effects on the material surface area. The higher rate of decomposition (corresponding to higher furnace temperatures) for a given phase, should produce more finely divided material. This decomposition is essentially a gas phase precipitation in which the particle size is controlled by the kinetics of the precipitation. Conversely, as temperature is increased, annealing facilitates inter-grain growth resulting in larger grain size and probably larger particle size. Currently, it is not possible to quantify both effects and their influence on particle size for the metal sulfides and nitrides prepared from the gas phase reactions. Generally, it appears that for the molybdenum nitride reactions decomposition kinetics dominate annealing effects, while the surface areas of the molybdenum/H_2S products decrease from 500 to 1100 °C due to a relatively strong annealing influence.

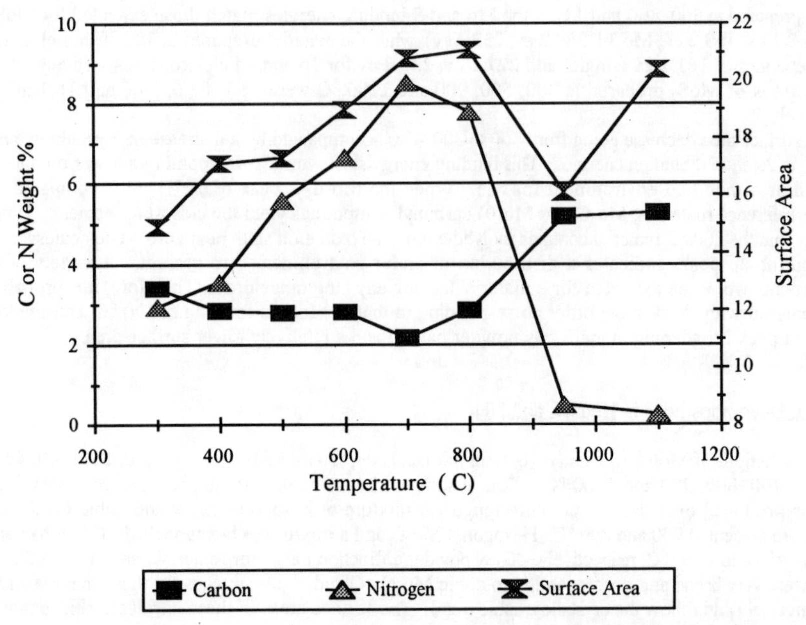

Figure 1. Graph of carbon and nitrogen percentages and surface areas for materials prepared by the reaction of $Mo(CO)_6$ and ammonia vapor at 300, 400, 500, 600, 700, 800, 950, 1100 °C.

$Fe(CO)_5$ Decompositions

$Fe(CO)_5$ was decomposed in ammonia at 300, 400, 500, 600, 700, 800, 900 and 1000 °C. For this reaction series, a crystallographic transformation occurs between 600 and 700°C. For materials formed between 300 and 600 °C, the X-ray powder patterns have been indexed to hexagonal $Fe_xN(C)$ (3>x>1), although the powder pattern of the material produced at 300 °C contained two additional peaks of unknown origin. At temperatures from 700 to 1000 °C, orthorhombic Fe_3C forms in addition to elemental iron. Generally, the carbon and nitrogen amounts vary inversely (similar to the $Mo(CO)_6$ result), with the maximum nitrogen and carbon percentages achieved at 500 and 700 °C, respectively (see Figure 2). The carbon and nitrogen amounts are essentially constant from 800 to 1000°C. The surface areas increase with temperature for a given phase and decrease with higher temperature phases. The direct variation of surface area with temperature for a given phase was also observed in the $Mo(CO)_6$/ammonia reactions. For the hexagonal Fe_3C phase, the surface area increases from 9.45 to 13.67 m^2/g going from 300 to 600 °C. The orthorhombic phase is predominant at 700 and 800 °C, corresponding to increasing surface areas of 6.69 and 7.59 m^2/g. Iron metal accounts for the majority of the material (both Fe and Fe_3C are present) at 900 and 1000 °C, corresponding to a lowered surface area of 2.24 and 2.05 m^2/g at 900 and 1000 °C, respectively.

Catalysis Studies

Six samples were evaluated for their activity towards alcohol production from synthesis gas. The binary compounds, hexagonal-MoN (A), hexagonal-Mo_2C (B) and hexagonal-Fe_3N (C) were produced using procedures detailed previously. Ternary carbonitrides, hexagonal-$(MoFe)_3C(N)$ (D), ortho-$(Mo_{1.3}Fe_{1.7})C(N)$ (E) and ortho-$(MoFe)_3C(N)$ (F) were prepared by thermally decomposing mixtures of

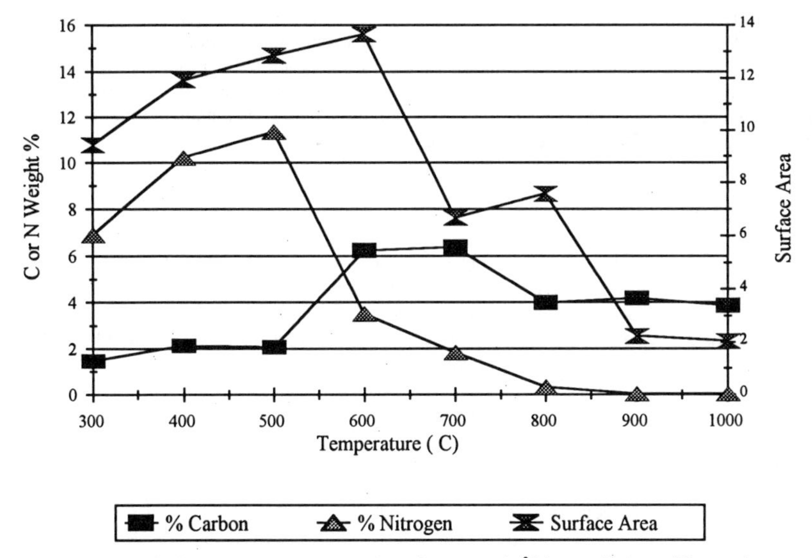

Figure 2. Carbon and nitrogen percentages and surface area (m^2/g) correlation with reaction temperature for iron carbonitrides and carbides produced from vapor phase thermolysis of $Fe(CO)_5$ and NH_3.

$Fe(CO)_5$, $Mo(CO)_6$ and ammonia vapors at 800 °C. Maxium methanol and methane production rates (g/kg/hr) are 40 (266 °C) and 1935 (486 °C), 109 (299 °C) and 1688 (386 °C), 245 (268 °C) and 1749 (375 °C), 440 (284 °C) and 2908 (396 °C), 100 (256°C) and 2004 (426 °C), 177 (224 °C) and 1924 (425 °C) for samples A-F, respectively. MoN and Mo_2C produce the lowest amount of methanol, while C and D produce the highest yield of methanol. The temperature/rate distributions are similar for hexagonal C and D and also for orthorhombic E and F. D and F have higher amounts of Mo and produce more alcohol than C or E, respectively. Based on these six experiments, molybdenum addition to the hexagonal iron nitride phase increases the selectivity to alcohol (methanol). Changing the iron carbonitride phase from orthorhombic to hexagonal shifts the maximum methanol production rate to higher temperatures. All six of these compounds are highly active catalysts and above 350 °C, they convert 95-98 % of CO to product Work on these materials is continuing to determine a phase and composition correlation with catalyst activity and selectivity toward the formation of higher alcohols.

CONCLUSIONS

MoS_2 surface areas can be varied from 17 to 82 m^2/g by varying the vapor phase reaction temperature. The surface areas of MoS_2 can be increased by lowering the reaction temperature from 1100 to 500 °C. From 500 to 300 °C the surface area decreases. This surface area/temperature trend reversal is partially due to the elimination of long range ordering and the chemical modification of molybdenum and sulfur environments. Hexagonal MoN, and relatively pure hexagonal Mo_2C can be produced with surface areas from 14.9 to 21.1 m^2/g, depending on the reaction temperature. By mixing the reactants on a molecular level prior to heating, diffusion controlled kinetic processes are minimized and thermodynamically controlled products are produced. MoN with a surface area of 21.1 m^2/g can be synthesized from this vapor reaction process. The purity of MoN relative to other molybdenum nitrides can be increased by lowering the reactor flow rate and increasing the partial pressure of ammonia.

Relatively pure Mo_2C can be made by varying the reactor temperature. A range of iron carbides and carbonitrides ($Fe_xN(C)$, $1.29 \leq x \leq 5.77$) can be produced by varying the reaction temperature in $Fe(CO)_5$ and ammonia gas phase reactions. The surface areas of iron carbonitrides produced from these vapor phase reactions vary from 13.7 to 2.1 m^2/g, as a function of increasing reaction temperature. The iron and/or molybdenum carbonitrides catalyze the conversion of CO/H_2 to methanol and methane at varying rates depending on crystallographic phase and Mo:Fe ratio.

ACKNOWLEDGEMENT

This work was funded by the Department of Energy under grant number DE-AC22-91PC91034. We thank Professor Edwin Kugler for BET surface area determinations and Dr. Ghaleb N. Salaita for ESCA/XPS analyses. Special appreciation to undergraduate students Michael Delancey and Damian Huff for preparation and X-ray analyses of the iron compounds. Also our appreciation goes to Professor John Renton for assistance with X-ray powder studies and Bob Romanoski and Don Floyd at Morgantown Energy Technology Center for CHN analyses.

REFERENCES

1. (a) C. N. R. Rao and J. Gopalakrishnan, Acc. Chem. Res. **20**, 228 (1987). (b) J. Livage, J. Solid State Chem. **64**, 322 (1986). (c) S. Shibata, T. Kitagawa, H. Okazaki, T. Kimuait, T. Murakami, Jap. J. Appl. Phys. **27**, L53 (1988). (d) J. Livage and J. Lemerle, Ann. Rev. Mater. Sci. **12**, 103 (1982).

2. (a) R. W. McHale, R. W. Schaeffer, A. Kebede, J. Macho, R. E. Salomon, J. Supercond. **5**(6), 511 (1992). (b) S. P. S. Arya, H. E. Hintermann, Thin Solid Films **193-194**(1-2), 841 (1990). (c) E. I. Cooper, E. A. Giess, A. Gupta, Mater. Lett. **7**(1-2), 5 (1988). (d) A. Gupta, G. Koren, E. A. Giess, N. R. Moore, E. J. M. O'Sullivan, E. I. Cooper, E. I. Appl. Phys. Lett. **52**(2), 163 (1988).

3. (a) P. Kumar, V. Pillai, D. O. Shah, Appl. Phys. Lett. **62**(7), 765 (1993). (b) Y. Zhang, M. Muhammed, L. Wang, J. Nogues, K. V. Rao, Mater. Chem. Phys. **32**(2), 213 (1992). (c) N. D. Spencer, Chem. Mater. **2**(6), 708 (1990).

4. (a) G. J. Quarderer and G. A. Cochran, European Patent No. EP-0-0119609, filed 3/11/84 (published 9/26/84), assigned to Dow Chemical Company. (b) X. Youchang, B. M. Naasz, G. A. Somorjai, Appl. Catal. **27**, 233 (1986).

5. (a) E. L. Kugler, L. E. McCandlish, A. J. Jacobson, R. R. Chianelli, U. S. Patent No. 5,071,813 (Dec. 10, 1981); Assigned to Exxon Research and Engineering Co. (b) Hee Chul Woo, Ki Yeop Park, Young Gul Kim, In Sik Nam, Jong Shik Chung, Jae Sung Lee, Appl. Catal. **75**(2), 267-80 (1991).

LASER SYNTHESIS OF IRON NITRIDE NANOPARTICLES FROM PYROLYSIS OF $FE(CO)_5$-NH_3 SYSTEM

X.Q. ZHAO*, Y. LIANG*, F. ZHENG*, Z.Q. HU*, G.B. ZHANG** AND K.C. BAI**
*Dept. of RS Crystallite and Laser Processing, Inst. of Metal Research, Academia Sinica, Shenyang 110015, PR.China.
**The Northeastern University, Shenyang 110015, PR.China.

ABSTRACT

By vapor-phase pyrolysis of $Fe(CO)_5$ and NH_3 induced by CW CO_2 laser, γ'-Fe_4N and ε-Fe_3N nanoparticles(<35nm) were prepared in Ar and N_2 atmospheres. In Ar, γ'-Fe_4N accompanied by a little α-Fe was formed at low temperatures below 650°C, while higher temperatures favored the formation of γ-Fe. In N_2 atmospheres, a mixture of Fe_4N and Fe_3N nanoparticles were obtained at high temperatures. In addition, the influences of synthesis parameters on the structure, morphology and magnetic properties of the nanoparticles were discussed.

Introduction

Currently nanometer-scale systems are of great interest since they possess many physical and chemical properties that differ substantially from those of conventional bulk materials. For example, as catalysts for Fischer-Tropsch reactions, iron and iron-based alloy ultrafine particles have shown excellent selectivity and activity due to their vast surface area and high surfacial activity [1,2,3,4].

To obtain iron nitride fine particles three methods have been developed: (I) evaporation of iron[4], (II) nitriding of iron ultrafine powders[5] and (III) nitriding of α-FeOOH fine powders[6]. The ultrafine particles prepared by the first method are a mixture of α-Fe and Fe_4N with a broad particle size distributions. Approaches (2) and (3) can produce pure Fe_4N acicular particles on a micrometer scale. In the present study, iron nitride(including Fe_4N and Fe_3N) nanoparticles were synthesized by laser-induced pyrolysis of $Fe(CO)_5$-NH_3 system.

Experimental

A 1000W CW CO_2 laser was used as the irradiation source with wavelength of 10.6μm. NH_3 absorbed the laser energy and transfered the absorbed energy to $Fe(CO)_5$ molecules. Meanwhile, $Fe(CO)_5$ molecules decomposed to iron atoms and carbon monoxide, and NH_3 decomposed to nitrogen atoms and hydrogen in the laser-induced reaction zone. Thus, the nitride ultrafine particles were synthesized. The as-prepared particles were collected in a glove-box with inert atmosphere such as Ar or N_2 and then put into alcohol(99.99%). The detailed description for this experimental was given in previous report[7]. By this method, 20-30 grams of ultrafine powder can be produced per hour. A D/Max-rA x-ray diffactometer with a Cu Kα source was used to study the structure of x-ray samples which were prepared

Mat. Res. Soc. Symp. Proc. Vol. 368

by mixing the powder with alcohol(99.99%) and then preparing a slurry on a glass slide. TEM micrographs were obtained in a Philips EM420 microscope. The particle samples were dissolved in alcohol and dispersed by ultrasonic vibration. The solution was then droped onto carbon-covered Cu grids. Mossbauer spectroscopy (Elscint Promeda II) with ^{57}Co source in Pb matrix were performed to investigate the magnetic state of the passivated particles, and then the magnetic properties of these samples were measured by a vibrating sample magnetometer(BHV-55) with a magnetic field of 12KOe.

Results and discussion

i) γ'-Fe_4N particles formed in Ar atmosphere

Table 1 summarizes the process parameters and the corresponding yields synthesized in Ar atmosphere, as revealed by the XRD examination.

Table 1. Synthesis parameters and the yields formed in Ar atmosphere

Samples	Temperatures (°C)	Laser intensities (W/cm^2)	Pressures (KPa)	NH_3 (SLM)	Yields
a	1150	3570	120	0.4	$\gamma'+\gamma$
b	980	2810	80	0.4	$\gamma'+\gamma$
c	850	2160	50	0.4	$\gamma'+\gamma$
d	600	1530	30	0.4	$\gamma'+\alpha$

Temperature of liquid $Fe(CO)_5$ is 50°C, while γ', γ and α denote Fe_4N, γ-Fe and α-Fe, respectively.

At lower temperatures such as 600°C, the yields mainly consist of γ'-Fe_4N and some α-Fe. The γ'-Fe_4N particles are approximately to spherically shaped and connect into particle-chains due to the magnetism as shown in Fig.1. The particles have a very uniform particle size distribution from 10nm to 25nm, with a average particle size of 16nm. In addition, two diffuse lines appearing in the electron diffraction pattern are attributed to the iron oxide film on the particle surfaces formed in the preparation of TEM samples.

As the volume of a ferromagnetic particle is reduced, a size is reached below which the anisotropy energy is smaller than the thermal energy. In that case, thermal fluctuations cause the magnetic moment of the domains to fluctuate randomly between their energy; hence the particles behave like a paramagnet, i.e superparamagnetic particles. Mossbauer spectroscopy indicated that 70% of the iron atoms are in paramagnetic state and 30% are in ferromagnetic state for sample d. The paramagnetic component can be attributed to reduction of particle size of γ'Fe_4N and to the iron oxide microcrystallite formed on the surfaces of the particles. Based on the particle size of 10-25nm, if 2-4nm iron oxide film on the surface of particles are taken into account, the core size range of the particles should be 2-21nm, with a average core size of 8-12nm in diameter. This value is in the regime of the critical particle sizes with superparamagnetic characteristics. The accurate measurement by X-ray diffraction indicated a mean grain size of 9nm for Fe_4N particles, which is in agreement with above analysis and suggest that most Fe_4N particles are single crystallite.

In spite of that, the γ'-Fe_4N particles with a little α-Fe possess a coercive force of 500 Oe

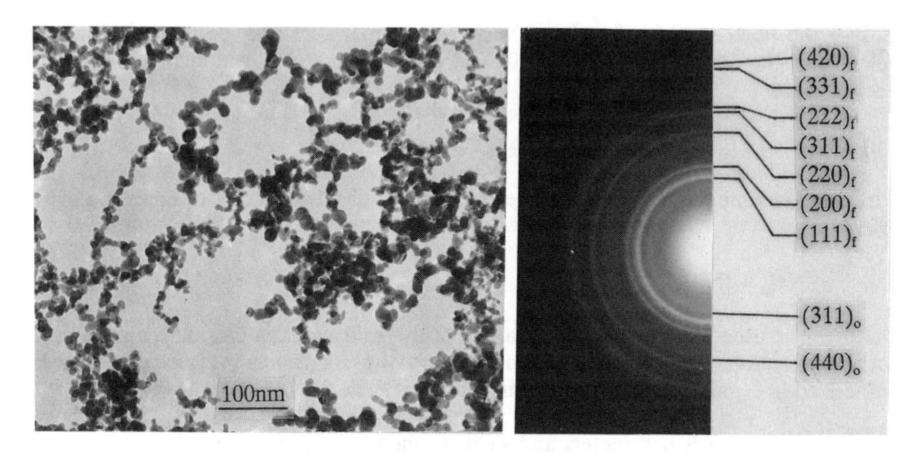

Fig.1.TEM micrograph of Fe_4N nanoparticles and the corresponding electron diffraction pattern.

and saturation magnetization of 58.6 emu/g. Consequently, if all the particles become larger than the critical size below which the particles are superparamagnetic, the saturation magnetization of the nanoparticles should increase to 195.3emu/g which is roughly consistent with the saturation magnetization of pure bulk Fe_4N material(184emu/g).

When synthesis temperature increased, α-Fe phase disappears and γ-Fe phase appears in the synthesized particles. The higher the temperature the more the component of γ-Fe phase becomes present in the yields. The γ-Fe particles have a larger size than γ'-Fe_4N and show clear crystal habits of polyhedra as shown in Fig.2. In contrast to the γ-Fe, γ'-Fe_4N

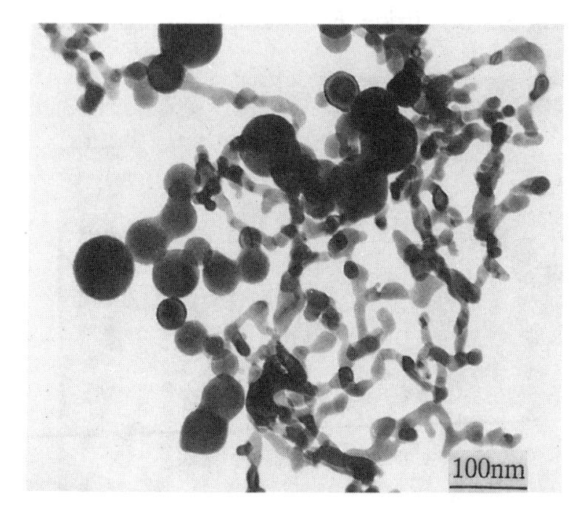

Fig.2. TEM micrograph of Fe_4N nanoparticles coexisted with γ-Fe particles.

particles show no crystal habit due to their very small particle size. It is worth noting that the formed Fe_4N particles become bar-like instead of spherically shape. This may be because the Fe_4N particles formed at very high temperature may be melted or surface-melted and then connected into bar-like particles along the direction of gas stream in reaction zone.

ii) Fe_4N and Fe_3N powders synthesized in N_2 atmosphere

In reaction zone induced by CO_2 laser irradiation, the dissociation of ammonia can be described as

$$NH_3 \rightarrow [N] + N_2 + H_2 \qquad (1).$$

If nitrogen is used as the environmental atmosphere, It would affect the balance of this pyrolysis reaction. The process parameters and the corresponding yields of powders synthesized in N_2 atmosphere were summarized in Table.2.

Table 2. The synthesis parameters and yields formed in N_2 atmosphere

Sample	Temperatures (°C)	Laser intensities (W/cm^2)	Pressure (KPa)	NH_3 (SLM)	Yields
m	650	1500	25	650	$\gamma'+\alpha$
n	850	1500	50	780	$\gamma'+\epsilon$
o	1000	1500	80	950	$\gamma'+\epsilon$
p	1150	1500	110	1100	$\gamma'+\epsilon$

The temperature of liquid $Fe(CO)_5$ is 50°C, while γ' and ϵ denote Fe_4N and Fe_3N phase, respectively.

From the above results, we can conclude that using nitrogen as atmosphere gas is favorable to the formation of iron nitride. For example, at high temperatures near 1150°C,

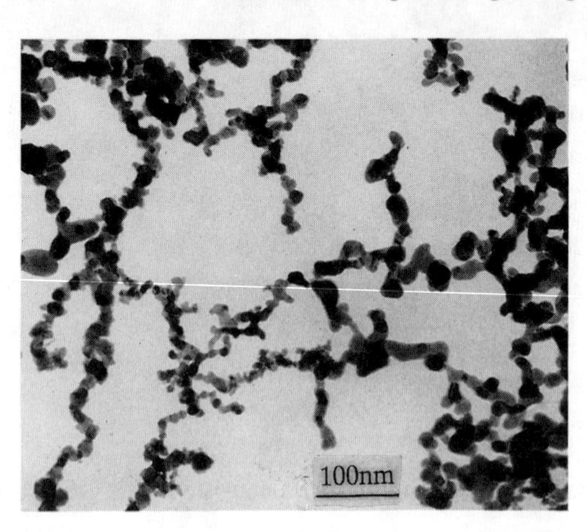

Fig.3. TEM micrograph of Fe_4N and Fe_3N nanoparticles formed at 1000°C and 80KPa.

there was no evidence of γ-Fe phase in the X-ray diffraction patterns. From a point of viewof chemical reaction equilibrium, the environmental atmosphere of nitrogen may give rise to an increase of active N atoms in the dissociation reaction of NH_3. More active N atoms favor the formation of Fe_3N, which is consistent with the Fe-N phase diagram[8]. Fig.3 shows the electron micrograph of a mixture of Fe_4N and Fe_3N ultrafine particles synthesized at 1000°C and 80KPa. The particles size ranges from 15 to 35nm, with an average size of 23nm in diameter. Based on accurate measurement of X-ray diffraction, the average grain size of Fe_4N and Fe_3N particles are 21.9nm and 17.3nm, respectively. These results suggest that most particles are single crystalline, and the individual particles only consist of one phase, either Fe_4N or Fe_3N phase. Nevertheless, the Fe_4N or Fe_3N individual particles are very difficult to determine due to their very small particle size. For this sample, Mossbauer spectra analysis indicated that the ferromagnetic and superparamagnetic components accounted for 57.6% and 43.4%, respectively. By comparison with the sample of Fe_4N prepared in Ar, the increase of ferromagnetic component may be assigned to the enlargement of the particle size. Meanwhile, the coercive force and saturation magnetization increased upto 580 Oe and 97.1 emu/g, respectively, which is roughly consistent with the estimation previously discussed. To enlarge or reduce the particle size and, accordingly, to have the all ultrafine particles be either ferromagnetic or superparamagnetic, the synthesis conditions and technology must be improved. As catalytic materials for the F-T synthesis, the iron oxide on the surfaces of the ultrafine particles have influences on the catalytic activities and selectivities[2]. Therefore, the treatment of as-synthesized ultrafine particles without oxidation is a important subject. Investigations on catalytic properties of the iron nitride ultrafine powders and the effects of the oxidation are currently under way.

Conclusions

Iron nitride nanoparticles (<35nm) were synthesized by laser-induced vapor-phase synthesis process. The particles are spherical, very small, and with a narrow size distribution.

The γ'-Fe_4N ultrafine particles coexisted with a little α-Fe were synthesized in Ar atmosphere and at lower temperature. With N_2 as atmosphere, a mixture of Fe_4N and Fe_3N particles were prepared at higher temperature and pressure. Despite large superparamagnetic component present, these particles revealed high magnetic properties.

References

1. A.Gupta and J.Yardley, SPIE, 458(1985) 131
2. I.Itoh, S.Nagano,T.Urata and E.Kikuchi, Appl. Catal. 67(1990) 1; 67(1991) 215; 77(1991)37
3. E.Yeh, N.Jaggi, J.Butt and H.Schwartz, J. Catal. 91(1985) 231
4. N.Saegusa, T.Tsukagoshi, E.Kita and A.Tasaki, IEEE Trans. on Magn. Mag.19,5(1983) 1629
5. K.Tagawa, E.Kita and A.Tasaki, Jpn. J. Appl. Phys. 21(1982) 1596
6. A.Tasaki, K.Tagawa and G.Kita, IEEE Trans. on Magn. Mag-17, 6(1981) 3026
7. X.Q.Zhao, Y.Liang, Z.J.Cui, K.S.Xiao, F.Zheng and Z.Q.Hu, Nanostructured Mater. 4(1994)397
8. H.Wriedt, N.Okcen and R.Nafziger, Bulletin of Alloy phase Diagrams, 8(1987) 355

SURFACE PROPERTIES OF MOLYBDENUM NITRIDE THIN FILMS

HYUEK JOON LEE, MANDAR S. MUDHOLKAR AND LEVI T. THOMPSON
Department of Chemical Engineering, The University of Michigan, Ann Arbor, MI 48105-2136

ABSTRACT

The interaction of NH_3 and CH_3NH_2 with the surfaces of β-$Mo_{16}N_7$, γ-Mo_2N, and δ-MoN films was investigated using thermal desorption spectroscopy. Ammonia temperature programmed desorption (TPD) spectra for the films were similar. Ammonia TPD (adsorption at ~280 K) produced an NH_3 peak at ~360 K. Some of the NH_3 decomposed into H_2 and N_2. Two H_2 desorption peaks were produced: a low temperature peak due to recombination of surface hydrogen and a high temperature peak due to hydrogen that emerged from the nitride subsurface. The N_2 desorption spectrum consisted of a peak at ~340 K and several peaks in the range 500-900 K. The desorption kinetics depended on the structure and composition of the film. $^{15}NH_3$ TPD experiments indicated that the low temperature N_2 desorption peak was due to NH_3 decomposition, while the origin of the high temperature peaks was the nitride itself. We believe that nitrogen desorption from the nitride was induced by the presence of hydrogen which altered the Mo–N bonding. Three different CH_3NH_2 decomposition processes were observed: complete decomposition of CH_3NH_2 into H_2, N_2 and C, partial decomposition into HCN, and simple C–N bond hydrogenolysis into CH_4 and NH_3. The decomposition pathways depended on the structural and compositional properties of the films. All three processes were observed for the β-$Mo_{16}N_7$(400), β-$Mo_{16}N_7$(203) and δ-MoN(002) films, while only trace amounts of HCN were detected for the γ-Mo_2N(200) film, suggesting that the partial decomposition into HCN was not favored on this surface. The β-$Mo_{16}N_7$(400) and δ-MoN(002) films had the high selectivities for simple C-N bond cleavage.

INTRODUCTION

Molybdenum nitrides have received a great deal of attention due to their competitive activities for commercially important reactions including hydrodenitrogenation [1], hydrodesulfurization [2], and ammonia synthesis [3]. For reactions such as hydrodenitrogenation and ammonia synthesis, the interaction of NH_3 with the catalyst surface is of particular importance. For example, Lee et al. [4] found that NH_3 suppressed hydrogenolysis and dealkylation during quinoline hydrodenitrogenation over high surface area Mo nitrides.

We have examined the interactions of several molecules of importance in hydrodenitrogenation catalysis with the surfaces of well-defined Mo nitride films. This paper describes our use of NH_3 and CH_3NH_2 temperature programmed desorption (TPD) to probe the surface properties of nearly phase-pure β-$Mo_{16}N_7$, γ-Mo_2N and δ-MoN films.

EXPERIMENTAL DETAILS

Phase pure β-$Mo_{16}N_7$, γ-Mo_2N and δ-MoN films were synthesized using Ion Beam Assisted Deposition (IBAD). Ion Beam Assisted Deposition involves simultaneous deposition of atoms and low energy ions. The molybdenum was deposited at a rate of 5 to 6 Å/s onto single crystal

Mat. Res. Soc. Symp. Proc. Vol. 368 © 1995 Materials Research Society

Si(100) substrates using an electron beam evaporator. The nitrogen ion flux was adjusted to achieve the desired film properties. Depositions were carried out using 500 eV nitrogen ions with a nitrogen-backfill pressure of 5×10^{-5} torr inside the deposition chamber. Additional details concerning the deposition of these films, and their compositions and structures can be found elsewhere [5].

The NH_3 and CH_3NH_2 TPD experiments were performed in an ultrahigh-vacuum chamber [6]. The chamber was equipped with a UTI quadrupole mass spectrometer for thermal desorption studies and an x-ray photoelectron spectrometer for surface analysis. The base pressure in the chamber was less than 8×10^{-10} torr. The films, mounted on a Mo platen, were heated to ~950 K using a heating rate of 6 K/s during the TPD experiments. Frequently the films were annealed at elevated temperatures prior to the TPD experiment. This treatment, which removed subsurface hydrogen, did not significantly alter the surface composition or structure. The films were cooled using liquid nitrogen. NH_3 and CH_3NH_2 were introduced directly to the film surface using a variable leak valve. The sample was turned to face the mass spectrometer during the TPD experiments to enhance the signal-to-noise ratio and minimize the effect of differing pumping speeds on the detection sensitivity for the various products. NH_3 and CH_3NH_2 were adsorbed at ~280 K, and the amounts of H_2, CH_4, NH_3, H_2O, HCN, N_2, CH_3NH_2 and CO_2 or N_2O were routinely measured. The exposure is reported in units of Langmuir based on the pressure in the chamber as determined using an ionization gauge uncorrected for gas composition. The films were cleaned when necessary by repeated sputtering with 500 eV Ar^+ ions at ambient temperature followed by an anneal at elevated temperatures.

RESULTS AND DISCUSSION

NH_3 TPD

Figure 1 shows a series of NH_3 TPD spectra which were collected for β-$Mo_{16}N_7$(400), γ-Mo_2N(200) and δ-MoN(101) films following a dose of 240 L. The low temperature peak appeared in a similar temperature range for each of the films. For the δ-MoN(101) film, the peak shifted to lower temperatures with increasing NH_3 exposure, indicative of second-order desorption. There are at least two mechanisms that could account for the second-order desorption of NH_3. Adsorbed NH_3 could have decomposed into NH_x and H species, and upon heating recombined to form NH_3. This type of desorption mechanism was suggested by Haddix, *et al.* [7] for γ-Mo_2N powders. Alternately, NH_3 could have bridge-bonded to two surface sites. For the β-$Mo_{16}N_7$(400) and γ-Mo_2N(200) film, the peaks did not shift indicating first-order desorption. According to Redhead [8], the energy for first order desorption can be estimated using the following equation

$$E = RT_p\,[\ln(\nu_o T_p/\alpha) - 3.64], \qquad (1)$$

where E is the desorption energy, R is the gas constant, T_p is the temperature corresponding to the maximum desorption rate (K), ν_o is the pre-exponential factor (sec^{-1}), and α is the heating rate (K/sec). We could not independently estimate the pre-exponential factor; therefore, we assumed a value of 10^{13} sec^{-1}, which is commonly used [8]. Typically, desorption energies are not very sensitive to order of magnitude changes in ν_o. For a pre-exponential factor of 10^{13} sec^{-1} with T_p = 360 K and α = 6 K/sec, the desorption energy was estimated to be 22 kcal/mol for the β-$Mo_{16}N_7$(400) and γ-Mo_2N(200) films.

Significant amounts of H_2 and N_2 also desorbed, indicating that some of the NH_3 decomposed completely into hydrogen and nitrogen. Figure 2 shows the H_2 desorption spectra at a dose of 240 L. The dominant feature in the spectra was a peak at ~360 K, which was coincident with that for NH_3 desorption. This desorption temperature was also similar to that for the low energy site observed during H_2 TPD from these films [6], which suggested that NH_3 decomposed shortly after adsorption and perhaps that these sites were different from those from which NH_3

desorbed. For γ-$Mo_2N(200)$ an additional peak was detected. This high temperature H_2 desorption peak is believed to be due to subsurface hydrogen sites [1,6]. For the δ-MoN(101) film hydrogen desorption appeared to be second-order as a shift in peak position was observed with increasing exposure.

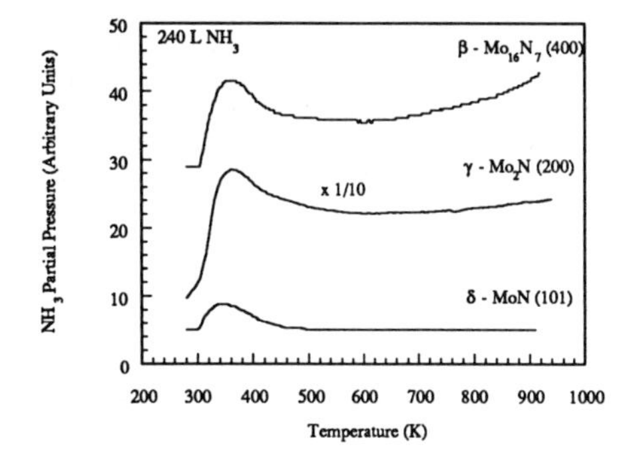

Fig. 1. NH_3 TPD spectra for β-$Mo_{16}N_7(400)$, γ-$Mo_2N(200)$ and δ-MoN(101) films following a dose of 240 L.

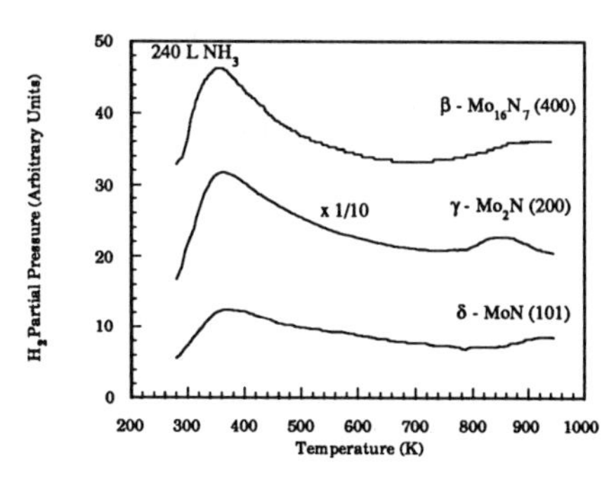

Fig. 2. H_2 spectra for β-$Mo_{16}N_7(400)$, γ-$Mo_2N(200)$ and δ-MoN(101) films following 240 L NH_3 dose.

Figure 3 compares the N_2 desorption spectra obtained for the β-$Mo_{16}N_7(400)$, γ-$Mo_2N(200)$ and δ-MoN(101) films following exposure to 240 L of NH_3. There were no significant differences between their characteristics at low temperature except for the amount of nitrogen that desorbed. The nitrogen desorption rate from the δ-MoN film was less than those from the β-$Mo_{16}N_7$ and γ-Mo_2N films. The high temperature desorption characteristics between 500 K and 900 K were complex and depended on the film structure.

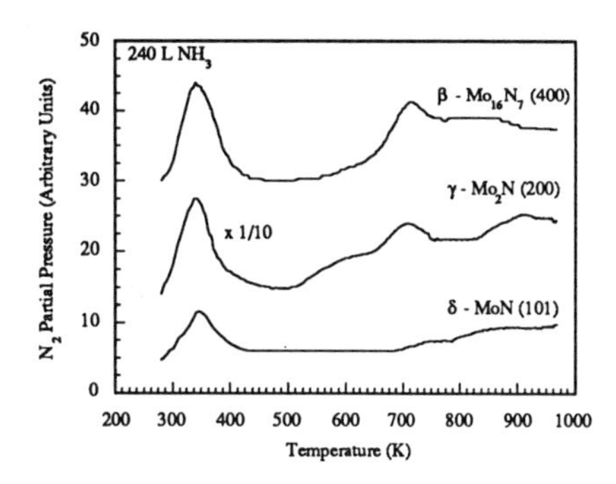

Fig. 3. N_2 spectra of β-$Mo_{16}N_7(400)$, γ-$Mo_2N(200)$ and δ-MoN(101) films following 240 L NH_3 dose.

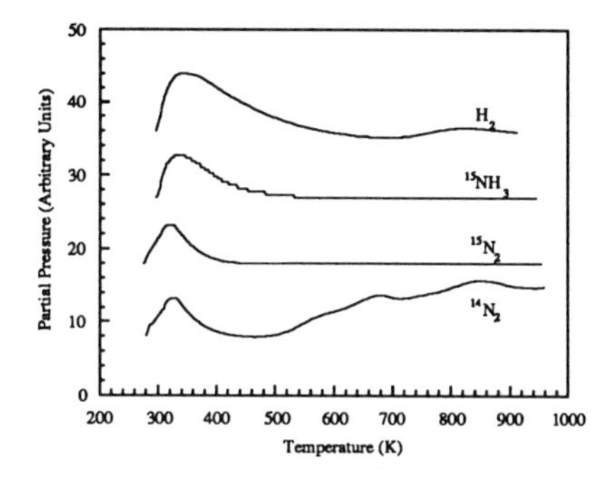

Fig. 4. $^{15}NH_3$ TPD spectra for γ-$Mo_2N(200)$ film.

To identify all the sources of N_2, we carried out a series of $^{15}NH_3$ TPD experiments on the films. The spectra were comprised of peaks for $^{14}N_2$ and $^{15}N_2$; however, there was no evidence of $^{14}N^{15}N$ (Fig. 4). Both $^{14}N_2$ and $^{15}N_2$ desorbed at low temperatures, while only $^{14}N_2$ desorbed at temperatures in excess of 450 K. The low temperature $^{14}N_2$ and $^{15}N_2$ peaks were similar to that observed during H_2 TPD [6]. The absence of high temperature $^{15}N_2$ desorption peaks suggested that the source of the high temperature nitrogen was the Mo nitride film itself. Subsurface hydrogen may have influenced the removal of nitrogen from the film. Results from a previous study suggested that hydrogen migrates into the subsurface layers of Mo nitride powders and films during H_2 TPD [1,6]. This hydrogen may have weakened the Mo–N bonds, facilitating the diffusion of nitrogen out of the subsurface region to the surface during heating.

Effects of orientation on the sorptive properties were determined by comparing NH_3 TPD spectra for the δ-MoN(101) and δ-MoN(002) films. Ammonia desorption was observed at a higher temperature for the δ-MoN(002) film (~420 K) than that for the δ-MoN(101) film. Furthermore, the H_2 and N_2 desorption spectra had very different characteristics.

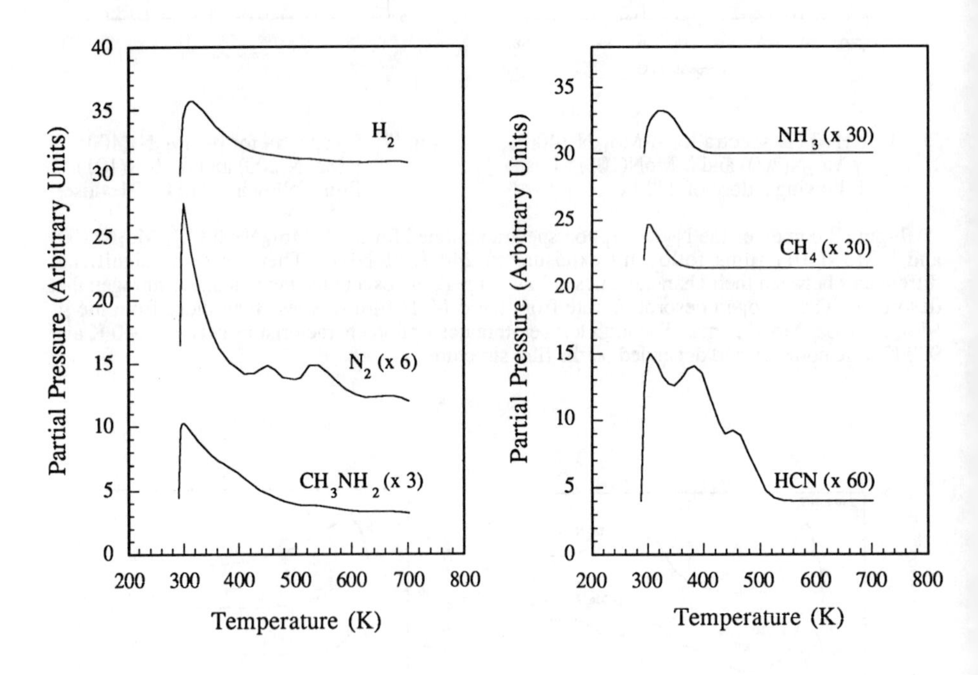

Fig. 5. CH_3NH_2 and products formed during CH_3NH_2 TPD from the β-$Mo_{16}N_7$(400) film (90 L CH_3NH_2 dose).

CH_3NH_2 TPD

Figure 5 shows the CH_3NH_2 TPD spectra for the β-$Mo_{16}N_7$(400) film at a dose of 90 L. The results suggested at least three different decomposition processes: complete decomposition of CH_3NH_2 into H_2, N_2 and C, partial decomposition into HCN, and selective C–N bond activation to produce CH_4 and NH_3. The low temperature H_2 and N_2 peaks were similar to those observed during the NH_3 TPD experiments, suggesting that part of the CH_3NH_2 decomposed shortly after adsorption. It is also interesting to note that the NH_3 peak appears to have similar characteristics to those observed during NH_3 TPD.

The influence of orientation can be assessed by comparing the spectra of β-$Mo_{16}N_7$(400) with those of β-$Mo_{16}N_7$(203) (Fig. 6). Although all six products were again observed, the peak characteristics were very different, suggesting different decomposition routes.

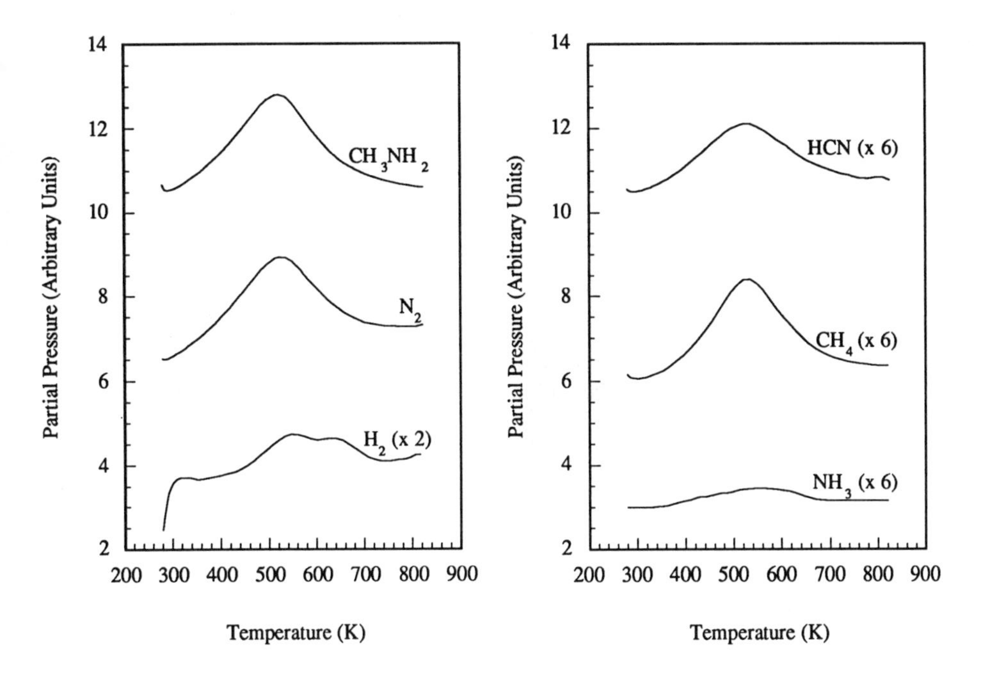

Fig. 6. CH_3NH_2 and products formed during CH_3NH_2 TPD from the β-$Mo_{16}N_7$(203) film (6 L CH_3NH_2 dose).

Spectra from the γ-Mo_2N(200) film showed completely different desorption characteristics. CH_3NH_2, N_2 and H_2 were observed, but the positions and shapes of these peaks were very different from those of the β-$Mo_{16}N_7$(400) film. Only trace amounts of HCN desorbed from this film. The partial decomposition via C–N bond activation was apparent from the presence of CH_4, though no NH_3 was detected. We believe that NH_3 further decomposed into N_2 and H_2, which is supported by an observation of NH_3 decomposition sites for this film.

Spectra from the δ-MoN(002) film resembled those of the β-$Mo_{16}N_7$(203) film except for some subtle differences. Only one H_2 desorption peak was detected at ~560 K, while two overlapping H_2 peaks were observed in the same temperature range for the β-$Mo_{16}N_7$(203) film. The differing relative intensities also indicated that the site distributions were different for these two films.

Comparison with Mo nitride powders

A series of unsupported Mo nitride powders with surface areas ranging up to 200 m^2/g was examined in an independent study [9]. The Mo nitride powders also decomposed NH_3 as we observed for the films. At low coverages only H_2 desorption was observed, indicating that the decomposition sites were populated first. A similar behavior was observed for the Mo nitride films. Two H_2 desorption peaks were produced for the low surface area Mo nitrides (< 20 m^2/g), while only one peak was observed for the higher surface area Mo nitrides. Recall that the γ-Mo_2N film produced two H_2 desorption peaks while the β-$Mo_{16}N_7$ and δ-MoN films produced one peak within the temperature range studied. This implied that the low surface area Mo nitrides possessed γ-Mo_2N-like species on the surface, and β-$Mo_{16}N_7$ and/or δ-MoN-like species were present for the higher surface area Mo nitrides.

SUMMARY

Temperature programmed desorption was used to investigate the interaction of NH_3 and CH_3NH_2 with molybdenum nitride thin films. The surface chemistry was very dependent on the phase and orientation of the films. Ammonia desorbed molecularly and decomposed into H_2 and N_2 during NH_3 TPD from the β-$Mo_{16}N_7$(400), γ-Mo_2N(200) and δ-MoN(101) films. No significant NH_3 decomposition was observed for the δ-MoN(002) film. Small amounts of nitrogen were also removed from the films in the form of N_2. There were a several different CH_3NH_2 decomposition pathways observed. Selectivity for simple hydrogenolysis of the C–N bond was greatest for the δ-MoN(002) film. Information derived from the films was useful in understanding the surface properties of Mo nitride powders, materials that have demonstrated activities for hydrodenitrogenation and hydrodesulfurization.

ACKNOWLEDGMENTS

Financial support for this work was provided by the National Science Foundation (NSF-CTS-8918107).

REFERENCES

[1] J.G. Choi, J.R. Brenner, C.W. Colling, B.G. Demczyk, J.L. Dunning and L.T. Thompson, Catalysis Today 15 (1992) 201.
[2] H. Abe and A.T. Bell, Catal. Lett. 18 (1993) 1.
[3] L. Volpe and M. Boudart, J. Phys. Chem. 90 (1986) 4874.
[4] K.S. Lee, H. Abe, J.A. Reimer and A.T. Bell, J. Catal. 139 (1993) 34.
[5] E.P. Donovan, G.K. Hubler, M.S. Mudholkar and L.T. Thompson, Surf. Coat. and Technol., in press (1994).
[6] J.G. Choi, H.J. Lee, and L.T. Thompson, Appl. Surf. Sci. 78 (1994) 299.
[7] G.W. Haddix, D.H. Jones, J.A. Reimer and A.T. Bell, J. Catal. 112 (1988) 556.
[8] P.A. Redhead, Vacuum 12 (1962) 203.
[9] C.W. Colling, J.G. Choi and L.T. Thompson, J. Catal. submitted (1994).

ON THE STABILITY OF Mo_2N DURING FIRST-STAGE HYDROCRACKING

J.A.R. VAN VEEN*, J.K. MINDERHOUD*, J.G. BUGLASS*, P.W. LEDNOR*
AND L.T. THOMPSON**
*Koninklijke/Shell-Laboratorium, Amsterdam, P.O. Box 38000, 1030 BN Amsterdam, The Netherlands
**University of Michigan, Department of Chemical Engineering, Ann Arbor, MI 48109-2136, U.S.A.

ABSTRACT

An unsupported sample of Mo_2N has been subjected to a first-stage hydrocracking test. The evolution of the HDS and HDN performance indicated a transformation of Mo_2N into MoS_2. This was substantiated by XPS and TEM, the latter technique showing that the transformation is limited to only a few surface layers.

INTRODUCTION

Molybdenum nitrides and carbides have, in recent years, emerged as hydroprocessing catalysts with interesting performance characteristics [1–11]. Under certain conditions, very high activities can be observed, and selectivities in HDS and HDN model reactions are often vastly different from those seen for traditional sulfidic catalysts. This is rather remarkable for feeds containing (some) sulfur in that, from thermodynamics, one would predict that even at fairly low H_2S partial pressures, conversion to MoS_2 should take place [1]. However, to our knowledge, clear analytical evidence for such a transformation has not been reported.

It has been surmised by Schlatter et al. [1] that their Mo_2C catalyst became covered by a surface layer of MoS_2 in the presence of H_2S because its performance tended towards that of MoS_2. However, although they made sure there was no bulk sulfidation (XRD), the existence of the sulfurised surface layer was not further substantiated. Remarkably however, Markel and Van Zee [3] did not find any evidence for MoS_2 with Raman spectroscopy — a very sensitive technique in this regard — upon sulfiding Mo_2N at 673 K. Also, Abe et al. [9] found evidence for strong adsorption of H_2S on Mo_2N, but did not suspect the actual formation of MoS_2. Lastly, Oyama [11] argued on the basis of XPS spectra that Mo_2N incorporates very little surface sulfur during coal liquids processing (containing 100–800 ppm S).

Intrigued by the above results, we decided to test the hydroprocessing capabilities of a molybdenum nitride catalyst under practical conditions and a first-stage hydrocracking test was selected. A preliminary study of a Mo oxynitride had already shown that it is liable to sulfidation under certain (model) test conditions, e.g. at 50 bar pressure, 3 %v H_2S, gas phase test [12]. Here we used an unsupported Mo_2N sample and its performance was compared to that of a typical sulfided NiMoP/alumina catalyst.

EXPERIMENTAL

As described before, the Mo_2N sample was prepared by reducing MoO_3 with ammonia [5]. Briefly, the sample was heated from room temperature to 623 K in about 30 minutes, further

Mat. Res. Soc. Symp. Proc. Vol. 368

heated to 723 K at 40 K/h and then to 973 K at 100 K/h where it was held for an hour. After cooling under flowing NH_3, the sample was purged with He and passivated with 1 %v O_2 in He. Its BET surface area was determined to be 18 m^2/g. XRD showed the bulk to consist of crystalline γ-Mo_2N, Mo and a trace of MoO_2. The surface composition was, according to XPS: N/Mo = 0.89, O/Mo = 1.13 and S/Mo = 0.0 (atomic ratios).

About 4 cm^3 of this material — as 30–80 mesh particles mixed with an equal volume of SiC — was loaded into a microflow unit and tested under typical first-stage HC conditions [13]. The feed, a Vacuum Gas Oil of Middle Eastern origin, contained about 1100 ppmw organic N and nearly 3 %w organic sulfur. A hydrogen partial pressure of slightly more than 100 bar was applied and weight-average-bed-temperatures were between 370 and 405°C. After the test, which took some 300 hours, the catalyst was flushed with toluene, unloaded and submitted for XRD, BET, XPS and TEM analysis.

RESULTS & DISCUSSION

After some 50 hours on stream, the molybdenum nitride catalyst had an HDN activity comparable to that of a commercial first-stage hydrocracking catalyst when the data are normalised to the same BET surface area and first-order kinetics are assumed. (So, actually its weight-based activity is about ten times lower, its volume activity less so in view of the very high bulk density, 2.5 g/cm^3). This is in itself an interesting finding, inasmuch as it shows that the activity of an 'unpromoted' material can be similar to that of a practical catalyst. However, during the run, the HDN activity gradually decreased and, at the end, the activity was only about one-third of that at the beginning.

The HDS activity, on the same basis as above but now assuming approximately second-order kinetics, is initially approximately ten times lower than that of commercial catalysts. In contrast to the HDN activity, however, it remains fairly constant over the run. This could be construed as indicating that the catalyst becomes more sulfide-like in that the k(HDN)/k(HDS) ratio becomes more 'normal'.

At the very end of the test, the hydrogen pressure was decreased, giving rise to a substantial loss in HDN activity, which at least is similar to what is observed for conventional sulfided NiMo/carrier catalysts.

As to the hydrogenation of aromatics, the activity observed is extremely low. Quantification in terms of a relative activity compared to conventional catalysts is, therefore, very difficult and was not attempted.

These results, it will be clear, may well point to a (partial) conversion of molybdenum nitride into the sulfide in the course of the run. This hypothesis, already enunciated by Schlatter et al. [1] for molybdenum carbide is, in fact, supported by the analytical data we will now discuss.

A rather instructive TEMicrograph of the spent catalyst is reproduced in Fig. 1. It is clearly seen that, while the bulk of this (typical) Mo_2N particle has remained intact, a layered continuous phase, some 1.2–2.4 nm 'deep', has formed on its surface. The layer separation of 0.6 nm is consistent with that of MoS_2. This assignment is in agreement with the XPS measurements which indicate the surface composition to be N/Mo = 0.38, O/Mo = 0.84 and S/Mo = 0.76. However, XRD only sees the unconverted bulk, Mo_2N and Mo. As one would expect from such a smooth surface transformation, the BET surface area is also unchanged (once the measured value, 10 m^2/g, is corrected for the fact that now the sample is a mixture of Mo_2N and SiC).

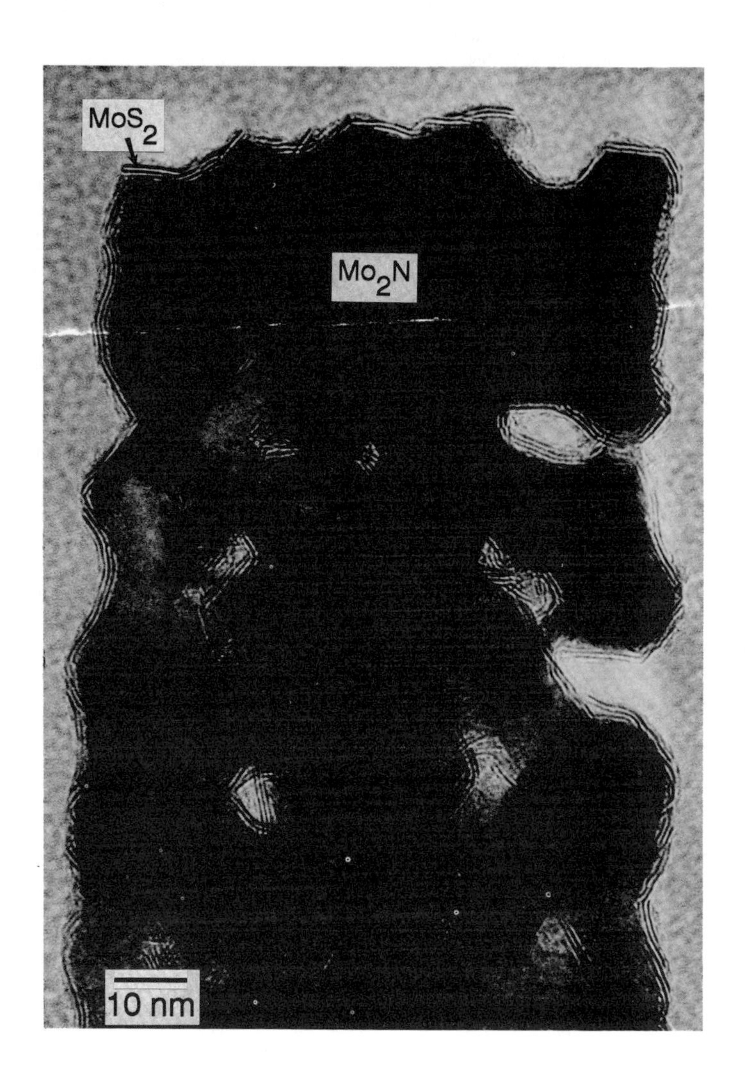

Fig. 1. Micrograph of the Mo_2N spent catalyst.

There is, therefore, no doubt that the thermodynamically expected transformation of molybdenum nitride into molybdenum sulfide has, in fact, taken place but, at the same time, it is clear that it is restricted to a rather thin surface layer. From the TEM result we would surmise that the MoS_2 layers are wrapped so well around the crystallites as to inhibit the further penetration of sulfur species into the nitride and/or the escape of nitrogen out of it. In all, the application of unsupported molybdenum nitride (and the same probably goes for the carbide) in first-stage hydrocracking as it is presently practised offers no advantage over $NiMoP/Al_2O_3$ as it quickly sulfides.

CONCLUSION

Unsupported molybdenum nitride is unable to withstand industrial first-stage hydrocracking conditions in that its surface is gradually covered with a thin MoS_2 layer. The activity and selectivity of the Mo nitride are lost once sulfided. That the surface is sulfided follows not only from the evolution of the HDN and HDS activity during the run, but is also proven by analytical data (TEM, XPS, XRD). Future work will show whether it is possible to stabilize the highly active Mo nitride phase, e.g. through the use of a support [10].

REFERENCES

1. J.C. Schlatter, S.T. Oyama, J.E. Metcalfe III and J.M. Lambert, Jr., Ind. Eng. Chem. Res. **27**, 1648 (1988).
2. D.J. Sajkowski, S.T. Oyama, Prepr. – Am. Chem. Soc., Div. Pet. Chem. **35**, 233 (1990); U.S. Patent No. 5200060 (26 April 1991).
3. E.J. Markel and J.W. van Zee, J. Catal. **126**, 643 (1990).
4. S.T. Oyama, R. Kapoor and C. Sudhakar, Prepr. Pap. – Am. Chem. Soc., Div. Fuel Chem. **37** (1), 156 (1992).
5. J.G. Choi, J.R. Brenner, C.W. Colling, B.G. Demczyk, J.L. Dunning and L.T. Thompson, Catal. Today **15**, 201 (1992).
6. M. Nagai and M. Toshihiro, Catal. Lett. **15**, 105 (1992).
7. K.S. Lee, H. Abe, J.A. Reimer and A.T. Bell, J. Catal. **139**, 34 (1993).
8. M. Nagai, M. Toshihiro and T. Takashi, Catal. Lett. **18**, 9 (1993).
9. H. Abe, T. Cheung and A.T. Bell, Catal. Lett. **21**, 11 (1993).
10. C.W. Colling and L.T. Thompson, J. Catal. **146**, 193 (1994).
11. S.T. Oyama, Catal. Today **15**, 179 (1992).
12. P. Moureaux, unpublished results (1992). C. Sayag, Ph.D. Thesis, Univ. Paris VI (1993).
13. J.K. Minderhoud and J.A.R. van Veen, Fuel Proc. Technol. **35**, 87 (1993).

PART II

High Surface Area Metal Carbides and Carbon

CHARACTERIZATION OF A CATALYTICALLY ACTIVE MOLYBDENUM OXYCARBIDE

MARC J. LEDOUX, PASCALE DELPORTE AND CUONG PHAM-HUU
Laboratoire de Chimie des Matériaux Catalytiques, EHICS-ULP, 1, rue Blaise Pascal, 67008 Strasbourg Cedex, France

ABSTRACT

A very active and selective molybdenum based catalyst for C_6^+ linear alkanes isomerization as been found. Physical characterization of this new active phase, such as XPS, XRD, MAS-NMR and HRTEM is described.

INTRODUCTION

In the chemical and petrochemical industry oil plays a crucial role as a raw material and catalysts are important in its processing. The increase in the gasoline octane number is achieved generally over active and selective Pt based catalysts supported on a zeolite or acidic alumina support. On these classical catalysts, isomerization with high selectivity is easily obtained with C_5 and C_6 as feedstock whatever the conversion, as opposed to C_7 or heavier hydrocarbons which give a high cracking selectivity when the conversion is increased. This has been identified as a challenge in catalysis for cleaner fuels, as non-aromatic octane enhancement of gasoline becomes more important [1].

The use of transition metal carbides as catalysts has received a great deal of interest since the discovery of their catalytic activity and their similarity with noble metals in heterogeneous catalysis [2,3]. However, clean surface carbides yield mostly hydrogenolysis products [4,5]. On tungsten carbides prepared by temperature-programmed reaction, according to the method of Boudart and co-workers, Iglesia and co-workers reported that oxidation gives a very selective surface for alkane isomerization [6-8]. Recently, experiments have been carried out by Ledoux and co-workers [9-12] concerning the isomerization of saturated hydrocarbons (C_6, C_7 and C_8) on a high specific surface area Mo_2C-oxygen-modified and MoO_3-carbon -modified material. They found that after an activation period under the reactant mixture at low temperature ($\leq$350°C) the activity becomes very significant and the isomerization selectivity remains very high even at high conversion; in addition, no deactivation occurs on the catalyst with time on stream up to several days. These results were attributed to the formation of an oxycarbide phase, formed by incorporating carbon atoms in the molybdenum oxide lattice [13]. But the presence of

Mat. Res. Soc. Symp. Proc. Vol. 368

several amorphous or poorly crystallized phases observed by TEM (encapsulated charcoal, carbidic or oxycarbidic phases), makes the crystallographic analyses, which would enable an easy characterization, extremely complicated.

The first aim of this article is to report briefly the isomerization activity obtained for C_6 and C_7 linear molecules on these systems and on a classical bifunctional catalyst, Pt/β-zeolite. The nature of the active phase and the characterization of its microstructure by the use of several techniques such as powder X-ray diffraction (XRD), surface area measurements by BET, ^{13}C MAS-NMR, X-ray photoelectron spectroscopy (XPS) and transmission electron microscopy (TEM) will constitute the second part of this article.

EXPERIMENTAL

MoO_3 (Strem Chemical) was used as received. A typical bifunctional catalyst was also used for comparison. Details of the preparation are given in [12].

n-Hexane and *n*-heptane supplied from Prolabo (puriss grade, >99.95%) were used as received. H_2 (Air Liquide, grade U) was purified by a copper catalyst and a molecular sieve drier before metering.

XRD measurements were performed on a Siemens Model D-500 diffractometer with CuKα monochromatic radiation (λ = 1.5406 Å).

The surface areas of the materials measured by the single point BET method using a Carlo Erba Ins. Model Sorpty 1750. The standard pretreatment consisted of heating the sample under dynamic vacuum at 573 K for 1h in order to remove the adsorbed water and other impurities. The measurement was made at liquid nitrogen temperature with nitrogen (Air Liquide, 99.95%) as the adsorbate gas.

The ^{13}C MAS-NMR was performed on a Brucker MSL400 spectrometer. In order to avoid contamination the sample was prepared and transferred via a glove-box under dry nitrogen and the analyse was performed in a sealed rotor.

XPS analyzes were performed in a Cameca Nanoscan-50 spectrometer using an X-ray source of Al Kα (1486.6 eV). The binding energies were calculated by taking the energy of the C 1s photoelectron at 284.8 ± 0.2 eV relative to the Fermi level. Only C 1s, O 1s and Mo 3d were recorded, and for a given emission the number of accumulations was selected in order to achieve a good signal-to-noise ratio. The samples were kept in the sealed reactor which was opened in a glove-box under a nitrogen atmosphere and then transferred onto the sample holder. Enclosed in a vessel, the mounted sample was transported to the spectrometer where it was transferred inside the XPS apparatus under flowing argon. The peak fit conditions used were: Full-Width at Half Maximum (FWHM) for metal = 1.0 eV and for oxide = 1.6 eV, ratio Mo $3d_{5/2}$ / Mo $3d_{3/2}$ = 1.5 and distance Mo $3d_{5/2}$ / Mo $3d_{3/2}$ = 3.2eV.

The High Resolution Transmission Electron Microscopy (HRTEM) was carried out with a Topcon Model EM-002B microscope working at 200 kV. The point to point resolution of this equipment could reach 1.7 Å. In order to avoid any artifact due to the solvent the samples were ground and deposited on a copper grid covered by a carbon membrane, transparent to electrons and able to withstand the beam energy.

RESULTS AND DISCUSSION

n-C_6 and *n*-C_7 isomerization on MoO_3-carbon -modified

It is well known in catalysis that as the reactant conversion increases, competition between the products and the reactants for the active sites becomes more and more important and therefore more secondary products are formed.

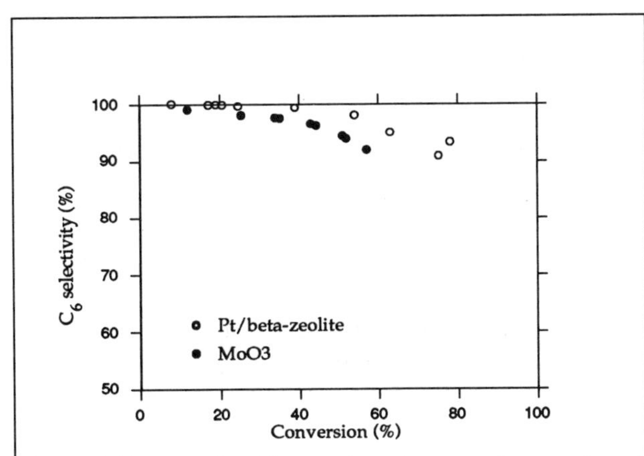

Figure 1. Isomer selectivity as a function of total conversion of n-hexane over MoO_3-carbon modified and Pt/β–zeolite at atmospheric pressure.

Figure 1 shows the variation of the isomer selectivity as a function of the total conversion of the MoO_3 - carbon - modified material and the Pt/β-zeolite for *n*-hexane . As expected, the isomerization selectivity slightly drops with increasing conversion but with *n*-hexane, the reaction can be carried out close to the thermodynamic equilibrium without affecting too strongly the selectivity (90%) (Fig. 1).

On a bifunctional catalyst one should expect for hydrocarbons having more than six carbon atoms, that thermodynamically the cracking of a branched alkane is much more favorable than the cracking of a linear reactant [14,15]. In figure 2, one observes a strong decrease of the selectivity when the conversion is increased on Pt/β-zeolite. However MoO_3-carbon-modified catalyst exhibits a remarkably high selectivity even at conversion as high as 80%. In addition, this catalyst showed a constant activity and selectivity after 52 days under reactive stream, which reflect a remarkable stability. Numerous re-activations did not affect neither this stability.

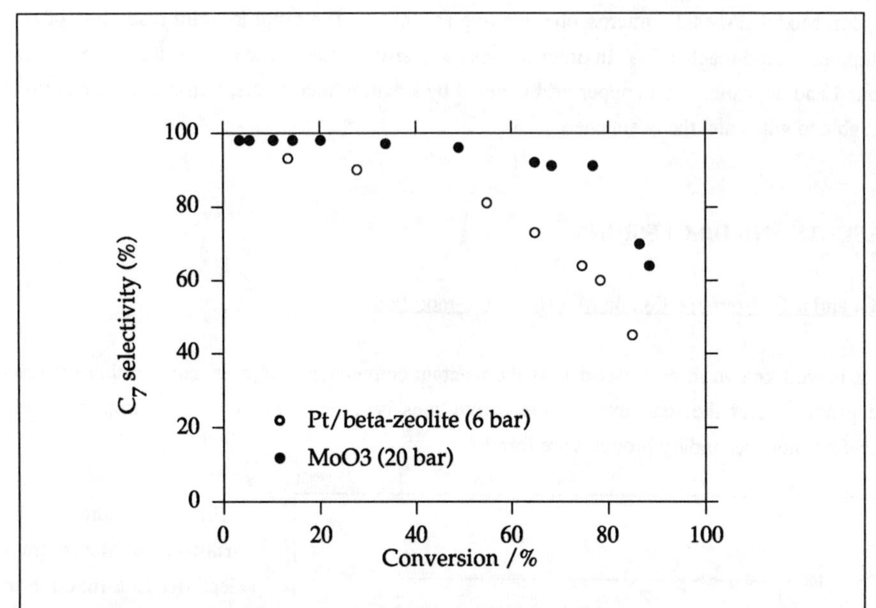

Figure 2. Isomer selectivity as a function of the total conversion for n-heptane isomerization over MoO_3-carbon modified and Pt/β-zeolite.

<u>Characterization results</u>

XRD. Unsupported MoO_3 was used as received and had a BET surface area of 4 m^2/g. For the starting oxide, only diffraction lines corresponding to MoO_3 are observed (Fig. 3a). The diffraction lines are narrow and correlate well with the low specific surface of the material. XRD pattern of the starting oxide also exhibits high intensities for all the <0k0>MoO_3 reflections. This is due to the structure of MoO_3 formed by layers of octahedra stacked along the [010]. After the activation period, all the MoO_3 diffraction lines have disappeared and the pattern only shows the presence of MoO_2 and another phase having interplanar distances, d, at 0.6198, 0.3057 and 0.2038 nm, which is unidentifiable from the literature data and is attributed to an oxycarbide phase. The structure of MoO_3 is made up of planes of MoO_6 octahedra bound together by Van der Waals interactions in the direction [010]. The new catalytically active unknown phase is formed by simultaneous reduction and substitution of some of the oxygen atoms by carbon atoms from the gas phase, resulting in the contraction of the distances between the planes of octahedra of the starting MoO_3. The transformation is accompanied by an important increase in the surface area of the catalyst from 4 to 140 m^2/g.

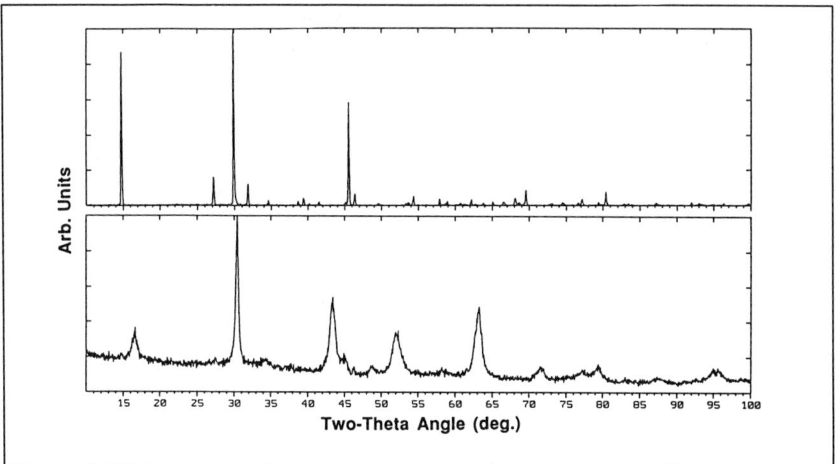

Figure 3. XRD patterns of unsupported MoO_3, (a) as received (4 m^2/g) and (b) after reaction with a hydrocarbon/hydrogen mixture at 350°C for 7h (140 m^2/g).

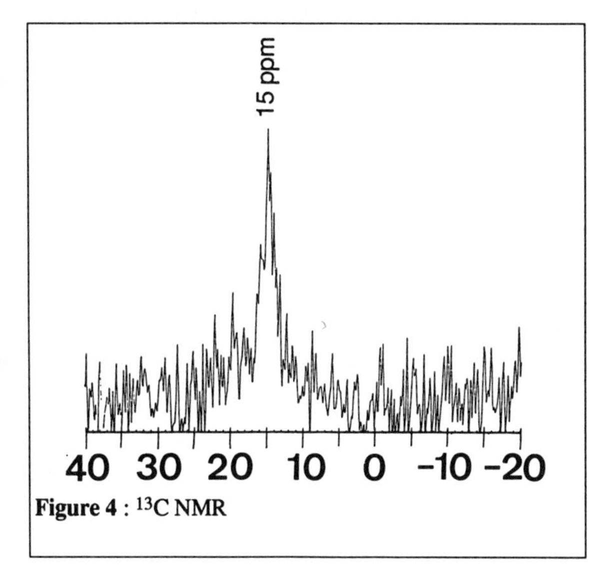

Figure 4 : ^{13}C NMR

NMR. Further evidence for the formation of an oxycarbide phase is shown in the ^{13}C MAS-NMR spectrum of the MoO_3-carbon-modified material presented in figure 4. n-Hexane was replaced by 1-^{13}C-hexane in the activation process. The spectrum shows a sharp peak located at 15 ppm which is attributed to the carbon present in the oxycarbide phase. This chemical shift is very different from that for the carbon present in an Mo_2C material, which appears at a shift value of 275 ppm.

XPS. The XPS spectra of Mo 3d and O 1s of the starting material MoO_3 is presented in Figs. 5a and b. The XPS spectrum of Mo 3d (Fig. 5a) shows only the presence of Mo^{6+}. The XPS spectrum of O 1s (Fig. 5b) shows the presence of a symmetrical peak centered at 531 eV which corresponds to the oxygen atoms linked to Mo in MoO_3 according to the results published in the literature [16].

The XPS spectra of Mo 3d, O 1s and C 1s recorded on the sample after reaction at 350°C are

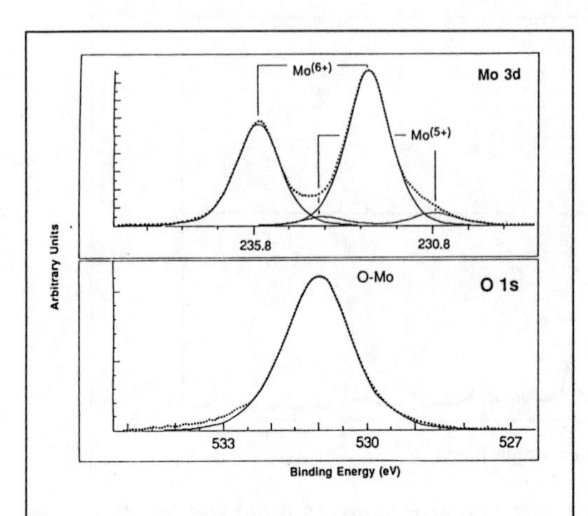

Figure 5 XPS spectra of Mo 3d and O 1s of the starting MoO_3.

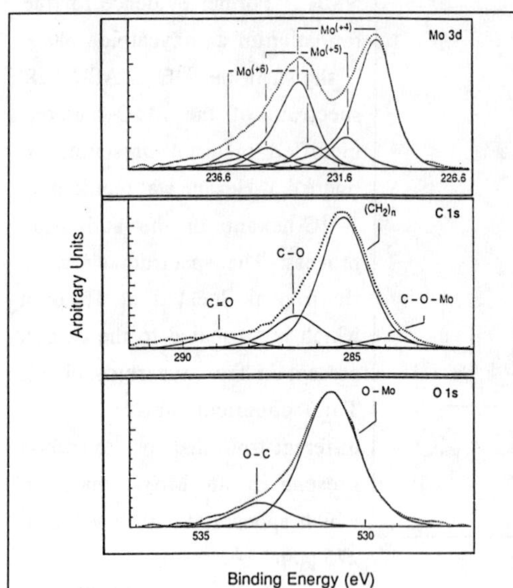

Figure 6. XPS spectra of Mo 3d, O 1s and C 1s of MoO_3-carbon modified catalyst.

presented in Figs. 6a, b and c. From the XPS spectrum of Mo 3d (Fig. 6a) there is no evidence for Mo carbide or Mo metal formation, but only a mixture of different oxidation states of Mo (+6, +5, +4).

The main oxidation state is Mo^{4+} (85%). It cannot be attributed only to a MoO_2 species (observed by XRD), because MoO_2 is completely inactive for isomerization, but to another phase which displays a similar binding energy; probably Mo^{4+} in the oxycarbide phase together with Mo^{5+}.

The O 1s spectrum (Fig. 6b) shows a peak due to the oxide at 530.6 eV. However, a small peak at around 532.7 eV is also present (although difficult to see due to its size and the small chemical shift); this is attributed to oxygen atoms engaged in the oxycarbide form.

The C 1s spectrum (Fig. 6c) shows the presence of four overlapping XPS peaks. The C 1s peaks located at 284.8, 286.2 and 288.5 eV can be attributed to $(CH_2)_n$, C-O and C=O, respectively. The first C 1s peak located at 283.5 eV is attributed to a carbon species bonded to an oxygen and a metal atom in the oxycarbide form (C-O-Mo) which is located at a higher binding energy compared to the C atoms involved in the carbide form (1.2 eV higher), in agreement with several publications devoted to silicon oxycarbide [17] and to aluminum oxycarbide [18].

HRTEM. TEM micrograph of MoO_3 is presented in Fig. 7a along with the microdiffraction pattern (Fig. 7b). TEM well illustrates the platelet form of the MoO_3 in the (010) plane.

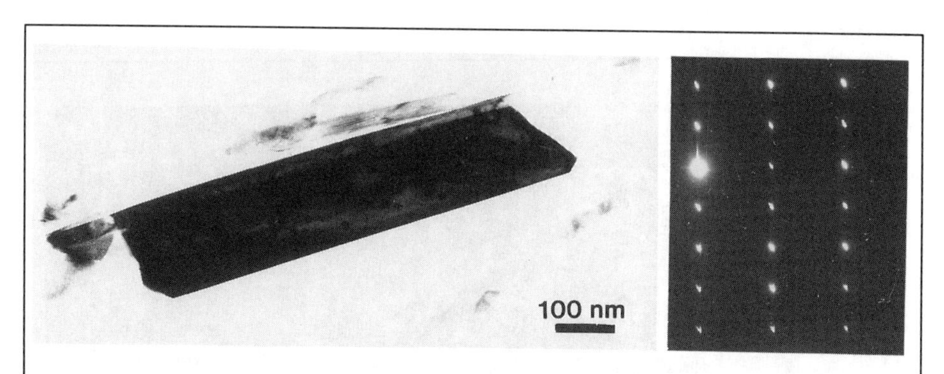

Figure 7. (a) Low magnification micrograph (x $107.9.10^3$) of MoO_3 platelet and (b) selected area diffraction pattern of $[100]_{MoO3}$.

The HRTEM image of MoO_3 (Fig. 8a) shows a crystalline structure only over a short distance, but this was due to the high sensitivity of the sample under the electron beam.

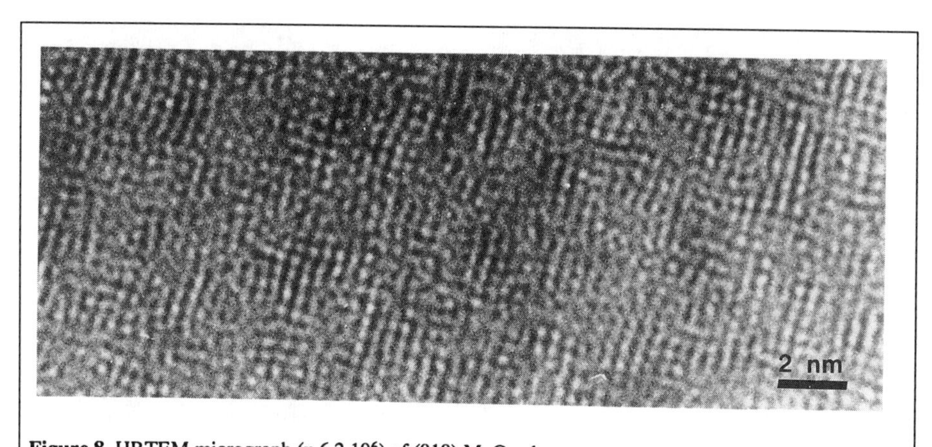

Figure 8. HRTEM micrograph (x $6.2.10^6$) of (010) MoO_3 planes.

HRTEM micrographs (Figs. 9a and b) of the sample after reaction at 350°C indicate the presence of well organized MoO_2 (Fig. 9a) which is confirmed by electron microdiffraction presented in an insert and a structure in a "chevron" shape and its corresponding electronic microdiffraction, characteristic of MoO_3-carbon-modified (molybdenum oxycarbide). In other samples this crystalline phase can co-exist with its pseudo or amorphous form [13].

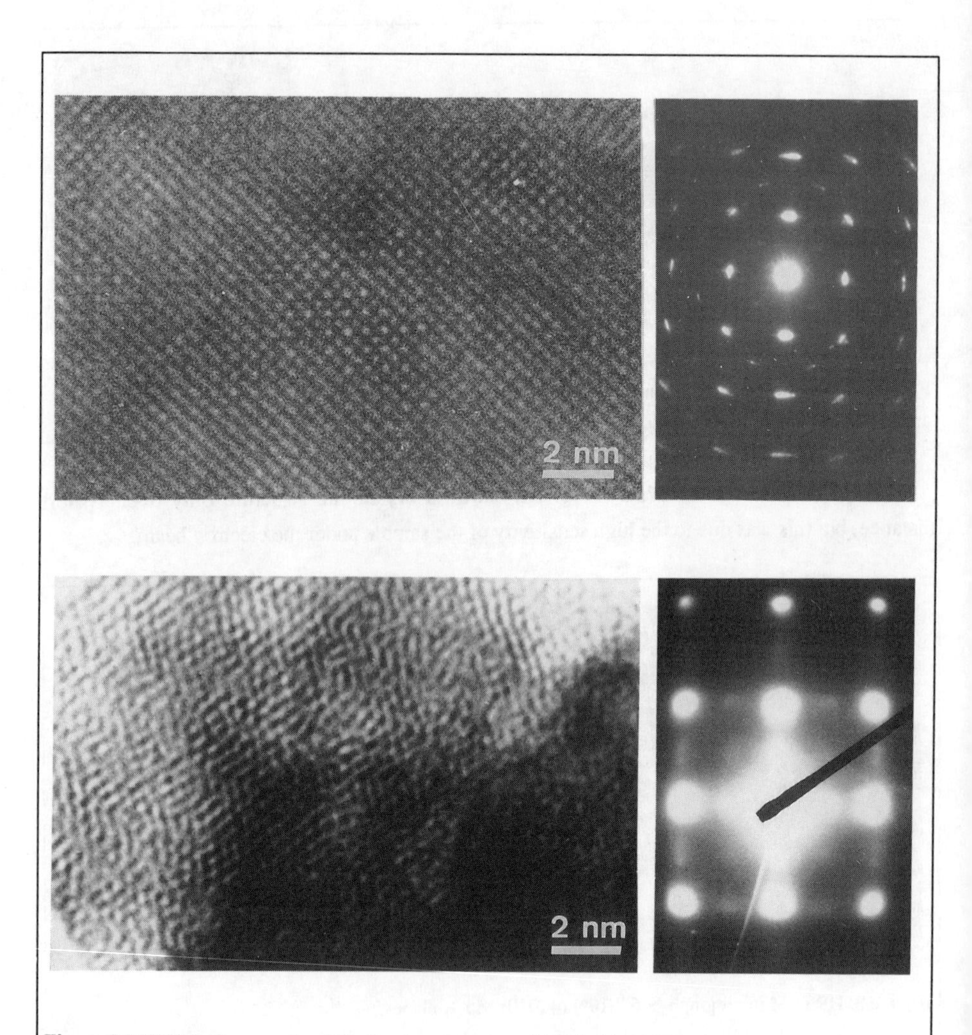

Figure 9 HRTEM micrographs of MoO_3 after reaction at 350°C. (a) HRTEM image (x 6.2.10^6) of (100) MoO_2 planes and a corresponding SADP of [100]MoO_2. (b) HRTEM image (x 6.2.10^6) of MoO_3-carbon modified planes and a corresponding SADP.

DISCUSSION

Reaction at low temperature of low specific surface area unsupported MoO_3 with a hydrocarbon and hydrogen mixture leads to the formation of a high specific surface area MoO_3-carbon-modified phase and MoO_2 as a side product, as shown by XRD (Fig. 3). MoO_2 under identical reaction conditions show no catalytic activity at all even after prolonged exposure: thus, the activity and selectivity observed here is due to the other phase.

The mechanism of the transformation $MoO_3 \rightarrow MoO_2$ + MoO_3-carbon-modified can be explained as follows: at low reaction temperature (≤350°C) hydrogen adsorbs on the (010) plane of MoO_3, i.e. on the oxygen atoms bonded to a single Mo atom (because of the layered structure), and the water formed by reaction of two adjacent OH groups desorbs, creating surface oxygen vacancies. These vacancies can be randomly distributed in the plane (010) at the beginning but a deeper and longer reduction produces a higher concentration of vacancies which will re-organize themselves along [101] because it is the most energetically favourable situation with each octahedron bearing a vacancy surrounded by saturated octahedra. The MoO_3 lattice may also rearrange to eliminate vacancies by local collapse and shears, which lead to a decrease of the [010] distance. Assuming that this ordered structure is the intermediate state of the reaction, two possible paths should be examined, depending on the nature and the composition of the reducing flow.

A reduction by pure H_2 finally forms MoO_2 *via* an intermediate phase MoO_xH_y. This phase is formed by insertion of H radical or H_2 molecule in the MoO_3 shear structure, and decreases to the benefit of MoO_2 with enough time on stream. A reduction-carburation under the mixture of hydrocarbon and hydrogen can introduce, in addition to the formation of MoO_2 by the previous mechanism, C atoms during the process and stabilize the intermediate phase. Indeed, Spevack et al. [19] considered that shear planes are preferential pathways for diffusion of S atoms to form molybdenum oxysulfide. In the same way, C atoms can fill O vacancies or replace O atoms in the lattice, to form the MoO_3-carbon-modified (MoO_xC_y) phase. The presence of carbon atoms bonded with oxygen and metal is shown by XPS and NMR. This active phase can also be obtained directly under pure H_2 activation leading firstly to the formation of the intermediate phase MoO_xH_y which, when hydrocarbon is added to the stream, is transformed into MoO_3-carbon-modified (MoO_xC_y) The size and the electronic properties of the H radical or H_2 molecule result in a quite unstable phase when compared to MoO_3-carbon-modified (MoO_xC_y) where C is not atomically very different of O. When the reaction temperature is higher than 350°C (i.e. 400 or 450°C) the transformation in MoO_2 (under pure H_2) or Mo carbide (under a mixture of hydrocarbon and hydrogen) is fast and the formation of MoO_3-carbon-modified (MoO_xC_y) cannot be controled. All these transformations can be summarized by the scheme presented in Fig. 10.

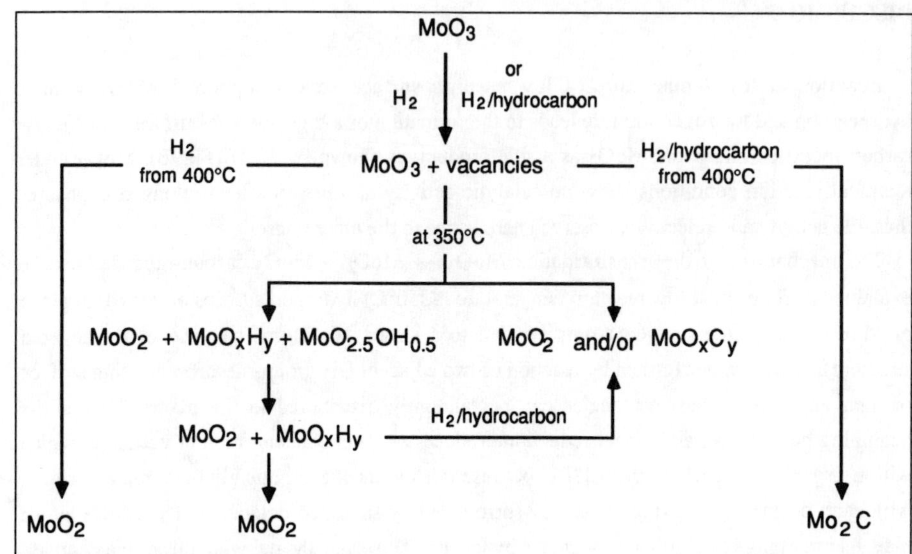

Figure 10. Mechanism of the transformation of MoO_3 as a function of the reaction temperature and the nature of the reactant.

The increase in the specific surface area during transformation can also be explained by the same mechanism. As localized reduction occurs in several domains of the starting oxide, the metal lattice contracts and fractures the crystal to produce fine pores. The surface area is also produced by the low temperature reaction. Similar results have already been reported by Volpe and Boudart during Mo_2N synthesis from MoO_3 [20].

CONCLUSION

The formation of the high specific surface area MoO_3-carbon-modified (oxycarbide) (140 m^2/g) phase from low specific surface area unsupported MoO_3 (4 m^2/g) can be explained as follows: creation of oxygen vacancies on MoO_3 planes (010) accompanied by a contraction of the distance between the layers through shears which minimize the constraints and fill some vacancies. If the number of vacancies is too high (high temperature reaction or time on stream long enough), then the solid reconstructs into MoO_2. If carbon is added to the flow and on condition that the temperature is not too high (between 300 and 400°C), some carbon atoms stop the contraction of the planes at a level of 14% along (010) allowing the formation of a new phase between MoO_3 and MoO_2 which is an oxycarbide phase.

The MoO_3-carbon-modified (MoO_xC_y) phase is an excellent catalyst for the isomerization of heptane and the mechanism operating allows a high total conversion of the hydrocarbon feed without drop in the isomerization selectivity. This is because the mechanism is not bifunctional but, as shown in a previous article [10], probably proceedes via a metallacyclobutane.

ACKNOWLEDGEMENTS

PECHINEY Co. is gratefully acknowledged for supporting this work.

References

1. I. E. Maxwell and J. E. Naber, Catal. Lett. **12**, 105 (1992).
2. J. M. Muller and F. G. Gault, Bull. Soc. Chim. Fr. **2**, 416 (1970).
3. M. Boudart and R. Levy, Science **181**, 547 (1973).
4. J.S. Lee, S. Locatelli, S.T. Oyama and M. Boudart, J. Catal., **125**, 157 (1990).
5. M. J. Ledoux, C. Pham-Huu, H. Dunlop and J. Guille, J. Catal. **134**, 383 (1992).
6. F. H. Ribeiro, R. A. Dalla Betta, M. Boudart, J. Baumgartner and E. Iglesia, J. Catal. **130**, 86 (1991).
7. F. H. Ribeiro, M. Boudart, R. A. Dalla Betta and E. Iglesia, J. Catal. **130** ,498 (1991).
8. E. Iglesia, J. Baumgartner, F. H. Ribeiro and M. Boudart, J. Catal. **131**, 523 (1991).
9. M. J. Ledoux, C. Pham-Huu, H. Dunlop and J. Guille, Proceedings 10th ICC, L. Guczi, F. Solymosi and P. Tétényi, Eds., Akadémiai Kiado, Budapest, Vol. B, 955 (1993).
10. C. Pham-Huu, M. J. Ledoux and J. Guille, J. Catal. **143**, 249 (1993).
11. M. J. Ledoux, J. Guille, C. Pham-Huu, E. A. Blekkan and E. Peschiera, Eur. Pat. Appl. No. 93-14199 (November 1993).
12. E. A. Blekkan, C. Pham-Huu, M. J. Ledoux and J. Guille, Ind. Eng. Chem. Res. **33**, 1657 (1994).
13. P. Delporte, F. Meunier, C. Pham-Huu, P. Vénnégues and M. J. Ledoux, Catal. Today, *in press.*
14. M. Steijns, G. Froment, P. Jacobs, J. Uytterhoeven and J. Weitkamp, Ind. Eng. Chem. Prod. Res. Dev. **20**, 654 (1981).
15. S. T. Sie, Ind. Eng. Chem. Res. **31**, 1881 (1992); **32**, 397 (1993); **32**, 403 (1993).
16. B. Brox and I. Olefjord, Surf. Inter. Anal. **13**, 3 (1988).
17. A. Julbe, A. Larbot, C. Guizard, L. Cot, T. Dupin, J. Charpin and P. Bergez, Eur. J. Solid State Inorg. Chem., **26**, 101 (1989).
18. M. Bou, J.M. Martin, Th. le Mogne and L. Vovelle, Appl. Surf. Sci., **47**, 149 (1991).
19. P.A. Spevack and N.S. MacIntyre, J. Phys. Chem., **96**, 9029 (1992).
20. L. Volpe and M. Boudart, J. Solid State Chem. **59**, 348 (1985).

STRUCTURAL CHARACTERIZATION OF NANOCRYSTALLINE Mo AND W CARBIDE AND NITRIDE CATALYSTS PRODUCED BY CO_2 LASER PYROLYSIS

Xiang-Xin Bi[1], K. Das Chowdhury[2], R. Ochoa[4], W. T. Lee[4], S. Bandow[5], M. S. Dresselhaus[1,3] and P. C. Eklund[4]
[1] Department of Electrical Engineering and Computer Science, MIT, Cambridge, MA 01239
[2] Department of Materials Science and Engineering, MIT, Cambridge, MA 01239
[3] Department of Physics, MIT, Cambridge, MA 02139
[4] Center for Applied Energy Research, University of Kentucky, Lexington, KY40511-8433
[5] Instrument Center, Institute for Molecular Science, Myodaiji, Okazaki, 444 Japan

Abstract Using both XRD and HRTEM lattice imaging, we have shown that CO_2 laser pyrolysis (LP) produces nanoscale transition metal carbide and nitride catalysts, including cubic Mo_2C, Mo_2N, and W_2N, which possess highly crystalline structures in their as-synthesized form. In contrast, LP-produced W_2C in its hexagonal phase is disordered. Clear lattice expansion, induced by the small crystallite size of the nanoparticles has been observed for LP-produced Mo_2C particles,which have a typical crystallite size of 2 nm. No carbon coating was observed in HRTEM for LP-produced Mo_2C particles. Furthermore, Mo=N and Mo=C bonding in Mo_2N and Mo_2C, respectively, were identified by an XPS measurement, which also reveals the presence of a thin oxide layer formed on the particle surface during the passivation process. Finally, the average crystallite sizes determined from HRTEM and XRD are in good agreement, indicating that the line broadening observed in XRD is due to the small crystallite size of the nanoparticles.

CO_2 laser pyrolysis (LP)[1] has been shown to provide a versatile, very rapid process for the production of nanoscale particles. Though a variety of nanoscale particles including Si, SiC, Si_3N_4[1-5], ZrB_2[6], TiO_2[4, 6], Mo_2C[7], and Fe_3C[8] were produced in the past using this technique, few studies were carried out to define the detailed crystal structures and degree of crystallinity of the LP-produced nanoparticles. In a recent study, we have shown that LP-produced α-Fe, Fe_3C, Fe_7C_3 nanoscale particles of 15 nm size are highly crystalline, based on the results from both X-ray diffraction (XRD) and high resolution TEM (HRTEM)[9]. In this work, we use these two techniques in combination with X-ray photoelectron spectroscopy (XPS) to characterize LP-produced nanoscale Mo_2C, W_2C, Mo_2N and W_2N particles. C in $Mo(W)_2C$ and N in $Mo(W)_2N$ come from C_2H_2 and NH_3 reactant gases, respectively, which both absorb CO_2 laser energy. The detailed synthesis process is addressed in a separate paper[10].

There is growing interest in nanoscale (< 100 nm) particles for their potential application as catalysts, sensor materials, battery electrodes, and dielectric media. Nanoscale catalysts are interesting for two reasons. First of all, if well dispersed, they have high specific surface areas. For example, a batch of 10 nm spherical carbon particles (ρ = 2.2 g/cm^3) would have a specific surface area of 1.4 x 10^2 m^2/g. Secondly, nanoscale particles may have different or improved catalytic activity and selectivity relative to their bulk form, because of the connection of catalytic activity to surface properties. The surface atoms in a nanoscale particle comprise a significant percentage of the total number of atoms (up to 50% for a 5 nm particle). Since these atoms likely arrange themselves differently from those in the bulk, they may present different physical and chemical properties, including catalytic activity.

Transition metal carbides and nitrides such as WC and Mo_2N have been shown to have catalytic activities similar to these of the noble metals[11, 12]. Therefore, these binary compounds have been considered to be economical alternatives to very expensive noble metal catalysts. To determine catalytic activity of these carbide and nitride compounds, it is important to obtain structural information about these catalysts. We employ XRD to identify the crystalline phase, extract *average* values of the lattice spacing and of the crystallite sizes of the nanoscale particles. However, XRD is unable to detect directly any surface coatings of a few monolayers or less, nor

Mat. Res. Soc. Symp. Proc. Vol. 368 © 1995 Materials Research Society

directly reveal atomic structures near the particle surface. In contrast, HRTEM lattice imaging has proven to be a powerful tool to probe atomic structures of an entire particle, especially when the particle has well ordered structure. The results extracted from XRD, HRTEM and XPS thus present a relatively complete picture of the structure of nanoscale particles.

In Fig. 1, we display XRD data (dots) for LP-produced nanocrystalline Mo_2N (Fig. 1a), W_2N (Fig. 1b), Mo_2C (Fig. 1c), and W_2C (Fig. 1d) particles. The data, corrected for the instrumental zero shift in 2θ, were collected in air using Cu (K_α) radiation and a Phillips 3100 powder diffractometer. The solid curve in each figure represents a sum of Lorentzian lineshapes. The values used for the diffraction peak positions and peak areas were taken directly from a standard powder diffraction file for these materials[13]. A single linewidth is used for all the peaks, and each curve is normalized to the intensity of the highest diffraction peak in the spectrum. The overall agreement between the data points and the calculated results for the standards indicates that our phase identification is correct[13]. Mo_2N, W_2N and Mo_2C are identified to have a cubic structure, whereas W_2C a hexagonal one. Diffraction peaks for these cubic compounds were indexed accordingly. A broad peak positioned around $2\theta \sim 27°$, not accounted by the calculated curves in Mo_2N, W_2N and W_2C, is attributed to MO_3 (M= Mo and W) formed during the oxygen passivation process[9][10]. Sharp lines in the observed W_2N XRD spectrum come from unreacted $W(CO)_6$ which recrystallized on the filters along with the particles. These carbonyl contaminants may be easily removed by subjecting the sample to a temperature of 150 °C (the sublimation point for $W(CO)_6$) in Ar.

We have fitted the doublet feature in the XRD pattern around 40° for each material with two Lorentzians, using a least square method to accurately extract the peak position $2\theta_{111}$ and the width at half maximum intensity A for the 111 diffraction line (LS in Fig. 1). Using the Debye-Scherrer formula ($D = 0.9\lambda/A\cos(\theta_{111})$) and the Bragg's law ($2d\sin(\theta_{111}) = \lambda$), we have determined the average crystallite size D and the lattice spacing d along the 111 direction for Mo_2N (Fig. 1a), W_2N (Fig. 1b), and Mo_2C (Fig. 1c), as given in Table 1. The XRD data for W_2C (shown in Fig. 1d) presents an amorphous pattern. Thus no effort has been made to extract lattice spacing information for W_2C. In the case of Mo_2C and W_2N, significant differences exist between the relative intensities of the standard powder diffraction pattern and our data. This may be attributed to Mo vacancies in the crystalline structures of our particles or to variations in the metal atom sublattice near the surfaces. Furthermore, clear shifts of the diffraction lines toward lower 2θ angles is seen for Mo_2C (Fig. 1c), indicative of a lattice expansion, presumably associated with the small crystallite size.

In Fig. 2, we show HRTEM lattice images for the same batch of Mo_2N (Fig. 2a) and W_2N (Fig. 2b), Mo_2C (Fig. 2c) and W_2C (Fig. 2d) nanoparticles, whose XRD data appear in Fig. 1. It can be seen that the lattice fringes are present in the majority of Mo_2C, Mo_2N, and W_2N particles, which indicates that the particles are crystalline. These images do not show significant surface oxide or carbides coatings, though the XRD results suggest the presence of oxide layers. We thus conclude that the oxide layers must be thin and disordered enough so that they are not resolvable under HRTEM. In particular, the Mo_2C particles do not appear to have a carbon coating on the surface, in contrast to LP-produced Fe_3C, where as many as 10 graphitic layers were often observed [9]. This is understandable, since Fe_3C is known to be a better catalyst than Mo_2C for producing graphitic carbons from the carbon source gas C_2H_4. We have identified most of the observable lattice fringes in Fig. 2 with {111} atomic planes, by measuring their interlayer lattice spacings. An average crystallite size has also been obtained. In the case of Mo_2N, however, significant particle agglomeration made it difficult to obtain an accurate value for the average crystallite dimension. These values including {111} lattice spacings d_{111} and crystallite size D are listed in Table 1. W_2C clearly displays an amorphous structure, consistent with the conclusion derived from the XRD analysis. It is clear from Table 1 that the crystallite sizes estimated from the finite-size-induced line broadening in XRD and HRTEM are in good agreement, indicating that within each small crystallite there is little disorder.

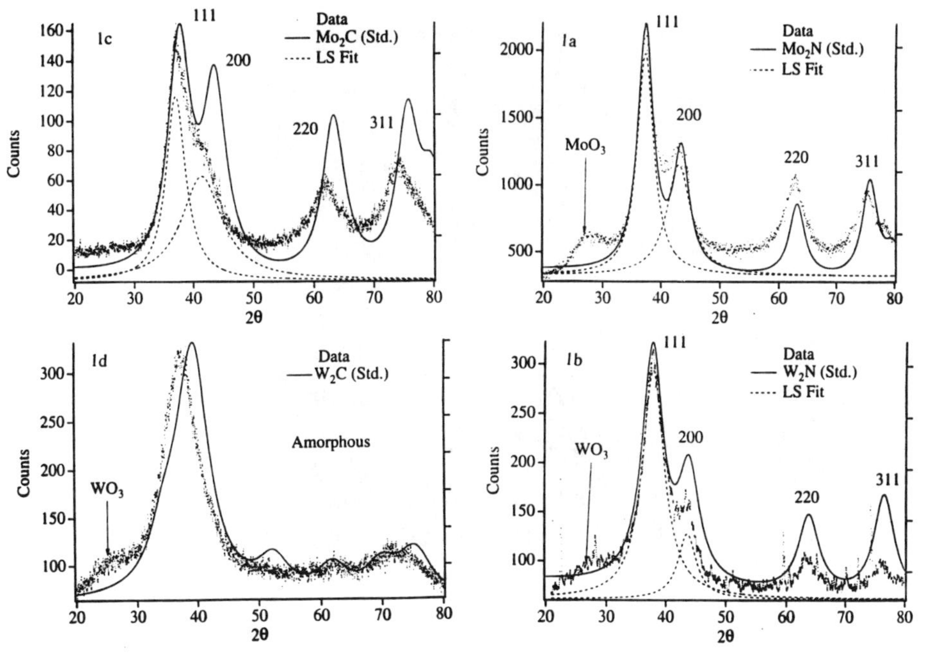

Fig. 1 XRD (Cu K_α, $\lambda = 1.5418$ Å) data (dots) for Mo_2N (1a), W_2N (1b), Mo_2C (1c), and W_2C (1d). The solid curve in each figure represents a sum of Lorentzian lineshapes whose peak positions and intensities were taken directly from a standard powder diffraction file for these materials[13]. Dashed lines are deconvoluted peaks from a least square (LS) fit to the doublet ({111} + {200}) around $2\theta \sim 40°$.

Fig. 2a HRTEM lattice images for Mo_2N W_2N (2b), Mo_2C (2c), and W_2C (2d). Lattice spacings along the {111} direction are indicated in the figures for Mo_2N, W_2N, and Mo_2C.

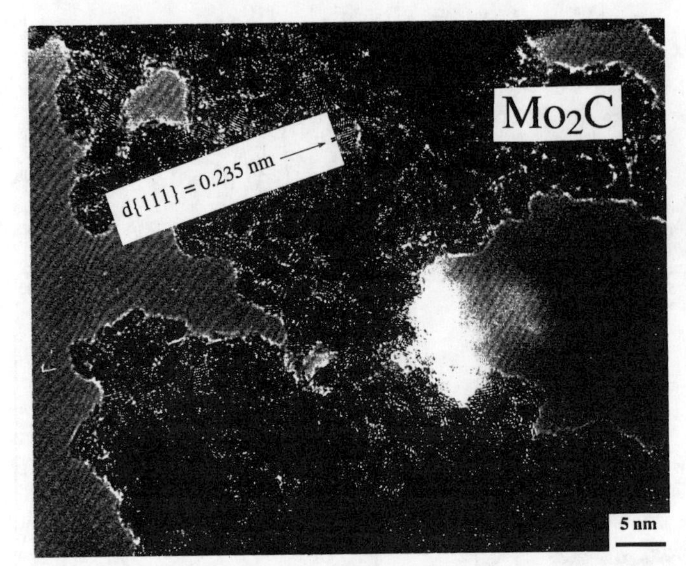

Fig. 2b HRTEM lattice images for Mo_2C Lattice spacing along the {111} direction is indicated.

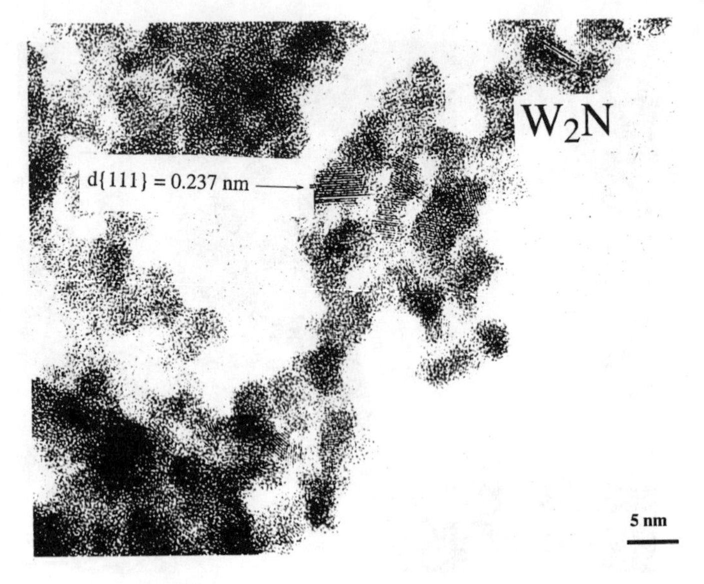

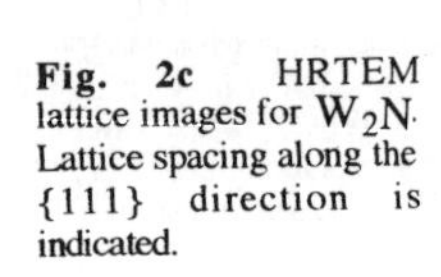
Fig. 2c HRTEM lattice images for W_2N. Lattice spacing along the {111} direction is indicated.

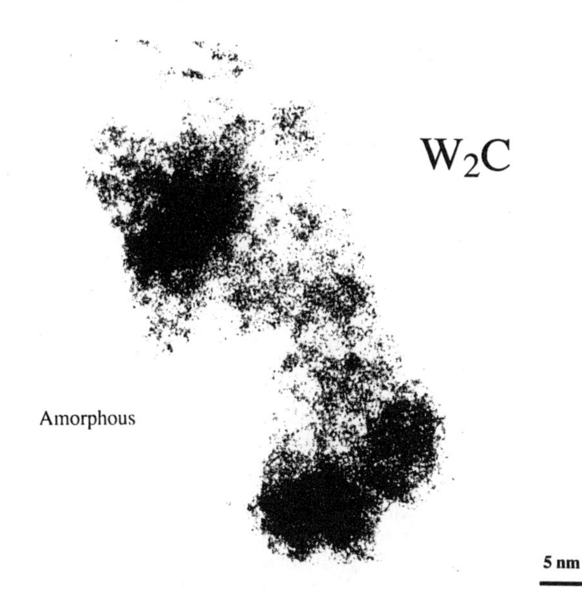

Fig. 2d HRTEM lattice images for W_2C, showing a amorphous structure.

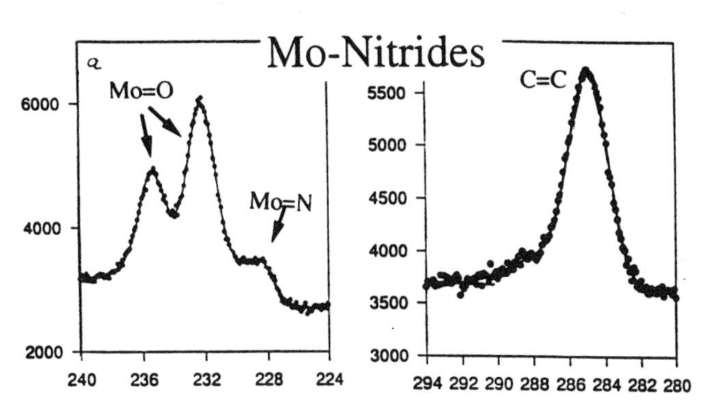

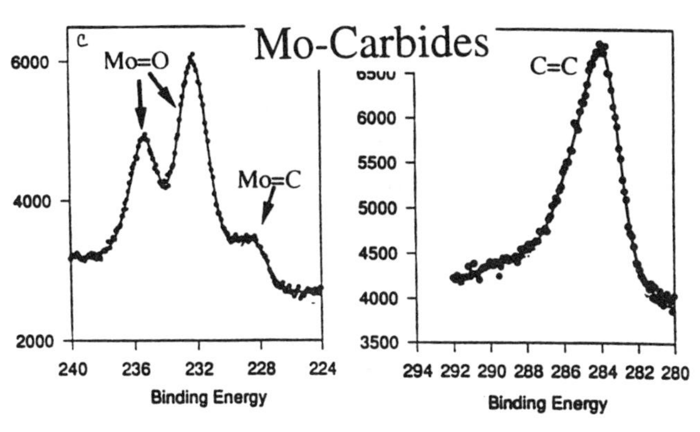

Fig. 3 XPS data (dots connected with solid lines) for Mo_2N (3a and 3b) and Mo_2C (3c and 3d). Peaks associated with Mo=N, Mo=C, Mo=O, and C=C are indicated.

Table 1 Average crystallite size (D) and lattice spacing (d) along the {111} direction, derived from XRD and HRTEM measurements.

		Mo_2N	W_2N	Mo_2C	W_2C
D_{XRD}	(Å)	25	19	21	Amorphous
D_{HRTEM}	(Å)	24*	27	20	Amorphous
d_{XRD}	(Å)	2.40	2.38	2.42	NA
d_{HRTEM}	(Å)	2.40	2.37	2.35	NA
$d_{standard}$	(Å)	2.40	2.38	2.38	NA

* The value is approximate, due to significant particle agglomeration.

Figure 3 shows the XPS spectra for Mo_2N (Fig. 3a and 3b) and Mo_2C (Fig. 3c and 3d). We assign peaks at 228.58 eV in Fig. 3a and 3c to the N=Mo and C=Mo local environment, respectively, whereas the peaks at 232.2 eV in Fig. 3a and 3c are from the +VI oxidation state of molybdenum, which suggests the presence of MoO_3 on the particle surface. This is consistent with the XRD results described previously, which exhibit a broad peak at $2\theta \sim 27$ °. Deconvolution of the data gives 90% of *surface* Mo in the form of MoO_3 for both Mo_2N and Mo_2C. The carbon 1s peak located around 284.5 eV in Figs. 3b and 3d corresponds to carbon deposited on the particle surface during the synthesis. The peak in Fig. 3d is slightly lower in energy than that in Fig. 3b because of an additional contribution from Mo=C bond. Note the surface MoO_3 and carbon are not resolved in HRTEM due to their poor crystallinity, which emphasizes the importance of employing multiple tools in the nanoparticle characterization. Further studies involving elemental analysis are under way to identify the incorporation of oxygen into the carbide or nitride crystalline lattice.

Acknowledgements

This work was supported, in part, by the NSF(9201878-DMR), University of Kentucky Center for Applied Energy Research, the U. S. Dept. of Energy and the NSF US-Japan Science Exchange Program.

References

1. J. S. Haggerty, "Sinterable Powders from Laser-Driven Reactions", in *Laser-induced Chemical Processes,* J.I. Steinfeld, Editor, 1981, Plenum Press: New York.
2. F. Curcio, G. Ghiglione, M. Musci, and C. Nannetti, Applied Surface Science, **36**: p. 52-58, 1989.
3. P. R. Buerki, T. Troxler, and S. Leutwyler, "Synthesis of Ultrafine Si_3N_4 Particles by CO_2-laser Induced Gas Phase Reactions" in *High Temperature Science*, Vol. 27, 1990, Humana Press Inc., pp 323.
4. F. Curcio, M. Musci, and N. Notaro, Applied Surface Science, **46**: p. 225-229, 1990.
5. R. Fantoni, E. Borsella, S. Piccirillo, and S. Enzo, SPIE, **1279**: p. 77, 1990.
6. G. W. Rice and R. L. Woodin, J. Am. Ceram. Soc., **71**: p. C181, 1988.
7. G. W. Rice, "Laser-Driven Synthesis of Transition-Metal Carbides, Sulfides, and Oxynitrides" in *Laser Chemistry of Organometallics*, 1993, p. 275.
8. G. W. Rice, R. A. Fiato, and S. L. Soled, United States Patent, **4,659,681**1987.
9. Xiang-Xin Bi, B. Ganguly, G. Huffman, E. Huggines, M. Endo, and P. C. Eklund, Journal of Materials Research, **8**(7): p. 1666, 1993.
10. Xiang-Xin Bi, R. Ochoa, W. Lee, and P. C. Eklund, unpublished.
11. S. T. Oyama, J. C. Schlatter, J. E. Metcalfe, and J. M. Lambert, Ind. Eng. Chem. Res., **27**: p. 1639-1648, 1988.
12. R. B. Levy and M. Boudart, Science, **181**: p. 547, 1973.
13. JCPDS, Standard X-ray Powder Diffraction Data File, 1991.

NICKEL CATALYZED CONVERSION OF ACTIVATED CARBON INTO POROUS SILICON CARBIDE

R. MOENE*, J. SCHOONMAN**, M. MAKKEE*, AND J.A. MOULIJN*
*Department of Chemical Process Technology, Section Industrial Catalysis,
**Laboratory of Applied Inorganic Chemistry,
Delft University of Technology, Julianalaan 136, 2628 BL Delft, The Netherlands

ABSTRACT

High surface area silicon carbide (SiC) of 31 m^2/g has been synthesized by the catalytic conversion of activated carbon. The thermal stability in non-oxidizing environments is shown to be excellent; no significant sintering has been observed after ageing in nitrogen for 4 hours at 1273 K. The presence of 2v% steam or the use of air results in SiC oxidation into SiO_2 and considerable sintering at 1273 K. Air oxidation of SiC is shown to cause substantial SiC conversion, *viz.* 60 % after 10 hours at 1273 K. Complete conversion is achieved at 1080 K in about 100 days. This rate of oxidation agrees with reports on the oxidation of non-porous Acheson SiC and Chemical Vapour Deposited SiC coatings. The use of SiC based catalysts is, therefore, limited to (1) high temperature gas phase reactions operating in the absence of oxidizing constituents (O_2 or H_2O) and (2) liquid phase processes at demanding pH. Syntheses of highly dispersed and highly loaded Ni/SiC catalysts are feasible by applying an ion-exchange technique, resulting in supported nickel particles of 4 nm.

INTRODUCTION

Silicon carbide's physical bulk properties (high thermostability, high mechanical strength, and high heat conductivity) have been claimed to enable the use of this material as catalyst support at extreme process conditions, *viz.* processes operating at high temperatures and oxidizing environments [1]. High surface area SiC has been synthesized by decomposing tetramethyl silane to arrive at surface areas of nearly 50 m^2/g [1], reaction of SiO with activated carbon (60 m^2/g) [2], and pyrolysis of an organosilicon polymer (172 m^2/g) [3]. Recently, Moene *et al.* [4] developed a new synthesis procedure for the preparation of high surface area SiC in which activated carbon is catalytically converted into SiC according to eq. 1.

$$C(s) + SiCl_4(g) + 2H_2(g) \overset{Ni}{\rightleftarrows} SiC(s) + 4HCl(g) \qquad (1)$$

The mechanism of conversion comprises the gasification of carbon by hydrogen followed by the intra-particle deposition of SiC from $SiCl_4$ and CH_4 *via* the Vapour Liquid Solid (VLS) mechanism, resulting in abundant whisker growth within the particle. The mechanism is outlined in Fig. 1. Both reactions are catalyzed by nickel, and the carbon acts as a source for CH_4 formation as well as a template for SiC deposition. The synthesis of SiC with surface areas ranging from 30 to 80 m^2/g has been achieved, according to this bi-functional catalysis mechanism. Application of a Fluidized Bed Chemical Vapour Deposition (FB-CVD) reactor allows a very reproducible and homogeneous conversion of several grams of activated carbon [5] which alleviates to a large extent the investigations for possible applications of SiC supports.

Mat. Res. Soc. Symp. Proc. Vol. 368 © 1995 Materials Research Society

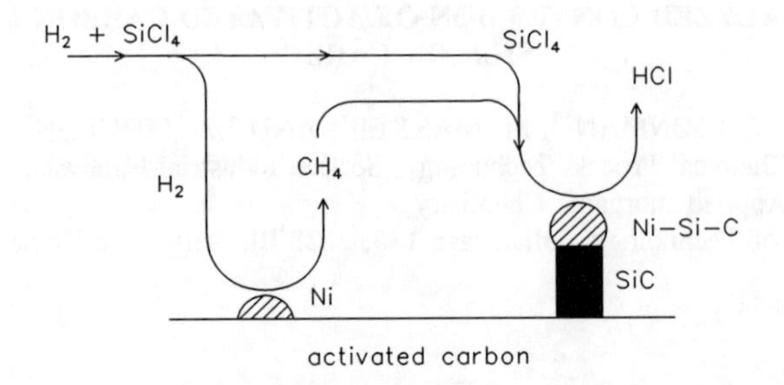

Fig. 1. The catalytic conversion of activated carbon into high surface area silicon carbide. Nickel catalyzes both carbon gasification and SiC formation. The activated carbon acts as carbon source and template

Ni/SiC catalysts prepared by impregnation have been shown to exhibit a high Metal Support Stability. Even after oxidation in air at 1273 K for 8 hours, an easily reducible nickel species is retained (d_{Ni}= 20 nm) [4]. Synthesis of highly dispersed SiC based catalysts might be laborious, due to the low amount of anchoring sites on SiC. This has led to the application of a sophisticated micro-emulsion technique in which impregnation is carried out with nano-sized metal particles [6]. Implementation of conventional catalyst synthesis procedures is, however, of great advantage for the application of highly dispersed SiC based catalysts. Automotive exhaust catalysis based on SiC powder [6] and porous SiC [7] has been investigated as a possible application in the field of high temperature processes. Here, the thermal stability of the silicon carbide itself and the limited reactions of the catalytically active components with the support are claimed to result in a superior operation compared to those of conventional alumina based catalysts. SiC has furthermore been suggested as a support for catalysts suitable for high temperature combustion [8]. Lednor and de Ruiter [9] report that the thermal stability of SiC exceeds that of SiO_2. Extensive sintering is, however, found for high surface area SiC in steam at 1273 K. In this paper the susceptibility towards sintering and oxidation are compared with that of other forms of SiC. Potential applications at extreme process conditions are investigated. The investigated properties comprise the thermal stability in N_2, N_2/H_2O mixtures, and air. Additionally, the application of ion-exchange has been investigated for the synthesis of highly dispersed Ni/SiC catalysts.

EXPERIMENTAL

High surface area SiC has been synthesized by FB-CVD using washed Norit Elorit carbon granulates (300 to 425 μm) loaded with 5w% nickel [4]. A mixture of 45 mol% H_2, 4.5 mol% $SiCl_4$, and 50.5 mol% argon is reacted with the carbon at 1380 K and 100 kPa for 40 minutes. The residual carbon present after conversion has been removed by oxidation at 1023 K in dry air. The resulting silicon carbide, with a surface area of 31 m^2/g and a pore volume of 0.2 ml/g will be referred to as SiC-5.

The thermal stability of the high surface area SiC-5 has been determined in a tubular reactor under flowing nitrogen at 1273 K for 4 hours. The hydrothermal stability has been determined at 1023 K and 1273 K by adding steam in an amount of 2v% to the nitrogen.

The rate of oxidation of the silicon carbide at 1080 K, 1180 K, and 1280 K has been determined by Thermal Gravimetric Analysis (TGA), measuring the weight increase during oxidation in dry air for 10 hours.

Ion-exchange has been carried out at 298 K in a jacketed and stirred glass flask filled with a 150 ml 1 M ammonium nitrate solution. Nickel nitrate is added (0.1 M) and the pH was set to 8.3 by bubbling gaseous NH_3 through the solution, after which silica (Engelhard Si-162-1, 20 m^2/g) or SiC-5 was added. After 24 hours the particles were subsequently filtered, washed, dried, and calcined at 773 K. The adsorbed amount nickel was determined by Atomic Absorption Spectrometry (AAS). The metal support interaction was investigated by Temperature Programmed Reduction (TPR).

RESULTS AND DISCUSSION

Stability of SiC-5

The surface areas and pore volumes of SiC-5 after the thermal and hydrothermal stability tests are shown in Table I. The nearly constant surface area after ageing in pure N_2 at 1273 K displays the excellent thermal stability of SiC-5. The stability in air and steam environments is less pronounced, which is evidenced by the decrease in surface area after ageing at 1273 K and partial oxidation of the silicon carbide into SiO_2. The SiC conversion into SiO_2 of 27% caused by oxidation at 1273 K in steam containing nitrogen results in a decrease in specific surface to 26 m^2/g; the remainder (to 19 m^2/g) is induced by sintering. Lednor and de Ruiter [9] describe that heat treatment in N_2 for 4 hours at 1273 K of SiC prepared by pyrolysis of a organosilicon polymer [3] resulted in a decrease in specific surface area of 15.7 %. SiC-5, however, displays a higher thermal stability; heat treatment at 1273 K in N_2 for 4 hours causes a decrease in surface area of only 6.5 %. Lednor and de Ruiter reported a surface area decrease for SiC after the hydrothermal treatment at 1023 K and 1273 K of 35 % and 84 %, respectively. Both are considerably higher than that of SiC-5. The physical rationale of these observed effects is probably the difference in initial surface area of the two different silicon carbides. It might be that their silicon carbide of 172 m^2/g is prone to more extensive sintering than SiC-5 (31 m^2/g).

The hydrothermal stability of SiC-5 is comparable with that of SiC synthesized by the reaction of SiO with activated carbon reported by Pham-Huu *et al.* [7]. Similar reductions in surface area were encountered after ageing in steam.

The conversion of SiC into SiO_2 in air is displayed in Fig. 2. The conversion of SiC-5 has been determined assuming that eq. 2 represents the stoichiometry of oxidation.

Table I. Surface area (S_{BET}), pore volume (V_{pore}), and SiC conversion into SiO_2 (ξ) of SiC-5 after ageing in air, N_2 and 2v% H_2O in N_2

properties	SiC-5	ageing process			
		air, 1 hour	N_2, 4 hours	2v% H_2O in N_2, 4 hours	
		1273 K	1273 K	1023 K	1273 K
S_{BET} (m^2/g)	31	17	29	29	19
V_{pore} (ml/g)	0.2	0.16	0.22	0.25	0.14
ξ (-)	-	0.20	0	0	0.27

$$SiC(s) + \frac{3}{2}O_2(g) \rightarrow SiO_2(s) + CO(g) \qquad (2)$$

The corresponding thickness of the formed SiO_2 layer is calculated assuming that the oxidation ensues via the unreacted shrinking core model and that the SiC-5 consists of agglomerated non-porous spheroids of 60 nm diameter (based on the BET surface). It is evidenced that both the rate and total amount of SiC-5 oxidation increase with increasing oxidation temperature. The rate of oxidation decreases in time. Oxidation at 1280 K for 10 hours causes abundant SiC conversion, *viz.* up to 60%. The surface area of SiC-5 decreases in the first hour of this oxidation from 31 m^2/g to 17 m^2/g (Table I).

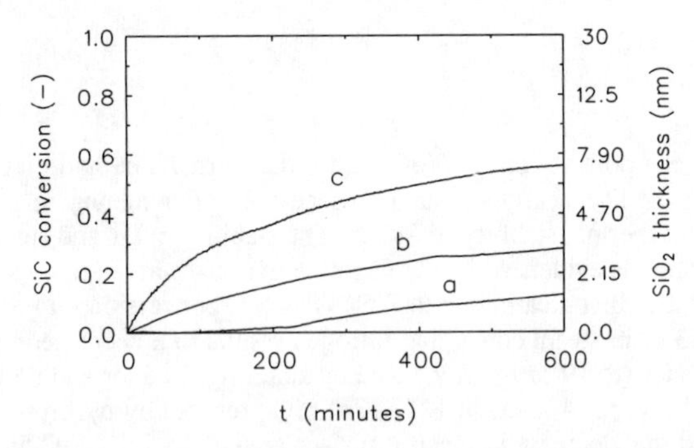

Fig. 2. Oxidation of SiC into SiO_2 in air at (a) 1080 K, (b) 1180 K, and (c) 1280 K

It is generally accepted that the resistance of SiC towards oxidation is caused by the so-called passive oxidation of SiC (*i.e.* the formation of a protective silica layer *via* eq. 2). The rate of oxidation is controlled by diffusion of the oxidant species through the growing amorphous oxide surface film which coats the surface [10]. The assumption that the rate of oxidation is inversely proportional to the oxidic layer thickness results in a linear relationship between the layer thickness and the square-root of the time, when assuming that the formed oxidic film is flat or much thinner than the total size of the particle. This relationship has indeed been reported as a suitable manner to describe the oxidation of non-porous SiC granules [10], single crystals [11], and coatings [12]. In the oxidation of high surface area SiC, however, the oxidic layer is certainly not flat while the conversion is extensive, implying that a shrinking unreacted core model has to be applied for the oxidation of the primary particles of the high surface area SiC. The rate equation that describes the fluid-solid reaction for fixed-size particles and diffusion through the ash as controlling step is

$$\frac{t}{\tau} = 1-3(1-\xi)^{\frac{2}{3}}+2(1-\xi) = 1-3\left[\frac{r}{R}\right]^2+2\left[\frac{r}{R}\right]^3 \qquad (3)$$

where τ represents the time necessary for complete oxidation, t is the duration of oxidation, ξ is the conversion, r denotes the radius of the non-converted part of the SiC, and R is the initial radius [13]. Fitting of eq. 3 with the oxidation data from Fig. 2 delivers τ and thus the time in which 100% conversion of the SiC can be expected. The values with corresponding 95% confidence regions are shown in Table II.

Table II. Time (τ) for 100 % conversion of SiC-5 into SiO_2 by oxidation in air at various temperatures.

	1080 K	1180 K	1280 K
τ (hours)	2232 ± 68	317 ± 2	62.9 ± 0.2

The small 95% confidence regions of Table II suggest that eq. 3 is a suitable description of the oxidation of high surface area silicon carbide. It can be concluded from Table II that high surface area SiC cannot be applied in oxidizing environments at high temperatures; total oxidation is to be expected in about hundred days at temperatures as low as 1080 K.

Hence, high surface area SiC transforms into silica in oxidizing environments. The formed silica layer reduces the rate of oxidation, but does not prevent further oxidation. The mechanism of oxidation suppression is diffusion limitation of the oxidizing compound through the silica layer, and, hence, this mechanism will only be effective for thick (>0.1 μm) silica layers. This prerequisite for oxidation protection makes the concept of oxidation inhibition *via* an oxidic protective coating inapplicable for SiC based catalyst supports in which the primary particles are one order of magnitude smaller than the required protective silica layer.

Synthesis of highly dispersed Ni/SiC catalysts by ion-exchange

The amount of nickel adsorbed on the SiO_2 and SiC surface is displayed in Table III.

Table III. The amount of adsorbed Ni on SiC and SiO_2, analyzed by AAS. TPR is used to determine the relative amount of NiO in moderate (NiO^{MMSI}, reduction at 700 K), and in strong interaction (NiO^{SMSI}, reduction above 800 K) with the support. (Bulk NiO reduces at 600 K)

	Ni^{AAS}		NiO^{MMSI}	NiO^{SMSI}
	(w%)	(at.Ni/nm^2)	(%)	(%)
SiC-5 after synthesis	0.27	0.92	100	0
SiC-5 ion-exchange	3.04	10.4	56	44
SiO_2 ion-exchange	0.57	2.92	4	96

It is shown that SiC-5 exhibits a 3-fold higher adsorption capacity for Ni^{2+} than silica based on surface concentration. Clause *et al.* [14] report the formation of nickel silicates during ion-exchange on silica under these applied conditions. High Resolution Electron Microscopy analysis of SiC-5 after ion-exchange discloses the presence of lamellar structures. After reduction of this NiO/SiC catalyst at 900 K, highly dispersed Ni is obtained (d_{Ni} = 4 nm). The origin of this high adsorption capacity is not entirely clear and necessitates further investigation. It is tentatively concluded that it is caused by the high reactivity of the partially oxidized SiC surface, which to some extent dissolves and re-adsorbs together with Ni^{2+} to form the nickel silicate. The Metal Support Stability of the ion-exchanged Ni/SiC catalyst is as high as the Ni/SiC ones prepared by impregnation [4]. The NiO is still easily reduced after calcination at 1273 K.

CONCLUSION

The potential of high surface area SiC prepared by the catalytic conversion of activated carbon as catalyst support has been studied. The thermal stability in non-oxidizing environments is excellent. No sintering has been observed in nitrogen atmospheres at temperatures of 1273 K. Utilization of air or the presence of steam in nitrogen at 1273 K results in SiC oxidation and considerable sintering. Oxidation in air at elevated temperatures is shown to cause substantial SiC conversion (60 % after 10 hours at 1273 K). Air oxidation of SiC with a surface area of 31 m^2/g at 1080 K will result in complete conversion in about 100 days. Because the conversion rate is proportional to the surface area, high surface area SiC can not be used as a catalyst support in processes operating in oxidizing environments at temperatures above 1080 K. Finally, highly dispersed Ni/SiC catalysts can be synthesized by exploitation of the reactivity of the partially oxidized SiC surface during ion-exchange. Thus, the application of sophisticated preparation techniques is not essential for arriving at high dispersions.

ACKNOWLEDGEMENT

This research was part of the Innovation-oriented Research Programme on Catalysis and was financially supported by the Ministry of Economic affairs of The Netherlands.

REFERENCES

1. M.A. Vannice, Yu-Lin Chao, and R.M. Friedman, Appl. Catal. **20,** 91 (1986).
2. M.J. Ledoux, S. Hantzer, C. Pham-Huu, J. Guille, and M.P. Desaneaux, *J. Catal.,* **114** (1988) 176.
3. D.A. White, S.M. Oleff, and J.R. Fox, Adv. Ceram. Mater. **2,** 53 (1987).
4. R. Moene, L.F. Kramer, J. Schoonman, M. Makkee, and J.A. Moulijn in Scientific bases for the Preparation of Heterogeneous Catalysts VI, Louvain-la-Neuve, preprints, vol 1, 1994, p. 379.
5. R. Moene, unpublished results.
6. M. Boutonnet-Kizling, P. Stenius, S. Andersson, and A. Frestad, Appl. Catal. B **1,** 149 (1992).
7. Pham-Huu Cuong, S. Marin, M.J. Ledoux, M. Weibel, G. Ehret, M. Benaissa, E. Peschiera, and J. Guille, Appl. Catal. B **4,** 45 (1994).
8. M.F.M. Zwinkels, S.G. Järås, P.G. Govind, and T.A. Griffin, Catal. Rev. —Sci. Eng. **35,** 319 (1993).
9. P.W. Lednor and R. de Ruiter, in Inorganic and Metal-Containing Polymeric Materials, J.E. Sheets, C.E. Carraher, C.U. Pittman, M. Zeldin, and B. Currel (eds.), Plenum, New York, 1990, p. 187.
10. R.J. Pugh, J. Coll. Interface. Sci. **138,** 16 (1990).
11. Z. Zheng, R.E. Tressler, and K.E. Spear, J. Electrochem. Soc. **137,** 854 (1990).
12. L. Philipuzzi, R. Naslain, and C. Jaussaud, J. Mater. Sci. **27,** 3330 (1992).
13. Perry's Chemical Engineer's Handbook, R.H. Perry and D. Green eds., 6th ed., 1984, McGraw-Hill, New York, p. 4-9.
14. O. Clause, L. Bonneviot, and M. Che, J. Catal. **138,** 195 (1992).

PALLADIUM CARBIDE FORMATION IN Pd/C CATALYSTS AND ITS EFFECT ON ADSORPTION, ABSORPTION AND CATALYTIC BEHAVIOR

NALINI KRISHNANKUTTY* AND M. A. VANNICE **
* Department of Chemical Engineering, The Pennsylvania State University, University Park, PA 16802 (Current Address: Materials Research Laboratory, The Pennsylvania State University, University Park, PA 16802)
** Department of Chemical Engineering, The Pennsylvania State University, University Park, PA 16802

ABSTRACT

The presence of PdC_x, a metastable palladium carbide phase, was detected in Pd dispersed on high surface area carbon blacks using a Pd acetylacetonate precursor. TEM and XRD measurements indicated well dispersed Pd after reduction at 573 K; however, hydride formation and chemisorption of gases was suppressed on these catalysts and this was accompanied by lattice expansion of Pd as seen from XRD. Cleaning pretreatments with O_2 slightly contracted the Pd lattice and enhanced hydride formation and chemisorption, while a 673 K reduction gave normal hydride ratios and Pd lattice parameters although chemisorption was still suppressed. Pd lattice expansion and suppression of hydride formation and chemisorption are attributed to carbon atoms present both in the bulk and on the Pd surface. The performance of these catalysts in probe reactions was consistent with this model of C contamination of the Pd surface and bulk.

INTRODUCTION

Carbon supports have many advantages, but the nature of surface functional groups and presence of small levels of impurities may influence the catalytic behavior of metals supported on them. The present study was initially aimed at elucidating the effect of carbon pretreatment prior to metal impregnation on the performance of C-supported Pd and examining the influence of pretreatment of these Pd/C catalysts on adsorption and catalytic behavior. Palladium was chosen because of its use in hydrogenation reactions; also, C-supported Pd has been reported to have lower specific activity compared to Pd on other supports for reactions like benzene and 1,3 butadiene hydrogenation[1-3]. Further, the existence of PdC_x in the presence of carbon containing gases and an amorphous carbon support has been reported[4-15]. Initial results indicated that carbon pretreatment did not have a significant effect on Pd dispersion with our preparative technique, but unexpected adsorption and catalytic behavior was observed; consequently, the effect of catalyst pretreatment on the behavior of C-supported Pd was emphasized. This paper discusses the contamination of the surface and bulk of Pd in these catalysts by C atoms, as shown by x-ray diffraction, transmission electron microscopy, gas adsorption and absorption.

EXPERIMENTAL

Black Pearls 2000 (BP2000, Cabot Corp.), a carbon black with a surface area of 1475 m^2/g, 0.36% ash, and 1.3 wt% sulfur, was given different pretreatments to change the chemical nature of the carbon surface. The untreated BP200 (C-AS IS), BP2000 which was heated in H_2 for 16 h

Mat. Res. Soc. Symp. Proc. Vol. 368

at 1173 K (C-HTT-H_2), BP2000 heated in Ar for 16 h at 1173 K (C-HTT-Ar) and the C-HTT-Ar which was boiled in HNO_3 for 6 h (C-HNO_3) were used as supports for 3 wt% Pd catalysts prepared anaerobically by an incipient wetness method from a Pd acetylacetonate precusor, $Pd(C_5H_7O_2)_2$ dissolved in tetrahydrofuran. The high temperature treatments given to the carbon support (HTT) decreased its S content to 0.1%. Catalysts and supports were handled without air exposure prior to characterization and probe reactions; further details are given elsewhere[16-18].

H_2, O_2 and CO uptakes at 300 K on these Pd/C catalysts were determined by volumetric techniques after the following pretreatment : heating to 393 K in 50 sccm He, holding at 393 K for 1 h before heating in He to either 573 or 673 K, reducing in 50 sccm H_2 for 2 h, then evacuating at the reduction temperature for 1 h and cooling under vacuum to the chemisorption temperature. Energy changes during adsorption of H_2, O_2 or CO at 300 K as well as during bulk β-hydride formation at 300 K were determined by a modified Perkin-Elmer DSC-2C differential scanning calorimeter. Some catalysts were heated in 2% O_2/98% He at 573 K for 30 min prior to reduction at 573 K and the gases evolved were monitored by a UTI mass spectrometer. XRD measurements were made ex-situ on reduced samples with a Rigaku Geigerflex diffractometer while TEM measurements were made on a Philips 420T transmission electron microscope. Vapor-phase benzene hydrogenation, CO hydrogenation and CO oxidation reactions at 1 atm were performed under differential conditions as described earlier[17].

RESULTS

Pd particle sizes, assuming spherical shapes and calculated from H_2 and CO uptakes using accepted 1:1 adsorption stoichiometries for H_{ad}:Pd_s and CO_{ad}:Pd_s where Pd_s represents a surface Pd atom, and from XRD and TEM are shown in Table I. Hydride ratios obtained from reversible H_2 uptakes and Pd lattice parameters obtained from XRD are also given below. XRD patterns and results from O_2 uptakes are presented elsewhere[16].

Table I : Pd particle sizes and bulk hydride ratios after different pretreatments of Pd/C catalysts

Pretreatment	Catalyst	Particle size, d, nm based on H_{irr}	CO_{ad}	TEM	XRD	H_{rev}/Pd_b	Pd (111) peak (2θ)	Pd lattice constant a, (Å)
573 K reduction	2.6% Pd/C-AS IS	18.8	-	4.0	4.0	0.28	39.41	3.961
673 K reduction	2.6% Pd/C-AS IS	28.3	-	22.7	-	0.40	-	-
573 K reduction	3% Pd/C-AS IS	86.9	11.3	9.6	-	0.01	-	-
O_2 pret before 573 K red.	3% Pd/C-AS IS	53.8	19.9	6.5	-	0.21	39.82	3.921
573 K reduction	2.8% Pd/ C-HTT-H_2	37.7	8.7	4.8	-	0.21	-	-
O_2 pret before 573 K red.	2.8% Pd/ C-HTT-H_2	10.3	8.8	-	3.5	0.38	39.52	3.950
673 K reduction	2.8% Pd/ C-HTT-H_2	37.7	10.3	-	7.1	0.47	39.81	3.922
573 K reduction	3.1% Pd/ C-HTT-Ar	28.2	12.5			0.23	39.23	3.978
573 K reduction	2.3% Pd/C-HNO_3	56.5	37.7	7.2	4.4	0.35	39.08	3.990
O_2 pret before 573 K red.	2.3% Pd/C-HNO_3	28.3	18.8	-	-	0.42	-	-
673 K reduction	2.3% Pd/C-HNO_3	56.5	30.5	-	7.2	0.64	40.17	3.888
REFERENCE	Pd (JCPDS # 6-681)	-	-	-	-	0.66	40.12	3.890

Spherical Pd particles of narrow size distributions were seen by TEM after a 573 K reduction, with peaks occurring between 3 - 10 nm. However, larger particle sizes were obtained from H_2 and CO chemisorption, indicating a substantial suppression of chemisorption. Further, for a given sample, the dispersions from H_2 and CO chemisorption do not agree well, in contrast with the behavior of Pd on oxide supports as discussed previously[16]. Pd forms α and β bulk hydride phases, and at 300 K the maximum H_{ab}/Pd_{bulk} ratio for bulk Pd is ~ 0.66, which can vary with the extent of alloying in Pd alloys, but is not influenced by particle size for particles above 1.5 nm[16,19]. After reduction at 573 K, low bulk hydride ratios of 0.2 - 0.35 along with a broad

Pd(111) peak, shifted from 40.11° to lower angles, was seen for most of the Pd/C catalysts, as listed in Table I and this shift is indicative of a lattice expansion.

Either exposure to reducing conditions at 673 K and higher or use of an O_2 pretreatment prior to reduction at 573 K affected hydride ratios and gas uptakes. Nearly normal hydride ratios were obtained after a 673 K reduction; however, chemisorption was still suppressed as shown by the large particle sizes from H_2 and CO chemisorption in comparison to TEM or XRD. This increased hydride formation was accompanied by lattice parameters near the value of 3.889 Å for bulk Pd. During the O_2 pretreatment, CO and CO_2 evolution from Pd supported on HTT carbon catalysts was detected by mass spectrometry, and evolution of SO_2 was also detected from the Pd/C-AS IS sample[16]. After O_2 pretreatment, H_2 chemisorption increased while CO uptakes either increased or remained unchanged. The Pd lattice contracted slightly and hydride ratios increased, but values similar to bulk Pd were still not attained as shown in Table I. Isothermal, integral heats of H_2 and CO adsorption on these catalysts were obtained to determine if weaker binding was associated with the suppressed chemisorption. As reported in detail elsewhere[16], Q_{ad} values for irreversible hydrogen adsorption on these Pd/C catalysts are closer to initial Q_{ad} values at low H coverage but are slightly higher than integral Q_{ad} values for similar size supported Pd crystallites. The Q_{ad} values for CO on these catalysts are lower than initial heats of adsorption on clean Pd surfaces and the isothermal, integral Q_{ad} values for supported Pd crystallites larger than 3 nm. The apparent heats of β-hydride formation, assuming all the reversible H_2 uptake is associated with this phase are close to the values of 9 ± 1 kcal/mole established previously[16].

The activity of Pd/C-HTT-H_2 and Pd/C-HNO_3 for benzene and CO hydrogenation and CO oxidation after a 573 K reduction are shown in Table II in comparison with Pd on oxide supports. The turnover frequencies (TOFs) calculated based on dispersions from CO adsorption are similar to those on Pd/oxide supports for benzene hydrogenation, lower in CO hydrogenation and similar or higher in CO oxidation. The results for other catalysts are presented elsewhere[17].

Table II : Turnover frequencies and activation energies for Pd/C catalysts in probe reactions

Catalyst	Benzene hydrogenation		CO hydrogenation		CO oxidation	
	TOF x 10^3 @ 413 K, s^{-1}	E_a kcal/mole	TOF x 10^3 @ 648 K, s^{-1}	E_a kcal/mole	TOF x 10^3 @ 400 K, s^{-1}	E_a kcal/mole
Pd/C-HTT-H_2	36	9	1.8	30	7.4	24
Pd/C-HNO_3	30	9.4	0.52	13	30.0	25
Pd/SiO_2	43	11.6	6.5	29	-	-
Pd/Al_2O_3	-	-	-	-	8.7	18

DISCUSSION

The adsorption behavior of Pd/C catalysts employing a clean turbostratic carbon support can be explained by C atoms occupying both surface and bulk interstitial Pd sites during the pretreatment used in this study. Additional surface contamination can occur when S is present in the support; however, with clean HTT carbons, no S-containing gases were observed during regeneration in O_2 and the inhibited adsorption capacities must be attributed to contamination by carbon. The contraction of the Pd lattice after an O_2 treatment and the increase in hydride ratios and H_2 chemisorption, shown in Table I, indicate that carbon is removed from the Pd particles during the O_2 treatment. The sharp decrease in CO and CO_2 evolution with time during O_2 treatment implies that their formation is associated mainly with the Pd, rather than the support[17].

Although it has been commonly accepted that second and third row Group VIII metals do not form bulk interstitial carbides[20], the existence of interstitial C atoms in Pd is well documented.

Exposure of Pd to C_2H_2, C_2H_4 and CO at 423-773 K produced C/Pd ratios up to 0.15, along with Pd lattice expansion up to a value of 0.399 nm and decreased hydride forming ability due to carbon atoms in octahedral sites[4-8,10]. Interstitial carbon was also indicated by a 2.8% expansion of the Pd lattice when Pd was deposited by vacuum evaporation on a carbon substrate[9]. The bulk PdC_x phase decomposed at 610-873 K in inert atmospheres and at 423-460 K in H_2 or O_2[5,6,11,12]. This phase decomposed at 700 K under vacuum for 2-5 nm Pd particles on amorphous carbon, accompanied by carbon deposition on the Pd surface[9]. The appearance of carbon on clean Pd surfaces during cooling from 1000 K, and its segregation at 400-600 K has been reported[13,14].

These studies clearly establish that Pd-C phases can exist, although single-crystal studies imply that little or no interstitial bulk carbon would be expected in these Pd/C catalysts due to the low solubility of C in Pd from 298 to 673 K. Regardless, the lattice expansion seen by XRD in the current study indicates that interstitial C is present in Pd after reduction and evacuation at 573 K, and the formation of this PdC_x phase may be facilitated in these Pd crystallites that are in continuous contact with a non-graphitic carbon support. Hamilton and Blakely used graphite[14], Stachurski[5], Ziemecki and Jones[7] deposited a finite amount of carbon from gases, while Lamber et al. evaporated Pd onto an amorphous carbon surface[9]; thus, there was limited contact between the carbonaceous layer and the Pd atoms. In the current study, Pd particles are present in the small pores of a turbostratic carbon, enhancing Pd-C contact which may make it more facile for carbon to migrate onto and into the Pd crystallites during the higher temperature reduction or evacuation periods. The enhanced Pd-C interface and a possible interaction between Pd and carbon at the Pd-C interface as suggested previously[21] could weaken C-C bonds and facilitate the formation of a solid solution at the Pd-C interface and in the bulk, similar to that in dispersed Ni, Pd and Pt on amorphous carbon during its catalyzed graphitization or gasification as discussed elsewhere[17]. The driving force for this mechanism is thought to be the free energy difference between the initial disordered carbon and the final graphitic form of carbon. The dissolution of C from an activated carbon into Ni crystallites, as well as C precipitation on the Ni surface during treatments in N_2 or Ar or H_2 above 723 K has been proposed to explain decreased catalytic activity of Ni for acetone hydrogenation[22] and the dissolution and diffusion of carbon species from hydrocarbons through metal particles leads to the formation of carbon filaments[23].

The source of the carbon in the PdC_x phase in the current study is partially from $Pd(AcAc)_2$. Pd/SiO_2 catalysts prepared from $Pd(AcAc)_2$ also had suppressed chemisorption of H_2 and CO along with suppressed hydride ratios after 573 K reduction[24]. However, stoichiometrically adsorbed H_2 and CO along with normal hydride ratios were obtained after O_2 pretreatment followed by H_2 reduction at 573 K for Pd/SiO_2, unlike the behavior of Pd/C catalysts after the same treatment, where chemisorption and hydride formation still remain suppressed. The decrease in CO uptakes after these sequential O_2 treatments (following reduction) suggests that C contamination occurs during either the reduction or, more likely, the evacuation at 573 K[16]. The possibility that decomposition of $Pd(AcAc)_2$ in these small pores under H_2 (or He) may leave residual C atoms cannot be dismissed, but different mechanisms exist on the two supports[24].

Five types of Pd particles can be proposed to coexist in these Pd/C catalysts[16]: Type A) Clean Pd particles, with no bulk or surface carbon, exhibiting normal chemisorption and hydride ratios; Type B) Pd particles completely encapsulated by carbon and incapable of either adsorbing gas or forming hydride; Type C) Pd particles partially covered by carbon and with interstitial carbon, having both suppressed chemisorption and hydride formation; Type D) Pd particles partially covered by carbon but with no interstitial carbon exhibiting suppressed chemisorption but normal hydride ratios; and Type E) Pd particles with clean surfaces but with interstitial carbon, resulting in normal chemisorption but lower hydride ratios. Type E particles cannot exist due to the thermodynamics of surface segregation of carbon in Pd, since C on the Pd surface is necessary to stabilize C in the bulk[13,14]. This is supported by our results showing that suppression of β-hydride formation was always accompanied by suppressed chemisorption in these Pd/C catalysts.

Two limiting models can be proposed as the simplest explanations of the behavior of these Pd/C catalysts. One (Model I) invokes a mixture of Pd particles with no interstitial carbon but with different extents of surface contamination - either clean or completely encapsulated Pd particles, i.e., Type A and B particles. This model can qualitatively explain both the suppression of chemisorption and hydride formation, but it predicts bulk hydride ratios 2 - 5 times lower than observed and cannot explain the expansion of the Pd lattice[16]. Consequently, the simplest model to explain the phenomena observed in this study is the proposal that Pd particles exist with both bulk and surface carbon contamination, i.e., all are Type C particles (Model II). The apparent bulk C/Pd ratios of 0.05 - 0.10, calculated assuming a H_{ab}/Pd_{bulk} ratio of 0.66 and that one interstitial C atom blocks four such sites for H atoms,[4] are presumably stabilized at 298 K by the presence of 0.4 - 0.8 monolayer of carbon present on the surface of these Pd particles[16].

The high ratios of CO_{ad}/H_{ad} can be explained by this model, since the larger CO uptakes compared to H_2 is consistent with the fact that H_2 needs two adjacent 3-fold or 4-fold hollow sites for dissociation, the population of which would decrease more rapidly than on-top sites due to occupancy of these hollow sites by C atoms, while CO could still adsorb linearly on single (on-top) sites. Further, in accordance with Model II, Q_{ad} values for H_2 relate more closely to the initial Q_{ad} values[16], which is consistent with the low surface coverages of H atoms that exist on these catalysts. In contrast, the lower Q_{ad} values for CO compared to small Pd crystallites may be a consequence of CO adsorption only on singlet (on-top) sites of lower energy[16].

Model II can explain the observed chemisorption behavior of Pd/C catalysts. Reducing conditions above 673 K or higher cause a shrinkage of the Pd lattice and provide normal hydride ratios but suppressed chemisorption. At these higher temperatures bulk C atoms diffuse to the Pd surface resulting in the contracted Pd lattice and normal hydride ratios. The maximum coverage of the Pd surface if all the carbon from the bulk migrated to the surface is 0.6 - 1.2 monolayers, which explains the continued suppression of chemisorption after this pretreatment[16]. Although some sintering may exist at higher reduction temperatures, the particle sizes from chemisorption are larger than from XRD, indicating substantial suppression of chemisorption. Hence, C still remains on the Pd surface after this pretreatment. This model is also consistent with the the evolution of CO and CO_2, the contraction of the Pd lattice, and the increase in both hydride ratios and H_2 uptakes after an O_2 regeneration step. These results establish that at least part of the CO and CO_2 evolved is associated with the Pd crystallites even though some of it may come from the support. The mild cleaning in O_2 followed by reduction in H_2 and evacuation at 573 K did not remove all the carbon, possibly due to diffusional limitations for interstitial carbon to migrate to the surface and react with O_2, or the presence of completely encapsulated Pd particles which cannot adsorb and dissociate O_2 at 573 K, or the migration of C atoms from the support to the Pd during the reduction and evacuation step in H_2 at 573 K following the O_2 pretreatment[16]. Consequently, Model II, which assumes the presence of Type C particles only, is the simplest one to explain all the observed phenomena, but the simultaneous presence of Type A, B, and D particles cannot be ruled out; however, Type C particles must be present. The Pd surface can also be contaminated by sulfur from the carbon support in the Pd/C-AS IS sample.

This model can also explain the observed catalytic behavior of these catalysts. TOFs based on dispersion from CO chemisorption in the structure-insensitive benzene hydrogenation should not be affected by partial C overlayers, presuming the absence of electronic effects[17] and this behavior is indeed observed. In contrast, these Pd/C catalysts had markedly lower activities for CO hydrogenation relative to other Pd catalysts and some catalysts prepared from $Pd(AcAc)_2$ had a TOF 25 times lower than typical Pd/SiO_2[17]. Severe deactivation occurred after a high temperature excursion and apparent activation energies varied from 13 to 30 kcal/mole, perhaps reflecting the deactivation process[17]. The low specific activity is attributed to the type of carbon on the Pd surface under reaction conditions. At the high reaction temperatures $\geq$ 673 K, all the C atoms in the Pd lattice can migrate to the Pd surface, achieving coverages close to a monolayer and they may be present as graphitic carbon[16,17]. Thus, the low activity and the deactivation could be due to the formation of this unreactive overlayer of graphitic carbon. Further, the rate determining step in CH_4 formation from CO and H_2 on Pd is thought to be the H-assisted dissociation of CO[17]. Coverage of the Pd surface by C atoms adsorbed in the 3-fold and 4-fold hollow sites typically occupied by H atoms or a graphitic overlayer would decrease the rate of this step. The TOFs for CO oxidation over Pd/C catalysts were comparable to or higher than those for Pd/Al_2O_3. It appears that carbon is removed from the Pd surface during the oxidizing conditions of the reaction, and Pd/C is similar to Pd on other supports during this reaction. The higher TOFs may be partially due to the initially suppressed chemisorption on the Pd surface, which did not count all the surface Pd atoms indicated by TEM.

CONCLUSIONS

A model invoking partial coverage of the Pd surface by carbon atoms and carbon atoms occupying interstitial positions in the Pd lattice can explain the suppressed chemisorption of H_2 and CO as well as the suppression of hydride formation in Pd/C catalysts. Carbon atoms from either the acetylacetonate group or the turbostratic carbon support appear to migrate onto and into Pd crystallites dispersed within the carbon black pore structure during reduction in H_2 and evacuation at 573 K. Exposure to reducing conditions above 673 K causes the carbon in the interstitial sites to migrate to the Pd surface, while a O_2 treatment at 573 K removes the carbon from the Pd, but it seems to reestablish itself during the subsequent reduction/evacuation step.

ACKNOWLEDGEMENTS

This study was funded by the National Science Foundation under Grant# CTS-9011275. We would like to thank Kevin Appold and Jin Li for conducting some of the O_2 pretreatment studies.

REFERENCES

1. Suh, D. J., Park, T. J., and Ihm, S. K., Carbon, **31(3)**, 427 (1993).
2. Chou, P. and Vannice, M. A., J. Catal., **107**, 129 (1987).
3. Ouchaib, T., Moraweck, B., Massardier, J., and Renouprez, A., Catal. Today, **7**, 191 (1990).
4. Ziemecki, S. B., Michel, J.B., and Jones, G. A., React. of Solids, **2**, 187 (1987).
5. Stachurski, J., J. Chem. Soc. Faraday Trans. 1, **81**, 2813 (1985).
6. Ziemecki, S. B., Jones, G. A., Swartzfager, D. G., and Harlow, R. L., J.Am.Chem.Soc., **107**, 4547 (1985).
7. Ziemecki, S. B., and Jones, G. A., J. Catal., **95**, 621 (1985).
8. Ziemecki, S. B.,Jones, G. A.,and Swartzfager, D. G.,J. Less-Common Metals, **131**,157 (1987).
9. Lamber, R., Jaeger, N., and Schulz-Ekloff, G., Surf. Sci., **227**, 15 (1990).
10. McCaulley, J. A., J. Phys. Chem., **97**, 10372 (1993).
11. Yamamoto, T. et al., Appl. Phys. Lett., **63**, 3020 (1993).
12. Maciejewski, M. and Baiker, A., J. Phys. Chem. **98**, 285 (1994).
13. Wolf,M., Goschnick,A., Loboda-Cackovic,J., Grunze,M., Unertl,W. N., and Block,J. H., Surf. Sci., 1**82**, 489 (1987).
14. Hamilton, J. C. and Blakely, J. M., Surf. Sci., **91**, 199 (1980).
15. Ratajczykowa, I., J. Vac. Sci. Technol., **A 1 (3)**, 1512 (1983).
16. Krishnankutty, N. and Vannice, M.A., J. Catal., Submitted for Publication, Aug. 1994.
17. Krishnankutty, N. and Vannice, M.A., J. Catal., Submitted for Publicationn, Aug. 1994.
18. Krishnankutty, N. and Vannice, M.A., Chem. Mater., Submitted for Publication, Nov. 1994.
19. Palczewska, W., Adv. Catal., **24**, 245 (1975).
20. Toth, L. E., Transition Metal Carbides and Nitrides, (Academic Press, New York, 1971), p.2.
21. Baetzold, R. C., Surf. Sci., **36**, 123 (1972).
22. Gandia, L. M., and Montes, M., J. Catal., **145**, 276 (1994).
23. Baker,R.T.K., Barber,M.A., Harris,P.S., Feates,F.S., and Waite,R.J., J. Catal., **26**, 51 (1972).
24. Krishnankutty, N., Li, J., and Vannice, M. A., In Preparation.

Carbon Fibrils
Mechanism of Growth and Utilization as a Catalyst Support

JOHN W. GEUS, A.J. VAN DILLEN, AND MARCO S. HOOGENRAAD
Department of Inorganic Chemistry, Debije Institute,
University Utrecht, The Netherlands

ABSTRACT

Especially with the production of fine chemicals catalytic reactions are being performed in which solid catalysts suspended in liquids are used. Carbon is stable in both acid and alkaline liquids and is not preferentially wetted by water. Carbon is therefore an attractive support for catalysts to be used in liquid-phase reactions. The relatively low diffusion coefficients in liquids call for catalyst bodies having wide pores and dimensions between 1 and 10 μm. Since particles smaller than 1 μm cannot be readily separated from liquids, attrition of carbon support bodies must be avoided. However, it is difficult to produce active carbon bodies of 1 to 10 μm of a high mechanical strength. Also with gas-phase reactions, the low mechanical strength of active carbon bodies limits the applicability of active carbon.

From supported metal particles carbon filaments of a diameter between 10 and 300 nm can be grown rapidly. The diameter is generally of the same order of magnitude as the size of the metal particles. The graphite layers within the fibrils, that have a circular cross-section, are oriented either parallel to the fibril axis or at an angle, which leads to cone-shaped graphite layers. The mechanism of growth of carbon fibrils from supported metal particles will be dealt with. The parallel orientation of the graphite layers leads to a very high mechanical strength, while the porous structure is extremely open. A representative value for the surface area is 225 m^2/g and for the pore volume 1.6 ml/g, which leads to an average pore dimension of 28 nm.

To apply active precursors, such as, palladium salts, the surface of the carbon fibrils must be oxidized, e.g., by treatment with nitric acid. A very high dispersion of palladium can thus be achieved. The dispersion of components active in hydrogenation can be assessed by the ability to react the support with hydrogen to methane. The activity of catalyst on carbon fibrils in the hydrogenation of nitrobenzene as well as the filterability of carbon fibril catalysts will be compared with that of commercial catalysts.

INTRODUCTION

Though solid catalysts are most frequently used in gas-phase reactions, reactions in liquid phases are becoming increasingly important due to the present interest in the production of fine chemicals without producing waste materials, such as, inorganic salts. Since with solid catalysts, the catalytic reaction proceeds at the surface of the solid, the catalytically active surface area per unit volume must be sufficiently large to ensure the high activity required in technical operations. A high active surface area calls for small catalyst bodies. To separate the catalyst from the liquid reaction products, however, the size of the catalyst bodies has to be at

Mat. Res. Soc. Symp. Proc. Vol. 368 © 1995 Materials Research Society

least about 3 μm. Since the surface area of bodies of 3 μm is too small to provide the required active surface area, porous bodies are required. Since it is difficult to produce porous bodies of a size of about 3 μm, often much larger catalyst bodies are utilized in technical reactions. The reacting molecules have to migrate through the pores to the catalytically active surface. The transport of species through liquids proceeds much more slowly than that in gases. With a fixed catalyst bed, in which bodies of about 5 mm are used, separation of the catalyst from the liquid proceeds smoothly, but the rate of the reaction is generally completely determined by the transport of the reacting molecules. With most catalytic reactions only the external edge of the large catalyst bodies used in a fixed catalyst bed are contributing to the reaction. With the much smaller catalyst bodies used in slurry reactors, transport is less often determining the rate of the reaction. An effect on the rate of the catalytic reaction does generally not present large problems, especially not with the production of fine chemicals. However, transport limitation can also very adversely affect the selectivity of catalytic reactions. Since the selectivity of catalytic reactions is almost invariably of paramount importance, transport limitations have usually to be avoided, which calls for catalyst bodies of 3 to 10 μm having wide pores. The size of the bodies determining the pore length is more important to avoid transport limitations than the width of the pores.

Besides the size of the catalyst bodies and the width of the pores, solid catalysts to be used in liquid phase reaction have to meet a large number of demands. The solid should be stable in acid or basic fluids, while the bodies must be mechanically strong. Disintegration of the catalyst bodies causes the separation of the catalyst from the reaction products to become very difficult. Not only loss of the often expensive catalyst, but also the presence of small amounts of the catalytically active material in the reaction products, often food additives or drugs, is not acceptable. Highly important is the wetting of the support by different liquids. When a liquid is processed which is not miscible with water, a catalyst which is preferentially wetted by water leads to problems, especially if water is produced in the reaction. Water present in the feedstock or generated in the reaction is filling the pores of the catalyst and is expelling the organic fluid out of the pores of the catalyst.

Since it is very difficult to process catalytically active materials alone to highly porous bodies of the required small dimensions, generally a support is utilized. The support consists of bodies of the desired size and porosity onto which the active component(s) are applied preferably as very small particles. For catalytic reactions in liquids, carbon is an attractive support; carbon is not affected by acids or bases and is not wetted preferentially by water. The interaction of carbon with water can be controlled; oxidation of the surface, which leads to te formaton of carboxyl groups at the surface, raises the wettability by water, whereas reduction, which removes the carboxyl groups strongly decreases the wetting by water. Frequently activated carbon is used as a support for liquid phase reactions. The properties of activated carbon supports are notoriously difficult to control, which brings about that the reproducibility of the production of carbon supports is often poor. Furthermore, the mechanical strength of active carbon is low, which causes much problems with suspended carbon supports. Raney metals, which does not involve a support, and can be readily separated from the liquid, are therefore often used in liquid phase reactions. When precious metals are desired for the catalytic reaction, a support is required, and usually carbon supports have to be used. Carbon supports of a higher mechanical strength and a higher porosity are, however, much desired, especially when properties, such as, the size of the bodies and the surface properties of the carbon can be well controlled.

Already in 1890 the first carbon fibrils were observed, but the fibrils did not draw much attention. Much later, in the seventies, Rostrup Nielsen[1] published about carbon fibrils formed in the usual nickel methane-steam reforming catalysts that had been exposed to gas flows containing not sufficient steam. Later Baker[2] published a number of spectacular results on the growth of carbon fibrils out of metal particles.

The carbon fibrils can exhibit a very high mechanical strength; catalyst bodies are disintegrated readily by the growing fibrils, and the fibrils easily fracture glass reactors. The mechanical strength, the high surface area, and the large pore volume cause the carbon fibrils to be excellent catalyst supports. With most gas phase reactions, activated carbon supports are not attractive due to the low mechanical strength. The carbon fibrils present, however, also interesting possibilities for gas phase reactions carried out within a fixed catalyst bed.

Highly intriguing is the mechanism which leads to the growth of carbon fibrils out of supported metal particles. The mechanism of the growth will therefore be considered as well as the structure of the carbon fibrils. The surface area and the porous structure of the fibrils will also be discussed. Subsequently the application of active components on the carbon fibrils will be dealt with using palladium as an example. The electrostatic charge on the surface of suspended carbon fibrils will be reported both before and after oxidation. Finally the activity of the palladium catalysts supported on carbon fibrils will be tested in the liquid-phase hydrogenation of nitrobenzene to aniline together with the filterability of the catalyst.

EXPERIMENTAL

Carbon fibrils were grown by passing either methane and hydrogen or carbon monoxide and hydrogen upwards through a catalyst bed containing alumina-supported nickel or iron. A number of other experiments has been performed with gases, such as, ethylene and toluene. The alumina support used was prepared by flame hydrolysis (Alon-C Degussa, Germany). When a broad size distribution of metal particles was desired, the support was loaded with iron or nickel by incipient wetness impregnation of iron(III) nitrate or nickel nitrate. To produce uniform metal particles deposition-precipitation from a homogeneous solution onto the suspended alumina was performed. For the preparation of the iron catalyst, which contained 20 wt.% Fe, an iron(III) nitrate solution was slowly injected into a suspension of the alumina support, the pH of which was kept at a level of 5 by simultaneous injection of ammonia. With the preparation of alumina-supported nickel catalysts the pH level was raised homogeneously to 8.

A weighed sample of the catalysts (about 100 mg) was placed onto a porous Pyrex plate mounted in a spherical reactor, which can accommodate the often rapidly increasing mass of carbon fibrils. The temperature was measured and controlled by a thermocouple placed a short distance above the initial catalyst bed. Previously the iron catalysts were reduced by raising the temperature to 700°C within two hours in a flow of 10% hydrogen in argon and keeping the catalyst at 700°C for another two hours. With the nickel catalysts the temperature was 600°C. In separate experiments the reduction of the catalysts had been investigated by thermogravimetry and by TPR. The supported iron and nickel catalysts could be reduced virtually completely at the temperatures used.

During the growth experiments a flow of 100 ml/min was passed through the catalyst bed. A back pressure controller kept the pressure in the reactor at 1.4 bar. Before and after passing through the reactor the composition of the gas flow was monitored by a hot-wire detector (HWD) and a gas-chromatograph equipped with a Porapack Q column. Water formed in the catalytic reaction was removed by a trap kept at -77°C. With the gas-chromatograph carbon monoxide, carbon dioxide, methane, and hydrogen could be measured. To control the rate of supply of carbon atoms to the surface of the metal particles, hydrogen was added to the carbon monoxide or the methane. The growth of carbon from carbon monoxide proceeded mainly by reaction to carbon dioxide, while some carbon monoxide reacted with hydrogen to methane. With a number of experiments the encapsulation of the carbon was carried out in a separate step by exposing the catalyst for the first 30 min to a carbon monoxide/hydrogen flow (usually 10% CO, 20%H_2); subsequently the gas flow was changed to, e.g., pure methane to achieve the growth of carbon fibrils. The total amount of carbon deposited was determined by determination of the weight of the reactor before and after the reaction.

Before and after the growth of carbon fibrils the catalysts were investigated in a Philips EM420 electron microscope operated at 120 kV. Specimens were ground in a corundum mortar and ultrasonically dispersed in alcohol. A drop of the suspension was brought onto a holey carbon film supported by a copper grid.

To measure the saturation magnetization a weighed amount of the catalyst was placed in an aluminum holder. The magnetization was measured with a vibrating sample magnetometer at magnetic field strengths up to 12,000 oe.

Palladium catalysts were prepared from carbon fibrils grown from supported iron particles. The surface area and the pore volume were measured using a Micromeretics ASAP 2400 apparatus. The loading of the fibrils was 2.5 wt.% Pd. To raise the surface reactivity of the carbon fibrils, 3 g of fibrils were suspended thus treated in 100 ml 65% nitric acid (Lamers & Pleugers very pure) and kept for 10 to 30 min at about 90°C. The carbon fibrils were subsequently thoroughly washed and dried at 120°C for 1 hour. The carbon fibrils were suspended in degased water under a flow of nitrogen, which was maintained during the precipitation and the filtering of the loaded support. Under vigorous agitation the Pd complex, $Pd(NH_3)_4Cl_2$ or $PdCl_2$ was added, after which an excess of a formaldehyde solution was injected to reduce the palladium. The loaded carbon fibrils were filtered and dried in the nitrogen flow at room temperature. Finally, the fibrils were kept at 80°C for 12 h and cooled to room temperature also in the nitrogen flow.

To determine the electrostatic charge of fresh and nitric acid-pretreated carbon filaments suspended in water as a function of the pH, the electrokinetic sonic amplitude (ESA) was measured using the Matec ESA-8000 system.

To assess the catalytic activity of the carbon-supported catalysts the hydrogenation of nitrobenzene to aniline was used. 50 mg of catalyst was suspended into 125 ml isopropylalcohol brought in a reactor of 175 ml kept at 25°C, after which 6.35 ml nitrobenzene was added. Through a gas recirculating stirrer (1900 rpm) 100 ml of a flow of 50% H_2 and 50% argon was passed through the solution. The consumption of hydrogen was measured with a HWD, and the conversion of nitrobenzene by determining the composition of the head space of the reactor by means of a gas chromatograph.

The filterability of a commercial activated carbon and the carbon fibrils was compared by measuring the time involved in filtering a suspension of 1 g of the support in an equal amount of water in the same filtering device.

STRUCTURE AND MECHANISM OF GROWTH OF CARBON FIBRILS

Fig. 1 shows an electronmicrograph of carbon fibrils grown out of supported iron particles. The tangled skeins can be seen, as well as the metal particles present at the end of the fibrils. The diameter of the carbon fibrils is generally about equal to that of the metal particles from which the fibrils have grown. With growth from large nickel particles (diameter about 25 nm or more), the metal particles have a cone shape with an almost flat basal plane. The cone is usually completely enclosed by the growing fibril. Smaller nickel particles from which a fibril has grown usually are more spherical. With iron the metal particles are often more spherical in shape and are situated on top of the fibril. The internal structure of fibrils grown from nickel and iron is different. Fibrils grown from small iron particles are exhibiting graphite layers oriented in parallel to the axis of the fibril, while fibrils grown from nickel are showing graphite layers oriented at an angle with the fibril axis. Schematically the internal structure of the fibrils is indicated in Fig. 2. The angle can vary appreciably, from about 40° to less than 10°; with larger nickel particles the angle is larger. The internal structure of the fibrils can be examined by lattice imaging of the graphite layers.

FIGURE 1. ELECTRON MICROGRAPH OF CARBON FIBRILS GROWN OUT OF SUPPORTED IRON PARTICLES. TANGLED SKEINS.

The distance between the layers agrees with the distance determined by X-ray diffraction, which is 0.342 nm. Since the layer distance apparent in X-ray diffraction of well ordered graphite is 0.335 nm, the graphite layer of the fibrils are badly ordered in the direction of the c-axis. The internal structure of the fibrils is also evident from the dark field image of the diffraction spot of the (002) diffraction of graphite. With a sufficiently large angle between the graphite layers the dark field

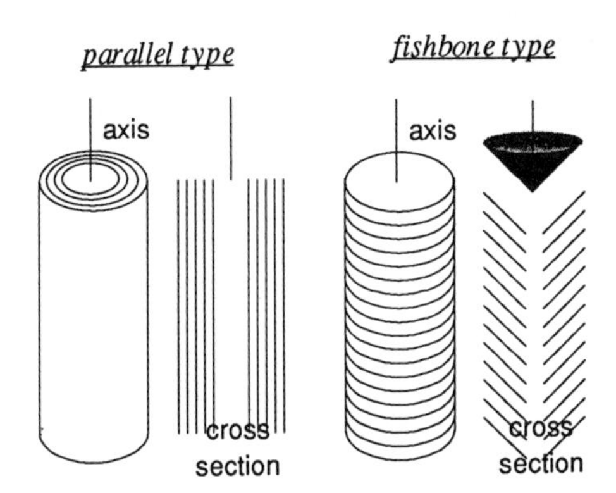

FIGURE 2. SCHEMATIC REPRESENTATION OF INTERNAL STRUCTURE OF THE TWO DIFFERENT TYPES OF CARBON FIBRILS . LEFT-HAND SIDE CYLINDRICAL. RIGHT-HAND SIDE FISHBONE.

image shows only one side of the fibril with fishbone-structured fibrils, whereas with parallel fibrils both sides are apparent in the dark field image. The fishbone structure as apparent in the electron microscope can very well result from a ribbon that has been wrapped up. Fig. 3 shows an electron diffraction pattern and dark-field images of the two pairs of diffraction spots. The nickel particle at the end of the fibril exposing a free nickel surface can be clearly seen.

Both from a scientific point of view and to control the properties of the carbon fibrils, it is highly important to elucidate the mechanism of growth of the filaments. We therefore consider the evidence on single crystal surfaces first. Experiments with single crystals have shown that upon exposure of nickel or iron to methane or carbon monoxide dissociative adsorption proceeds. With methane formation of a surface carbide is clearly evident from the Auger spectrum. With carbon monoxide, more elevated temperatures are required to induce desorption of the initially adsorbed oxygen by reaction with carbon monoxide as carbon dioxide. Other hydrocarbons, which are more reactive than methane, are also dissociating upon adsorption on nickel, iron, or cobalt surfaces. The dissociative adsorption on nickel is plane-specific, i.e., on the closest packed nickel (111) surface no dissociative adsorption takes place, whereas on iron all crystallographic planes are dissociatively adsorbing carbon monoxide.

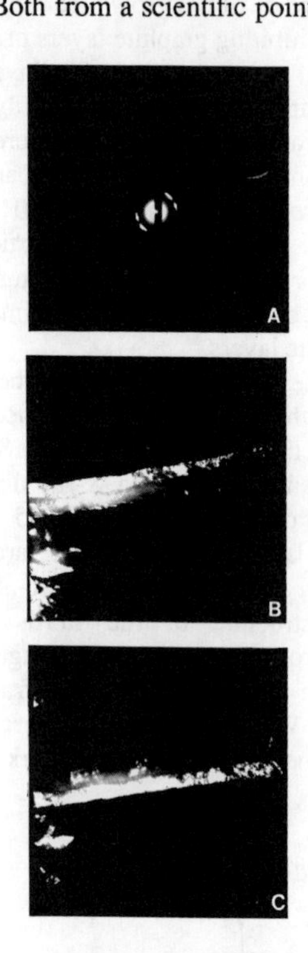

FIGURE 3. ELECTRON DIFFRACTION PATTERN (A) AND DARK-FIELD IMAGES (B AND C) OF CARBON FIBRIL GROWN OUT OF ALUMINA SUPPORTED NICKEL PARTICLE

Upon prolonged exposure to the above carbon containing gas molecules the carbon atoms set free in the metal surface cause reaction of subsurface layers of metal atoms to metal carbide. The formation of metal carbide can be appreciated best from the magnetization measured on supported metal particles. Since nickel carbide is not ferromagnetic, a pronounced drop of the ferromagnetism is evidence for the reaction to the carbide. The saturation magnetization of iron carbide is still considerable, though is clearly smaller than that of metallic iron. A drop in saturation magnetization is measured upon exposure of iron to, e.g., carbon monoxide. The reaction to iron carbide is evident from the much lower Curie temperature; the different iron carbides are exhibiting much lower Curie temperatures than metallic iron.

Generally metal carbides are less stable than a system containing the metal and graphite. However, nucleation of graphite is impeding the establishment of the thermodynamic equilibrium. Whether the metal particles have reacted completely to the metal carbide before nucleation of graphite on the surface of the metal particles proceeds, is not completely sure. The experimental evidence indicates that the period of time involved in the nucleation of graphite at the surface of the metal particles depends upon the momentary concentration of carbon within the metal surface. At low concentrations nucleation does not proceed readily, whereas at high concentrations the nucleation is more rapid. The nucleation of graphite depends furthermore on the temperature and on the crystallographic plane of the metal. On Ni(111), e.g., nucleation occurs relatively rapidly in view of the almost complete fit of the lattices. Once graphite has

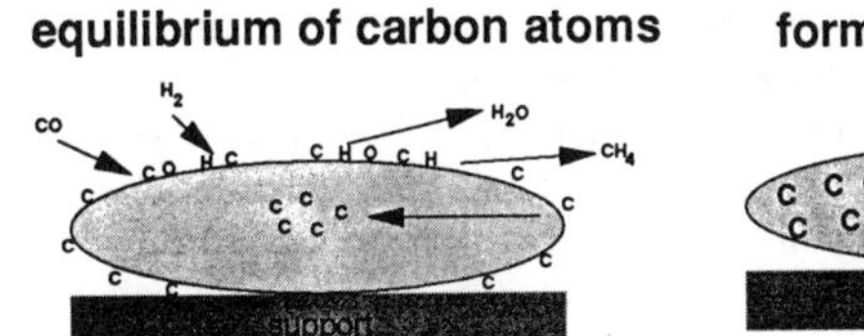

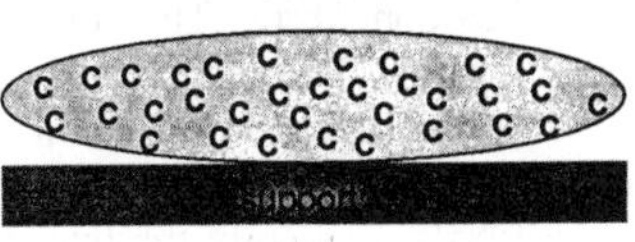

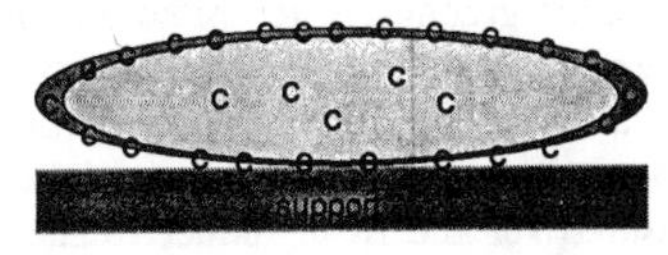

FIGURE 4. SCHEMATIC REPRESENTATION OF REACTION OF SUPPORTED METAL PARTICLES TO METAL CARBIDES AND SUBSEQUENT ENCAPSULATION BY GRAPHITE LAYERS.

nucleated, a rapid growth over the surface of the metal particles proceeds. The particles are deformed by the strong graphite layers, which are bent to surround completely the metal particles, that are encapsulated in graphite. Since the curvature that graphite layers can exhibit is limited, very small particles cannot be encapsulated. It has been found experimentally that very small nickel particles (diameter about 3 nm) are not covered by graphite layers at all, which indicates that the critical nuclei of graphite layers are of the order of 3 nm. Usually the process ends with the encapsulation. We could show that encapsulated metal particles are no longer active in the decomposition of methane; reaction of CH_4 to carbon and hydrogen therefore does not proceed on graphite surfaces. The encapsulation of a supported metal is schematically indicated in Fig. 4.

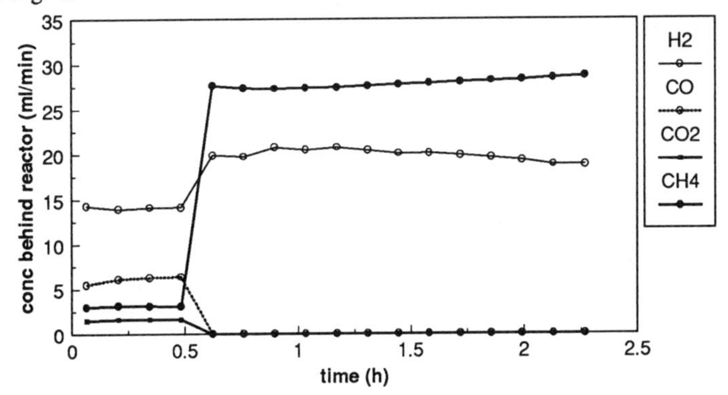

FIGURE 5. COMPOSITION OF GAS FLOW BEHIND THE REACTOR LOADED WITH ALUMINA-SUPPORTED NICKEL CATALYST PREPARED BY IMPREGNATION WITH $Ni(NO_3)_2$. FIRST 30 MIN 10%CO AND 20%H_2, SUBSEQUENTLY 40%CH_4 IN ARGON

When pure methane or carbon monoxide are passed through a supported iron or nickel catalyst, the metal particles larger than about 3 nm are merely encapsulated and the consumption of methane or carbon monoxide stops. Investigation in the electron microscope indeed indicates that the metal particles are completely encapsulated. When the rate of supply of carbon atoms to the surface of the metal particles is decreased by addition of hydrogen to the gas flow, the amount of carbon taken up by the metal particles is lower. After passing a flow of hydrogen and carbon monoxide for 30 min through the catalyst bed, switching the gas flow to pure methane leads to growth of carbon fibrils. The composition of the gas flow after passage through the reactor is shown Fig. 5. The reaction of CO to CH_4 and CO_2 can be seen as well as the consumption of methane and the production of hydrogen during the subsequent growth of the carbon filaments. An initial treatment with 10% CH_4 and 20% H_2 also brings about subsequent growth of carbon fibrils. Evidently the thickness of the graphite layer encapsulating the metal particles is determining whether subsequently growth of carbon fibrils will proceed.

Segregation of graphite out of the metal particles puts the metal particle under a considerable compressive stress, since the graphite is deposited within the strong graphite layers surrounding the metal particle. When the graphite layer is relatively thick, segregation of more graphite is not possible. The segregation of graphite is evident from the magnetization of the supported metal. With encapsulation and complete segregation of the carbon, the magnetization approaches the original level measured for the fresh reduced catalyst. When the segregation cannot proceed further, the magnetization does not recover completely. With relatively thin graphite envelopes, however, the graphite layers can be locally broken by the metal and the metal is being squeezed out of the graphite envelope. To minimize the surface energy the metal migrates completely out of the graphite enclosure and leaves an empty envelope. When the metal is being pushed out, a fresh metal surface is being exposed to the gas atmosphere, while the broken graphite layers at the bottom are providing nucleation sites for the graphite layers. As a result carbon atoms are being transported from the free metal surface at the top to the graphite layers at the bottom of the particles. The thus growing graphite layers are pushing the metal particle up, which produces the well known situation of metal particles situated at the top of a growing carbon fibril. The process of the initiation of the growth of carbon fibrils is schematically represented in Fig. 6.

In the fibril growth process indicated above, the metal particles are lifted from the surface of the support. With some supports, the metal particles remain fixed on the support and the direction of growth of the fibrils is not random, but initially oriented perpendicularly to the support.

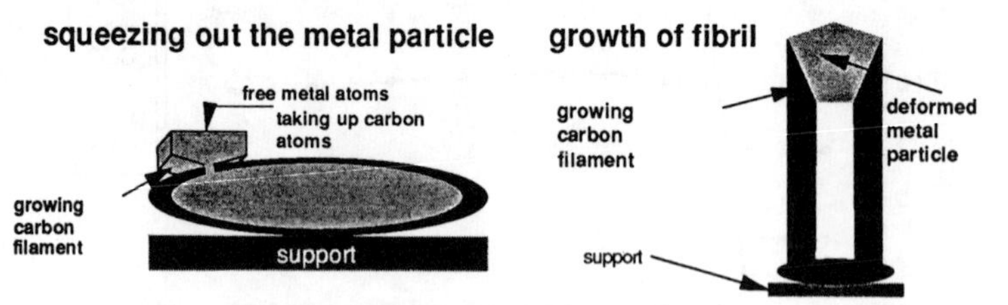

FIGURE 6. INITIATION AND GROWTH OF A CARBON FILAMENT OUT OF A METAL PARTICLE SQUEEZED FROM A GRAPHITE ENVELOPE

When the rate of addition of carbon atoms to the metal surface is high and, hence, the rate of nucleation of graphite elevated, encapsulation of the pushed out metal particles can proceed immediately. As a result a chain of empty graphite envelopes results. Fig. 7 shows an electronmicrograph in which some of the processes discussed can be seen. The pushing out of a metal particle is evident, as well as the some chains of empty graphite capsules. In some instances a metal particle is completely encapsulated after some growth of a fibril and after encapsulation pushed back into the empty inner core of the fibril. Also some well grown fibrils can be seen.

The saturation magnetization is indicating the above processes. As an instance Fig. 8 shows the saturation magnetization and the increase in weight of a nickel-on-alumina catalyst due to the growth of carbon. At the right-hand side of the figure, the saturation magnetization and the increase in weight measured on samples of the catalysts are represented as a function of the logarithm of the exposure time. It can be seen that the magnetization passes through a minimum. Inspection of the catalyst after the experiment indicated that the smaller nickel particles had been encapsulated. The recovery of the magnetization is due to the decomposition of the nickel carbide initially formed. The permanent drop in magnetization is due to the metal particles from which the filaments are growing, that are partly or completely present as nickel carbide.

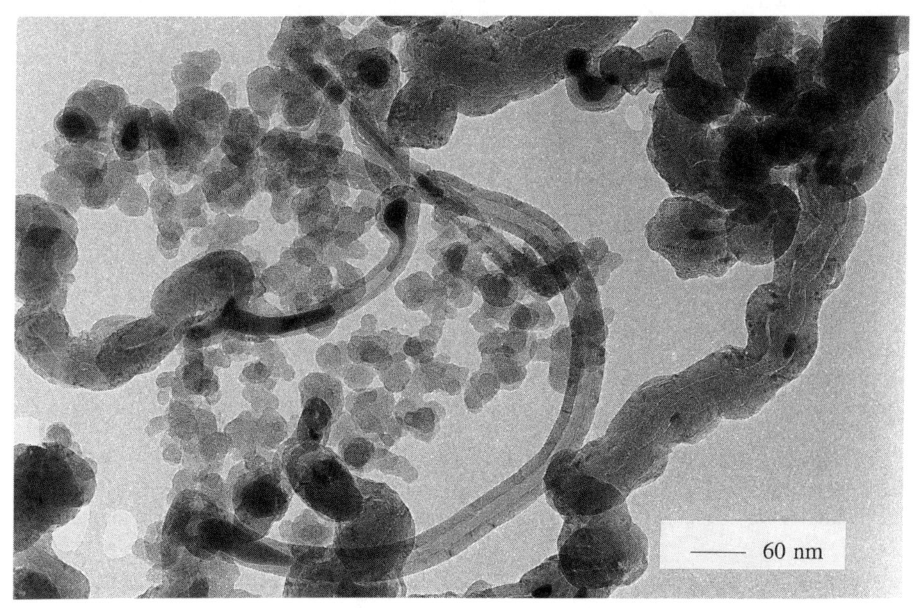

FIGURE 7. ELECTRON MICROGRAPH OF CARBON GROWN OUT OF SUPPORTED IRON PARTICLES.

The drop in magnetization is difficult to explain unambiguously. It is possible that only a fraction of all metal particles is present as nickel carbide, but it can also be that on one moment some particles are completely encapsulated and have decomposed to nickel and graphite, whereas other particles are growing filaments and are completely carbidic. During the growing process metal particles can proceed from an encapsulated state to a growing state and reversely. Baker has seen the stop in the growth and the sudden re-initiation of the growth in a different direction of the fibril in a controlled atmosphere electron microscope.

The above mechanism for the growth of carbon fibrils out of supported metal particles has been established for the growth out of iron, nickel, and cobalt particles and, less extensively, from palladium particles[3,4]. A crucial point in the mechanism is the relatively difficult nucleation of graphite. Consequently initially a metal carbide results upon exposure of metal particles to decomposing gas molecules containing carbon. The graphite layers encapsulating the metal particles have grown by segregation of carbon out of the metal. In separate experiments we established that encapsulated metal particles exposing exclusively graphite surfaces do not exhibit any activity in the decomposition of methane or carbon monoxide. We feel that the presented mechanism can explain the complicated growth phenomena of carbon fibrils quite satisfactorily.

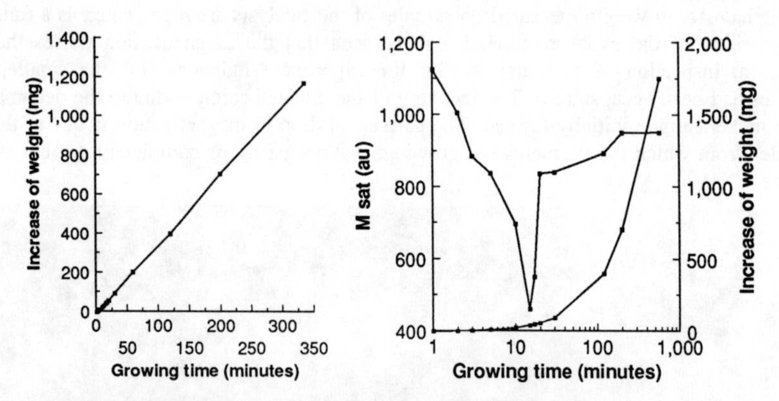

FIGURE 8. INCREASE IN WEIGHT AND SATURATION MAGNETIZATION OF AN ALUMINA-SUPPORTED NICKEL CATALYST EXPOSED TO CH_4. MAGNETIZATION CALCULATED PER GRAM OF NI. NOTE LOGARITHMIC SCALE AT RIGHT-HAND SIDE.

APPLICATION OF PALLADIUM ONTO CARBON FIBRILS

The growth of the mechanically strong carbon fibrils completely disintegrates the original catalyst bodies. Tangled skeins of fibrils result. The size of the skeins is of the order of 6 µm, which is very suitable for liquid-phase reactions. If desired the support used in the fibril

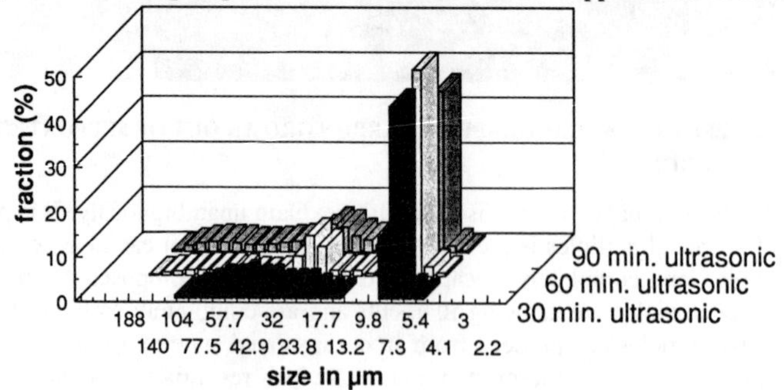

FIGURE 9. PARTICLE SIZE DISTRUTIONS OF CARBON FIBRILS (ULTRASONICALLY TREATED FOR 90 MIN) AND A COMMERCIAL ACTIVATED CARBON (ULTRASONICALLY TREATED FOR 150 MIN). PARTICLE SIZE DISTRIBUTION DETERMINED BY LIGHT SCATTERING.

producing catalyst can be removed by treatment with alkali or with acid. Also the metal particles from which the carbon fibrils have grown can be removed by treatment with acid, provided the metal particles are not completely encapsulated with graphite layers owing to which the particles exhibit the same chemical inertness as graphite. The surface area of the fibrils is typically 225 m^2 per g and the pore volume 1.6 ml per g, from

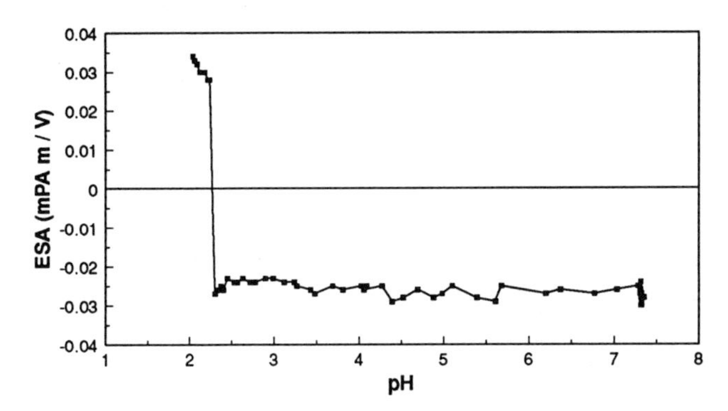

FIGURE 10. SIGN OF ELECTROSTATIC CHARGE OF SKEINS OF CARBON FIBRILS AS DETERMINED FROM ESA. ABOVE PH 2.3 NEGATIVELY CHARGED FIBRILS

which a mean pore diameter of 28 nm can be calculated. Wide pores are very favorable with liquid-phase catalysts. The size as determined by light scattering of suspended carbon fibrils was compared with the size of a commercial activated carbon. Fig. 9 shows that the size distribution of the skeins of carbon fibrils is much more favorable than that of commercial activated carbon. By controlling the size of the supported metal particles and the temperature and the exposure to carbon containing gas molecules, the porosity and surface area of the fibrils can be controlled within fairly wide ranges.

To apply catalytically active precursors highly dispersed on the carbon fibrils, a sufficiently strong interaction with the precursor is required. To assess the properties of the carbon fibrils as a catalyst support, the ability to take up active precursors uniformly over the surface of the fibrils has to be investigated. Usually carboxyl groups on the surface of activated carbon provide anchoring points for active precursors. Carboxyl groups were applied onto the surface of carbon fibrils with graphite layers parallel to and at an angle with the fibril axis. Oxidation with nitric acid generated the carboxyl groups.

The presence of carboxyl groups can be established most easily from the electrostatic charge on the surface of the fibrils. The ESA signal of fibrils treated with nitric acid as a function of the pH is shown in Fig. 10. Without treatment in nitric acid, it was not possible to apply the palladium almost atomically dispersed on the carbon fibrils. Application of a positively charged complex of palladium was found to be a prerequisite; addition of $PdCl_2$ did not result in a good distribution. Also exclusion of oxygen was found to be highly important. Since large Pd particles are more stable than small Pd particles, oxidation leading to a low solubility of Pd results in Ostwald ripening.

TESTING OF CARBON FIBRIL-SUPPORTED PALLADIUM CATALYSTS

The activity of a Pd catalyst supported on carbon fibrils grown from supported iron particles is compared with that of a commercial catalyst based on activated carbon in Fig. 11. It can be seen that the catalyst based on carbon fibrils which has the same Pd loading is distinctly more active. More important is the filterability, which is compared in Fig. 12. It is evident that the catalyst based on carbon fibrils filters much more easily.

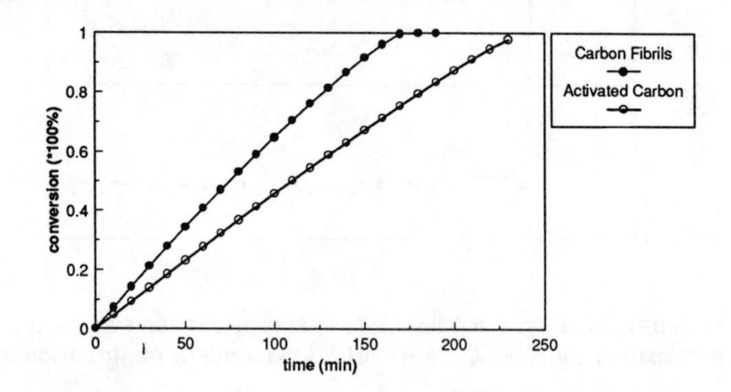

FIGURE 11. ACTIVITY OF 5 WT.% PD-ON-CARBON IN LIQUID PHASE HYDROGENATION OF $C_6H_5NO_2$. COMPARISON OF COMMERCIAL ACTIVATED CARBON SUPPORT AND CARBON FIBRILS.

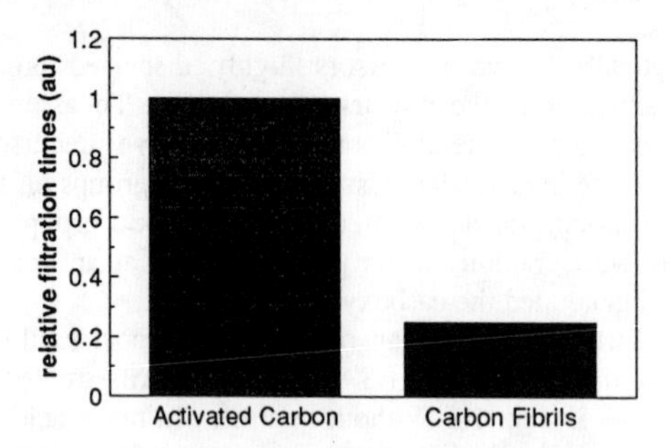

FIGURE 12. COMPARISON OF RATE OF FILTRATION OF COMMERCIAL ACTIVATED CARBON AND CARBON FIBRILS GROWN OUT OF SUPPORTED IRON PARTICLES

---

[1] J.R. Rostrup-Nielsen, J. Catal. **27** (1972) 343

[2] R.T.K. Baker and P.S. Harris, Chemistry and Physics of Carbon : A series of advances, New York (1978), Chapter 2, p.83

[3] P.K. de Bokx, A.H.J.M. Kock, E. Boellaard, W. Klop and J.W. Geus, J. Catal. **96** (1985) 454

[4] E. Boellaard, P.K. de Bokx, J.W. Geus, J. Catal. **96** (1985) 455

CARBON NANOFIBERS AS A NOVEL CATALYST SUPPORT

MYUNG-SOO KIM , NELLY M. RODRIGUEZ AND R. TERRY K. BAKER,
Catalytic Materials Center, Materials Research Laboratory, The Pennsylvania State University, University Park, PA 16802, USA

ABSTRACT

Catalytically grown carbon nanofibers have been prepared by the thermal decomposition of carbon containing gases over copper-nickel and iron surfaces. This material is found to be highly graphitic in nature when prepared from certain catalysts and gaseous reactants. In the as-grown state, carbon nanofibers have surface areas in the range 200 to 300 m^2/g, and by following careful activation procedures this value can readily be increased to ~700 m^2/g. Electrical measurements indicate that the material has a conductivity approaching that of single crystal graphite. This material combines the attributes of active carbon and graphite and in addition, the physical form of carbon nanofibers offers some interesting opportunities for the design of unique catalyst systems.

INTRODUCTION

Although carbon is only used as a support medium in certain commercial catalytic hydrogenation processes [1,2], there has been a growing interest in exploring its potential application for other reactions. Vannice and coworkers [3-5] have carried out extensive studies on the use of various types of carbon as supports for transition metals in the CO-H_2 synthesis reactions. This kind of support appears to generate high dispersion of the metallic component and frequently gives rise to unusual catalytic properties [6-8]. Studies carried out by Webb and coworkers [9] suggested that catalysts derived from palladium decorated graphite specimens exhibited completely different selectivity patterns for various hydrocarbon hydrogenation reactions to those found from palladium deposited on amorphous carbon substrates. More recently Gallezot and coworkers [10] have confirmed the advantages of utilizing graphite rather than disordered forms of carbon such as charcoal as a support for platinum catalysts. They used a combination of X-ray diffraction and high resolution transmission electron microscopy to show that the metal clusters prepared by the decomposition of a platinum-dibenzylidene acetone complex were selectively located on the basal plane of graphite. Furthermore, the platinum crystallites produced from this source had a raft-like morphology and formed in an epitaxial mode on the support. The strong interaction of the metal with graphite was thought to account for the difference in selectivity for the hydrogenation of cinnamaldehyde compared to that found when platinum was supported on charcoal, a system where only weak forces would exist between the metal and the support. Furthermore,when carbon is in the highly crystalline form of graphite there is the possibility of modifying the chemistry of the metal due to electron transfer reactions with the support. Under normal circumstances, however, graphite has an extremely low surface area and is therefore not a suitable material for a catalyst support.

The existence of carbon nanofibers (sometimes known as carbon filaments) has been known for decades and such deposits are produced from the interaction of carbon-containing gases with certain hot metal surfaces [11]. It is only in recent years, however, that we have learned how to control the mode of growth and tailor the structural characteristics of the material by the use of carefully selected metal catalyst particles and gas environments [12]. The arrangement of atoms at the surface of the particle in contact with the gas phase is a crucial factor in determining the adsorption and rupture of the precursor molecule [13,14]. On the other hand, the geometric alignment of graphite platelets (parallel, perpendicular or at an angle with respect to the fiber axis) precipitated at the rear of the catalyst particle is controlled by the interfacial properties between the metal and carbon [15,16]. The diameter of the nanofiber formed is

Mat. Res. Soc. Symp. Proc. Vol. 368 © 1995 Materials Research Society

directly dependent on the size of the catalyst particle and this parameter can easily be controlled by careful choice of the pre-treatment conditions. The typical dimensions of catalytically grown carbon nanofibers vary from 5 to 100 μm in length, and 5 to 1000 nm in diameter. It is possible to produce relatively large amounts of carbon nanofibers possessing surface areas in excess of 300 m^2/g and having structures consisting of an ordered array of graphitic micro-crystals in a stacked arrangement, which gives rise to a high density of active edge sites.

In the current investigation we have used a combination of techniques to examine the properties of highly graphitic carbon nanofibers which are important to the use of this material as a catalyst support medium. Attention is focussed on structural characteristics, methods of increasing the surface area of the structures, determination of pore size distribution and the electrical resistivity of the nanofibers. Finally, a discussion is given highlighting the advantages afforded by the use of carbon nanofibers over activated carbon as a catalyst support.

EXPERIMENTAL

The carbon nanofibers used in this study were produced by two different methods; in the first preparation samples were grown from the copper-nickel (3:7) catalyzed decomposition of ethylene at 600°C, and in the other procedure, the material was formed from the interaction of iron powder with a CO/H_2 mixture at 600°C. Following the growth process, the deposit was cooled to room temperature in helium and then passivated in 2%O_2/He prior to exposure to air. When necessary, the metal catalyst particles associated with the carbon nanostructures could be removed by dissolution in 1.0 M HCl.

The structural features of the carbon nanofibers were established from a combination of high resolution examination, selected area electron diffraction and lattice fringe imaging techniques, which were performed in a 2000 EX II transmission electron microscope. The lattice resolution of this instrument is estimated to be 0.14 nm. Suitable transmission specimens were prepared by ultrasonic dispersion of the deposit in iso-butanol, and then application of a drop of the supernate to a holey carbon film. BET surface areas measurements of carbon nanofibers were carried out by nitrogen adsorption at -196°C, using a Coulter Omnisorp 100CX instrument. It was also possible to determine the pore size distribution at the same time.

The overall degree of crystallinity of the nanofibers was assessed by comparison of the oxidation characteristics of the various samples with that of high purity active carbon and single crystal graphite, which start to undergo oxidation in 10% CO_2/Ar at 690 and 860°C, respectively. These experiments were performed in a Cahn 2000 micro-balance at a heating of 5°C/ min. The gases used in this work, ethylene, carbon monoxide, hydrogen and helium were all 99.999% purity and were supplied by MG Industries and used without further purification. Reagent grade nickel and cupric nitrates used in the preparation of catalyst powders were obtained from Fischer Scientific. Iron powder (200 mesh) was obtained from Johnson Matthey, Inc. (99.99% purity), and had a BET surface area of 0.3 m^2/g at -196°C.

RESULTS AND DISCUSSION

Examination of the carbon nanofibers by transmission electron microscopy showed that while both types of material were highly graphitic in nature their respective structures were quite different. The appearance of fibers produced from the copper-nickel catalyzed decomposition of ethylene is presented in Figure 1. These fibers had been formed by a bidirectional mode from two opposite faces of a twinned catalyst particle, which remained embedded within the structure throughout the growth process. High resolution studies and electron diffraction analysis indicated that the nanofibers consisted of graphitic platelets oriented at an angle to the fiber axis in a "fish-bone" stacking arrangement. Examination of the nanofibers produced from the interaction of iron with CO/H_2 also indicated that the majority of the fibers were formed by a

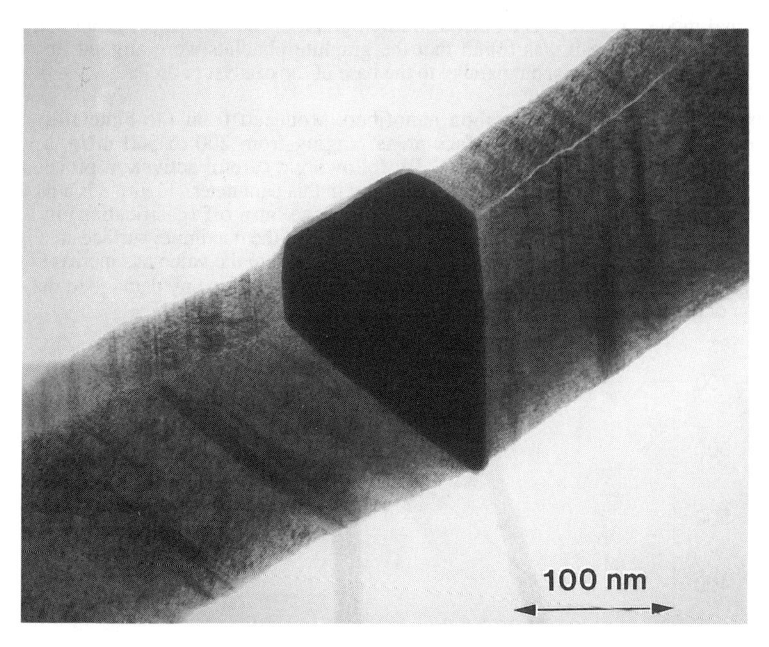

Figure 1. Carbon nanofiber formed from the Cu-Ni catalyzed decomposition of C_2H_4 at 600°C.

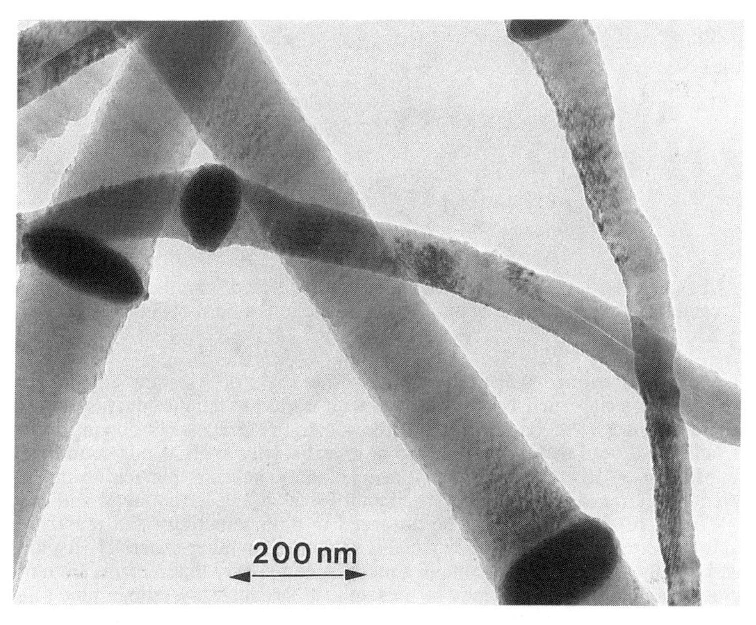

Figure 2. Carbon nanofibers formed during the interaction of a CO/H_2 (4:1) mixture with a powdered iron catalyst at 600°C.

bidirectional mode. Figure 2 is an electron micrograph of the nanofibers generated from this reaction and in this case it was found that the graphite platelets were aligned in a direction perpendicular to the fiber axis, but parallel to the base of the catalyst particles.

In the "as grown" state, carbon nanofibers produced from the bimetallic catalyzed decomposition of ethylene have surface areas ranging from 200 to 300 m^2/g, a value that depends on their method of preparation. By following a careful activation procedure it is a relatively simple task to obtain a significant increase in this parameter. Figure 3 is a plot showing the change in surface areas as a function of percentage burn off (gasification) in a CO_2/He mixture at 850°C. From this dependence it is apparent that the maximum surface area is attained for this batch of nanofibers at 40% burn off, a condition where the value has increased to almost 700 m^2/g. It should be stressed that this treatment did not result in any damage to the structural perfection of the material.

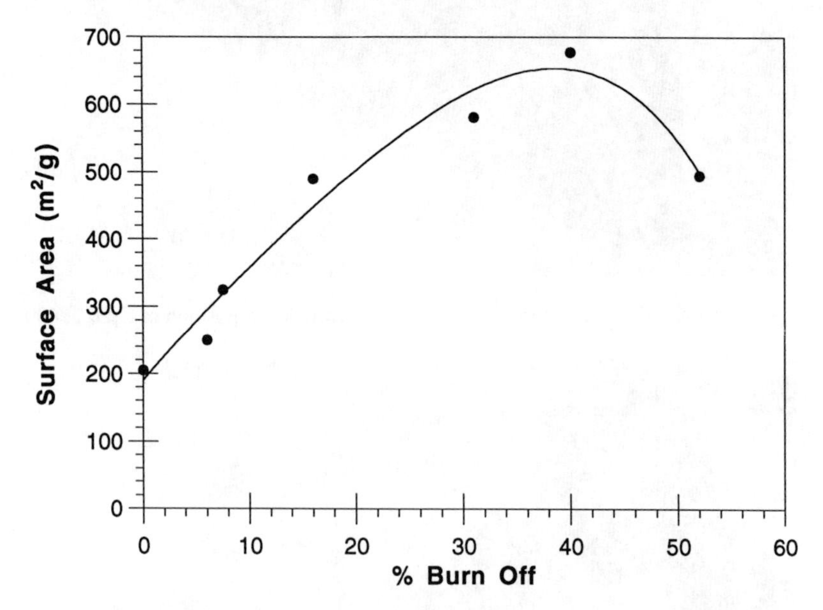

Figure 3. Change in surface area of carbon nanofibers produced from the copper-nickel catalyzed decomposition of ethylene as a function of burn-off in CO_2/He (1:1) at 850°C.

The structural arrangement of carbon nanofibers also offers many advantages over the more disordered materials such as active carbon with respect to the introduction and subsequent growth characteristics of metal particles on the substrate surfaces. Following impregnation, metal particles will tend to become located at specific sites such as edges and steps on the graphitic platelets. In addition, such particles may acquire particular morphological characteristics as a result of the strong interfacial forces between the metal and the graphitic support. Therefore, differences will be observed in both selectivity and reactivity patterns compared to those found when the same metal is supported on other materials. It should also be appreciated that the nanofibers can contain a metallic component that remains from the original growth process and these particles may be present in a particular crystallographic orientation at the tip of the structures. This latter feature is clearly evident in Figure 4, a micrograph taken of a carbon nanotube produced from the interaction of iron with a CO/He mixture at 600°C showing the formation of an octagonal shaped metal particle at end of the carbon structure.

It has been demonstrated that a reactant gas will exhibit unusual adsorption and decomposition characteristics on such a catalyst particle. In a recent study we have compared the behavior of iron particles supported on carbon nanofibers and γ-alumina for the conversion of $CO/C_2H_4/H_2$ mixtures at 600°C [17]. While the maximum amount of gas and solid phase products arising from the decomposition of CO and C_2H_4 were similar from both catalyst systems, the oxide supported iron particles exhibited severe deactivation after about 30 minutes on stream. In contrast, when the metal particles were supported on carbon nanofibers, the maximum level of activity was maintained for a prolonged period of time. It was suggested that the rapid deactivation observed with the Fe/alumina system was related to reconstruction of the metal crystallites into an arrangement that did not favor decomposition of the carbon-containing gases, a phenomenon that did not occur on the graphitic support material.

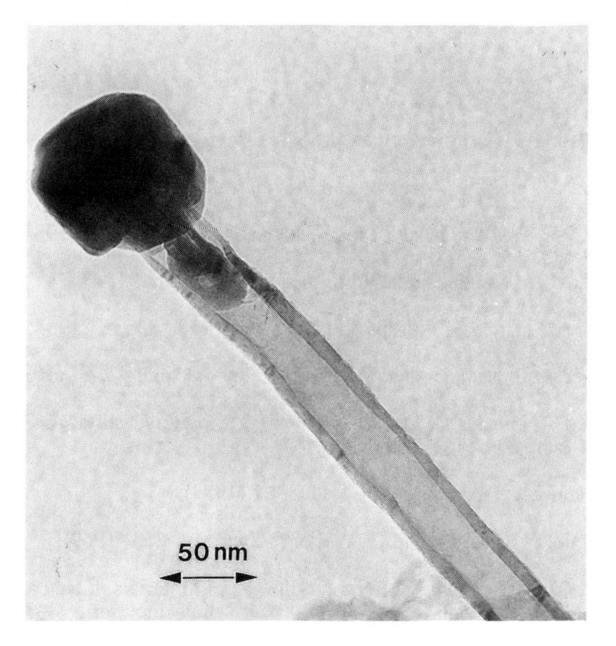

Figure 4. Transmission electron micrograph of a carbon nanotube created from the interaction of iron/graphite with a CO/He (4:1) mixture at 600°C showing the existence of an octagonal shaped catalyst particle.

A further feature that should be taken into consideration concerns the selective adsorption ability of this material which can allow for discrimination between reactant molecules. It can be argued that because of the non-polar nature of the graphite basal plane, adsorption in these regions will be restricted to organic molecules, which will establish a very strong interaction with the carbon. This aspect should give rise to an increase in the lifetime of the catalyst, since polymeric residues that are frequently responsible for causing deactivation will tend to adsorb on areas of the support which are remote from the location of metal particles. In addition, if a small fraction of polar groups are present at the graphite edge sites, enhanced wetting of the carbon by aqueous solutions of metal salts will lead to improved catalyst preparation procedures. This behavior is to be contrasted with the situation encountered with active carbon where simultaneous adsorption of both inorganic and organic species occurs at the same sites.

Finally, there are the subtle changes in the wetting characteristics on a given type of carbon substrate as the functionality of the surface is progressively altered from acidic to basic. An acidic surface will tend to withdraw electrons from the metal, whereas a basic surface will donate electrons to the particles. This induction effect could modify the chemisorption characteristics with respect to a particular adsorbate gas molecule. In future studies we intend to explore how the interplay between these factors and those exerted by a gaseous reactant ultimately dictates the particle morphology, crystallographic orientation and chemical state for various carbon nanofiber supported metal samples, since these features have a direct impact on the catalytic behavior of the system.

ACKNOWLEDGMENTS

Financial support for this work was provided by the National Science Foundation under grant CTS92-22572.

REFERENCES

1. M. Smiseck and S. Cerny, "Active Carbon", Elsevier, N. Y., 1970.

2. C.L. Thomas, "Catalytic Processes and Proven Catalysts", Academic Press, N. Y., 1970.

3. H.-J. Jung, P.L. Walker, Jr. and M.A. Vannice, J. catal. 75, 416 (1982).

4. A.A. Chen, M.A. Vannice and J. Phillips, J. Phys. Chem. 91, 67 (1985).

5. J.J. Venter, A.A. Chen, J. Phillips and M.A. Vannice, J. Catal. 119, 451 (1989).

6. P. Ehrburger, O.P. Mahajan and P.L. Walker, Jr., J. Catal. 43, 61 (1976).

7. F. Rodriguez-Reinoso, J.D. Lopez-Gonzalez, C. Moreno-Castilla, A. Guerrero-Ruiz and I. Rodriguez-Ramos, Fuel 63, 1089 (1984)

8. J. Phillips and J.A. Dumesic, Appl. Catal. 9, 1 (1984).

9. I.C. Brownlie, G.R. Fryer, and G. Webb, J. Catal. 14, 263 (1969).

10. P. Gallezot, D. Richard and G. Bergeret, in "Novel Materials in Heterogeneous Catalysis" edited by R.T.K. Baker and L.L. Murrell, ACS Symposium Series 437, p. 150, 1990

11. R. T. K. Baker., Carbon 27, 315 (1989).

12. N. M. Rodriguez., J. Mater. Res. 8, 3233 (1993).

13. R. J. Koestner, J. C. Frost, P. C. Stair, M. A. Van Hove, and G. A. Somorjai, Surf. Sci. 116, 85 (1982).

14. X. Y. Zhu, and J. M. White., Surf. Sci. 214, 240 (1989).

15. R.T. Yang and J.P. Chen, J. Catal. 115, 52 (1989).

16. M. S. Kim, N. M. Rodriguez, and R. T. K. Baker, J. Catal. 134, 253 (1992).

17. N. M. Rodriguez, M. S. Kim, and R. T. K. Baker, J. Phys. Chem. in press.

THERMAL CONDITIONS FAVORING BUCKYTUBE GROWTH ON THE ANODE

NIKOLA KOPRINAROV*, MIKO MARINOV**, GEORGE PCHELAROV*,
MARIANA KONSTANTINOVA*, RADOSLAV STEFANOV*

* Bulgarian Academy of Sciences, Central Laboratory for Solar Energy & New Energy Sources, 72 Tzarigradsko shose, 1784 - Sofia, Bulgaria
** Bulgarian Academy of Sciences, Institute of Physicalchemistry, Academic Bonchev str., Bl.11, 1040 - Sofia, Bulgaria

ABSTRACT

The most widespread methoid for buckytube synthesis is by arc discharge in an inert gas ambient. At arc discharge the major kind of charge carriers are the electrons generated from the cathode by thermal emission. For this reason up until now the deposit was always obtained on the cathode. By supplementary electrode heating and cooling we studied the effects of the anode-cathode temperature difference and found that it plays a major role in defining the quality and quantity of the deposit obtained. Given appropriate heat-up or cool-down of the electrodes, a state can be reached such that the temperature at the cathode is higher than the temperature at the anode. A deposit was obtained on the anode, observed by a TEM. In our view employing this method yields the most favorable conditions for buckytube growth. Our results show that except buckytubes with a high ratio of their length with respect to their width, a multitude of small size polyhedral carbon systems closed in by spatial carbon nets and thin flat interfering carbon plates were also obtained. When using electrodes of different cross-section, with the deposition of an ever thicker deposit on the anode, its temperature regime changes. The lower initial temperature at the anode begins to grow steadily with time tending towards the temperature at the cathode. Thus, buckytube growth proceeds at the lowest possible rate, at a small temperature difference between the vapor source and growing structure. In our view employing this method yields the most favorable conditions for buckytube growth.

INTRODUCTION

The principle requirement in the utilization of an arc discharge is that the electric current flowing through the discharge chamber should be so high that, the the carriers it generates are sufficient to further self-maintain a stable flow of current. At that the temperatures of the electrodes can reach such high values that they may start to melt or even evaporate. The production of vapor by this means is very intense and has a number of advantages. The method is preferable when employing hard-melting metals or conductive materials, the other methods for heat-up of these are difficult to implement. For these reasons the arc discharge method persevered as the method most often employed in carbon sublimation for fullerene production [1, 2, 3, 4, 5, 6].

Mat. Res. Soc. Symp. Proc. Vol. 368 © 1995 Materials Research Society

In the utilization of an arc discharge for the fabrication of fullerenes, the temperatures on the electrodes' surfaces are of special importance. These heat-up to the highest temperatures and depending on the values reached can be a vapor source or spots of carbon vapor condensation. The precise surface temperature reached at heat-up can be determined with the aid of Fig.1, when the heat flow between the cathode and anode forefront surfaces and also the heat flow from these surfaces towards the cool regions within the discharge space are calculated.Heat energy is released along the two electrodes due to the current flow. The heat energies within the cathode and anode respectively $Q_{RC}(I, l_C, S_C, \rho_C)$ and $Q_{RA}(I, l_A, S_A, \rho_A)$ are functions of the electric current I, the geometric size of the respective electrode (the length l and cross-section $S=2\pi r$) and the specific resistance ρ. The heat is dispersed by irradiation, the energies at the cathode and anode respectively being $Q_{LC}(l_C, S_C, T_C)$ and $Q_{LA}(l_A, S_A, T_A)$, these are temperature and irradiation area dependant. Also they depend on the heat conduction through the electrodes $Q_{CC}(l_C, S_C, a_C, T_C)$ and $Q_{CA}(l_A, S_A, a_A, T_A)$, the geometric size, temperature of the electrodes and thermal conduction of electrode material **a**. The thermal transport through the gas in the discharge chamber $Q_{GC}(l_C, S_C, T_C)$ and $Q_{GA}(l_A, S_A, T_A)$ is dependant on the area and temperature of the electrodes.

There are additional reasons for changing the electrode surface energy [7], when an arc discharge is employed. Given that the required charge carriers are generated from the cathode by means of thermionic emission as in this case, then each electron that leaves the cathode carries with-it an energy equal to:

(1) $q = W_m + 2kT$, where W_m is the effective work function for the cathode, k is Boltzman's constant and T the temperature of the cathode. The total energy taken away by the cathode is:

(2) $Q_e = I(W_m + 2kT)/e$, where e is the charge of the electron.

This energy is transported to the anode by means of the electrons and is mostly shed as heat energy at the anode. Thus we obtain the equalities:

(3) $Q_{RC} - Q_e - Q_{LC} - Q_{GC} - Q_{CC} = 0$

and

(4) $Q_{RA} + Q_e - Q_{LA} - Q_{GA} - Q_{CA} = 0$

Thus the temperatures of the forefront electrode surfaces of the anode T_A and cathode T_C can be determined. Normal values for Q_e at the equal or approximately equal electrode parameters l,s ,a and always lead to $T_C < T_A$ and that is why the carbon of the hotter anode sublimates and deposits on the cooler electrode.

A multitude of parameters connected with the electrodes yield the possibility such that by changing these the values for T_C and T_A are also changed.

EXPERIMENT

The case when $T_A > T_C$ is most easily realized, it is utilized most often in the fabrication of buckytubes, hence it will not be considered here. In accordance with what has been stated above regardles of the route by which the inequality $T_A < T_C$ was obtained, the sublimation of the cathode and the growth of a deposit on the anode must be expected. We chose the easiest and quickest way for realizing this - by changing the geometric size of the electrodes.

By increasing the cross-section of the anode to 6X8mm and length to 15mm and decreasing the cross-section of the cathode to 1x1mm and length to 35mm, we were able to obtain the inequality $T_A < T_C$ and the deposition of growth on the anode at the expence of material sublimating from the cathode. The power supply current was 50A while the voltage was 17V, the distance between the electrodes was at the same time maintained at 0.2mm. The work pressure was 3.10^4Pa while the gas used was Argon. The electrodes were made from spectrally pure carbon.

RESULTS AND DISCUSSION

The principle aim of the experiment is not to determine if carbon atoms or clusters will deposit on the cooler electrode surface. Usually these are widely deposited in the form of soot, but rather to determine wether a deposit will grow on the anode that has the characteristic peculiarities of a deposit obtained on the cathode at $T_C < T_A$. It is important to understand to what extent the plasma within the arc space, which is in contact with the growing deposit most of the time, as well as the electric field will all influence nanotube growth.

A phtograph of the obtained anode deposit is shown in Fig.2. What is most impressive at the time of deposition is that the speed at which the cathode is spent is very low. In our experiment it is about 20 times lower than the speed at which the anode spent given the same conditions and the same electrode sizes although the supply voltage is with reversed polarity.

The crater formed in the anode deposit is quite deep, with a wall of pyrolytic graphite that is of a silvery color in appearance. The black color deposit standing on the inside of the crater does not have a well defined column structure characteristic of the cathode deposit.

After mechanically grinding and scraping of some material from the inside of the crater a sample was taken which was then observed by a TEM. If we compare the forms of the various Carbon structures that are obtained on the cathode also at AC discharge [8] it can be seen that the anode deposit structures have a better symmetry and therefore the observed nanotubes Fig.3 are longer.

In equations (1) and (2) the temperature at the surface of the electrodes is expressed as a function of the initial sizes of the electrodes. With the deposition of a growth on the cooler electrode, the spots of carbon vapor deposition no longer lay on the electrode surfaces but are drawn further away from them as the deposit grows. The larger the size of the deposit becomes the larger the changes in temperature of the growing surface become. These start to increase with respect to the initial values. In order to calculate theses values, it is required that in the equation for the cooler electrode the fact that the length of the electrode has changed so that we now have the initial length plus the constantly increasing length of the deposit will have to be considered, while the cross-section of the electrode is also constantly changing.

In the case of a cold anode the growing deposit has a small cross-section which is close to that of the thiner cathode and induces the heat-up of the anode. As a result of this, T_A grows, while at the same time the difference T_C - T_A decreases. When the value of T_A becomes close to the value of T_C, this will bring about an ever slower growth of the deposit and this will lead to a slower growth in temperature and finally to an overall decrease in the rate of deposit growth. Such a slow rate of growth at which the temperature of the vapor source is close to the temperature of the growing layer should lead to the fabrication of orderly carbon structures of a high quality.

We belive that the conditions considered above are the main reason why amorphous carbon has not been observed in the anode deposit and the growth of a multitude of layers around the buckytubes is quite favorable so that they grow in thikness as shown on Fig.4. The thickened regions are shown on Fig.4 with arrows. On Fig.5 we can observe the same carbon structure by means of the DF method. It can be seen that the material deposited on the buckytubes forms ordered atomic structures. It is also characteristic for the anode deposit that small size almost defect free polyhedral structures Fig. 6. The synthesis of flat carbon forms-probably graphite crystals Fig. 7, have been observed in this deposit for the first time [9]. With the above in mind we conclude that the slow deposit growth at a temperature of T_A close to T_C should supply the best conditions for a perfect growth of the ordered carbon structures.

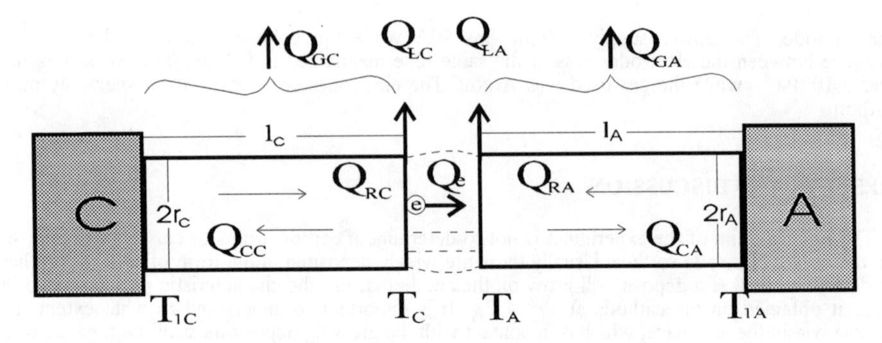

Fig. 1 Heat energy exchange between the anode and cathode

Fig. 2 A photograph of the deposit on the anode

Fig. 3 Buckytubes grown on the anode

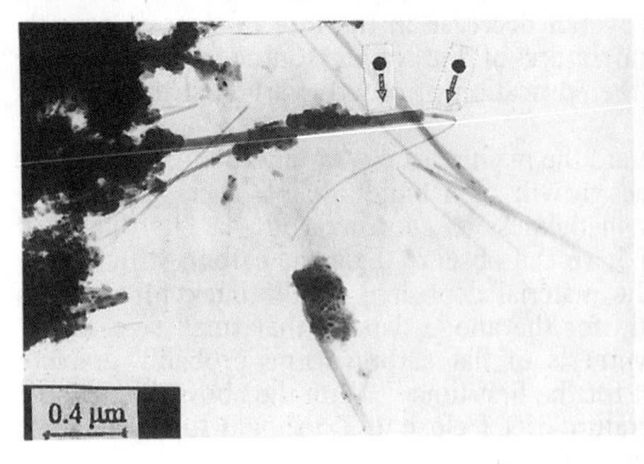

Fig. 4 A buckytube covered by incomplete concentric layers. - Bright Field

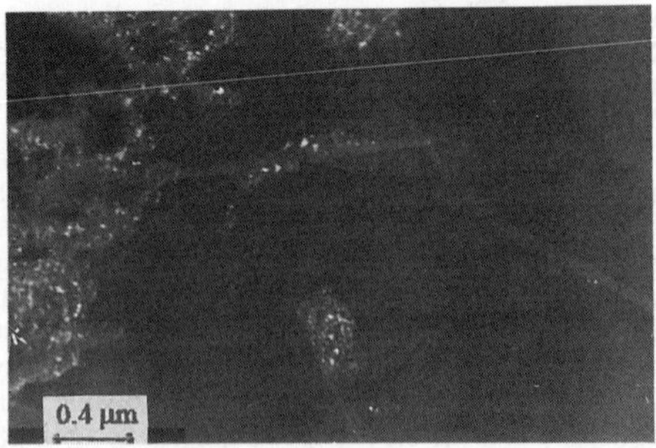

Fig. 5 A buckytube covered by incomplete concentric layers. - Dark Field

Fig. 6 Polyhedral closed Carbon nets

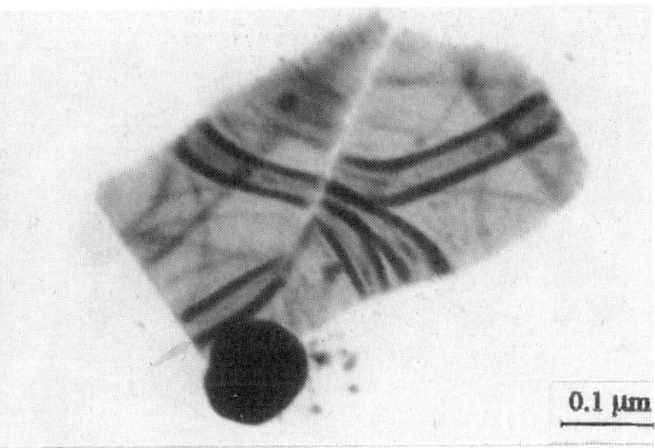

Fig. 7 A plane Carbon flake grown on the anode

CONCLUSION

A multitude of factors determine the temperature T which the electrodes will reach in the arc discharge process. Its variation allows us to obtain and study carbon deposits under different conditions. The deposition of a growth on the anode at the expense of the cathode is especially interesting as here within the plasma that surrounds the cathode. The presence of carbon micro and nano structures of the same nature as those found in the cathode deposit allow us to conclude that the field and differences in the properties of the plasma that surrounds the growing deposit do not play a leading role in the nucleation and build-up of carbon structures. The small difference in the temperatures of the anode and cathode at deposition on the anode together with our observations of the structures grown on the anode allow us to conclude that as the temperature of carbon condensation approaches the temperature of sublimation of carbon, the carbon structure quality improves.

REFERENCES

1. Iijima S., Nature 354, 56-58 (1991).
2. Iijima S., Nature 356, 776-778 (1992).
3. Iijima S., Phys. Rev. Lett.,69, 3100-3103 (1992).
4. Iijima S., Nature 363, 603-605 (1993).
5. Ebbesen T.W. & Ajayan P.M., Nature 358, 220-222 (1992).
6. Taylor G.H., Fitz Gerald J.D., Pang L. and Wilson M.A., J. Crystal Growth, 135, 157-164 (1994).
7. Granowski W.L., Der Elektrische Strom im Gas, Akademie Verlag, Berlin, 1955.
8. Koprinarov N., Marinov M., Pchelarov G., Konstantinova M., Stefanov R., J. Phys. Chem., in press.
9. Koprinarov N., Marinov M., Pchelarov G., Konstantinova M., Stefanov R., Zeitsch fur Physik B, submitted for publication.

Acknowledgment: The sponsorship of the Bulgarian Science Foundation is much appreciated under contract No. X-323.

PART III

Supported Metal Catalysts

SYNTHETIC DESIGN OF COBALT FISCHER-TROPSCH CATALYSTS

STUART L. SOLED*, JOSEPH E. BAUMGARTNER*, SEBASTIAN C. REYES*, AND ENRIQUE IGLESIA**
*Exxon Research and Engineering Co., Rt. 22 East, Annandale, NJ 08801
**Department of Chemical Engineering, University of California, Berkeley, CA 94720

We describe synthetic strategies for designing cobalt Fischer-Tropsch catalysts in order to maximize synthesis rates and C_5^+ selectivity. The support cobalt-precursor interaction must be balanced to maximize dispersion yet allow facile low temperature reduction. The cobalt time yield is independent of support and varies directly as the cobalt dispersion for dispersions in the 0.02 to 0.12 range [1].

Diffusional constraints can dramatically alter apparent turnover rates and selectivities [1-6]. Transport restrictions become increasingly important when large catalyst pellets (1-3mm) are used in packed-bed reactors in order to avoid substantial pressure drops. As previously shown by Iglesia et al. [4-6], at typical FT synthesis conditions, two types of reaction-diffusion couplings occur: (a) diffusion-limited product removal from catalyst pellets and (b) diffusion-limited reactant arrival at catalytic sites. In the first regime, diffusion-enhanced readsorption of α-olefins leads to higher product molecular weight and paraffin content as pellet size or active site density increase. In the second regime, catalyst pellets become depleted of CO, which favors formation of lighter products and decreases the desirable C_5+ selectivity.

Growing chains that desorb as olefins can readsorb and initiate chain growth, leading to desirable higher molecular weight products. As pellet size, site density or olefin carbon number increase, the probability of readsorption increases. With increasing transport restrictions, the catalytic sites are exposed to higher effective H_2/CO ratios, which produce undesirable lower molecular weight products. Intermediate levels of transport restrictions lead to optimum product distributions.

We have prepared large SiO_2 pellets with uniform and eggshell Co distributions. The imbibition of a liquid into a sphere was shown to follow Washburn's equation [7] and depends on both solution properties (viscosity and surface tension) and solid properties (pore radius, pore tortuosity, and contact angle). The eggshell catalysts contain Co near the outer support surface when prepared by impregnating 2 mm silica spheres with high viscosity molten cobalt nitrate. The local cobalt content approaches 50% wt. in a 0.1mm external shell; yet, we can obtain relatively high Co dispersions (0.05-0.10) by directly reducing the nitrate precursor at a slow heating rate.

Extensive catalytic tests on small and large uniformly impregnated pellets and on the eggshell catalysts of this study show that maximum C_5+ selectivities are obtained at intermediate levels of transport restrictions [8]. Eggshell catalysts have the optimum C_5+ selectivity. In effect, eggshell catalysts decouple the pellet diameter from the characteristic diffusion distance, which is then controlled independently by varying the eggshell thickness. Within these eggshell catalysts, moderate transport restrictions retard the removal of reactive olefins, which then readsorb and initiate surface chains leading to higher molecular weight products. Transport restrictions, however, are not sufficiently severe to introduce CO concentration gradients that reduce catalyst effectiveness and lead to undesirable lighter hydrocarbons.

Mat. Res. Soc. Symp. Proc. Vol. 368

ACKNOWLEDGMENTS

We thank Dr. Rocco A. Fiato for many helpful discussions and Ms. Hilda Vroman and Mr. Bruce DeRites for the synthesis, characterization, and catalytic evaluation of some of these materials. We also thank Dr. Eric Herbolzheimer and Ms. Dee Redd for the viscosity and surface tension measurements.

REFERENCES

1. E. Iglesia, S.L. Soled, and R.A. Fiato, J. Catal. **137**, 212 (1992).
2. E. Iglesia, S.C. Reyes, and R.J. Madon, J. Catal. **129**, 238 (1991).
3. R.J. Madon, S.C. Reyes, and E. Iglesia., J. Phys. Chem. **95** 7795 (1991).
4. E. Iglesia, S.C. Reyes, and S.L. Soled, in *Computer-Aided Design of Catalysts and Reactors*, edited by E. R. Becker and C. J. Pereira (Marcel Dekker, New York, 1993), p. 199.
5. E. Iglesia, S.C. Reyes, R.J. Madon, and S.L. Soled, in *Advances in Catalysis and Related Subjects*, edited by D.D. Eley, H. Pines, and P.B. Weisz (Academic Press, New York, 1993), vol. 39, p. 239.
6. R.J. Madon, S.C. Reyes, and E. Iglesia, in *Selectivity in Catalysis*, edited by S.L. Suib and M.E. Davis (ACS Symposium Series, 1992).
7. E. Washburn, Phys. Rev. **17**, 273 (1921).
8. E. Iglesia, S. Soled, S.C. Reyes and J.E. Baumgartner, J. Cat., submitted for publication (1994).

PREPARATION AND PRETREATMENT EFFECTS ON Co/SiO_2 CATALYSTS FOR FISCHER-TROPSCH SYNTHESIS

KENT E. COULTER AND ALLEN G. SAULT
Fuel Science Department, Sandia National Laboratories, Albuquerque, NM 87185-0710

ABSTRACT

Catalyst drying procedures are often given little attention in the experimental section of papers on supported metal catalysts. In general, drying appears to be regarded as a method to remove water and other volatile components prior to calcining or reduction, but not as a method to affect the surface properties of the catalyst. This study uses x-ray photoelectron spectroscopy (XPS) to examine the surface properties of silica supported cobalt catalysts, prepared using incipient wetness impregnation of cobalt nitrate hexahydrate, and finds a wide range of cobalt distributions, extent of nitrate decomposition, and reducibility for various drying procedures. After UHV annealing and subsequent reduction, the final cobalt surface properties are found to depend on the length of heating and the environment during the drying process. Maximum cobalt metal surface area is obtained for samples exposed to limited amounts of air and dried under conditions where gas phase species generated during the precursor decomposition are rapidly removed from the surface of the sample.

INTRODUCTION

Supported cobalt catalysts are often prepared by incipient wetness impregnation using a cobalt nitrate hexahydrate precursor[1-4] followed by drying at ~100°C and calcination in air at 400-500°C. In general, this procedure produces an easily reduced but poorly dispersed (<5%), Co_3O_4 surface phase. While several studies have focused on the effects of cobalt precursors[2], calcination conditions[4], and cobalt dispersions[3], little information is provided in the literature on the effects of drying. Examination of the literature[2,5] suggests that drying variables such as time, temperature, heating rate, and catalyst environment can significantly alter the cobalt surface properties.

In this letter we report on the effects of conventional and novel drying procedures. Following reduction, cobalt dispersions and reducibilities strongly depend upon the initial conditions utilized in the drying process, demonstrating the importance of catalyst drying, and the need to carefully control conditions during this important, but often neglected, stage of catalyst preparation. Although the chemical processes occurring during the different drying procedures are not well understood at this time, the observed variations in cobalt surface properties are reported and possible chemical mechanisms are proposed. The maximum concentration of reducible cobalt surface area is obtained by vacuum drying the cobalt nitrate precursor at 110°C followed by vacuum annealing at 400°C and reduction at 400°C.

EXPERIMENTAL

All Co/SiO_2 samples were prepared by incipient wetness impregnation[1] of SiO_2 (Davisil grade 643: 300 m^2/g; 150Å pores; 115 cm^3/g pore volume) with aqueous solutions of $Co(NO_3)_2 \cdot 6H_2O$ (Fischer ACS reagent grade). Cobalt loadings of 18.8 and 26.4 wt% were measured by atomic absorption after calcining in air at 400°C. Following impregnation, the bright pink samples were dried in air at room temperature until visibly dry and portions of each were saved for analysis. Subsequent drying at ~100°C was performed in either a static air oven for 16 hours (26.4 wt%) or in a vacuum oven for 24 hours (18.8 wt%). For clarity, samples air dried at room temperature are identified in

Mat. Res. Soc. Symp. Proc. Vol. 368

the text as air dried samples. Samples air dried at 110°C are identified as oven air dried and samples vacuum dried at 100°C are identified as vacuum dried.

Surface analysis was performed in an ultrahigh vacuum (UHV) chamber ($3x10^{-10}$ Torr) coupled to an atmospheric gas-phase reactor. The samples were analyzed by X-ray photoelectron spectroscopy (XPS) using Al Kα radiation and a VG Microtech CLAM2 analyzer operating at an analyzer pass energy of 50 eV, and 4-mm entrance and exit slits. Scans of the Co 2p, Si 2p, O 1s, C 1s, and N 1s regions were taken with charging corrections made by referencing to the reported Si 2p value of 103.4 eV for SiO_2[6]. All XPS spectra are normalized to the Si 2p intensity and the cobalt weight loading of each respective sample to ensure relevant comparisons are made.

RESULTS AND DISCUSSION

Drying

Fig. 1 shows the Co 2p region for a Co/SiO_2 sample (a) after air drying at room temperature, (b) vacuum drying at 100°C and (c) oven air drying at 110°C. The air dried sample exhibits a Co $2p_{3/2}$ binding energy of 782.0 eV and large shakeup features associated with highly coordinated cobalt nitrate[2]. Analysis of the N 1s region indicates the presence of nitrate on the surface of the air dried sample. Visual inspection of the sample once in the UHV chamber shows a color change from pink to purple and the N 1s/Co $2p_{3/2}$ intensity ratio, corrected for known sensitivity factors[7] indicates a N/Co atomic ratio of 0.43, which is well below the expected value of 2 for $Co(NO_3)_2$. The color change and low N/Co XPS intensity ratio suggests some precursor decomposition has occurred on the air dried sample during evacuation.

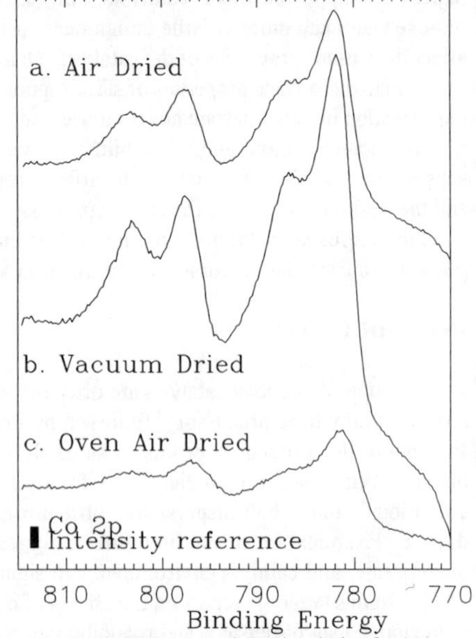

Figure 1. Co 2p XPS spectra following drying for (a) an air dried sample, (b) a vacuum dried sample, and (c) an oven air dried sample. Co 2p intensity reference shows equivalent counts for comparison between figures.

The Co 2p region for the vacuum dried sample (fig.1b) exhibits peak positions and shapes very similar to the air dried sample, although the Co $2p_{3/2}$/Si 2p intensity ratio for the vacuum dried sample is a factor of two larger than the air dried sample. This increase is attributed to spreading of the nitrate precursor during the vacuum drying at 100°C and is supported by a large N 1s peak at 407.2 eV assigned to nitrate[8]. The N 1s/Co $2p_{3/2}$ atomic intensity ratio is 0.76 which is twice as high as observed for the air dried samples. Apparently vacuum drying promotes the spreading of the cobalt precursor but not the decomposition of the nitrate as evidenced by the N 1s peak position and the pink color of sample. It is intriguing that the vacuum dried sample has a larger concentration of nitrate remaining than the air dried sample considering no treatment was carried out on the air dried sample A possible explanation is the length of time each sample was exposed to air prior to high temperature drying. The air dried sample was stored in air for three weeks prior to introduction into the UHV chamber, whereas the vacuum dried sample was first dried in air at room temperature overnight and immediately placed in the vacuum furnace. Extended exposure to air possibly alters the sample through water and/or CO_2 adsorption.

As compared to the previous samples, the Co 2p region for the oven air dried sample shown in fig. 1c is markedly different. The Co $2p_{3/2}$ peak is located at 780.8 eV and the Co $2p_{3/2}$/Si 2p intensity ratio is considerably lower than any previously discussed samples. The Co $2p_{3/2}$ binding energy is higher than the reported values for cobalt oxides and lower than the reported values for cobalt silicate[9]. In addition the N 1s region for the oven air dried sample shows only an extremely small peak at 399.4 eV attributed to residual nitrogen adsorbed on the sample. The lack of nitrate features and the position of the Co $2p_{3/2}$ binding energy are indicative of a fully decomposed cobalt nitrate precursor that has formed a mixed cobalt oxide and cobalt silicate phase on the surface[10]. Another striking difference between the oven air dried sample and all other samples is that the oven air dried sample is black after removal from the drying oven and no color change is observed after introduction into the UHV chamber. The black color of the oven air dried sample is similar to the color of all samples following calcining at 400°C in air and is indicative of complete precursor decomposition. Total decomposition of the cobalt nitrate hexahydrate precursor for the oven air dried sample supports the proposal that air drying promotes precursor decomposition more than vacuum drying. In addition, drying in air appears to promote cobalt agglomeration. This agglomeration is attributed to the presence of NO_x species generated during the decomposition process oxidizing the Co^{2+} to Co^{3+} and promoting migration of the precursor over the silica surface to form large particles. In a study on the effects of preparation conditions on the reduction behavior of Co/Al_2O_3 catalysts[4], the concentration of gas phase NO_2 present that can oxidize Co^{2+} to Co^{3+}, determines the probability of cobalt-aluminate formation. X-ray diffraction indicates that after drying in air at 65°C, cobalt is well dispersed as small droplets of $Co(NO_3)_2$. Upon calcining at 300°C in the absence of NO_2 these nitrate droplets decompose to form Co^{2+} ions that disperse over the surface and diffuse into the support lattice to form a surface aluminate. In contrast, in the presence of NO_2 Co^{2+} oxidizes to Co^{3+}, preventing this dispersion and forming large Co_3O_4 crystallites. The same argument may explain the trends observed in this study of silica supported cobalt catalysts.

Vacuum Annealing

After the initial analysis, each sample was annealed in the UHV chamber. Fig.2 shows the Co 2p region for each sample following an UHV anneal at 250°C, which is above the reported decomposition temperature for cobalt nitrate hexahydrate[11]. The spectrum for the air dried sample shows a dominant Co $2p_{3/2}$ peak at 781.8 eV and the emergence of a low binding energy shoulder assigned to a cobalt silicate phase that is in contact with electrical ground and does not exhibit the charging typically associated with supported catalysts[10]. In addition, the cobalt intensity is 30% greater for the vacuum annealed air dried sample as compared to the air dried sample. The N 1s region for the air dried sample annealed to 250°C is dramatically different from the air dried only sample. The major nitrogen peak is 60% lower in intensity and is

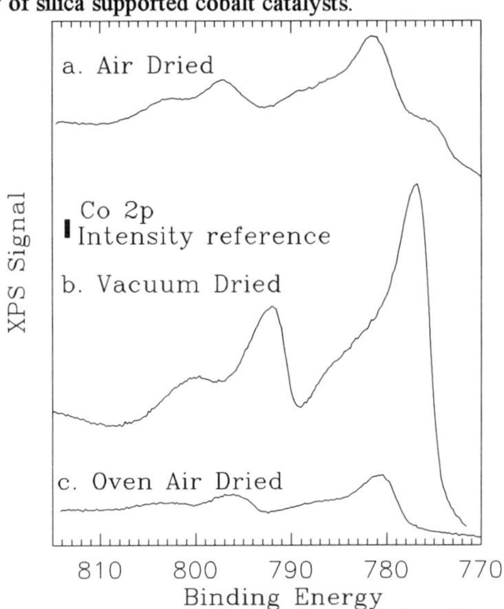

Figure 2. Co 2p XPS spectra following UHV annealing at 250°C for (a) an air dried sample, (b) a vacuum dried sample, and (c) an oven air dried sample. Co 2p intensity reference shows equivalent counts for comparison between figures.

located at 399.8 eV, with only a small shoulder at the nitrate position of 407.2 eV. In addition, the sample color has changed from purple to black. Obviously, the vast majority of the nitrate is decomposed and the high $Co2p_{3/2}$ binding energy indicates the remaining cobalt phase formed is a silicate phase[9]. Vacuum annealing the vacuum dried sample increases the cobalt intensity by 50% (fig. 2b) with the observed increase occurring in the main Co $2p_{3/2}$ and $2p_{1/2}$ peaks and not in the shakeup features. In addition, the vacuum dried sample in fig. 2b exhibits a very unexpected shift in the Co 2p features after the 250°C anneal. The major Co $2p_{3/2}$ peak has a charge corrected binding energy of 777.0 eV which is well below any reported binding energy for Co^{2+}. Without charge correction, the binding energy is 780.4 eV which is assigned to a continuous, conductive, cobalt silicate phase[10]. A portion of the silicate could be well grounded through direct contact with the tungsten mesh supporting the sample, provided the layer is both continuous and conductive. The conductive nature of the cobalt phase and the cobalt intensity, which is the highest observed for any sample, suggests that vacuum drying followed by vacuum annealing at 250°C produces a thin-film of cobalt over the silica and thereby maximizes cobalt surface area.

UHV annealing of the oven air dried sample to 250°C does not significantly alter the Co 2p features as shown in fig. 2c. The Co $2p_{3/2}$ peak remains at 780.8 eV and the intensity increases by only 10%. Air drying at 100°C decomposes the cobalt nitrate precursor and the resulting mixed cobalt oxide/silicate phase is stabilized against modifications introduced by UHV annealing at 250°C.

<u>Reduction</u>

Fig. 3 shows the Co 2p region for each respective sample following vacuum annealing to 400°C and reduction in 630 Torr of hydrogen at 350, or 400°C. The air dried sample reduced at 350°C (fig 3a) has a dominant Co $2p_{3/2}$ peak at 782 eV that is assigned to a bulk cobalt silicate[9] and a small shoulder at 777.6 eV that is assigned to metallic cobalt. The cobalt intensity decreases slightly after reduction possibly indicating sintering of cobalt particles. After reduction of the vacuum dried sample at 400°C (fig. 3b) the metallic Co 2p peaks are larger than for any other sample examined. Furthermore, a majority of the cobalt probed by XPS is metallic and parallel hydrogen chemisorption measurements[10] verify that the metallic cobalt surface area is higher following this treatment than for any other treatment. The high metallic cobalt surface area of the vacuum dried sample following reduction at 400°C is a surprising result considering high reduction temperatures are often associated with sintering of the metal and loss of surface area[10]. Reducing the oven air dried sample at 400°C, produces a metallic cobalt peak at 777.8 eV but a large silicate peak at 781.8 eV also remains. In contrast to the vacuum annealed samples discussed in this paper, samples that are dried at 100°C, and calcined at 400°C in air are easily reducible at

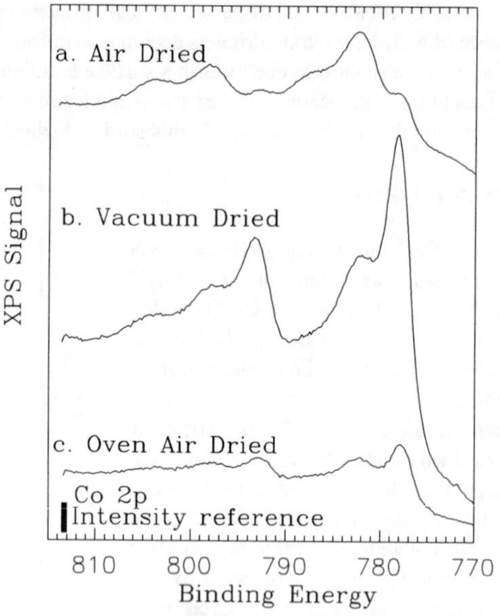

Figure 3. Co 2p XPS spectra following reduction in H_2 for (a) an air dried sample (350°C), (b) a vacuum dried sample (400°C), and (c) an oven air dried sample (400°C). Co 2p intensity reference shows equivalent counts for comparison between figures.

temperatures as low as 250°C, although the metallic cobalt surface area is half that observed for a vacuum dried sample reduced at 400°C[10].

CONCLUSIONS

The initial drying procedures strongly influence the distribution and reducibility of cobalt supported on silica prepared by incipient wetness impregnation using cobalt nitrate hexahydrate. Differences in preparation procedures such as the length of time the freshly prepared sample is exposed to air, the rate of removal of decomposition products during drying, and drying temperature have a profound effect on the surface properties of a silica supported cobalt catalyst. For example, neither drying in air at room temperature nor vacuum drying at 100°C fully decomposes the nitrate precursor, although vacuum drying increases the cobalt nitrate dispersion. In contrast, drying at 110°C in air decomposes the nitrate precursor and forms large particles of a mixed cobalt oxide/silicate phase. Results presented here demonstrate that under carefully controlled conditions drying can exert a profound influence on metal distribution and reducibility, and result in increased metal surface area compared to conventional drying and calcining procedures.

ACKNOWLEDGMENTS

The authors wish to thank Elaine Boespflug, Mark Harrington, Nancy Jackson, Steve Kohler, and Calvin Bartholomew for various technical contributions. This work was performed at Sandia National Laboratories for the U. S. Department of Energy under contract DE-AC04-76DP000789 was supported by the National Renewable Energy Laboratories under contract RAC 3-13250.

REFERENCES

1. A. Hoek, M. F. M. Post, J. K. Minderhoud and P. W. Lednor, US Patent 4,499,209 (1985).

2. M. P. Rosynek and C. A. Polansky, Appl. Catal. **73**, 97 (1991).

3. E. Iglesia, S. L. Soled and R. A. Fiato, J. Catal. **137**,212 (1992).

4. P. Arnoldy and J. A. Moulijn, J. Catal. **93**, 38 (1985).

5. J. M. Zowtiak and C. H. Bartholomew. J. Catal. **83** ,107 (1983).

6. M. Klasson, A. Berndtsson, J. Hedman, R. Nilsson, R. Nyholm and C. Nordling, J. Electron Spectrosc. **4**, 427 (1974).

7. C. D. Wagner, L. E. Davis, M. V. Zeller, J. A. Taylor, R. M. Raymond and L. H. Gale, Surf. Interface Anal. **3**, 211 (1981).

8. C. D. Wagner, W. M. Riggs, L. E. Davis, J. F. Moulder and G. E. Muilenberg, Handbook of XPS, (Perkin Elmer Corporation, Eden Prairie, MN, USA, 1978) 40.

9. C. D. Wagner, L. H. Gale and R. H. Raymond, Anal. Chem. **5**, 466 (1979).

10. K. E. Coulter and A. G. Sault, submitted J. Catal.

DESIGN AND PREPARATION OF HETEROGENEOUS CATALYSTS BY CONTROLLED CHEMICAL REACTIONS WITH OXYGEN AND HYDROGEN

DAVID L. COCKE*, DONALD G. NAUGLE** AND THOMAS R. HESS**
*Department of Chemistry, Gill Chair, Lamar University, Beaumont, TX 77710
**Department of Physics, Texas A&M University, College Station, TX 77843

ABSTRACT

Chemical reactions of metals and strongly interacting alloys such as Cu-Mn, Ni-Ti, Ni-Hf and Ni-Zr with oxygen and hydrogen play important roles in the preparation, activation, and regeneration of many important heterogeneous catalytic systems involving supported and unsupported metals and alloys. Recent advances in the understanding of metal and alloy oxidation is bringing new insight into the reactive design and activation of bi- and multi-metallic catalysts. By surface studies of oxidation, thermal annealing and reduction of selected alloys and their thin films and reaction layers and products we have been able to delineate the factors which are most important to the oxide formation processes and the oxide reduction processes. Reaction models developed from these results are permitting the design of new catalyst systems and providing long sought understanding to explain specific aspects of well established metallic catalysts.

INTRODUCTION

In 1970-80's considerable work on the preparation of new active catalysts from alloys and in particular from strongly interacting alloys occurred [1-56]. With this work came the realization that reaction with oxygen [1-8] or hydrogen [9-28] of one or more of the alloy's components played a major role in the ultimate activity and selectivity of the catalyst. Work in our laboratory centered on those strongly interacting alloys that could produce planar model catalysts and amorphous alloy catalysts [34,37,43]. The reactions of these alloys with oxygen and hydrogen assumed paramount importance in this work. In the years since these early attempts to make new catalysts from alloys, our efforts have concentrated on explaining the complicated reaction behavior of alloys with hydrogen and, in particular, oxygen [7,29-49,53,58-61]. Catalysis by strongly interacting alloys remains an active area of research [50-52]. This paper will briefly review the modeling progress and show its implications in catalyst design. The discussion will be limited to strongly interacting alloys for catalyst development. Recent work by Cocke and co-workers [see ref. list] has delineated the parameters that control alloy oxidation, and has shown that the modified Cabrera-Mott model [68] has merit for qualitative discussion of the oxidation process. In addition, this model has been extended to include compositions of the alloy.

THEORY AND MODELS

Previous surface studies of the post-chemisorption oxidation of binary polycrystalline alloys such as Ni-Zr, [8,29,30,38-42,46,52,53,58] Cu-Mn, [43-45], Cu-Ti [48,49], Ni-Ti, [7,8,52,55-57] Ni-Hf [59] Ti-Al [49,60] and Ag-Mn [61] and other Ni-M alloys (M= Nb[62], Al[63-65], Co[66], and U[67]) have shown that the parameters that control alloy oxidation can be delineated and that the selective oxidation of the components combined with ion movement are the key components of the surface oxide layer growth and subsequent structure. At low temperatures the Al, Zr, Mn, Nb, Ti, U, and Hf are all preferentially oxidized in the above alloys. Only at higher temperatures and oxygen activities are the Ni, Cu, and Ag and similar metals oxidized in the presence of the other metals [1,2]. The early oxidation process of these compounds, and of alloys in general, can be discussed in terms of the thin film oxidation model of Cabrera and Mott [see 68 and, et al.] in which the potential across the oxide that controls the oxidation process can be expressed as:

Mat. Res. Soc. Symp. Proc. Vol. 368 © 1995 Materials Research Society

$$\Delta\Phi = \frac{-\Delta G^{o}_{f}}{2e} + \frac{kT}{2e} \ln\left[\frac{4e^{2} N_{s} a_{O_2}^{1/2} x}{kT\varepsilon\varepsilon_{o}}\right] \tag{1}$$

here the potential, $\Delta\Phi$, across the growing oxide is determined by the free energy of formation of oxygen anions at the surface, $-\Delta G^{o}_{f}$, plus a term related to the temperature, T, the concentration of the surface oxygen, N_s, the oxygen activity, a_{O_2}, the film thickness, x, the electronic charge, e, the electron permittivity, ε, and the electron permittivity in vacuum, ε_o.

In the binary systems mentioned above, where one of the metals has a relatively large $-\Delta G^{o}_{f}$ for oxide formation and the other has a relatively small $-\Delta G^{o}_{f}$ for oxide formation, the alloy oxidation can be considered in two steps. The metal with the relatively large $-\Delta G^{o}_{f}$ will oxidize first and either segregate to the surface or contain unoxidized more noble metal particles such as Ni. This differentiation of oxidation of various alloy components and the means to control it is suggested from Equation 1 have allowed several methods to be developed for producing various planar model and granular catalysts. These are: a) compositions and oxidation reaction conditions have be found that directly produce highly dispersed metal particles on a porous oxide substrate; b) controlled oxidations have been found to produce supported metal catalysts with layered structures; c) oxidation and reduction schemes have been developed that produce a number of different catalyst structures such as a layered oxide catalyst particle with an alloy core or fine granular catalysts directly from bulk alloy.

The thermodynamic driving force for the oxidation reaction of alloys can be considered as the change in the standard free energy resulting from the formation of the oxide from the reactants and is negative for all metals. For a $M\text{-}M_2$ system, the chemical reactions between each metal component and gaseous oxygen is:

$$aM(s) + {}^{b}/_{2}\, O_2(g) \dashrightarrow M_aO_b(s) \tag{2}$$

$$cM_2(s) + {}^{d}/_{2}\, O_2(g) \dashrightarrow M_{2c}O_d(s) \tag{2a}$$

where each reaction can be divided into two major steps, an anodic and the a cathodic reaction at the oxide-metal and oxide gas interfaces, respectively. By considering the oxide overlayer structure as an electrochemical cell, the total potential can be written as the sum of the potentials for each metal components' oxide, $\Delta\Phi(\text{total}) = \Delta\Phi_M + \Delta\Phi_{M_2} + \ldots$, such that a new equation has been proposed:

$$\Delta\Phi = \frac{-\Delta G^{0}_{f_M}}{2be} + \frac{kT}{2be} \ln\left[\frac{(2be)^{b} N_s^{b} a_{O_2}^{b/2} a_M^{a} X^{b}}{kT\,\varepsilon^{b} \varepsilon_0^{b} a^{a}_{M^{2b/a+}}}\right] + \frac{-\Delta G^{0}_{f_{M_2}}}{2de} + \frac{kT}{2de} \ln\left[\frac{(2de)^{d} N_s^{d} a_{O_2}^{d/2} a_{M_2}^{c} X^{d}}{kT\,\varepsilon^{d} \varepsilon_0^{d} a^{c}_{M_2^{2d/c+}}}\right] + \ldots \tag{3}$$

which follows directly from the consideration of alloy composition on the anodic reaction. Many of the terms in equation (3) have been defined previously. The free energies of oxide formation per mole of O^{2-} for alloy component M and M_2 are $-\Delta G^{o}_{fM}$ and $-\Delta G^{o}_{fM_2}$, respectively. The activity of M, M_2 and their oxidized forms are a_M, a_{M_2}, $a_{M^{2b/a+}}$ and $a_{M_2^{2d/c+}}$, respectively. The lower case letters a, b, c and d, often found as exponents, are the stoichiometric factors used in equations (2) and (2a). We are currently testing this model in gas phase and electrochemical studies on a range of binary and ternary alloys (primarily amorphous alloy thin films since we have demonstrated that the lack of grain boundaries, large scale homogeneity and planar geometry are ideally suited for fundamental studies of oxidation). These studies are providing information that are proving useful in the design of catalysts or in the improvement of catalyst preparation and activation or regeneration procedures.

PHENOMENA AND PROCESSES

Numerous surface phenomena and reactive processes can be used to design and prepare catalysts from alloys. These include, selective oxidation, oxide segregation, oxide layering, mixed oxide formation, hydride fractioning and redox dispersion. These are briefly surveyed below.

Surface Compound Formation

Limited reaction with oxygen or hydrogen can produce catalytic surface compounds. These can be mixed oxide compounds such as spinels or layered structures with different oxides in different layers. A common behavior of all the strongly interacting alloy systems is that the oxide overlayer of the more electropositive element (MO_x) is formed first upon oxidation. The thickness, structure, stoichiometry, etc. of these layers depend on the oxidation conditions; although, in general, this first oxide layer acts as a protection barrier against the oxidation of the more noble metal. The models described above explain this behavior. Other more catalytically interesting phenomena can be made to occur if conditions are used to force the oxidation of the more noble metal and/or if oxidation-reduction schemes are pursued.

Mixed Compound Oxide: Our first attempt to design a catalyst from alloy oxidation was the planar Hopcalite catalyst [45,69]. In this work the oxidation of various composition Cu-Mn alloys under various conditions of temperature and oxygen activity were performed in surface analysis systems to follow the mechanisms of compound oxide formation on the surface of the alloy. Insight into the mechanism of oxidation of the Cu-Mn alloy was obtained and a set of composition, temperature and oxygen activity conditions were determined to produce the $CuMn_2O_4$ Hopcalite catalysts on the surface of the alloy [69]. However a model for the oxidation of an alloy did not emerge until the study of the Ni-Zr alloy where layered oxide structures developed.

Layered Structures: One of the most interesting occurrences of oxygen reacting with strongly interacting alloys is the formation of layered structures that have the most noble metal's oxide on the exterior of the more reactive metal's oxide. This occurs under conditions that provide a potential that is sufficiently high to oxidize not only the more reactive metal but the more noble component as well. Equation (1) shows that this requires elevated temperatures with sufficient oxygen activity. By subsequent reduction of the outer oxide with hydrogen or by reducing the oxygen activity planar supported metal catalysts are produced [48,49]. Ni-Zr alloys have been most extensively studied and exemplary of strongly interacting alloys. Reaction of Ni-Zr alloys with oxygen resulted in Zr- and Ni-oxide layered structures at sufficiently high temperatures and oxygen activities. If the oxidation conditions were less severe, a ZrO_2 supported Ni catalyst [see Cocke and Owens and co-workers].

Supported Metal Catalyst: Oxidizing Ni-Zr under conditions such that the Zr oxidizes and the Ni does not, (i.e. from Equation 1 at low T and oxygen activity) produces a Ni/ZrO_2 supported metal catalyst. The phenomenon occurs in the thin oxide overlayer of the metal alloy and produces a planar model catalyst. If the alloy is powdered prior to oxidation, the supported metal catalyst produced has a metallic core.

Bulk Oxygen and Hydrogen Absorption

Strongly interacting alloys generally have substantial oxygen and hydrogen solubility. Absorption of oxygen or hydrogen into the bulk can change the catalytic properties of the alloy and these absorbed gases can become active participants in surface catalytic reactions.

Bulk Oxide and Hydride Formation

Complete oxidation or reduction of bulk alloys can produce mixed oxide catalysts, supported metal catalysts or supported oxide catalysts. Reaction of a powdered alloy can result in the total conversion of the bulk alloy to oxides or hydrides. In the latter case, the formation of a hydride causes physical integrity to be destroyed and reduction in particle size occurs. Oxidation of these tiny hydride/metal particles can form mixed oxide compounds or layered structures as discussed above. However, oxidation of materials must be done with care. Several studies on ZrNi emphasize that hydride formation is seen in this intermetallic compound when it is used as a hydrogenation catalyst under reducing conditions [9]. The hydrided alloy can cause hydrogenolysis in hexane better than pure Ni or Co [11,12]. The presence of this hydride phase was invoked to explain decreased sintering and coke and resin formation on SiO_2 supported binary alloys [14]. Numerous other works by Lunin and co-workers [15-28] document the unique activity obtained by hydrided strongly interacting alloys.

Redox Cycles

Numerous studies [1-6] as well as our own work [29-49] have observed that oxidation/reduction cycles increase catalytic activity. Crystalline and amorphous alloys such as Pd-Zr, Pt-Zr, Ti-Ni and Ni-Zr have low activity in the virgin state but cycles of oxidation and reduction substantially increases their catalytic activity [52]. Oxidation at lower temperatures will preferentially oxidize the more electropositive metal, such as Zr and Ti, and only partially oxidize or not oxidize the more noble metal component. Reduction treatments will reduce any of the latter that has been oxidized as well as, possibly, absorb hydrogen into the bulk causing fracturing of the structure. Very recent pulsed field desorption work on the oxidation of Ni-Zr alloys containing absorbed hydrogen has shown that the sorbed hydrogen reacts at the surface through coupled reactions to produce new surface compounds [58].

Coupled Reactions

Coupled reactions have been discovered that can efficiently produce remote reduced metals and surface carbides. The latter at temperatures far below those required for normal carbide formation. The coupling agent is the bulk alloy. Heating a preformed oxide layer containing carbon in vacuum (very low oxygen activity) causes the metal to decompose with absorption of the oxygen into the bulk [41]. The active metal released by the reaction, MO_2 + (Alloy) = M + $2O_{absorbed}$, allows metal atoms react with carbon by the reaction, $C_x + M_{atomic} = MC_x$, to form dispersed or surface metal carbides. Another coupled reaction of interest is where the noble metal oxide (such as NiO) layer is on the outside of a reactive metal oxide (ZrO2)layer on the bulk can be reduced from above (oxide-gas interface) and below (oxide-oxide interface) via a hydrogen treatment (a low oxygen activity situation). In this case, (Zr-Ni), the NiO species at the oxide-oxide interface decompose and O^{2-} ions are transported across the ZrO_2 layer to form ZrO_2 at the alloy-ZrO_2 interface and electrons are transported to reduce the more noble metal to form Ni. While the NiO is reduced at the oxide-oxide interface, hydrogen reduces the NiO from above. This remote displacement reaction between the metals is common and occurs in processes that change oxygen activity.

CONCLUSIONS

Strongly interacting alloys show complex reactions with hydrogen and oxygen. However using our current understanding of the reaction mechanisms and in particular the experimental (process) variables of temperature, pressure, alloy composition and reaction cycles, new catalysts are being designed.

ACKNOWLEDGMENTS

The authors are grateful for the financial support of the Robert A. Welch Foundation (Houston TX), the National Science Foundation (Grant DMR-89-03135), the Texas A&M Interdisciplinary Research Initiative and the Texas Advanced Technology Program (Grant 3606).

REFERENCES

1. H. Imamura and W. E. Wallace, J. Phys. Chem. **83**, 2009 (1979).
2. H. Imamura and W. E. Wallace, J. Phys. Chem. , **83**, 3261 (1979).
3. H. Imamura and W. E. Wallace, J. Catal. **64**, 238 (1980).
4. H. Imamura and W. E. Wallace, J. Catal. **65**, 127 (1980).
5. R. Chin, A. Elattar, W. E. Wallace and D. M. Hercules, J. Phys. Chem. **84**, 2895 (1980).
6. W. E. Wallace, Chemtech, **12**, 752 (1982).
7. D. L. Cocke, M. S. Owen and R. B. Wright, Langmuir **4** , 1311 (1988).
8. C. M. Chan, S. Trigwell and T. During, Surf. Interface Anal. **15**, 349 (1990).
9. V. Nefedov, Y. Salyn, A. Chertkov, L. Padurets, Russ. J. Inorg. Chem. **19**, 785 (1974).
10. V. V. Lunin, Russian J. Inorg. Chem. **20**, 1279 (1975).
11. V. V. Lunin, V. I. Deineka, A. F. Plate, Dokl. Akad. Nauk SSSR , **229(2)**, 353.(1976).
12. V. V. Lunin, V. I. Nefedov, E. K. Zhumadilov, B. Y. Rakhamimov, and P. A. Chernavskii, Dokl. Akad. Nauk SSSR **240**, 114, (1977).
13. V. V. Lunin, A. E. Ogronomov, Y. M. Bondarev and L. K. Denisov, Vestn. Mosk. Univ., Ser. 2: Khim. **18(2)**, 218 (1977).
14. V. V. Lunin and B. Y. Rakhamimov, Prevrashch. Uglevodorodov Kislotno-Osnovn. Getertogennykh Ktal., Tezisy Dokl. Vses. Knof.; Grozn. Net. Inst. Im. Akad. M. D. Millionshchikova: Grozny, USSR; pp. 122-3 (1977).
15. V. V. Lunin, V. I. Nefedov, E. K. Zhumadilov, B. Y. Rakhamimov, and P. A. Chernavskii, Dokl. Akad. Nauk SSSR, **240(1)**, 114, (1978).
16. V. V. Lunin, Kh. N. Askhabova, T. V. Bychkova and E. N. Anisochkina, Neftekhimiya **20(3)**, 365 (1980), (Russian); Petrol. Chem. USSR **20(20)**, 91, (1980), (English).
17. V. V. Lunin, V. I. Nefedov, B. Y. Rakhamimov, and L. A. Erivanskaya, Zh. Fiz. Khim. **54(7)**, 1853 (1980).
18. V. V. Lunin, V.A. Galafeev, A. F. Plate, Neftekhimiya, **21(1)**, 92 (1981).
19. V. V. Lunin, Y. I. Solovetskii and P. A. Chernavskii, Dokl. Akad. Nauk SSSR **266(6)**, 1417 (1982).
20. V. V. Lunin and A. Z. Kahn, J. Molec. Catal. **25**, 317 (1984).
21. V. V. Lunin, O. V. Kryukov, I. A. Bruk, and A. L. Lapidus, Khimiya Tverdogo Topliva **18**, 84 (1984).
22. V. V. Lunin, O. V. Kryukov, S. O. Kozhinskii, I. A.Bruk, M. M./ Savel'yev, and A. L. Lapidus, Neftekhimiya **24**, 233 (1984).
23. V. V. Lunin and Y. I. Solovetskii, Kinet. Katal. **26**, 694 (1985).
24. R. M. Frak, J. Less-Common Metals **109**, 279 (1985).
25. A. L. Lapidus, A. I. Bruk, E. Z. Gildenberg and V V. Lunin, Izv. Akad. Nauk SSSR, Ser. Khim. **ll**, 2452 (1980).
26. Y. I. Solovetskii, P.A. Chernavskii and V. V. Lunin, Zh. Fiz. Khim. **56(7)**, 1634 (1982).
27. A. A. Lokteva, L. A. Erivanskaya, and V. V. Lunin, Kinetika i Kataliz **22**, 644 (1981).

28. I. L. Tseitlin, P. A. Chernavskii and V. V. Lunin, Zh. Fiz. Khim. **56(1)**, 122 (1982).
29. R. B. Wright, J. B. Jolly, M. S. Owens and D. L. Cocke, J. Vac. Sci. Technol. **5A**, 586 (1987).
30. R. B. Wright, M. R. Hankins, M. S. Owens and D. L. Cocke, J. Vac. Sci. Technol. **5A**, 593 (1987).
31. G. K. Chuah and D. L. Cocke, J. Trace Microprobe Tech. **4 (1/2)**, 1 (1986).
32. D. L. Cocke and K. A. Gingerich, J. Chem. Phys. **57(9)**, 3654 (1972).
33. D. L. Cocke, E. E. Johnson, and R. P. Merrill, Catal. Rev.-Sci. Eng.**26(2)**, 163 (1984).
34. D. L. Cocke, J. Metals, **70** (February 1986).
35. D. L. Cocke and S. Veprek, Solid State Commun. **57(9)**, 745 (1986).
36. D. L. Cocke, D. E. Mencer, Jr., and D. G. Naugle, Mat. Chem. and Phys. **2**, 17 (1987).
37. D. L. Cocke and D. E. Halverson, Thin Solid Films **155**, 133 (1987).
38. D. L. Cocke, M. S. Owens and R. B. Wright, Appl. Surface Sci. **31**, 341 (1988).
39. D. L. Cocke, M. S. Owens and R. B. Wright, J. Colloid and Interface Sci. **19**, 166 (1989).
40. D. L. Cocke, M. S. Owens and R. B. Wright, Ind. Eng. Chem. Res. Submitted (1994).
41. D. L. Cocke and M. S. Owens, Appl. Suface Sci. **31**, 471 (1988).
42. D. L. Cocke, M. S. Owens and R. B. Wright, 11th International Symposium on the Reactivity of Solids, June 19-24, 1988, Princeton, New Jersey, USA; edited by M. Stanley Whittingham, (1988).
43. C. H. Yoon and D. L. Cocke, J. Non-Crystalline Solids, **79**, 217 (1986).
44. C. H. Yoon and D. L. Cocke, J. Electrochem. Soc., **134**, 643 (1987).
45. C. H. Yoon and D. L. Cocke, Appl. Suface Sci. **31**, 118 (1988).
46. D.L. Cocke, G. Liang, M. Owens, D. E. Halverson and D. G. Naugle, Mater. Sci. Eng. **99**, 497 (1988).
47. D. L. Cocke, M. S. Owens and R. B. Wright, Appl. Surf. Sci. **31**, 341 (1988).
48. D. L. Cocke, T. R. Hess, D. E. Mencer, T. Mebrahtu and D. G. Naugle, Solid State Ionics, **43**, 119 (1990).
49. D. E. Mencer, T. R. Hess, T. Mebrahtu, D. L. Cocke and D. G. Naugle, J. Vac. Sci. Tech. **A9**, 1610 (1991).
50. T. Takahashi, S. Higashi, T. Kai, H. Kimura and T. Masumoto, Catal. Let. **26**, 401(1994).
51. A. F. L. Shammary, I. T. Caga, A. Y. Tata, J. M. Winterbottom and I. R. Harris, J. Chem. Tech. Biotechnol. **55**, 361, 369, 375 (1992).
52. J. P. Espinós, A. Fernández and A. R. González-Elipe, Surf. Sci. **295**, 402 (1993); A. R. González-Elipe, A. Fernandez, J. P. Espinós and G. Munuera, J. Catal.**131**, 51, (1991).
53. E. M. Perry, D. L. Cocke and M. K. Miller, Appl. Surf. Sci **44**, 321 (1990).
54. B. Walz, P. Oelhafen, H. J. Güntherodt and A. Baiker, Appl. Surf. Sci **37**, 337 (1989).
55. P. H. McBreen and M. Polak, Surf. Sci **179**, 483 (1987).
56. J. H. Thomas and T. T. Hitch, Surf. Interface Anal. **15**, 85 (1990).
57. A. Petri, A. Neumann and J. Küppers, J. Vac. Sci. Technol. **A8**, 2576 (1990).
58. T. Hess, G. Abend, J. H. Block and D. L. Cocke, unpublished results.
59. D. L. Cocke, W. E. Daulet, M. S. Owens and R. B. Wright, Solid State Ionics **32/33**, 930 (1989).
60. D. Mencer, D. L. Cocke and C. Yoon, Surf. and Interfac. Anal., **17**, 31 (1991).
61. D. L. Cocke, R. Campbell, K. Balke and M. Owens, Appl. Surf. Sci **40**, 227 (1989).
62. U. Bardi, A. Atrei and G. Rovida, Surf. Sci. **268**, 87 (1992).
63. A. Atrei, U. Bardi and G. Rovida, J. Electron Spectrosc. Relat. Phenom. **57**, 99 (1991).
64. U. Bardi, A. Atrei and G. Rovida, Surf. Sci. Lett. **239**, L511 (1990).
65. E. W. A. Young, J. C. Rivière and L. S. Welch, Appl. Surf. Sci. **28**, 71 (1987).
66. J. C. Rivière, F. P. Netzer and G. Rosina, Surf. Interface Anal. **18**, 333 (1992).
67. T. Gonder, C. A. Colmenares, J. R. Neagle, J. C. Spirlet and J. Verbist, Surf. Sci. **265**, 175 (1992).
68. A. Atkinson, Rev. Mod. Phys. **57**, 437 (1985).
69. C. Yoon and D. L. Cocke, J. Catal. **113**, 267 (1988).

ENCAPSULATED SILVER CLUSTERS AS OXIDATION CATALYSTS

R.A. CRANE, L.C. CHAO, AND R.P. ANDRES
School of Chemical Engineering, Purdue University, West Lafayette, IN 47907

ABSTRACT

A novel supported metal catalyst has been prepared, that consists of Ag clusters a few nanometers in diameter which are encapsulated in an amorphous silica coating only a few Ångstroms in thickness and are deposited on an oxide support. The synthesis of this catalyst and its physical characterization are described. Its potential use as a selective oxidation catalyst is discussed.

INTRODUCTION

Because nanometer scale metal clusters exhibit size dependent physical properties, they offer interesting possibilities as catalysts if their size can be controlled within a narrow size range. An even more interesting possibility is presented by nanometer scale metal clusters coated with a semiconducting layer of controlled thickness. The important catalytic consequences of metal-support interactions (MSI) in the case of TiO_2-supported metals, platinum in particular, are well documented. The explanation most often proposed for this behavior is the creation of special sites at the metal-support interface [1]. G-M. Schwab has presented experimental data and theoretical arguments that support a continuum electronic model for such effects [2]. Schwab argues for even larger consequences in cases when a thin layer of a semiconducting material is supported on a metal substrate rather than the usual MSI situation in which a thin layer of metal is supported on a semiconducting substrate [3].

Recent advances in the synthesis of nanometer diameter clusters make it possible to produce heteronuclear clusters having a "fish-eye" structure in which a metal sphere is encapsulated in a skin consisting of a semiconducting oxide [4, 5]. We have developed techniques at Purdue by which these encapsulated clusters can be isolated as dispersed particles on an inert support and have produced sufficient amounts of such novel materials that their catalytic activity can be determined [6, 7].

In what follows the synthetic methods used to produce these new catalytic materials are described for the specific case of nanometer diameter Ag clusters encapsulated by an amorphous SiO_2 coat and supported on fumed silica. TEM micrographs of the encapsulated clusters and physical characteristics of the supported catalyst are presented. Finally, the potential catalytic consequences of encapsulated-cluster-based-catalysts are discussed.

CATALYST SYNTHESIS

A cluster source known as the Multiple Expansion Cluster Source (MECS) has been developed at Purdue [4]. This is an aerosol reactor in which nanometer scale clusters of

Mat. Res. Soc. Symp. Proc. Vol. 368

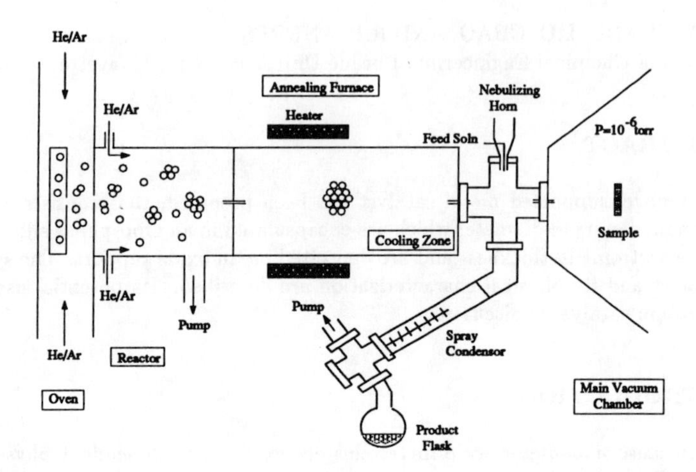

Figure 1: Schematic Drawing of Multiple Expansion Cluster Source and Metal Cluster Colloid Reactor

controlled size and composition are grown in the gas phase. Clusters are captured from the gas phase using a Metal Cluster Colloid Reactor (MCCR) [6]. The entire apparatus is shown schematically in Figure 1. A superheated mixture of Si and Ag atoms is created by evaporating silver shot (Surepure Chemetals, 99.99%) and silicon powder (Aldrich Co., 99.999%) from carbon crucibles held in a resistively heated carbon oven. An inert gas, typically helium (BOC Group, Inc., 99.995%), transports the metal/silicon vapor through a sonic orifice in the oven wall. The stream is quenched by mixing with a cool inert gas stream, either helium or argon (Ar, BOC Group, Inc., 99.997%). Cluster formation then occurs in a fast-flow tubular reactor. Cluster size is controlled by oven temperature, type of quench gas, and quench to oven gas flow ratio.

The clusters are annealed in the gas phase by flowing the aerosol through an alumina tube (3/4 in. ID, 30 in. length, Vesuvius McDaniel Refractory, Beaver Falls, PA) which passes through a Lindberg furnace (General Signal, Watertown, WI) held at 950°C. The length of the heating zone is about 12 inches and is followed by a cooling zone which is approximately 8 inches in length. The total residence time in the alumina tube is about 1 second. On exiting from the annealing region the Ag/Si clusters have the "fish-eye" structure shown in Figure 2.

From the annealing region, the clusters can be expanded through a capillary orifice into a vacuum chamber maintained at 10^{-5} torr and deposited onto 5 nm thick amorphous carbon films supported on 400 mesh copper grids (Ernest Fullam Inc., Latham, NY) for TEM (JEOL 2000FX) analysis.

Alternatively, the clusters can be collected as a colloidal solution. A low volatility solvent, typically dodecane (Aldrich Co., 99+%), containing 500 ppm of a surfactant such as palmitic acid (Fluka Co., 99+%), is sprayed into the MCCR by means of an ultrasonic nebulizer. As the cluster aerosol passes through the mist, the clusters are captured in the

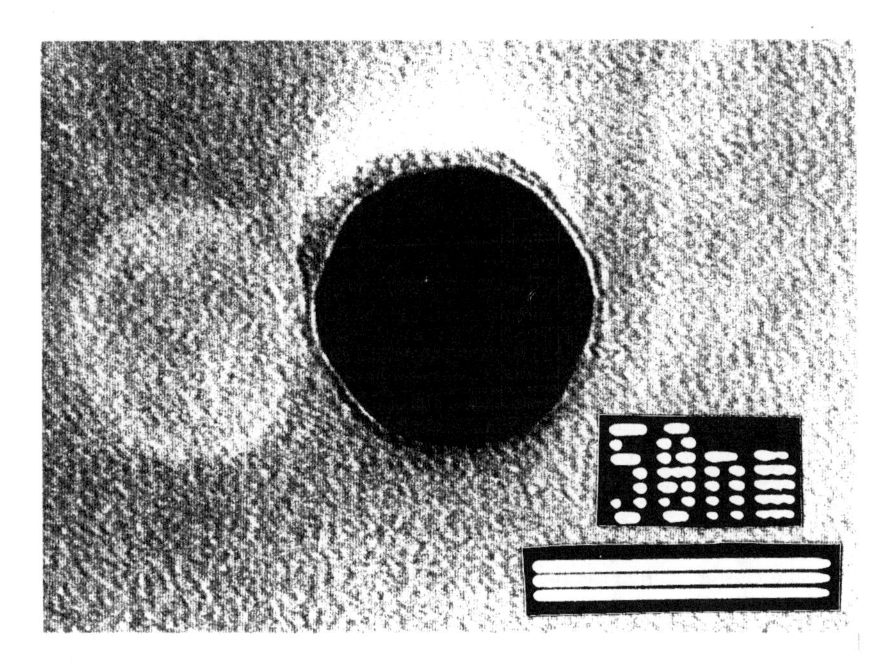

Figure 2: Transmission Electron Micrograph of "Fish-Eye" Cluster: Silver Encapsulated by Silica

spray. A spray impactor is used to condense the liquid and deliver it to a product flask. With dodecane, 96% recovery of the solvent is possible. No attempt is made to keep the clusters from exposure to O_2 and their Si outerlayer is oxidized to SiO_2 in the product flask. The resulting colloid (˜ 6 ppm Ag/SiO_2) is a bright yellow color.

The colloid obtained is very stable with time. The Ag/SiO_2 clusters are uniformly dispersed onto fumed silica (Cabot Co., Cab-O-Sil M-5, 200 m^2/g) with the aid of ultrasonic agitation. The fumed silica is allowed to adsorb the clusters and flocculate for several hours. The catalyst, which settles to the bottom of the flask, is then removed, washed with heptane, and air-dried.

The catalyst is oxidized in oxygen (Matheson, 99.999%) at 250°C and 1 atm for 2 h and then reduced in hydrogen (Matheson, 99.999%) at 300°C and 1 atm for 3 h to eliminate organic residues.

CATALYST CHARACTERIZATION

Silica-encapsulated silver clusters with a mean diameter of 50 Å are produced in the MECS/MCCR using the conditions given in Table 1. A TEM showing the "fish-eye" structure is displayed in Figure 2. This TEM micrograph is actually of a relatively large Ag/SiO_2 cluster produced using Ar as quench gas and a low quench gas flow. In the colloid,

Table 1: Typical Conditions in the MECS/MCCR

Oven Gas	He	
Quench Gas	He	
Quench to Oven Gas Ratio	3	
Total Gas Flow	1.5	g-mol/s x10^3
Oven Temperature	1300	°C
Reactor pressure	50	torr
Annealing Reactor Pressure	40-48	torr
Annealing Temperature	950	°C
Feed solution upstream pressure	50	torr
Feed solution rate	10	mL/min
Product flask temperature	20	°C
Feed solvent	dodecane	
Surfactant, palmitic acid	500	ppm

the mean cluster diameter is the same as in the aerosol, without significant broadening of the size distribution. Dispersed on the silica support, the particle size distribution is broadened only slightly. A TEM of the final catalyst is shown in Figure 3. The weight loading of silver for this catalyst is 1.39%, as determined by atomic absorption. XPS confirms that all of the Si is in the +4 oxidation state.

ENCAPSULATED CLUSTERS AS CATALYSTS

There are several ways in which encapsulated clusters might function as unique catalytic materials. The encapsulating oxide may serve to protect the metal core from poisons, as depicted in Figure 4a. The encapsulating layer may also serve to protect the metal particles from sintering, as portrayed in Figure 4b. In terms of exhibiting unique behavior toward oxidation reactions, however, the most interesting possibilities are those shown schematically in Figure 4c-e. The first possibility assumes the encapsulating layer acts as a diffusion barrier. The second possibility postulates that the overlayer is not continuous and that the silica is attached to selective sites on the Ag surface. The third idea invokes the Metal-Support Interaction (MSI) concept. In this case, the underlying Ag imparts novel catalytic activity to the SiO_2 outerlayer through electronic interaction.

An oxide film serving as a diffusion barrier, as in Figure 4c, would behave similarly to a spillover catalyst. Assuming that hydrocarbons cannot penetrate the oxide layer, catalysis would occur away from the metal. In this scheme, molecular oxygen diffuses through the oxide film, is activated by the metal, and then diffuses back to the surface of the oxide film. The activated species might be atomic oxygen, an ion, or a radical. In terms of selective oxidation, this scheme is attractive since the acidic metal sites that promote combustion are not accessible to the hydrocarbon.

When only selective sites are blocked by the silica overlayer (Figure 4d), sites which

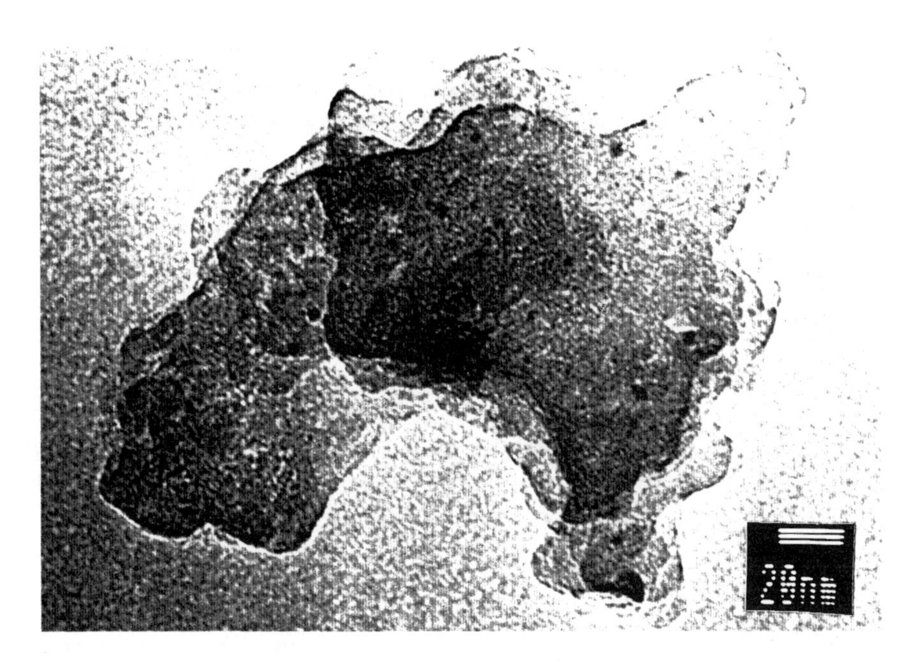

Figure 3: Transmission Electron Micrograph of Encapsulated Silver Clusters Supported on Silica

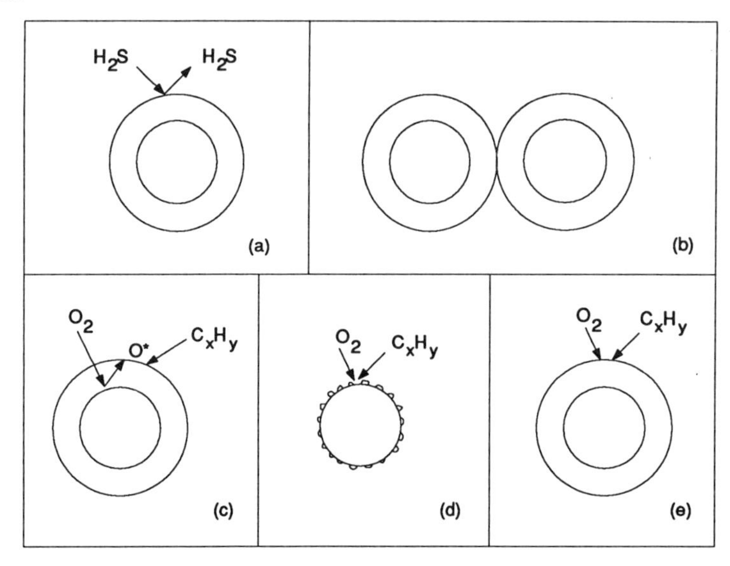

Figure 4: Schematic Representations of the Possible Catalytic Behaviors of Encapsulated Clusters

promote combustion may be inaccessible. In this scheme, catalysis occurs by a more conventional mechanism, i.e. Langmuir-Hinshelwood kinetics on metal sites. Any selectivity promotion occurs by redistribution of the active sites. Ideally, acidic sites will be blocked by the silica, and sites responsible for promoting partial oxidation will be available to the hydrocarbon.

In the third scheme (Figure 4e), the surface of the SiO_2 layer is modified electronically. In this scheme, the inert SiO_2 is activated by the underlying Ag. Ordinary Langmuir-Hinshelwood kinetics may be invoked, but here the active sites have changed. This allows for quite different kinetics. For example, the heats of adsorption and activation energies for reaction may now be quite different.

CONCLUSIONS

Silver clusters encapsulated in SiO_2 offer numerous possibilities for unique selective oxidation behavior. The development of techniques by which these clusters and other clusters with similar structures can be synthesized makes possible a whole new class of catalysts.

ACKNOWLEDGEMENTS

This research was partially funded by the National Science Foundation under grant ECS-9117691. The assistance of Atul Patil in synthesizing the Ag/SiO_2 cluster shown in Figure 2 is gratefully acknowledged.

REFERENCES

[1] M.A. Vannice, Catal. Today **12**, 255 (1992).

[2] G.-M. Schwab, Mémoires Société Royale des Sciences de Liège, 6e sèrie **1**, 31 (1971).

[3] G.-M. Schwab, Adv. in Catalysis **27**, 1 (1978).

[4] S.B. Park, PhD thesis, Purdue University, West Lafayette, IN, 1988.

[5] A.N. Patil, N. Otsuka, and R.P. Andres, J. Phys. Chem. **98**, 9247 (1994).

[6] L.-C. Chao, PhD thesis, Purdue University, 1994.

[7] L.-C. Chao and R.P. Andres, J. of Colloid and Interface Sci. **165**, 290 (1994).

DIRECT EVIDENCE OF SURFACE ROUGHNESS IN SMALL METALLIC PARTICLES

MIGUEL JOSE-YACAMAN, SAMUEL TEHUACANERO, CRISTINA ZORRILLA AND GABRIELA DIAZ
Universidad Nacional Autónoma de México,Instituto de Física, P.O. Box 20-364, México 01000.

ABSTRACT

The characterization of nanoparticles is of prime importance in catalysis. High Resolution Electron Microscopy coupled with image processing has produce a lot of new information on the detailed structure of the particles. In this paper we discuss the possibility of applying these techniques to the study of surface roughness.

A deep understanding of catalytic phenomena requires a good knowledge of the surface atomic structure of the catalyst. Heterogeneous catalysts are mostly composed by small metal particles dispersed over a support. High Resolution Electron Microscopy (HREM) combined with image processing techniques are actually the current trend in the study of particles. Image processing can improve the information extracted from the image. In practical terms the processing involves several steps which are shown in Figure 1.

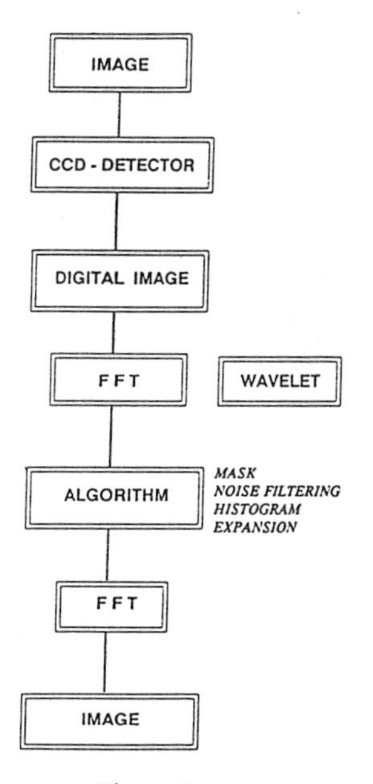

Figure 1.

The most common substrates, γ-Al_2O_3, SiO_2 and graphite used in catalysts, are crystalline at the nanoscale and therefore scatter the electrons coming out of the metal particle. Even if the substrate is amorphous it reduces the intensity signal coming from the particles. The theory for this case has been worked out by Yacamán and Avalos [1]. A crucial point is the orientation relationship between the particle and the substrate. It is always desirable to orient the electron-beam in such a way that the particle is strongly diffracting and the substrate is not. This produces the optimum conditions for particle imaging. However this kind of orientation is very difficult to achieve and in many cases impossible because of the orientation relationship between the particles and the support. However, in the case of High Resolution Images of catalysts the image processing can be invaluable. Figure 2 shows the example of a Pt particle oriented along <110> direction and supported on amorphous carbon.

Mat. Res. Soc. Symp. Proc. Vol. 368 © 1995 Materials Research Society

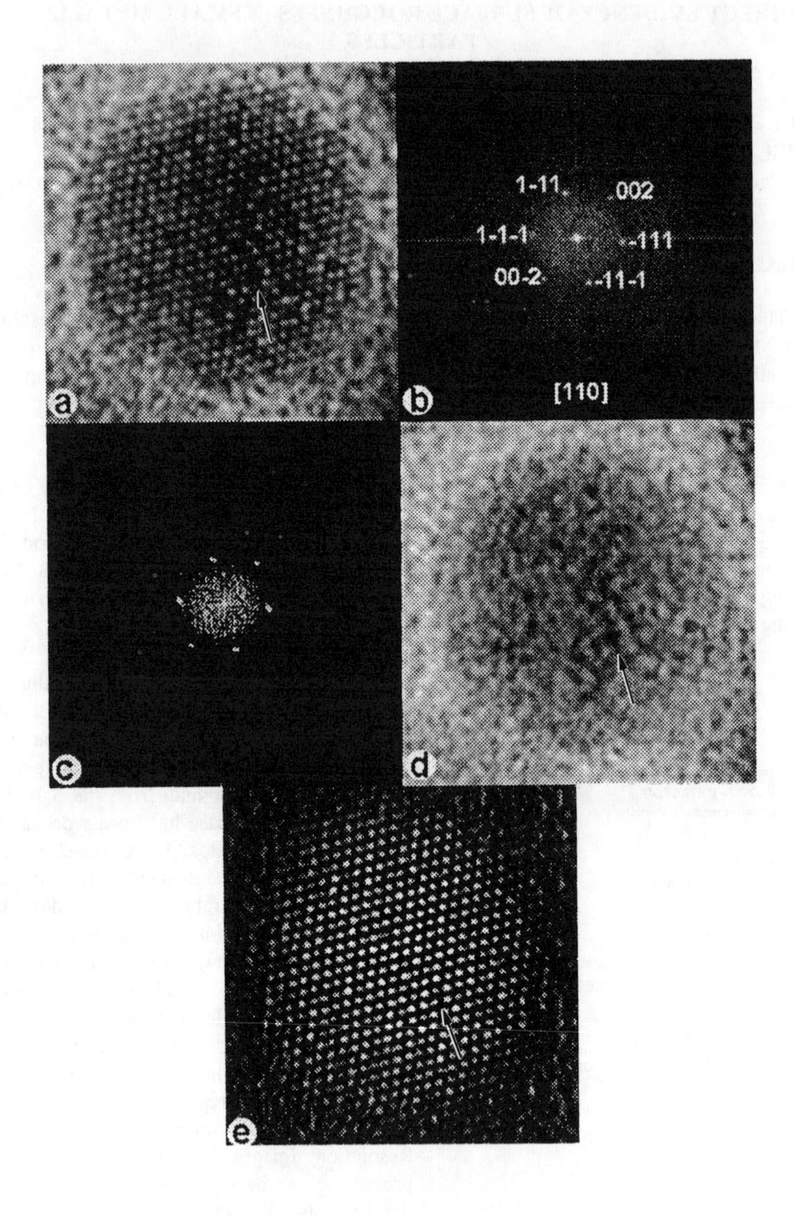

Figure 2. Image processing procedure applied to a Pt catalyst particle.

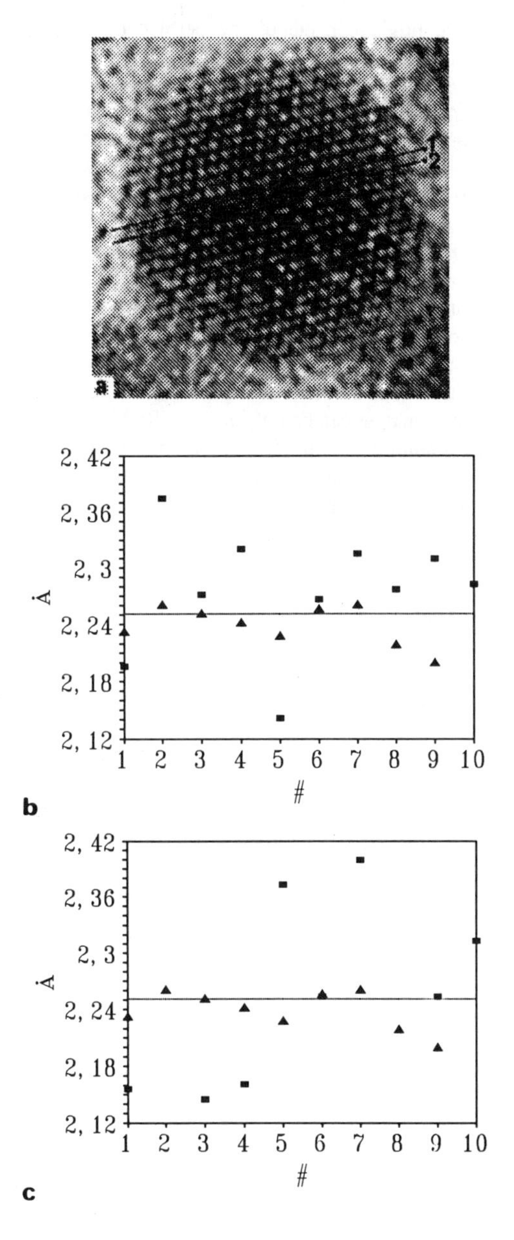

Figure 3. Plot of the distance between spots along a line crossing a Pt particle oriented in the <110> direction vs. the position of the atomic column. Fig. 3b) for line 1 and Fig. 3c) for line 2.

Figure 2a) shows the original image of the particle obtained at the optimum focus condition. The Fourier transform of the particle (figure 2b)) shows the spots corresponding to the particle which appear well defined plus the background signal due to the support.

We now apply a mask to the spots of the particle as shown in figure 2c). The spots should be identified by indexing the pattern as for an electron diffraction pattern. By applying a Fourier transform again, we obtain an image corresponding to the support as shown in figure 2d). This image although basically structureless contains some interesting contrast variations. If we now produce an image by substracting the figure 2d) from the image of figure 2a) we obtain a new image that corresponds to the particle without support. As can be seen in figure 2e), the particle looks much cleaner. In fact many of the irregular features that are apparent in the figure 2a) have now disappeared. For instance, in the region marked with an arrow in figure 2a) it appears that the particle contains a distorted region. The same region in figure 2d) is shown by the arrow and a black feature can be seen. In figure 2e) the feature has disappeared. Therefore, we can conclude that this irregularity was not real and resulted on the image only as an artefact of the substrate. This shows that erroneous information can be obtained in HREM pictures of supported catalysts if caution is not taken. In addition image processing can do the trick for us by eliminating the blurring of information on the catalysts due to unwanted contrast due to the substrate. Recently an extensive study of the effect of the support on HREM images has been published by Yao and Smith [2].

Effect of thickness and roughness in small particles

We have a particle with 3-D such as a cubo-octahedron. The atomic columns have a different number of atoms i.e.; the thickness of each column is different. This thickness variation has a strong effect on the image, the image spots shift with respect to the real position. This effect was first discussed theoretically by Fluei et al. [3] and confirmed on experimental images by M. José-Yacamán et al. [4].

This effect can be used to analyse the roughness on nanoparticles. We refer again to the particle in figure 2. In this case we have measured the distance between spots along a line crossing the particle along the $<1\ \bar{1}1>$ direction, figure 3a). In figures 3b) and 3c) we show the plot of the distance between spots along the line. The expected average value is shown in the figure by a solid line. The corresponding experimental points are indicated by ■ . The plots also show the theoretical values indicated by ▲ obtained for an undistorted particle with the same orientation. It is clear from the figure that there are very large variations on the distance between spots up to 10%. Analysing a large number of pictures we found similar results. In order to obtain the figures 3b) and 3c) a very careful calibration of the magnification of the microscope was necessary. On the other hand magnification errors alone will not explain variations on the same picture. The observed variations are well above the experimental error. A second effect that we considered was the possibility the variations were due to the image processing. In order to assess this effect we used a calculated image of a particle with cubo-octahedral shape oriented in the <110> direction. It was processed using the same algorithm that was applied to the real image and the distances were measured. We found that the distances between dots were not altered by the image processing.

In order to reduce the errors in the localisation of the peaks of figures 3b) and 3c) we used the algorithm developed by Beltrán del Río et. al. [5] which takes advantage of the gaussian nature of the peak and that locates a peak with a high accuracy. We also consider the possibility that the variations observed in figure 3 were the result of the dynamical effects produced during image formation. Indeed it is conceivable that the complex interaction between the electrons and the atoms of the particle might result in random variations on the distance. In the figure 3b) the distances between atomic columns in an ideal undistorted <110> oriented particle indeed show variations. The variations are due to the dynamical nature of the diffraction through a particle, however, these oscillations are below the level observed experimentally as shown in figure 3b) and 3c) (Theoretical ▲).

We are left only with three possibilities; the observed variations are due to a true distortion of the lattice, they are due to surface roughness, or they are due to the substrate. The first possibility is unlikely, remembering that each bright spot represents an atomic column, therefore the whole column will have to be expand or contract. The fact that the substrate has an effect on the images of atomic columns and, in particular, that it affects the apparent positions, can be estimated. A study of the effect of the support (in the case of an amorphous) was performed by Paciornik et al. [6] and Saxton and Smith [7]. The shift on the images is smaller than the ones observed in Figure 3. In order to further isolate the effect of the support on the shift on the images, we performed some measurements on particles that were located on a region near a hole on the support film. In that case the effect of the support is eliminated. In those samples we also observe shifts of the same order of magnitude as the ones observed in Figure 3.

Therefore, surface roughness is the most likely effect. In figure 4 we show a model of a rough particle and the resulting images at different defocus conditions.

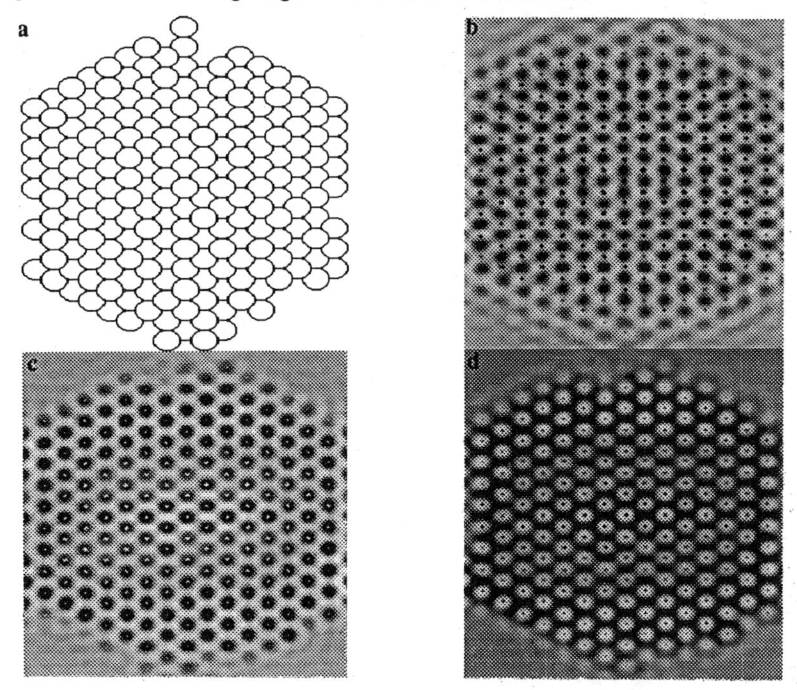

Figure 4. Model of a rough particle with 1474 atoms oriented in the <110> direction, at different defocus conditions. Fig.4b) 0 A, Fig. 4c) -400 A and Fig. 4d) -700 A.

As it is possible to see, a shift between image dots and real positions is observed. A further confirmation of the roughness effect can be obtained from the images of the particle edges.The calculations of figure 4 show that because the roughness is more pronounced at the edges of the particle, atomic columns appear lighter (even fading away) and the edge irregularities become apparent.

We conclude that all the theoretical and experimental evidence shows that small particles are rough. Most of the work in catalysis by small particles was based on the idea of correlating atomic sites in flat low index surfaces or in corners of regular polyhedral shapes. The classical work of Van Handervelg and Hartog [8] made countings of sites of different atomic coordination. In view of our results the nature of the sites is more complex that originally expected. Indeed it seems that the roughness effect is the dominant factor, with the shape and even the crystal structure of the particle being less relevant for catalysis.

References

1. M. José-Yacamán and M. Avalos, Ultramicroscopy , **10,** 211 (1983).

2. M.H. Yao and D.J. Smith, J. of Microscopy, **175**, 252 (1994)

3. M. Fluei, PhD thesis No.796 EPFL, Laussanne, 1989.

4. M. José-Yacamán, P. Schabes, R. Herrera, A. Gómez and S.Tehuacanero,Surf. Sci.,**237,** 248 (1990).

5. L.M. Beltrán del Río, A. Gómez and M.José in Proceedings of ICEM-13,**1**, 443 (1994).

6. S. Paciornik, R. Kilaas, U. Dahmen and M. O'Keefe, Proceedings of 51st Annual EMSA Meeting ,p. 458, Ed. by G.W. Bailey and C.L. Riedel, San Fco. (1993)

7. W. O. Saxton and D.J. Smith, Ultramicroscopy, **18**, 39 (1985).

8. R. Van Hardeveld and F. Hartog, Surf. Sci., **15,** 189 (1969).

PERFORMANCE CHARACTERISTICS OF LOW-TEMPERATURE CO OXIDATION CATALYSTS

GAR B. HOFLUND*, STEVEN D. GARDNER*, DAVID R. SCHRYER**, BILLY T. UPCHURCH**, JACQUELINE SCHRYER** AND ERIK J. KIELIN**
*Department of Chemical Engineering, University of Florida, Gainesville, FL 32611
**NASA Langley Research Center, Hampton, VA 23665

ABSTRACT

A series of Au/MnO_x catalysts with the Au content ranging from 0 to 10 at%, a 19.5 wt% Pt/SnO_x catalyst and a commercially available 2 wt% Pt/SnO_x catalyst have been examined for activity toward low-temperature CO oxidation. A 1 atm, 10 sccm stoichiometric mixture of CO (1 vol.%) and O_2 in He were reacted at 55°C over 100 mg of each catalyst for time periods as long as 20,000 min. Under these conditions a 10 at% Au/MnO_x catalyst exhibits the highest activity and the lowest decay. Outgassing experiments indicate that CO_2 retention is primarily responsible for the activity decay of the Pt/SnO_x catalysts and that the 10 at% Au/MnO_x catalyst is not significantly affected by this process.

INTRODUCTION

Long used as a research tool to study heterogeneous catalysis, the catalytic oxidation of CO is becoming increasingly important in practical applications. As concerns about the deterioration of air quality continue to intensify, there is a growing demand for CO oxidation catalysts which are more effective and versatile because these catalysts are often an integral component of pollution-control devices designed to reduce industrial and automotive emissions. Air-purification devices for respiratory protection and CO gas sensors commonly exploit CO oxidation catalysis.

Recently, low-temperature catalytic oxidation of CO has been utilized in closed-cycle CO_2 lasers[1-5] in order to maintain high power output. The function of the catalyst is to recombine stoichiometric concentrations of CO and O_2 that are produced during the lasing process. These closed-cycle CO_2 lasers are used for monitoring atmospheric pollutants or weather from a remote location such as earth orbit. Obviously, these lasers require a catalyst which maintains high activity for an extended period of time (3 years).

Research directed towards the development of long-life, sealed CO_2 lasers[5,6] has produced several new materials which actively oxidize CO below 50°C. Among the most active of these materials are gold supported on manganese oxide (Au/MnO_x) and platinized tin oxide (Pt/SnO_x). CO oxidation in CO_2 lasers is difficult since CO and O_2 are present in small, stoichiometric concentrations, the temperature is low and the presence of CO_2 lowers the activity. For example, a Au/Fe_2O_3 catalyst has been shown to exhibit excellent low-temperature CO oxidation activity in air[7,8], but its performance is unacceptable in a CO_2 laser environment[6]. The present study examines and compares the long-term performance characteristics of Au/MnO_x and Pt/SnO_x catalysts for reaction times approaching 20,000 minutes.

EXPERIMENTAL

Details regarding sample preparation and the CO oxidation reactor have been described previously[4-6]. The samples prepared include MnO_x, 2 at% Au/MnO_x, 5 at% Au/MnO_x, 10 at% Au/MnO_x and 19.5 wt% Pt/SnO_x. A 2 wt% Pt/SnO_x catalyst was obtained from Engelhard Industries. The BET surface areas of these catalysts are given in table 1.

Mat. Res. Soc. Symp. Proc. Vol. 368 © 1995 Materials Research Society

The experiments were conducted using 100 mg of catalyst and a reaction temperature of 55°C. Unless noted otherwise, the reactor feed contained 1 vol% CO, 0.5 vol% O_2 and 2 vol% Ne in helium flowing at 10 sccm and a total pressure of 1 atm. Different pretreatments were

Table I

BET Surface Areas of Catalysts Tested

Catalyst	Surface Area (m^2/g)
10 at% Au/MnO_x	64
5 at% Au/MnO_x	58
2 at% Au/MnO_x	65
MnO_x	43
2 wt% Pt/SnO_x	6.4
19.5 wt% Pt/SnO_x	127

used for the different types of catalysts. The MnO_x-based catalysts were exposed to 10 sccm of pure helium for one hour as the reaction temperature stabilized. The Pt/SnO_x catalysts underwent the following stepwise pretreatment to optimize their activity[9]: (a) heating to 125°C for one hour in 10 sccm of helium, (b) exposure to a 5% CO/He mixture (10 sccm) for 180 minutes, and (c) cooling to the reaction temperature in 10 sccm of helium for one hour. In each case the helium flow was replaced with the reaction gas mixture and product sampling was initiated after the reactor had stabilized at 55°C. An automated sampling valve directed a one ml sample of the reaction products to a gas chromatograph for analysis of % CO_2 yield at predetermined time intervals. The results were plotted versus time to yield the CO oxidation activity curves for the various catalysts.

RESULTS AND DISCUSSION

The CO-oxidation activity of 10 at% Au/MnO_x, 19.5 wt% Pt/SnO_x and 2 wt% Pt/SnO_x catalysts as a function of time are shown in figure 1. The activities of both Pt/SnO_x catalysts have been optimized by the reductive pretreatment, and the performance of the 19.5 wt% Pt/SnO_x catalyst has been optimized further with respect to Pt content. Although a reductive pretreatment enhances the overall activity of Pt/SnO_x, there are disadvantages associated with the pretreatment process itself. It is costly and inconvenient and completely incompatible with some applications in which the catalyst must be stored in or exposed to air. An interesting consequence of such a pretreatment is the appearance of an induction period wherein the CO oxidation activity exhibits an initial decrease and increase before going into a slow decay. As indicated in figure 1, the activity of the 2 wt% Pt/SnO_x catalyst exhibits an induction period. Schryer and coworkers[9] have studied this induction period for this catalyst and found that its shape and duration are dependent upon the pretreatment conditions. They also found that the addition of moisture to the reactor feed gas eliminates the dip in activity. Based on this observation they suggest that the activity dip is caused by surface dehydration during the pretreatment. As the surface adsorbs moisture and forms surface hydroxyl groups, the activity increases.

The performance of the 10 at% Au/MnO_x catalyst has not been optimized with respect to pretreatment procedures, but it exhibits superior CO oxidation activity compared to both of the Pt/SnO_x catalysts under the conditions of this test. A very important property of the Au/MnO_x catalyst is negligible activity decay. Even after 16,000 minutes the 10 at% Au/MnO_x catalyst does not exhibit a loss in activity whereas the activities of both Pt/SnO_x catalysts are diminished. While efforts continue to probe the mechanisms responsible for the decay of Pt/SnO_x catalysts, CO_2 retention plays an important role[5,10]. Surface characterization studies of Pt/SnO_x catalysts before and after pretreatment[11,12] have found that a Pt/Sn alloy forms during the reductive pre-

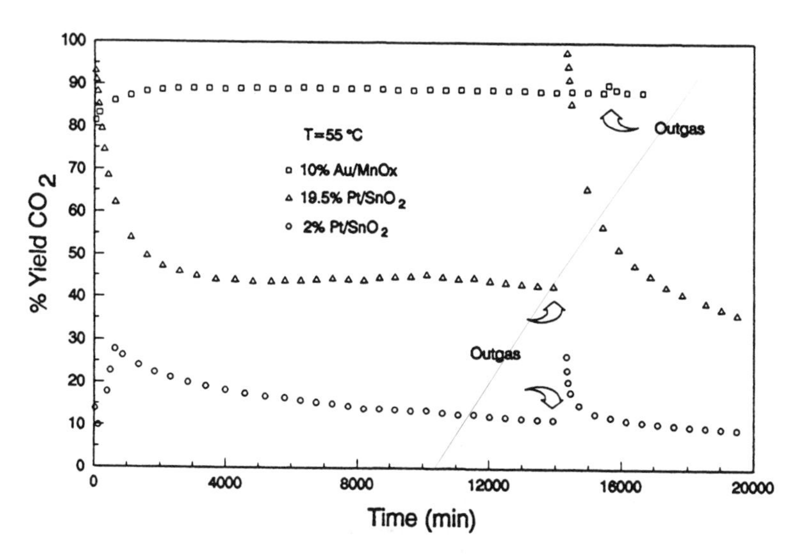

Figure 1. CO oxidation activity of 10 at% Au/MnO_x, 19.5 wt% Pt/SnO_2 and 2 wt% Pt/SnO_2 at 55°C as a function of time.

treatment and that significant amounts of $Pt(OH)_2$ are present on the catalytically active surface. A mechanism formulated from the characterization studies and reactor studies[10] proposes that an adsorbing CO reacts with a surface hydroxyl group and a surface oxygen to form a surface bicarbonate species which decomposes to form desorbing CO_2 and a surface hydroxyl group. More CO and O_2 adsorbs and the cycle continues. If the desorption of CO_2 is rate limiting, then the active site concentration decreases resulting in a decay in activity. Several days into the reaction, each catalyst was outgassed in flowing helium at the reaction temperature of 55°C, and the CO_2 concentration was monitored with the gas chromatograph. CO_2 desorbed for about 2 hours, and then the reaction was continued. As a result, the activities of both Pt/SnO_x catalysts were briefly regenerated followed by a rapid decay. This experiment demonstrates that CO_2 retention is a primary process contributing to activity decay. However, the outgassing procedure had little effect on the Au/MnO_x catalyst suggesting that CO_2 retention is not significant on this surface under the conditions used in this study. This may explain the low-decay rate for the Au/MnO_x catalyst.

Optimization of a catalyst with regard to all of the preparative and pretreatment variables is an important process. Optimization of the Au/MnO_x catalyst is in the initial stages. The effect of varying gold loading on the CO oxidation activity of unpretreated Au/MnO_x is shown in figure 2. Each catalyst was prepared and tested under similar conditions so direct comparisons are valid. Metallic Au surfaces do not exhibit appreciable CO oxidation activity below 300°C[8], and the catalytic activity exhibited by MnOx in figure 2 is quite low. Therefore, a synergistic effect between the Au and MnO_x is evident. Increasingly large Au loadings not only enhance the overall activity but decrease the rate of activity decay as well. The effect of outgassing becomes more prominent with decreasing catalyst performance, which provides further support to the hypothesis regarding surface CO_2 retention and activity decay. As indicated in figure 3, however, the Au loading is not the only variable influencing overall CO oxidation performance. Two 5 at% Au/MnO_x samples were prepared which differ only in the number of distilled-water washes prior to drying and calcining. The extra washing step produces an improved activity profile. This is believed to be due to more complete removal of surface Cl that originates from the precursor used in preparing the catalyst. Surface Cl is known to inhibit catalytic low-temperature CO oxidation.[5,13]

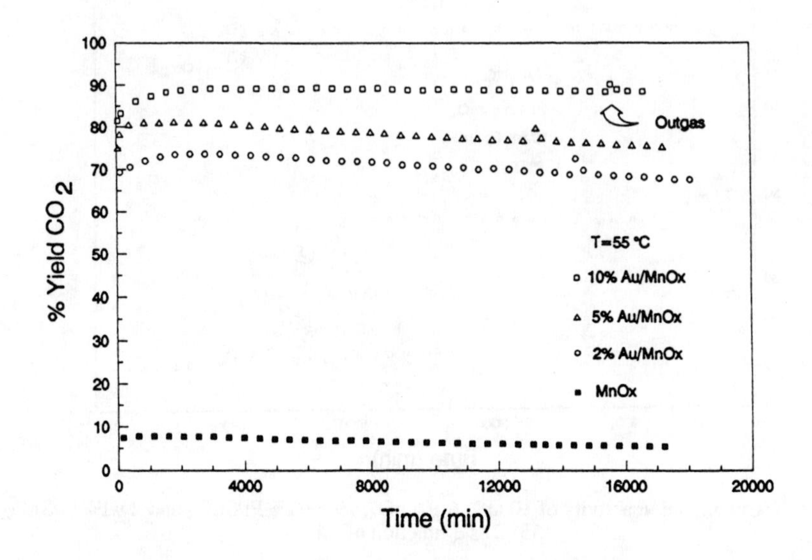

Figure 2. CO oxidation activity of 10 at% Au/MnO_x, 5 at% Au/MnO_x, 2 at% Au/MnO_x and MnO_x at 55°C as a function of time.

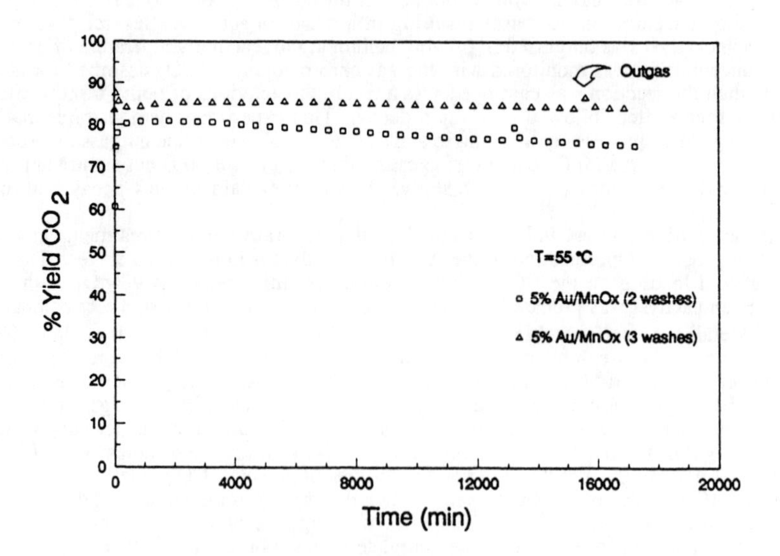

Figure 3. Effect of washing with distilled water on the CO oxidation activity of 5 at% Au/MnO_x at 55°C as a function of time.

ACKNOWLEDGMENT

Support for this research was provided by the National Science Foundation through grant #CTS-9122575.

REFERENCES

1. D.S. Stark, A. Crocker and G.J. Steward, J. Phys. E: Sci. Instrum. **16**, 158 (1983).

2. D.S. Stark and M.R. Harris, J. Phys. E: Sci. Instrum. **16**, 492 (1983).

3. D.S. Stark and M.R. Harris, J. Phys. E: Sci. Instrum. **21**, 715 (1988).

4. C.E. Batten, I.M. Miller, G.M. Wood, Jr. and D.V. Willetts, Eds., "Closed-Cycle, Frequency-Stable CO_2 Laser Technology," Proceedings of a workshop held at NASA Langley Research Center, Hampton, VA, June 10-12, 1986, NASA Conference Publication 2456.

5. D.R. Schryer and G.B. Hoflund, Eds., "Low-Temperature CO-Oxidation Catalysts for Long-Life CO_2 Lasers," Collected papers from an international conference held at NASA Langley Research Center, Hampton, VA, October 17-19, 1989, NASA Conference Publication 3076.

6. S.D. Gardner, G.B. Hoflund, D.R. Schryer, J. Schryer, B.T. Upchurch and E.J. Kielin, Langmuir **7**, 2135 (1991).

7. M. Haruta, T. Kobayashi and N. Yamada, Chem. Lett. 405 (1987).

8. M. Haruta, N. Yamada, T. Kobayashi and S. Iijima, J. Catal. **115**, 301 (1989).

9. D.R. Schryer, B.T. Upchurch, J.D. Van Norman, K.G. Brown and J. Schryer, J. Catal. **122**, 193 (1990).

10. D.R. Schryer, B.T. Upchurch, B.D. Sidney, K.G. Brown, G.B. Hoflund and R.K. Herz, J. Catal. **130**, 314 (1991).

11. S.D. Gardner, G.B. Hoflund, D.R. Schryer and B.T. Upchurch, J. Phys. Chem. **95**, 835 (1991).

12. J.E. Drawdy, G.B. Hoflund, S.D. Gardner, E. Yngvadottir and D.R. Schryer, Surface Inter. Anal. **16**, 369 (1990).

13. "Catalysts for Long Range CO_2 Laser System," a progress report from a catalyst development program carried out by GEC Avionics, UK Atomic Energy Authority Harwell and UOP Ltd. under the auspices of the Royal Signals and Radar Establishment, November, 1988. Contract nl. SDIO 84-87-C-0046.

THE PREPARATION AND CHARACTERIZATION OF Pt/Fe ELECTROCATALYST FOR DIRECT METHANOL OXIDATION

Chaoying Ma*, A. D. Kowalak*, Changmo Sung#
*Chemistry Department , University of Massachusetts at Lowell, Lowell, MA 01854
#Center for Advanced Materials, University of Massachusetts at Lowell, Lowell, MA 01854

ABSTRACT

Pt/Fe electrocatalysts (Pt/Fe) supported on Vulcan XC-72 carbon black were prepared by chemical method. The formation of Pt/Fe clusters was confirmed by transmission electron microscopy (TEM) and energy dispersive x-ray spectrometry (EDXS). The electrochemical efficiency and durability of the catalysts for direct methanol oxidation were compared with the catalyst containing Pt only (Pt) under the same reaction conditions of 1 M CH_3OH in 0.1 M H_2SO_4 electrolyte solution. The current density for the Pt/Fe is three times of that of the Pt catalyst. Effects of various factors on the behavior of the catalysts, such as reaction time during preparation, order of the addition of the chemicals into the reaction system, methanol concentration, activation per unit Pt, scan rate for cyclic voltammograph, were examined.

INTRODUCTION

Direct methanol oxidation fuel cell is an ideal energy source for the zero emission, low pollution vehicles. Methanol is used as a fuel at relatively low temperature [1, 2]. Unlike gasoline-powered engines or electric power plants, fuel cells do not burn their fuel. Instead, energy is extracted chemically in a process that is much more efficient than burning the fuel and produces no combustion byproducts. One of the major difficulties for a commercial fuel cell comes from a suitable anode which has the advantages of high efficiency and low cost. The highly dispersed Pt colloidal particles have been proved to be the most effective electrocatalyst for direct methanol oxidation in fuel cell. Electrocatalysts incorporating Pt are the most active catalyst for the direct electro-oxidation of methanol in acid solutions and the electrochemical oxidation of methanol on Pt surfaces is known to be strongly enhanced by the presence of a second metal. The incorporation of Ru, Pd, Sn, Mo, Fe into Pt catalysts by various methods has been investigated to increase the efficiency and lower the cost of the electrocatalyst as anode materials in fuel cells.

Pt is used as the main material since it has the highest catalytic activity among single metals over a wide potential range. The modification of Pt metal is presently a major subject for the development of a catalytically active anode. The surface or bulk composition of a Pt electrode was modified by a second metal through mechanical mixing [3], submonolayer electrodeposition [4], alloy formation [5], and sputtering (or ion plating). Here in this research, a new Pt/Fe electrocatalyst was prepared by a chemical method for the direct methanol oxidation in fuel cells.

EXPERIMENTAL

Sample Preparation

H_2PtCl_6 (Aldrich) solution was reduced by $NaHSO_3$; HCl gas was driven off from the solution [6]. Hydrogen peroxide (H_2O_2, 35 wt%) was used to oxidize the platinum complex. Platinum oxide was combined with iron oxide. The resulting solution was allowed to settle down and combine with Vulcan XC-72 (carbon black, Cabot Co., Billerica, MA). The paste was reduced by H_2 bubbling for 15 minutes. The final catalysts were air dried and stored for further measurements.

Mat. Res. Soc. Symp. Proc. Vol. 368 © 1995 Materials Research Society

Characterization by TEM and EDXS

Transmission electron microscopy (TEM) was used to characterize the formation of Pt/Fe mixtures. TEM samples were prepared by placing a drop of the solution to be tested, which was a mixture of the reaction solution and ethanol/water, onto a copper support grid. The copper grid was placed on some filter paper to adsorb excess solution. Microscopic examination of the samples was carried out on a Philip 400T analytical electron microscope operating at 120 keV with a LaB_6 energy source.

The chemical composition analyses were carried out on a Noran Instrument 5,500 low-Z energy dispersive x-ray spectrometry system (EDXS).

Electrochemical Measurements

The catalytic activity for the electro-oxidation of methanol was measured at 25 ^oC using an EG&G Princeton Applied Research Potentiostat/Galvanostat Model 273, which was connected to an IBM Instruments 7424 MT x-y-T recorder. The reaction system was 1 M CH_3OH in 0.1 M H_2SO_4 electrolyte solution. The solution was deoxygenated with a constant stream of Ar gas for 20 minutes before the measurements. The counter-electrode was a platinum wire. A reversible Pd/H_2 was employed as the reference electrode. A glassy carbon electrode with a diameter of 6 mm, which was coated with the catalyst paste by using Nafion as a proton exchange membrane (PEM), was used as the working electrode.

RESULTS

Electron Microscopy

TEM samples were prepared from the resulting colloidal solutions after the addition of H_2O_2 followed by a period of reaction time from 30 minutes to 48 hours. TEM bright field images of the samples showed regular shape and monodispersed colloidal particles. Figure 1 shows the Pt/Fe colloidal particles size distribution. From preparation to preparation, the diameter of particles varied in the range of 50 nm to 90 nm. Figure 2 shows the TEM bright field image of monodispersed Pt/Fe particles. Energy dispersive x-ray analyses of the samples indicate that the metal particles were composed of platinum, iron and oxygen (Figure 3). Table 1 lists the peak intensities of elements Pt, Fe, O, Cu in different particles a, b, c of the same catalyst. In EDXS measurements, the peak intensity is proportional to the chemical composition of the elements. Since copper grids were used as the support for the TEM samples, Cu background peaks are observed in the EDXS spectra.

Table 1. Peak intensities of various elements in different particles from EDXS measurements

	Intensity		
Element	a	b	c
Pt	1	1	1
Fe	1.2	1.8	5.2
O	0.6	0.5	1.4
Cu	2.4	3.9	5.8

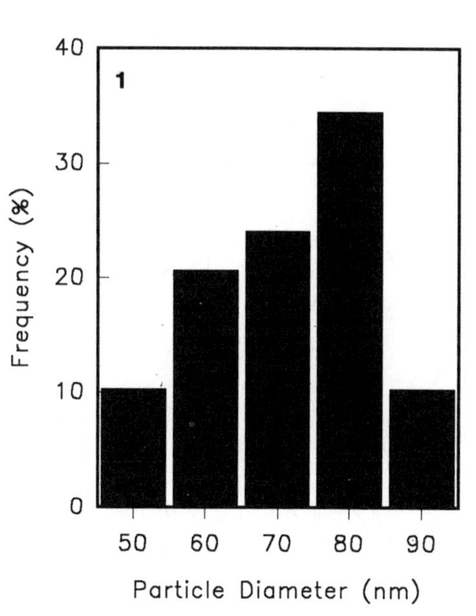

Figure 1. Size distribution of the Pt/Fe colloidal particles.

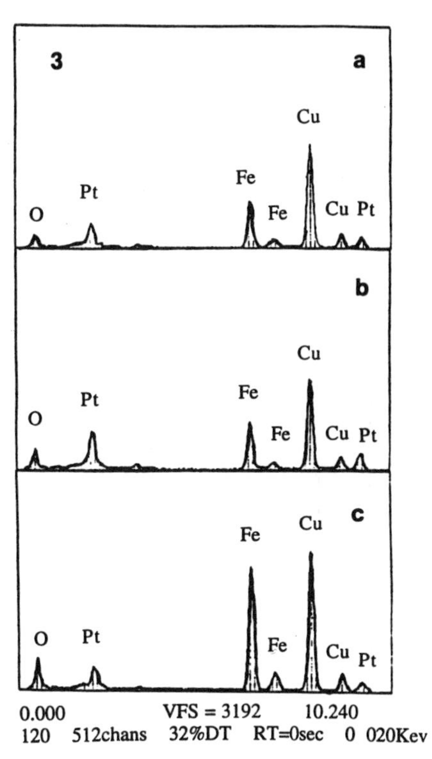

Figure 3. EDXS analyses of the Pt/Fe colloidal particles.

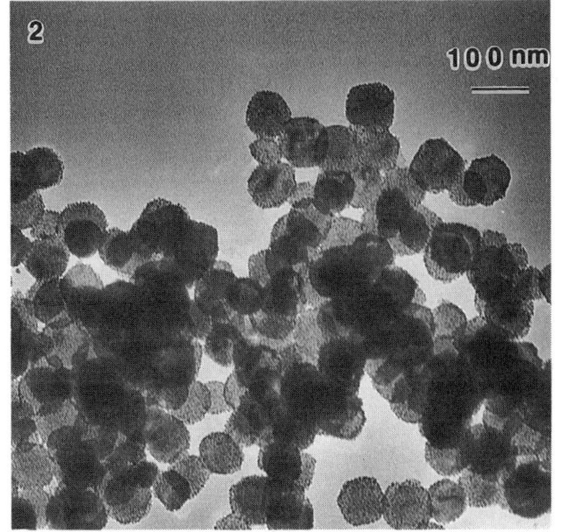

Figure 2. The TEM bright field image of the Pt/Fe colloidal particles.

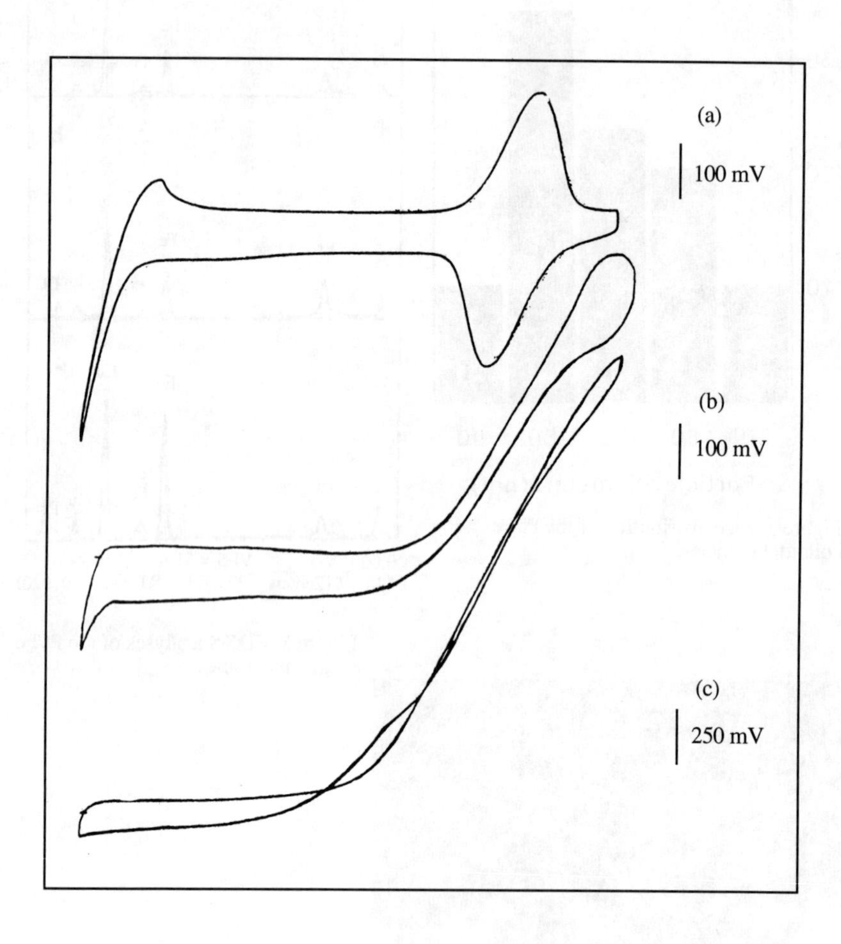

Figure 4. Cyclic voltammetric curves of catalysts on a glassy working carbon electrode, for 1 M methanol oxidation in 0.1 M H_2SO_4 electrolyte solution at a scan rate of 50 mV/s. Scan range: 0.1 V - 1.1V; counter electrode: platinum wire; reference electrode: Pd/H_2; (a) iron oxide/Vulcan XC-72 carbon black; (b) platinum oxide/Vulcan XC-72 carbon black; (c) platinum/iron oxide/Vulcan XC-72 carbon black.

Electrochemical Studies

20 μl of Pt/Fe electrocatalyst/Nafion mixture was coated onto the surface of a glassy carbon working electrode. The platinum content of the electrode was calculated to be 0.06 mg Pt/cm^2. The cyclic voltammograph of the Pt/Fe catalyst for 1M CH_3OH in 0.1 M H_2SO_4 electrolyte solution is shown in Figure 4. The cyclic voltammographs of the Pt and the Fe as electrode materials are compared with that of the Pt/Fe catalyst. Pt content of the Pt catalyst coated on the surface of the glassy carbon working electrode was also 0.06 mg Pt/cm^2, which is the same as that of the Pt/Fe catalyst.

DISCUSSION AND CONCLUSION

Figure 3 shows EDXS spectra of different Pt/Fe particles with different ratios of platinum/iron at the surface. It is assumed that the platinum oxide or platinum deposits on the surface of iron oxide after the addition of H_2O_2 during the preparation. The generated platinum oxide particles grow on the iron oxide nucleating centers. Since the size of the iron oxide colloidal particles can be controlled, the size of the resulting Pt/Fe particles can be controlled. Experimental results reveal that the reaction time after the addition of H_2O_2 is critical for the growth of platinum oxide on the surface of iron oxide particles [7] . This reaction time ranged from 30 minutes to 48 hours. The product with a reaction time of 2 hours only has half of the current density of the product with a reaction time of 48 hours under the same reaction conditions.

The electrochemical efficiency measurements show some promising results. With the similar Pt content coated onto the working glassy carbon electrode, the current density of Pt and Pt/Fe catalysts are compared. Generally, Pt/Fe catalysts show three times higher current density than that of the Pt catalyst, where Pt/Fe catalyst can reach a current density of 0.1 A/mg Pt while Pt catalyst has a current density of 0.03 A/mg Pt. Another feature of the Pt/Fe catalyst is its relatively stable, long lasting working efficiency. The Pt/Fe working electrode was continuously cycled in 1 M CH_3OH with electrolyte concentration of 0.1 M H_2SO_4 at the potential range of 0.1 V to 1.1 V for 5 hours. There was no significant change observed in the current density for the catalysts. A high electrochemical efficiency still was observed with the same Pt/Fe catalyst coated electrode used repeatedly for 4 times. However, the catalyst coated on the surface of glassy carbon electrode eventually flaked off. This is one of the major reasons that the catalyst finally failed. The electrochemical efficiency of the catalyst is approximately proportional to the Pt content in the electrode. The efficiencies of catalysts with different amounts of coatings, 10 μl, 20 μl, 30 μl, are compared. A higher scan rate of cyclic voltammograph ended up with a higher efficiency, though the change is not very large. Iron oxide particles themselves are not stable in H_2SO_4 solution. However, the Pt/Fe catalysts prepared by this method are stable in 0.1 M H_2SO_4 solution for several months.

The structure of the Pt/Fe particles and the interaction between Pt and Fe will be an interesting subject for further research. Whether this compound is a bimetallic particle or simple mixture of platinum and iron depends on the structure determination. The extremely low loading and tiny size of Pt in the catalyst makes it very difficult to determine the Pt-Fe structure by the x-ray diffraction (XRD) and x-ray photoelectron spectrum (XPS). Gauthier [8] has studied the surface segregation of Pt-M alloys (where M is a 3d transition metal including Co, Ni and Fe), where LEED has been used to study the relaxation of interlayer spacing between PtCo and PtNi alloys. The surface segregation phenomenon of the Pt/Fe catalyst in this research needs further study.

REFERENCES

1. P.N.Ross, Electrochim. Acta **36**, 2053 (1991).

2. R. Parson and T. VanderNoot, J. Electroanal. Chem. **9**, 257 (1988).

3. M. Watanabe, M. Uchida, S. Motoo, J. Electroanal. Chem. **229**, 395 (1987).

4. A. Aramata, T. Kodora and M. Masuda, J. Appl. Electrochem. **18**, 577 (1988).

5. J. O'M Bockris, and H. Wroblowa, J. Electroanal.. Chem. **7**, 428 (1964).

6. H.G.Petrow and R.J. Allen, U. S. Patent No. 4,044,193 (23, August, 1977).

7. Chaoying Ma, A. D. Kowalak, Changmo Sung, in Determination Nanoscale Physical Properties of Materials by Microscopy and Spectroscopy, edited by Mehmet Sarikaya, H. Kumar Wickramasinghe, Michael Issacson (Mater. Res. Soc. Proc. Vol. **332**, Boston, MA, 1993).

8. Y. Gauthier, R. Baudoing-Savois, J.J.W.M. Rosink and M. Sotto, Surface Science. **297**, 193(1993).

ELECTRONIC AND STRUCTURAL REQUISITES FOR CATALYTIC ACTIVITY IN COPPER-RUTHENIUM CATALYSTS

M. Malaty, D. Singh, Noel Vadel, M.L. Gomez, M. Palmieri, S.A. Jansen; Department of Chemistry, Temple University, Phila. PA 19122; S. Lawrence; Saginaw Valley State University, University Center, MI 48710.

ABSTRACT:

A system made by combining two non-alloying metals, ruthenium and copper, using alumina as a support was studied. This bimetallic supported catalyst has been used mainly in hydrogenolysis, dehydrogenation and oxidation reactions of hydrocarbons. The samples were characterized by Electron Paramagnetic Resonance Spectrometry (EPR) and X-ray Diffraction (XRD). These two molecular techniques are ideal for studying the electronic and structural changes of the samples at different temperatures and concentrations. Catalytic reactions were performed using the catalyst in the reduced and non-reduced forms on a series of catalytic hydrogenations. A correlation between the electronic, structural and catalytic properties has been made. A correlation of catalytic process to molecular phenomena has yield a better understanding of the catalytic site.

INTRODUCTION:

In the last twenty years there have been multiple reports on mixed metal catalysts. The general motivation for the preparation of such materials was to effect the selectivity of a highly active metal by inclusion of a second less active metal. The metallic components were selected based on certain electronic and packing characteristics including metals capable of forming alloys in the metallic state such as nickel and copper, copper and gold, ruthenium and copper. In the latter case, it was determined that copper grew pseudomorphically atop the (0001) surface of ruthenium with very little tensile strain. Catalytic application required that these materials be supported on oxide substrates, usually silica or alumina. As was expected the catalytic properties of the mixed metal systems varied widely depending on the preparation, support characteristics and catalytic application, i.e. varying effects on activity and selectivity were observed for oxidation and reduction.[1] Applications in ceramics have been directed at materials of similar composition. Electronic ceramics based on nickel and/or copper aluminates are continually of study. Resistive ceramic systems based on ruthenium dioxide "encapsulated" in alumina or silica glasses are also of topical interest. The effect of a secondary metal in the electronic oxides has not been well studied, but clearly the effect of a secondary metal on the resistive or materials properties is an area that needs further development.

Mat. Res. Soc. Symp. Proc. Vol. 368 © 1995 Materials Research Society

In this report, we will focus on a Cu-Ru/ g-alumina. The catalytic properties of several related silica and alumina supported systems have been evaluated extensively. The metal were chosen for several reasons. (i) Copper aluminates which adopt the spinel structure are known to form in preparation of ceramic materials, inclusion of a secondary metal may effect the formation of the stable spinel structure. (ii) Formation of ruthenium oxides or related mixed metal ruthenium-copper oxides are expected to exhibit modified electrical properties. (iii) The structural data obtained will be of utility for the assessment of catalytic properties of mixed metal systems. To date, there has been no attempt at correlation of the catalytic or electronic properties with the structural phases produced during preparation. This is in part because direct characterization of the structural phases has been limited. This work will provide a extensive characterization of the electronic properties of a copper-ruthenium system supported on alumina.

EXPERIMENTAL:

I. Preparation of materials:

The materials were prepared by the incipient wetness method.[2] There were two different incipient wetness methods employed. (i) Simultaneous Deposition: This procedure has the impregnating solution containing the two salt metals added to the alumina simultaneously. (ii) Step-wise Deposition: This method divides the impregnating solution into two steps. First the addition of one metal solution to the alumina and then dried. The second metal solution was then added and the final heat treatments were performed(Fig 1). The calcined materials were characterized by XRD and EPR. XRD measurements were performed on samples which ranged from 1%-35% in copper and ruthenium.

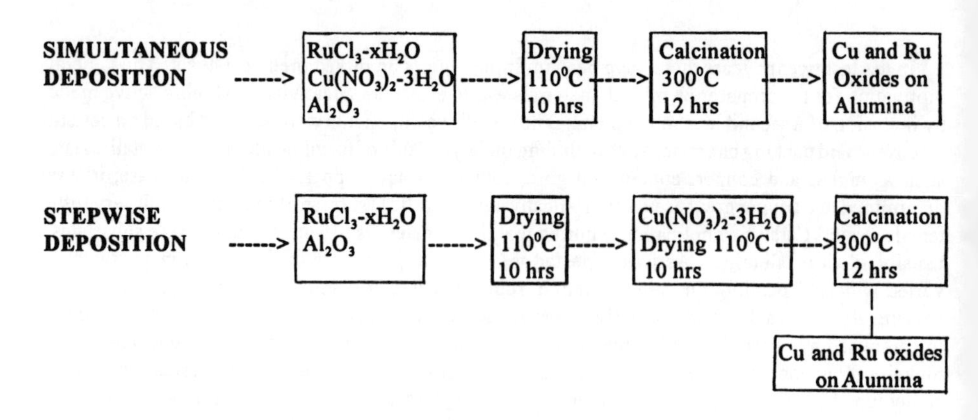

Figure 1: Methods of Preparation

II. Electron Paramagnetic Resonance (EPR)

EPR measurements were performed on a Bruker ER-200D spectrometer, fitted with an ER035 gaussmeter to provide accurate magnetic field measurement to 0.001G/1000 G. For variable temperature measurements, temperature control was maintained by a VT-4114 unit which provides temperatures within 0.5°C. A series EPR measurements were made on materials in which the overall loading was relatively low, 1-3% by weight. For these samples, the theoretical coverage limits were less than a tenth of a monolayer and such, these studies provide the potential for analysis of isolated metal - oxide interaction. Potential for aggregation can also be addressed as any cooperative metal-metal/oxide interaction will be pronounced in the EPR characterization at such low loadings and coverages. A complementary study at high loading and 1/2 and full monolayer, i.e. 20-40% by weight was also undertaken to determine the structural limits effecting catalytic and electronic properties.

III. Catalysis:

The catalytic activity and selectivity of these systems were studied using the hydrogenation reaction of o-xylene(Fig 2). Since this reaction is well understood, it was employed to probe the electronic characteristics and structure of the catalysts through their catalytic behaviors. This allowed us to make a correlation between the catalytic activity and selectivity to the electronic structure, which was studied by EPR. These catalysts were studied by a Gas-Solid reaction which was carried out under continuous-flow conditions in a glass tubular reactor.[3] The reactor tube was mounted in a tube furnance and the feed stream was 100 percent hydrogen. Since the products were in the gas phase, GC was the method employed to study the reaction.

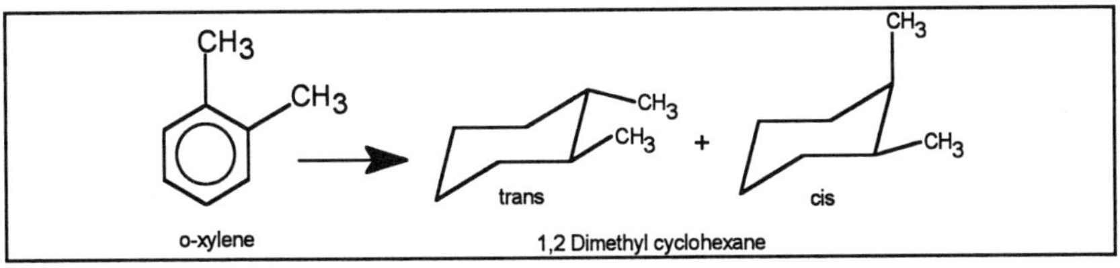

CH_3 CH_3 → CH_3 CH_3 + CH_3 CH_3

o-xylene trans cis

1,2 Dimethyl cyclohexane

Figure 2: o-xylene reaction

IV. Experimental Conditions:

The reaction conditions used for the hydrogenation reaction was a temperature of 100°C, and a hydrogen flow of 10 and 25ml/min. The catalyst used was run (i) unreduced and (ii) reduced(catalyst was treated for 3 hrs at 350°C:H_2-25ml/min).

RESULTS AND DISCUSSION:

(a) Low loading/low coverage. Multiple EPR analyses were performed to determine how the copper and ruthenium interacts with the oxide surface of alumina. This description will directed at the calcined materials only. The "g-value" as a function of metal composition showed a rather profound effect due to deposition conditions. In general, the "g-value" of a material represents a ratio between the resonant field and the magnetic transition energy. Therefore it serves as a sensitive indicator of processes occurring at or near the Fermi level. Deposition of a copper onto alumina at moderate temperatures has been shown to produce both surface oxides and copper aluminates. Production of these species is effected by the order of deposition in the mixed metal system. This can be assessed by analysis of variation in one of the components of the g-tensor. The resolved spectrum arises from isolated copper oxide states. Figure 3 shows the EPR spectra for the different step-wise deposition samples. Figure 4 shows the average "g" as a function of preparation. Here, it is important to note the rather striking difference between samples prepared when the copper and ruthenium were co-deposited versus those prepared by step-wise deposition. The values of the "g" for the co-deposition and the step-wise deposition in which copper precedes the ruthenium provide the same trend in "g". However when ruthenium is deposited prior to copper, the magnitude and range of "g" is limited. We believe this is due to a reduction in the contribution of the aluminate species. The XRD data which supports this conclusion is shown in figure 5. A small amount of ruthenium dioxide is observed with the spinel form of the copper aluminate obscured by the profile of the g-alumina.

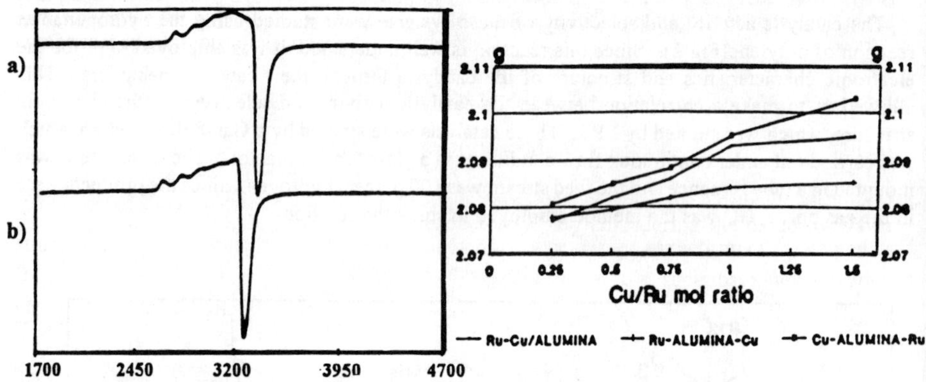

Figure 3: EPR spectra of (a) Cu-Al_2O_3-Ru and (b) Ru-Al_2O_3-Cu

Figure 4: average "g" values as a function of preparation

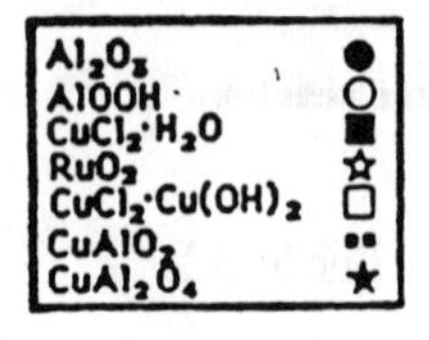

10 20 30 40 50 60 70 80

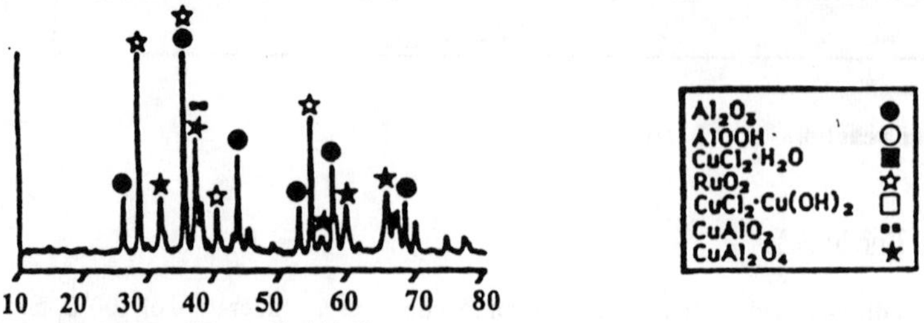

Figure 5: XRD of Ru + Cu/Al_2O_3

Catalysis:

The hydrogenation reaction of o-xylene was used as a probe to study the chemistry of the Ru-Cu/Al_2O_3 catalysts. We were able to make a correlation between the electronic, structural and catalytic properties of these catalysts. In understanding the relationship between these properties a better understanding of the chemistry of the catalyst on a molecular level can be developed and applied to similar reactions(e.g. hydrogenation of benzene). The active species in this reaction is ruthenium dioxide which is primarily been used for oxidation of hydrocarbons and for hydrogenation and dehydrogenation of alkanes. The η^6 coordination of the aromatic ring is understood to be the preferred binding site at the ruthenium(RuO_2) and binding does not occur on CuO. This is important for the reaction process because the first step is thought to be the adsorption of the o-xylene on to the catalyst and the ruthenium(RuO_2) is that site. The second step is the adsorption of hydrogen on the catalyst in the radical form(H·) and then migrates to the o-xylene, hydrogenating the molecule and forming the two products: 1,2 cis-trans, dimethyl cyclohexane.

Copper was added to the ruthenium catalysts in order to effect the selectivity of a highly active metal(Ru) with a less active metal(Cu). Different mechanisms appear to be active for low and high coverages. In the highly loaded pure ruthenium catalysts(Ru/Al_2O_3) the ruthenium(IV) becomes reduced to ruthenium(0) during the catalytic process, when using temperatures greater than 150^oC. This was due to the decomposition of the o-xylene, which deposited carbon onto the catalyst causing reduction of the ruthenium. When the ruthenium is in its reduced form(Ru^0) the activity of the catalyst is lost. Futhermore, when copper is added to the highly loaded samples the catalytic material becomes completely inactive. This suggests the copper is covering the surface of the catalyst and blocking the ruthenium sites or the copper is occupying the ruthenium sites in RuO_2 at the surface. In a typical ruthenium-copper aggregate, the ruthenium forms an inner core, while the copper is present at the surface.[4-5] In our highly loaded samples we suspect integrated oxides form because the electrical conductivities of the ruthenium/copper catalysts are significantly greater than the pure ruthenium catalysts. It has been determined both experimentally and theoretically that the π system of the aromatic ring will not bind to copper, therefore copper at the surface will deter the adsorption process.

In the low loaded samples the addition of copper changed the activity and selectivity of the catalysts. The preparation method of the low loading materials plays an important role in their catalytic properties. The stepwise deposition samples of Ru-Al_2O_3-Cu showed no catalytic activity. The "g" values for the two stepwise deposition samples are different(Fig 4). The stepwise deposition in the reverse order Cu-Al_2O_3-Ru showed activity. As more copper was added to the catalyst, there was a decrease in the percent spin active copper(Table I). The activity of these catalysts are following a linear response to the amount of spin active copper(Table I). This data suggests that the presence of copper oxides inhibits the activity in the low loaded catalysts.

These catalysts have copper located at the surface as well as in the bulk; the surface copper is considered the reducible species and the bulk copper is considered the ireducible species. When the catalyst is pretreated under reducing conditions the copper(II)oxide at the surface is reduced to copper(I) oxide and copper(0) metal, where as the copper in the bulk cannot be reduced. In figure 6 we show a correlation between the percent reducible copper and the percent of product formed. The samples which are first prepared under reducing conditions and then used in a reaction show little or no catalytic activity. In the EPR spectra we see a great reduction in the percent spin active copper in the reduced samples and this correlates to the catalytic data. This is due to the copper(II) oxide being reduced to copper(I) oxide and copper metal along the surface. This suggests that the copper oxide is occupying specific sites at the surface.

Table I: Percent Spin Active Copper vs. Percent Product for Cu-Al_2O_3-Ru Catalysts

Ru- %wt	Cu-%wt	Mole Ru-Cu	%Spin active Cu	%Reducible[a]	%Irreducible[b]	%Product[c]
1.00	0.00	-----	-----	-----	-----	35.6
1.00	0.30	1-0.50	65	49	16	9.3
1.00	0.46	1-0.75	62	47	15	8.8
1.00	0.62	1-1.00	45	34	11	6.1
1.00	1.11	1-1.50	25	19	6	2.2

[a] amount of copper that can be reduced[CuO--Cu_2O--Cu(0)]
[b] amount of copper that cannot be reduced
[c] percent of total product: cis + trans

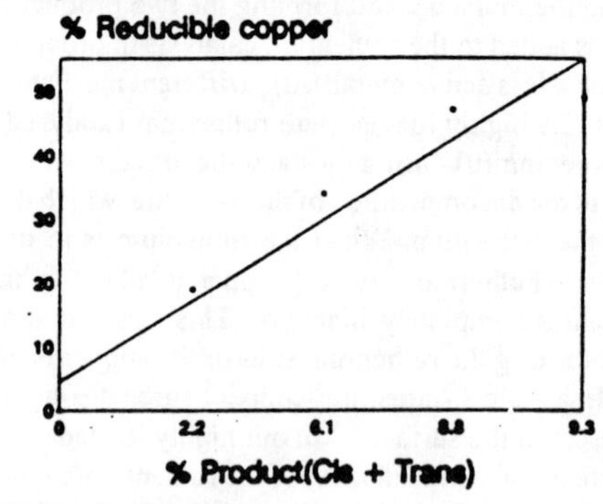

Figure 6 : % Reducible Copper vs. % Products(cis + trans)

CONCLUSION:

When copper was added to the highly loaded catalysts the reaction was completely inhibited. In the low loaded samples the preparation was important for the activity and selectivity. The catalysts prepared with copper(Ru-Al_2O_3-Cu) last in the deposition order showed no activity, where as the samples prepared with copper(Cu-Al_2O_3-Ru) first in the order were active. The activity of the Cu-Al_2O_3-Ru catalysts can be linearly correlated to the percent spin active copper. When these catalysts are reduced the catalytic activity is lost. Though the correlation between spin active copper sites and RuO_2 active centers is not yet well understood, it seems that there is some miscibility between the copper and ruthenium oxides even at the low loadings and thus a pronounced electronic effect is produced when ruthenium must "compete" for surface sites or co-ad to the copper treated support. If ruthenium is added first, the stable oxide, RuO_2 is formed and thus the cuprate appears to mask the active sites.

References:

1. J.H. Sinfelt. Y.L. Lan, J.A. Cusumano, and A.E. Barnett, J. Cat., 42, 227 (1977).
2. J.H. Sinfelt, J. Cat., 29, 308 (1973).
3. J.B. Black, J.D. Scott, E.M. Serwicka, and J.B. Goodenough, J. Cat., 106, 16-22(1987).
4. J.H. Sinfelt, Acc. Chem. Res., 10, 15 (1977).
5. J.C. Conesa, P. Esteban, H. Decpert, and D. Bazin, Spectroscopic Characterization of Heterogenous Catalysts, 57A, 265, Ed. J.L.G. Fierro (1990).

NOVEL MICROFABRICATED Pd-Au/SiO_2 MODEL CATALYSTS FOR THE HYDROGENATION OF 1,3-BUTADIENE

ANTHONY C. KRAUTH AND EDUARDO E. WOLF
University of Notre Dame, Department of Chemical Engineering, Notre Dame, Indiana 46556

ABSTRACT

Model Pd-Au/SiO_2 catalysts were prepared via a microfabrication methodology and their catalytic activity was evaluated using the hydrogenation of 1,3-butadiene as a probe reaction. The catalytic particles of the microfabricated catalyst were Pd-Au squares, ranging in composition, each approximately 4 μm in size and separated by 4 μm on a silica/silicon support. The reaction was carried out at atmospheric pressure with a hydrogen to hydrocarbon ratio of 125. The bimetallic catalysts exhibited lower apparent activation energies than the pure Pd catalyst with a minimum occurring at ~65 at% Pd. Corresponding to the decrease in activation energy, there is an increase in the turnover frequency over the monometallic catalyst at low temperatures, i.e. room temperature. At high temperatures (above 200°C) the pure Pd catalyst has the highest conversion (0.61) and the largest pre-exponential factor ($4.71x10^{16}$ cc/g Pd/s), while the 73 at% Pd catalyst has the lowest conversion (0.45) and the smallest pre-exponential factor ($5.59x10^6$ cc/g Pd/s). The bimetallic catalysts also showed a higher selectivity towards 1-butene than the monometallic catalyst; however, at lower conversions (less than 25%) the pure Pd catalyst was more selectivity towards 1-butene. AFM was used to characterize the surface morphology of each of the samples.

INTRODUCTION

Bimetallic catalysts have found use in a wide variety of industrial applications; however, there is still much that is unknown about the true nature of these catalysts. Most bimetallic catalysts are very sensitive to the preparation procedure and phenomenon such as alloying, surface segregation and particle composition distribution are factors that need to be clarified. The metals chosen for this study were palladium and gold. Some of the reasons for selecting these two metals are that they have been extensively studied, as witnessed by the number of reviews [1-4], their solid solutions are almost completely miscible [5] and the system should not exhibit surface segregation. This last point has been the source of some controversy. Some authors state that they were not able to detect any significant surface enrichment in the system [6-8], while others have stated that there is a significant enrichment in Pd or Au, depending on how the sample was treated, whether it was cleaned and if the bulk is Pd-rich or Au-rich [9-11]. A method of preparation currently being tested to facilitate in the study of bimetallic catalysts involves the use of microfabrication methodology [12], which has evolved from the microelectronics industry. The advantage of designing a catalyst via microfabrication techniques is that the particles will have a uniform size, shape, spacing and composition. This paper briefly describes the use of microfabrication technology to prepare supported Pd-Au/SiO_2 model catalysts.

The hydrogenation of 1,3-butadiene was used as a probe reaction to characterize the bimetallic catalyst's activity. Two examples exist in the literature of previous studies using the same probe reaction for Pd-Au supported catalysts. The work done by Joice, et al. [13] used pumice-supported Pd-Au bimetallic catalysts. Their results indicate that the activation energy remained constant as the composition was varied, except for a sharp increase of ~4.5 kcal/mol between 65-70% gold content. There was also a corresponding increase in the yield of 1-butene in the 60-70% gold range. The later work done by Miura, et al. [14] focused on an egg-shell type Pd-Au/Al_2O_3 catalyst. The main conclusion of their study was that a higher reaction temperature was required for the Pd-Au catalyst to achieve the same conversion as the monometallic Pd catalyst; however, a higher selectivity of 1-butene was achieved at high conversions.

Mat. Res. Soc. Symp. Proc. Vol. 368 © 1995 Materials Research Society

CATALYTIC PREPARATION

The silica support was obtained by oxidizing a clean, n-type, <100> silicon wafer. Oxidation conditions were 1200°C in a dry oxygen environment for twenty-four hours, which yielded a silica layer with a thickness of 1 μm.

The next step was to transfer the desired pattern for the catalytic particles layout to the silica layer. First the surface of the substrate needed to be primed to ensure good adhesion between it and the photoresist. Five milliliters of a solution composed of 50 vol% hexamethyldisilazane (HMDS) and 50 vol% acetone was applied to the surface of the wafer and spun at a high speed until dry. A positive photoresist (Shipley Microposit S1400-27 Photoresist) was then applied to the surface of the wafer and spun at a speed of 5000 rpm for 30 seconds, which resulted in a layer approximately 1.2 μm thick. The substrate was then "soft" baked at 90°C for 30 minutes to evaporate a portion of the solvents in the photoresist. The wafer was next brought into contact with a glass photomask, which had the pattern of the catalytic particles imprinted on it, using a contact mask aligner. The pattern was 4 μm transparent squares separated by 4 μm opaque spaces. The wafer was exposed to an ultraviolet light source (350 W mercury arc lamp, i-line, 365 nm) through the photomask and then immersed in a developing solution of one part water and one part Shipley Microposit Developer Concentrate. Positive photoresist depolymerizes when exposed to UV light; therefore, the resist underneath the transparent square areas washed away when immersed in the developer. The resist was then "hard" baked at 130°C for 30 minutes to remove any remaining solvents and completely harden the resist. Prior to metal deposition, the wafer was immersed in 5% buffered hydrofluoric acid for 30 seconds to etch part of the exposed silica and transfer the pattern for the catalytic particles to the silica layer.

The catalytic metal, Pd (99.997% pure, Johnson Matthey AESAR/ALFA), was deposited onto the wafer by vacuum thermal deposition. A deposition using 30.2 mg resulted in an estimated film thickness of approximately 200Å. Then the required film thickness of Au to yield the desired concentration was deposited using a vacuum electron-beam evaporator. Afterwards, the wafer was immersed in acetone to lift off the remaining photoresist and the metal on top of the resist. The metal that remained filled the squares previously opened in the silica layer.

ACTIVITY MEASUREMENTS

The hydrogenation of 1,3-butadiene was carried out in a 3/8" ID quartz tube at atmospheric pressure. The tube was heated using a flexible, electrical heating jacket (300 W/120 V, BriskHeat) and the temperature was kept constant by a temperature controller (Eurotherm Series 818).

The catalysts were reduced in a hydrogen flow of 20 cc/min for 6 h at 450°C. After the treatment, the temperature profile was programmed for a series of ramp and isothermal steps. The flow of reactants introduced into the reactor consisted of hydrogen (partial pressure of 500 torr), 1,3-butadiene (4 torr) and nitrogen to produce a constant flowrate of 110 cc/min. Ultra high purity grade nitrogen and hydrogen were supplied by Mittler Supply, Inc. and the 1,3-butadiene (99.5% purity) was supplied by Matheson Gas Products. The catalyst was exposed to hydrogen flow until the desired temperature was achieved to measure the activity. Then the flow was switched to the reaction mixture and a sample of the product stream was taken after 60 seconds. After sampling, the catalyst was again exposed to pure hydrogen flow to prevent deactivation until the next data point was to be taken. The reaction products were analyzed using a Hewlett Packard (Series 5710A) FID detector and separated by a 7 foot chromatographic column packed with 0.19% picric acid/graphpac packing (Alltech Associates, Inc.).

RESULTS AND DISCUSSION

Six samples were prepared for study. One sample was a monometallic gold catalyst that exhibited no catalytic activity at any temperature below 250°C. The preparation parameters of the other five samples are shown in Table I as well as the actual surface compositions as determined by

Table I: Compositional Data for the Five Microfabricated Pd-Au/SiO_2 Samples.

Pd Thickness (Å)[a]	Au Thickness (Å)[b]	Bulk Comp. (% Pd)[a]	Surface Comp. (% Pd)[c]
177	613	25	44
197	231	50	57
204	80	75	73
192	37	85	87
137	0	100	100

[a]Estimated [b]from calibration curve of E-beam evap. [c]from XPS

X-ray Photoelectron Spectroscopy (XPS). The goal was to hold the Pd thickness approximately constant, while adjusting the Au thickness to achieve the desired bulk composition. For the Pd-rich samples, there does not seem to be any evidence of significant surface segregation, but as the bulk becomes richer in Au, the surface exhibits a higher enrichment in Pd.

Figure 1 shows that the apparent activation energy for the reaction drops to a minimum around

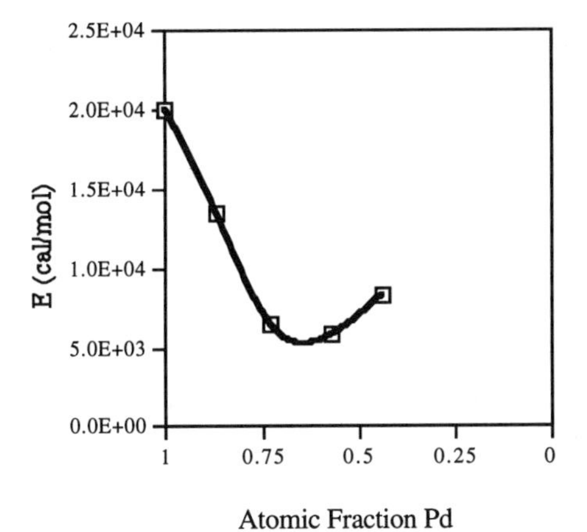

Fig. 1: Activation Energy as a Function of Composition for the Pd-Au Microfabricated Catalysts.

a Pd surface concentration of 65 at% and that the turnover frequency (TOF) at room temperature is greater for the bimetallic catalysts than the monometallic catalysts. Figure 2 displays the TOF for the five samples between 20°C and 60°C. The TOF was calculated using the number of Pd surface atoms based on the surface composition. It is of interest to note that measurements made through application of the de Haas-van Alphen effect [15], and later confirmed [16, 17], show that there are actually ~0.36 electrons in the s-band, or 0.36 holes in the Pd 4d-band, which roughly corresponds to the composition where the activation energy is at a minimum (35 at% Au). This would suggest that the filling of the Pd 4d-band leads to the decrease in activation energy from pure Pd. This effect of composition is also reflected in the TOF data. One would expect that at low enough temperatures, i.e. low conversions, the sample with the lowest activation energy would have the highest TOF, and this is true. At low temperatures the reaction is rate limiting and a relatively small percentage of reactant molecules possess enough energy to react. By comparing

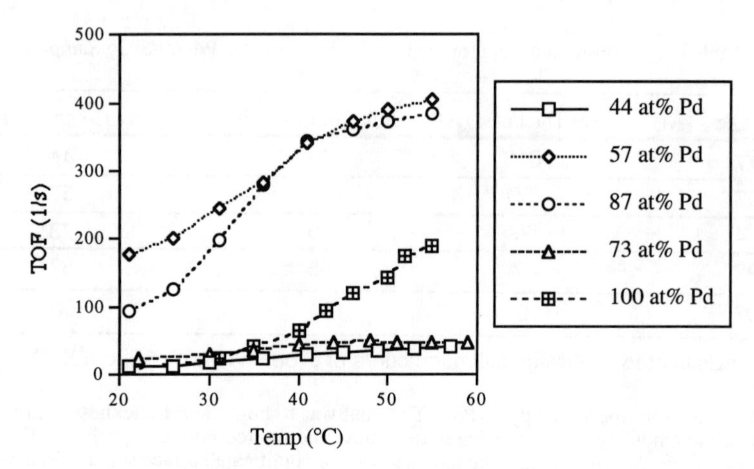

Fig. 2: Turnover Frequency as a Function of Temperature for the Pd-Au Microfabricated Catalysts.

the TOF at 21°C against the activation energy, the data shows a trend of a lowering of the activation energy increases the TOF, except for one sample. The 87 at% Pd sample has a much higher TOF than would be expected as compared to the other samples. The explanation for this is that at room temperature this sample already has a relatively high conversion (12 %) and the data taken may not reflect the regime where the reaction is rate controlled, i.e. the rate step and the mass transfer step are of comparable orders. All of the aforementioned evidence suggests that gold has an electronic effect on the Pd catalyst.

Along with the electronic effect from the presence of a second metal, there is also a structural one. There are many possible factors that can contribute to a structural effect, such as ordering, phase separation, surface morphology changes (surface roughness, particle size and particle shape), and simple dilution. By studying the catalytic activity at high temperatures, the overall rate is determined by the rate of mass transfer to and from the catalyst. The conversion for each sample as a function of temperature is displayed in Figure 3. The data suggests that the pure Pd sample has the largest number of active sites, which seems logical since it is undiluted, but the trend does not follow that the higher the Pd composition, the greater the conversion. One variable that seems to reflect the trend in conversions is the pre-exponential factor. Of the five samples, the pure Pd catalyst has the highest pre-exponential factor of 4.71×10^{16} cc/g Pd/s, while the 73 at% Pd catalyst, which displays the lowest conversion at 220°C, also has the lowest pre-exponential factor (5.59×10^{6} cc/g Pd/s). This general trend also holds for the other samples. Explaining why the 73 at% Pd catalyst seems to have the least number of active sites is difficult without information about the arrangement of the metals on the surface. Possible explanations include the presence of any long-range order at the surface, which might decrease the number of "ensembles" present, or the concentration distribution among particles.

One goal of this study was to determine if any relationship exists between the surface morphology and the performance of the catalyst. Using an atomic force microscope (AFM), it was possible to image the surface of the catalytic particles for each sample. The two most interesting samples are pure Pd and 73 at% Pd. The pure Pd sample has a columnar grain structure with an average particle size of 140 nm. The surface roughness for this catalyst (1.16 nm^2/nm^2 of geometric area) is higher than any other sample as determined using the Nanoscope II Surface Roughness Program. On the opposite end, the least overall active catalyst, 73 at% Pd, exhibits a completely different surface morphology. The particle size of this catalyst is the largest of any of the bimetallic catalysts at 160 nm and has a surface comprised of hills and valleys. The three remaining catalysts have very similar surface structures with no unique feature to distinguish between them. An important point demonstrated by the AFM images is that the pattern of the

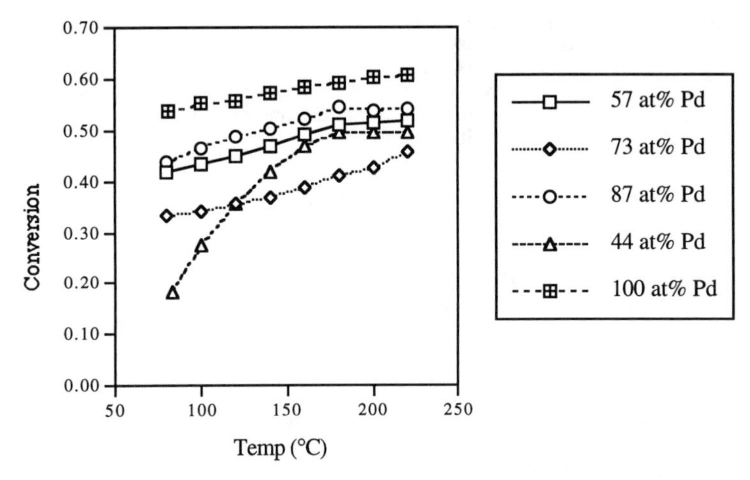

Fig. 3: Conversion as a Function of Temperature for the Pd-Au Microfabricated Catalysts.

microfabricated catalyst must be made smaller to extract the full advantage of fabricating a catalyst, because each particle size is not well defined. In essence the particles fabricated for this study are particles within a larger particle. The next step is to generate particles on the order of 100 nm by switching to electron-beam lithography (EBL) and allow us to better define the size of each particle.

Besides altering the overall rate of reaction and conversion, the presence of a second metal can also effect the product selectivity. As was mentioned earlier, other researchers had seen increases in 1-Butene selectivity with the addition of gold. This is true for our study also, but only in certain cases. A plot of 1-butene selectivity versus conversion (Figure 4) shows an increase in 1-butene yields, but only at higher conversions. As the conversion decreases below 0.25, the pure Pd catalyst actually has a higher preference towards 1-butene than does the bimetallic catalysts.

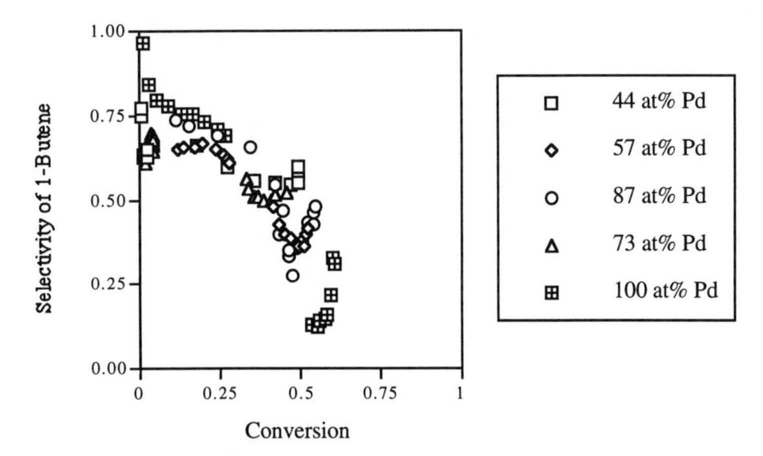

Fig. 4: Selectivity of 1-Butene as a Function of Conversion for the Pd-Au Microfabricated Catalysts.

Conversely, the Pd catalyst is more selective towards butane at higher conversions and less selective at lower conversions. The reaction scheme originally suggested by Phillipson, et al. [18] and later supported by Pradier, et al. [19] for the hydrogenation of 1,3-butadiene on a Pt catalyst states that butane is formed by the hydrogenation of 1-butene and not the 2-butenes. The data from our study indicates that this mechanism is also valid for our catalysts. The formation of 2-butenes is not significantly affected by the addition of gold.

CONCLUSIONS

The addition of a second metal (Au) has both an electronic effect and a structural effect on the Pd microfabricated catalyst for the hydrogenation of 1,3-butadiene. The electronic effect is a lowering of the apparent activation energy, which leads to an increase in the TOF at low temperatures, when the overall rate of reaction is controlled by kinetics. At high temperatures the Pd catalyst is most active, presumably due to it having the largest number of surface sites, while the 73 at% Pd catalyst has the lowest conversion. A larger pre-exponential factor seems to indicate a greater number of active surface sites are present. From the AFM images, parameters such as roughness and particle size may be important in relating the morphology to other factors, including the number of surface sites or the pre-exponential factor, but no single relationship was possible between two parameters. Finally, the presence of Au also has an affect on one of the reaction pathways (hydrogenation to 1-butene and further hydrogenation to butane), but not a significant one on the other (hydrogenation to the 2-butenes).

ACKNOWLEDGMENT

Financial support from an NSF grant (NSF CTS 92-15339) and the Graduate Assistance in Areas of National Need Program (GAANNP) through Notre Dame's Center for Bioengineering and Pollution Control is gratefully acknowledged. Furthermore, we would like to thank the National Nanofabrication Facility at Cornell for technical assistance and for supplying the photomasks.

REFERENCES

1. D. D. Eley, J. Res. Inst. Catal., Hokkaido Univ., **16** (1), 101 (1968).
2. E. G. Allison and G. C. Bond, Catal. Rev., **7** (2), 233 (1972).
3. R. L. Moss and L. Whalley, Adv. Catal., **22**, 115 (1972).
4. J. Schwank, Gold Bull., **18** (1), 2 (1985).
5. H. Okamoto and T. B. Massalski, Binary Alloy Phase Diagrams, Vol. 1, 2nd ed., editor-in-chief T. B. Massalski (ASM International, 1990), p. 409.
6. B. J. Wood and H. Wise, Surf. Sci., **52**, 151-160 (1975).
7. Y. L. Lam and M. Boudart, J. Catal., **50**, 530-540 (1977).
8. D. F. Ollis, J. Catal., **59**, 430-433 (1979).
9. G. Maire, L. Hilaire, F. G. Gault and A. O'Cinneide, J. Catal., **44**, 293-299 (1976).
10. A. Jablonski, S. H. Overbury and G. A. Somorjai, Surf. Sci., **65**, 578-592 (1977).
11. D. D. Eley and P. B. Moore, J. C. S. Faraday I, **76**, 1388-1390 (1980).
12. A. C. Krauth, K. H. Lee, G. H. Bernstein and E. E. Wolf, Catal. Lett., **27**, 43-51 (1994).
13. B.J. Joice, J.J. Rooney, P.B. Wells and G.R. Wilson, Disc. Faraday Soc., **41**, 223 (1966).
14. H. Miura, M. Terasaka, K. Oki and T. Matsuda, New Frontiers in Catalysis: Proceedings of the 10th International Congress on Catalysis, edited by L. Guczi et al., (Elsevier Science Publishers, New York, 1993), p. 2379.
15. J. J. Vuillemin and M. G. Priestley, Phys. Rev. Lett., **14**, 307 (1965).
16. F. M. Meuller and M. G. Priestley, Phys. Rev., **148**, 638 (1966).
17. H. Kimura, A. Katsuki and M. Shimizu, J. Phys. Soc. Japan, **21** 307 (1966).
18. J. J. Phillipson, P. B. Wells and G. R. Wilson, J. Chem. Soc., 1351 (1969).
19. C. M. Pradier, E. Margot, Y. Berthier and J. Oudar, Appl. Catal., **31**, 243 (1987).

PART IV

Molecular and Cluster Derived Catalytic Materials

FINE CHEMICALS SYNTHESIS BY HETEROGENEOUS ASYMMETRIC CATALYSIS

Kam T. Wan and Mark E. Davis

Division of Chemistry and Chemical Engineering
California Institute of Technology
Pasadena, CA 91125

ABSTRACT

The application of heterogeneous asymmetric catalysis to fine chemicals synthesis is discussed. The asymmetric synthesis of naproxen, a non-steroidal anti-inflammatory drug, using a novel, heterogeneous, chiral catalyst is used as an example to illustrate the feasibility to industrial application.

INTRODUCTION

The combination of environmental and health related concerns are causing governmental regulatory boards to pay more attention to the importance of the optical purity of chiral drugs [1]. The Food and Drug Administration recently published a new policy statement for the development of drugs comprised of stereoisomers [2]. It states that adverse reactions could be attributed to one enantiomer and recommends the study of individual isomers if toxic effects are found in the racemate. It also states that the racemic mixture is not expected to be the optimal ratio of isomers for therapeutic value. Also, the U.S. Patent Office is encouraging the development of single enantiomeric drugs. A new patent can be granted to inventors who first isolate the pure enantiomer and show its benefits. Such incentives help convince pharmaceutical companies that single-enantiomer drugs can be profitable. In fact, the most important trend in the chiral drug industry is the steady, rising flow of single-isomer forms of chiral drugs onto US and world markets [3] which creates an increasing demand for enantiomeric intermediates and bulk active compounds as well as enantioselective technology. At present, over 20% of the top selling drugs are single enantiomers. In 1993, a worldwide market for chiral drugs

Mat. Res. Soc. Symp. Proc. Vol. 368

totalled over $30 billion. For instance, in the case of a "mega-block-buster" prescription drug; e.g., for AIDS treatment, production may reach commodity-chemical scale, where the cost of production can be a crucial factor in commercialization [4]. In the future, a similar trend may also occur with other best-selling "block-buster" prescription drugs. Hence, a need continues for successfully development of large scale, economical production of pure enantiomers.

It should be noted that the large and rapidly growing pharmaceutical industry produces the largest amount of by-products per kilogram of desired product. This is due to the following reasons: (i) typical processes involve multi-step synthesis with low overall yields; (ii) the complexity of the final product; (iii) very high purity required and difficult separation of by-products, products, and catalyst.

Two strategies are available for the production of single enantiomers. The first, and more elegant strategy, involves the direct manufacture of the pure enantiomer through asymmetric synthesis. The second is to produce a racemic mixture through traditional routes. The desired product must then be separated from the racemic mixture using techniques that include crystallization, chromatography, and kinetic resolution. These methods are often labor intensive and expensive. Despite this, these procedures are used on a regular basis because the desired enantiomer is so valuable. The undesired enantiomer will then be recycled, discarded, or destroyed. The large amounts of by-products (or so-called "biological pollutant") and the high volumes of solvents used in the resolution make these systems targets for more economical and environmentally benign replacements. A recent breakthrough in chiral separation techniques is the emergence of liquid chromatographic methods which has risen from relative disfavor to become highly promising for numerous separations. New developments include simulated moving bed chromatography and the countercurrent extraction technique [3]. However, in terms of metric-ton-scale production, direct manufacture of the pure enantiomer through asymmetric catalysis is always preferred.

Homogeneous catalysis, based on soluble metal complexes, is now used extensively in the production of fine chemicals [5,6]. The extreme selectivity of homogeneous catalysts underlies their extensive use. This is especially true for the synthesis of chiral compounds [6-8]. The highly organized coordination sphere found in soluble organometallic complexes gives homogeneous, asymmetric catalysis numerous advantages over heterogeneous catalysis. Both the activity and the enantioselectivity of homogeneous catalysts are superior to heterogeneous catalysts. However, one of the shortcomings of homogeneous catalysis is the need for separating the catalytic species from reaction products. Depending on the system: (i) the catalytic species can be very expensive [9,10], e.g., rhodium and phosphine ligand; (ii) the catalytic species may not be robust enough for a particular separation and/or the separation may not be as complete or cost-effective as desired; and (iii) contamination of pharmaceutical products by the catalyt must be minimized. Therefore, there continues to be a need for the immobilization of the catalytically active organometallic species in order to enhance the economic viability of the production process by eliminating separation and/or catalyst recovery steps and to minimize the possible toxicity hazards by contamination of trace transition metal complexes.

Some progress has been made in the development of asymmetric heterogeneous catalysts. One approach to the preparation of heterogeneous catalysts involves attachment of a homogeneous catalyst onto a suitable support. In this way, the catalyst can in principle acquire the property of insolubility and at the same time, retain the same reactivity and selectivity exhibited by its homogeneous analogue. In attempts to do this, homogeneous catalysts have been attached to a variety of supports including cross-linked polymers [11-16]. Cross-linked polystyrenes are one of the most widely used polymer supports for the attachment of phosphine ligands through chemical modifications. However, results with polystyrenes containing optically active phosphine ligands aimed at effecting asymmetric syntheses have been disappointing [17]. In polar solutions of the substrates, the polystyrene beads collapse preventing entry of the substrate to the catalytic site. Problems of polymer shrinking in polar solvents and leaching of catalytic species from the support have always been obstacles [17]. Additionally, homogeneous catalysts covalently bound to the support are in general much less active than their unbound analogues. This has been attributed to the restricted degree of motional freedom of the metal complex on the support surface, leading to undesirable interactions or changes in re-organization energy, thus affecting both the activity and the selectivity [11,18-19].

Another approach to heterogenizing homogeneous catalysts involves modification of the catalyst with suitable hydrophilic groups, e.g., sulfonate or quaternary ammonium, that make it possible to transfer the catalyst into the aqueous phase, and thus leaving the product in a second phase, i.e., a two-phase system [20-22]. Product separation is simple for two-phase systems incorporating water-soluble catalysts. The hydrophilic catalyst, which is insoluble in the organic phase is molecularly well-defined, like conventional homogeneous catalysts. These catalysts accomplish the asymmetric catalytic reaction in the aqueous phase or at the phase boundary, and are removed from the desired product at the end of the reaction by a simple phase separation. Thermal separation processes, which can have detrimental effects on the active lifetime of the catalyst, are thereby avoided. Because of the high polarity of the water-soluble catalyst and its insolubility in the organic phase, leaching of the water-soluble catalyst into the organic phase is often below the detection limit. Unfortunately, the reaction rates achievable with two-phase systems are strongly dependent on the polarity of the substrate and/or the limiting interfacial area between the two phases. In general, the reaction activities are orders of magnitude less active than the homogeneous catalysts [23]. However, it is the poor enantioselectivity that really hinders the application of two-phase systems in asymmetric catalysis. Usually, poor enantioselectivities are found in aqueous media [24-29]. Hence, new genuine heterogeneous, asymmetric catalysts has yet to be developed for practical asymmetric synthesis of optically pure compounds.

Hydrogenation is arguably the most important synthetic application of asymmetric catalysis because of its potential to produce a wide variety of chiral functional groups. It also leaves no reagent to recover and no boron or aluminum wastes to dispose of as is the case of reaction with $NaBH_4$ or $LiAlH_4$. It is therefore not surprising that the largest number of catalytic systems has been described for this reaction type. However, a closer inspection of the literature reveals that there are only two families of synthetically useful heterogeneous, asymmetric catalytic systems: (i) a nickel catalyst modified with tartrate/NaBr [30] and (ii) Pt(Pd) catalysts modified with cinchona alkaloids [31-32]. Though many functionalized olefins can now be hydrogenated with optical yields over 95% using homogeneous catalysts [33], the performance of many heterogeneous systems are not that impressive. Instead of randomly searching for new heterogeneous asymmetric hydrogenation catalysts, heterogenization of known homogeneous systems still appears to be a better and more effective approach to the solution of this problem. This is especially true for asymmetric synthesis of chiral pharmaceutical products which demands extremely enantioselective catalysts with very high e.e.'s. Other than using the traditional immobilization techniques our strategy is to preserve a microscopically, homogeneous environment within a heterogeneous system, i.e., the catalytic site is an organometallic complex and the product and catalyst are phase separated. In this way, we could hopefully retain the selectivity/activity upon heterogenization of the homogeneous catalyst. The following is an illustration of our design and engineering of a new, heterogeneous chiral catalyst.

NEW HETEROGENEOUS ASYMMETRIC CATALYST

Based on the concept of supported-aqueous-phase-catalysis (SAPC) [34], we have developed a new heterogeneous asymmetric hydrogenation catalyst, designed specially to convert liquid-phase reactants [35]. This solid catalyst consists of a thin film that resides on a high-surface-area hydrophilic support (controlled-pore-glass), and is composed of water-soluble, chiral organometallic complex and ethylene glycol as the hydrophilic phase (Figure 1). Reaction of liquid-phase organic reactants takes place at the film-bulk organic interface [35]. This new heterogeneous catalyst differs from its SAP predecessor in that no water is used, and is therefore not considered as a supported-aqueous-phase catalyst [19].

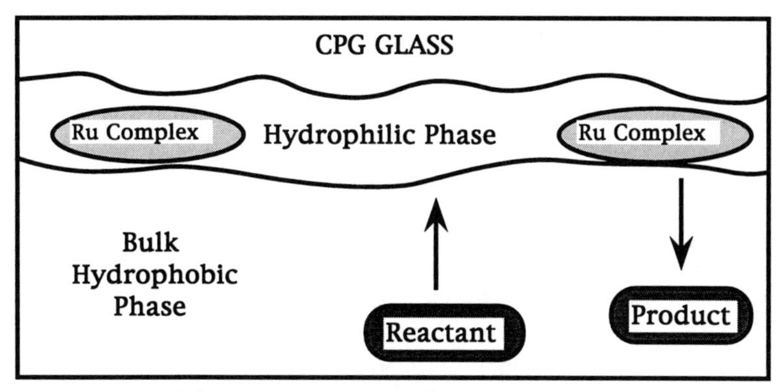

Figure 1 -- Schematic diagram of the heterogeneous catalyst

The efficacy of this new catalyst was proven in its application to the asymmetric synthesis of naproxen, a commercially important, non-steroidal, anti-inflammatory drug [35]. Annual sale of naproxen totalled over $ 750 million in 1992. With the approval of its over-the-counter version (Aleve) by Food and Drug Administration in early 1994, the sale of naproxen is expected to increase dramatically. Procter & Gamble is projecting first year retail sales for Aleve of $200 million. To keep up with the production in an economical way, new and more efficient chiral synthesis is highly desirable. In the Monsanto naproxen process [36], naproxen is prepared through homogeneous asymmetric hydrogenation of dehydronaproxen, [2-(6'-

methoxy-2'-naphthyl)-acrylic acid], as shown in Figure 2; which is in turn prepared by electrocarboxylation of 2-acetyl-6-methoxynaphthalene. One advantage of this process is that the key "process reagents", the electron and hydrogen, are environmentally friendly. However, it would be even more desirable to convert the hydrogenation step into a heterogeneous reaction, thus making the whole process clean and more economical.

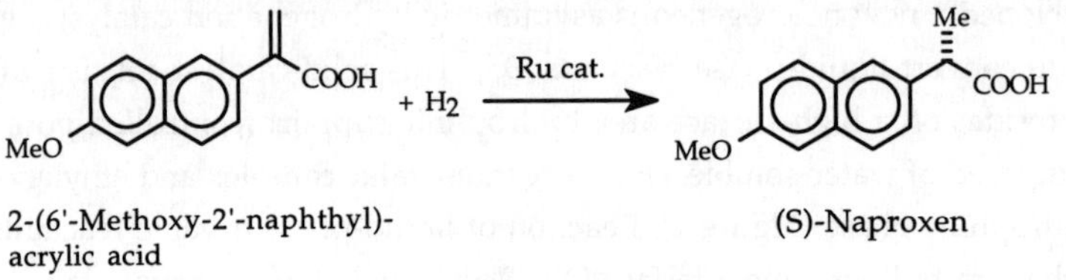

Ru cat.

MeO COOH + H_2 → MeO COOH Me

2-(6'-Methoxy-2'-naphthyl)-acrylic acid

(S)-Naproxen

Figure 2 -- Asymmetric synthesis of naproxen

We began our work on the sulfonation of 2,2'-bis(diphenylphosphino)-1,1'-binaphthyl (BINAP) with sulfur trioxide in concentrated sulfuric acid [37]. Instead of placing two sulfonate groups onto the binaphthyl rings (which most likely would have provided insufficient hydrophilicity), four sulfonate groups were added to the four phenyl rings in the meta positions. As a result, the highly hydrophobic BINAP molecule was converted into a hydrophilic ligand with a solubility of ~ 1 g/cm^3 in water. Extreme solubility in water is required in order to prevent the metal complex from leaching out into the non-polar, bulk organic phase. Our previous work on sulfonation of triphenylphosphine has shown that sufficient solubility could be achieved with one sulfonate group for every phenyl ring [18]. A corresponding rhodium sulfonated BINAP complex was prepared, and asymmetric hydrogenations of amino-acid precursors were performed in aqueous media [37]. Surprisingly, enantioselectivities in neat water were found to be as high as that of its parent Rh-BINAP system observed in non-aqueous solvents. More importantly, this rhodium catalyst was the first ever reported water-soluble asymmetric hydrogenation catalyst with enhanced enantioselectivity in neat water (Table I). In contrast to its parent Rh-BINAP system, enantioselectivity of the present system is independent of the substrate concentration. Formation of 1:2 catalyst: substrate adduct in the un-modified parent system has been suggested to be responsible for the decline in enantioselectivity [38]. The presence of four bulky sulfonate groups in the present system might have prevented the formation of any 1:2 adduct. These findings suggest the possible scaling-up to practical level, and they also prove that homogeneous asymmetric catalysis is indeed feasible in water - an environmentally benign solvent.

Table 1 Reduction of 2-acetamidoacrylic acids with rhodium catalyst†

Substrate	Solvent	e.e. (%)‡
I	H_2O	70.4
I	MeOH	58.0
I	EtOH	58.6
II	H_2O	69.0
II	MeOH	47.8
II	EtOH	56.0

†: Reacted at room temperature and one atm. of H_2
‡: Determined at 100% conversion
Catalyst: $[Rh(BINAP\text{-}4SO_3Na)(solvent)_2]ClO_4$
Substrate I: 2-Acetamidoacrylic acid
Substrate II: Methyl-2-acetamidoacrylate

While the BINAP ligand has been proven useful for a wide range of asymmetric reactions, the variety of the Rh (I)-catalyzed reactions is rather limited. Corresponding ruthenium complexes are more versatile, and catalyze a broader range of reactions. In an attempt to combine the utility of sulfonated BINAP with the reaction chemistry of ruthenium, we prepared a ruthenium sulfonated BINAP complex [39]. It indeed revealed superior enantioselectivity and stability to the corresponding rhodium complex thus making it a promising candidate for heterogenization onto a solid support. However, in the ruthenium system, the enantioselectivity dependence on solvent was found to be substrate dependent. For the case of dehydronaproxen, the e.e. observed in 1:1 v/v mixture of water/methanol was almost 10% lower than that in neat methanol. Similar results were also found with the hydrated supported-aqueous-phase catalyst (Table II) [19]. Nevertheless, the SAPC is only seven times less active than its homogeneous analogue, but 50 times more active than an ethyl acetate/water two-phase reaction mixture in the asymmetric hydrogenation of dehydronaproxen. This is due to the much larger interfacial area resulting from the controlled-pore-glass (CPG) support. The similar enantioselectivities found in hydrated SAP and two-phase systems (Table II) suggest that the active complex in the thin film has close resemblance to the one dissolved homogeneously in two-phase configuration. In fact, the upper limit in e.e. (75%) of SAPC is bounded by the intrinsic enantioselectivity limit of the homogeneous catalyst in water, not by any constraint from the SAP configuration [19].

Table II Reduction of dehydronaproxen in the presence of water with Ru catalyst†

System	Solvent	e.e. (%)‡
Homogeneous	*MeOH*	*86.0*
Homogeneous	1:1 MeOH/H_2O	78.9
Two-Phase	1:1 AcOEt/H_2O	75.0
Hydrated SAP	AcOEt	70.0

†: Reacted at room temperature and 1,350 psig of H_2
‡: Determined at 100% conversion

Therefore, high enantioselectivity is achievable with the SAP configuration, providing the solvent effect is solved. The origin of the solvent effect was found to be related to the aquation of Ru-Cl bond [40]. The Ru-Cl bond is crucial in providing the necessary molecular geometry for a more enantioselective binding with the substrate. Upon replacement of the water phase by anhydrous ethylene glycol, the ruthenium complex was solubilized without the cleavage of the Ru-Cl bond. The same high e.e. (Table III) as obtained in neat methanol is observed in 1:1 ethylene-glycol/methanol solvent mixture. In all cases, lowering of reaction temperature to 3-4°C resulted in even higher enantioselectivities. Since ethylene glycol is immiscible with most common, non-polar organic solvents, it was used as a substitute for water in the SAP system to create a new heterogeneous, chiral catalyst. The heterogeneous, asymmetric hydrogenation of dehydronaproxen was accomplished at 96% e.e. (Table III) and at one-third of the rate observed from the homogeneous catalysis (40.7 hr^{-1} vs. 131.0 hr^{-1} at room temperature) [35]. After the reaction, the catalyst could be recovered by simple filtration, leaving the product in the bulk organic phase.

Table III Reduction of dehydronaproxen in the presence of ethylene glycol EG with ruthenium catalyst†

System	Solvent	e.e. (%)‡
Homogeneous	*MeOH*	*88.2*
Homogeneous	*MeOH*	*96.1*[a]
Homogeneous	1:1 MeOH/EG	89.1
Heterogeneous	1:1 $CHCl_3$/Cyclohexane	88.4
Heterogeneous	1:1 $CHCl_3$/Cyclohexane	95.7[b]

†: Reacted at room temperature and 1,350 psig of dihydrogen
‡: Determined at 100% conversion
a: Reaction temperature = 4°C
b: Reaction temperature = 3°C

With the proper choice of the hydrophilic and hydrophobic phases, we were able to sustain a "homogeneous-like" environment within a macroscopically heterogeneous system. This unique configuration contributes to the extreme enantioselectivity of this heterogeneous catalyst. No metal leaching was found at a detection limit of 32 ppb and recycling (7 times) of the used catalyst is possible without any loss in enantioselectivity. A self-assembly test was employed to assure the long-term stability of this heterogeneous catalyst [35, 40]. The ruthenium complex dissolved in ethylene glycol was mixed with substrate in 1:1 chloroform/cyclohexane. Blank CPG support was added, then the reactor pressurized with H_2 and stirred for two hours. Complete conversion was observed, while less than 2% conversion was found in the control experiment for which no CPG was added. These results indicate that, under the reaction conditions, the individual components of the heterogeneous system self-assemble into a more thermodynamically stable supported-phase catalyst configuration. Therefore, the reverse, i.e., the separation of the hydrophilic catalytic phase from the support, is unlikely to happen under reaction conditions. The use of CPG support in this work is not mandatory; other high-surface-area hydrophilic supports can also be used, as was shown in previous work [18]. Additionally, the solid support can be reused by washing with either water or methanol, and so the waste from this process is minimal.

CONCLUSIONS

Although only the application to asymmetric hydrogenation is demonstrated, we strongly believe that the concept of supported phase catalysis can be applicable to other areas of asymmetric catalysis. As the phase separation of the catalyst from the reactants/products is based on differential hydrophilicity between the organometallic complex and the reactant/product, reactions with small changes in the hydrophilicity between the reactant and product are the most promising. Possible applications of this type of catalysis include epoxidation, hydroformylation, and probably reductions of ketones and imines. Dihydroxylation of small molecules may result in significant increases in hydrophilicity of the molecule, and thus causing catalyst leaching. However, this concept may still be applicable in the dihydroxylation of large, hydrophobic molecules.

ACKNOWLEDGEMENT

We thank Dr. Forster (Monsanto Company) for the generous gift of 2-(6'-methoxy-2'-naphthyl)-acrylic acid. This work was supported by the US NSF Alan T. Waterman Award (M.E.D.).

REFERENCES

[1] S. C. Stinson, Chem. & Eng. News, May 16,10 (1994).

[2] Food and Drug Administration, Chirality **2**, 338 (1992).

[3] S. C. Stinson, Chem. & Eng. News, September 19, 38 (1994).

[4] S. C. Stinson, Chem. & Eng. News, May 16, 6 (1994).

[5] G. W. Parshall, Homogeneous Catalysis (Wiley: New York, 1980).

[6] G. W. Parshall, and W. A. Nugent, Chemtech, March, 184 (1988).

[7] G. W. Parshall, and W. A. Nugent, Chemtech, May, 314 (1988).

[8] G. W. Parshall, and W. A. Nugent, Chemtech, June, 376 (1988).

[9] H. B. Kagan, Asymmetric Synthesis, J. D. Morrison, Ed. (Academic Press, Orlando, 1985), **5**, p.1.

[10] A. Karim, A. Mortreaux, and F. Petit, J. Organomet. Chem. **312**,375 (1986).

[11] F. R. Hartley, Supported Metal Complexes, R. Ugo and B. R. James, Eds. (Reidel, New York, 1985).

[12] C. U. Pittman, and G. O. Evans, Chemtech, 560 (1973).
[13] J. Manassen, Catalysis Progress in Research, F. Basolo and R. L. Burwell, Jr., Eds. (Plenum Press, New York, 1973), p.117.
[14] J. C. Bailer, Catal. Rev. **10**, 17 (1974).
[15] C. Lezenoff, Chem. Soc. Rev. **3**, 65 (1974).
[16] J. Manassen, Catalysis, Heterogeneous and Homogeneous, B. Delmon and G. James, Eds. (American Elsevier, New York, 1975), p.293.
[17] W. Dumont, J. C. Poulin, T. P. Dang, and H. B. Kagan, J. Am. Chem. Soc. **95**, 8295 (1973).
[18] J. P. Arhancet, M. E. Davis, J. S. Merola, and B. E. Hanson, J. Catal. **121**, 327 (1990).
[19] K. T. Wan and M. E. Davis, J. Catal. **148**, 1 (1994).
[20] F. Joo, and Z. Toth, J. Mol. Catal. **8**, 369 (1980).
[21] D. Sinou, Bull. Soc. Chim. Fr. 480 (1987).
[22] P. Kalck, and F. Monteil, Adv. Organomet. Chem. **34**, 219 (1992).
[23] W. A. Herrmann, C. W. Kohlpaintner, H. Bahrmann, and W. Konkol, J. Mol. Catal. **73**, 191 (1992).
[24] I. Toth, B. E. Hanson, and M. E. Davis, Tetrahedron: Asymmetry **1**, 913 (1990).
[25] Y. Amrani, and D. Sinou, J. Mol. Catal. **24**, 231 (1984).
[26] D. Sinou, and Y. Amrani, J. Mol. Catal. **36**, 319 (1986).
[27] R. Benhanza, Y. Amrani, and D. Sinou, J. Organomet. Chem. **288**, C37 (1985).
[28] L. Lecomte, and D. Sinou, J. Organomet. Chem. **370**, 277 (1988).
[29] G. Oehme, E. Paetzold, and R. Selke, J. Mol. Catal. **71**, L1 (1992).
[30] A. Tai, and T. Harada, Taylored Metal Catalysis, Y. Iwasawa, Ed. (Reidel, Dordrecht, 1986), p.265.
[31] H. U. Blaser, H. P. Jalett, and J. Wiehl, J. Mol. Catal. **68**, 215 (1991).
[32] H. U. Blaser, S. K. Boyer, and U. Pittelkow, Tetrahedron: Asymmetry **2**, 721 (1991).
[33] R. Noyori, and M. Kitamura, Modern Synthetic Methods, R. Scheffold, Ed. (Springer-Verlag, Berlin, 1989), p.115.
[34] J. P. Arhancet, M. E. Davis, J. S. Merola, and B. E. Hanson, Nature **339**, 454 (1989).
[35] K. T. Wan and M. E. Davis, Nature **370**, 449 (1994).
[36] A. S. C. Chan, Chemtech, March, 46 (1993).
[37] K. T. Wan and M. E. Davis, J. Chem. Soc., Chem. Commun., 1262 (1993).
[38] A. Miyashita, H. Takaya, T. Souchi, and R. Noyori, Tetrahedron **40**, 1245 (1984).
[39] K. T. Wan and M. E. Davis, Tetrahedron: Asymmetry **4**, 2461 (1993).
[40] K. T. Wan and M. E. Davis, accepted by J. Catal. (1994).

SURFACE ASSEMBLED HYBRID POLYMERS

PHILIP L. ROSE*, LARRY MANZIEK* AND MICHAEL K. GALLAGHER*
*Rohm and Haas Company, Research Laboratory, 727 Norristown Road, Spring House, PA, 19477

ABSTRACT

Novel copolymer composite resins have been synthesized using the surface assembled hybrid technology. Monodisperse polystyrene/DVB microspheres with diameters of 5.3 and 7.5 μm and polydisperse carbonaceous and acrylic microspheres with diameters ranging from 5-30 μm have been assembled onto the surface of styrene and acrylonitrile based gel and macroreticular resins.

Carbonaceous adsorbents with holes of controlled diameter on their surfaces have been prepared using the surface assembling technique. Core-shell carbonaceous adsorbents with divergent surface chemistries have been prepared using this technique

Novel catalytic materials have been made and are envisioned using the assembling technique to form short-diffusion path reactive systems. The range of surface functional groups include: strong acid, strong base, palladium, silver, gold and reductive amine-borane.

INTRODUCTION

The ability to create materials with two or more different functionalities and control their location in a single particle is of great interest for separations and catalytic applications. Surface assembled hybrid polymers offer a synthetic route to such novel polymeric, carbonaceous and inorganic materials. These materials can be prepared easily via suspension polymerization and then post-functionalized using conventional chemical techniques[1-3]. This approach allows the incorporation and controlled use of expensive materials, since they can be confined to the surface of a support. The structure of surface assembled hybrid materials permits a short diffusion path to the active site and reduced pressure drop in packed bed applications versus conventional resins. Moreover, with the appropriate pyrolysis conditions, these cross-linked polymer particles can be converted into high surface area carbonaceous materials with controlled dimensional holes on the surface.

Background

Even though surface assembled hybrid polymers are not new, as evidenced by earlier work from Dionex[4-5], Dow[5-6], Rohm and Haas[7-10], Yokogawa Hokushin Electric Corporation[11], we believe that our current method is more general and overcomes many of the earlier synthetic limitations. Two previous methods relied on either "irreversible" ion salt pair formation or hydrophobic interaction. While the latter approach, developed at Rohm and Haas, lends itself to the assembly of a wider array of materials than either the Dow or Dionex approach, it still suffers from sloughing of the latex particles. In addition, the pellicular composite resins are restricted largely to use in systems where the latex particles are insoluble in the eluent phase. An alternate approach was developed at Yokogawa Hokushin Electric Corporation, where in a

Mat. Res. Soc. Symp. Proc. Vol. 368

multi-step process, a binder system is used to "glue" small particles to a support surface. However, it is not clear to what degree the binder fouls the supported particles.

The technology described in this paper can be applied to virtually any combination of support and supported material. The composite is obtained in one step. The resulting material is physically robust and the functional groups and active sites contained in the supported material are readily accessible. The technology of surface assembled hybrid polymers takes advantage of the suspension polymerization process to place organic, carbonaceous, and inorganic particles primarily on a resin's outer surface.

Polymeric Composites

We have found that suspension polymerization can be adapted to create composite structures where, for the most part, the polymer, carbonaceous or inorganic microparticle ends up on the surface of the new polymer macroparticle. Key to obtaining the desired product is first forming a good dispersion of the microparticle in the organic phase. Thus for the resin sample shown in Figure 1, the polystyrene/divinylbenzene (DVB) microspheres, 7.5 µm diameter, are mixed with a porogen. Acrylonitrile, DVB and an initiator, are then added to the stirred solution. Once the microparticles are well dispersed in the organic phase, the entire mixture is added to an appropriate aqueous phase. The organic phase containing the suspended microparticles is broken up into droplets. Each suspended droplet contains monomer, cross linker, porogen and microparticles. Polymerization results in the entrapment of the microparticles primarily on the surface of the larger resin particle.

Figure 1. SEM of the outer surface of a composite resin resulting from the suspension polymerization of acrylonitrile in the presence of 7.5 µm diameter polystyrene/DVB microparticles.

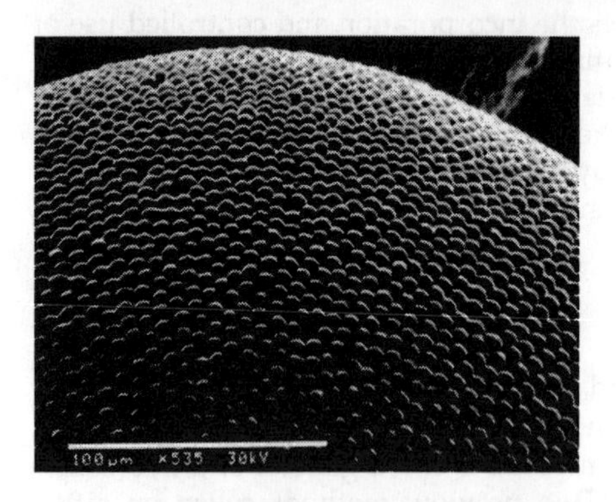

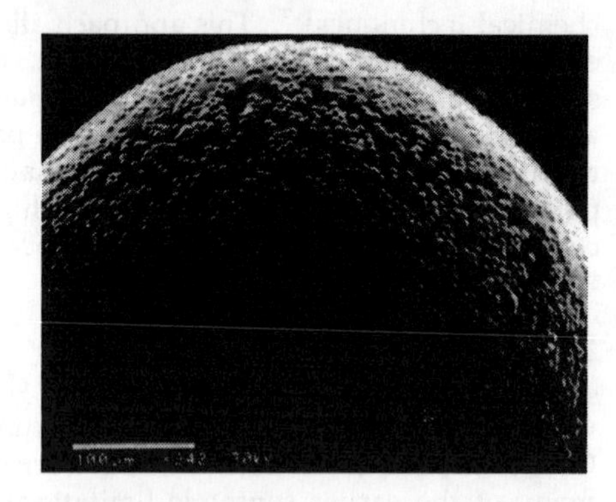

The resins in both Figure 1a and 1b are from the same suspension polymerization batch. The variation in the concentration of microparticles on the surface is related to the initial concentration of these microparticles in the individual monomer droplets once they have been dispersed in the aqueous suspension medium and appears to be a random event. Cross sections of composite resin beads reveal very few microspheres in the interior of the resin particle (see Figure 2). It is estimated that greater than 90% of the microspheres are on the composite resin's outer surface.

We have found that there are a number of factors that can effect the surface assembly process. For example, reducing the particle size of a microparticle, we observed a more ordered array of particles on the surface of the polyacrylonitrile particle (Figure 3.). The smaller particle size, 5.3 versus 7.5 μm diameter, may be responsible for the enhancement of hydrophobic interactions. If the aggregation of the 5.3 μm microbeads results from the minimization of non-polar/polar interactions, then we reasoned that decreasing the polarity of the bulk organic phase should disperse the microparticles on the resin surface.

Figure 2. SEM cross sections of the composite resin in Figure 1. The a) edge and b) interior of the composite resin.

a)

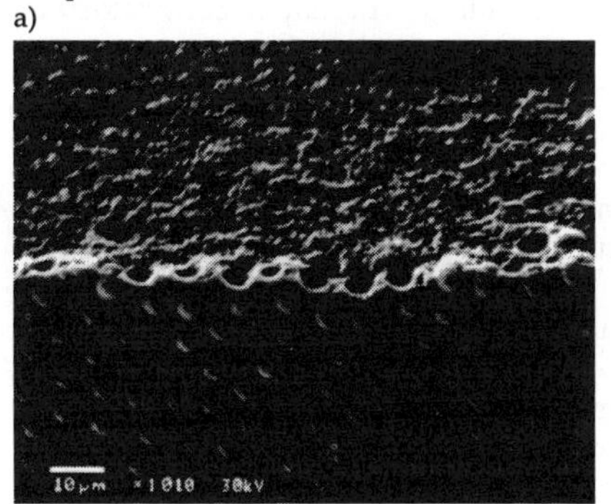

b)

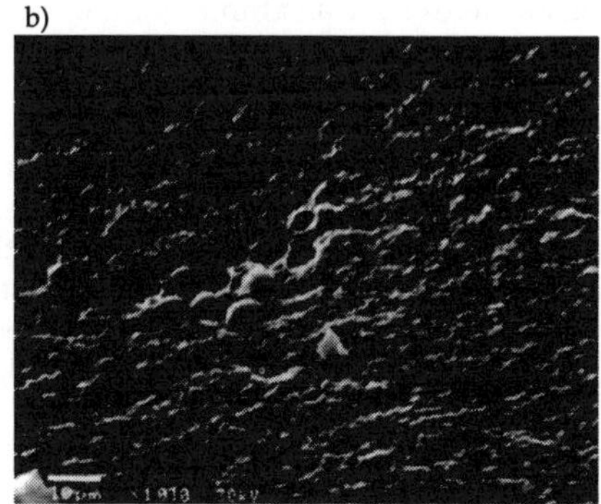

We repeated the experiment used to make the composite resins in Figure 3, substituting the ketone containing porogen for a less polar aliphatic hydrocarbon. After the polymerization we found that the styrenic spheres no longer formed ordered arrays on the surface but were randomly dispersed on the surface of the macroparticle. Reducing the polarity of the porogen increased the number of particles in the interior of composite resin, however, the surface still contained the majority of microparticles, irrespective of porogen and microparticle diameter.

Figure 3. SEM of the surface of a composite resin detailing the closepack ordered arrays of microparticles, monodispersed 5.3 μm diameter polystyrene/DVB.

a)

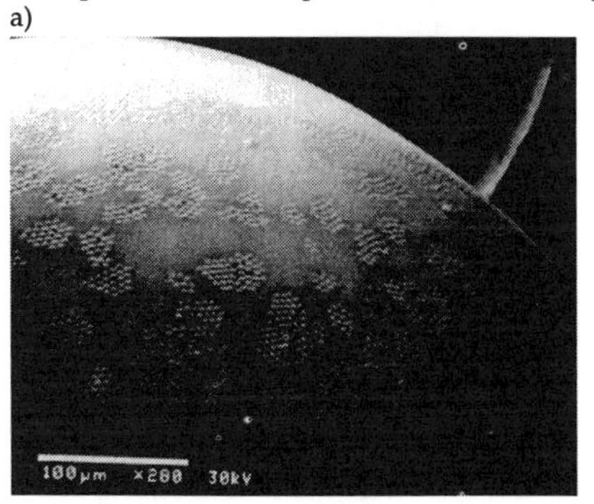

b)

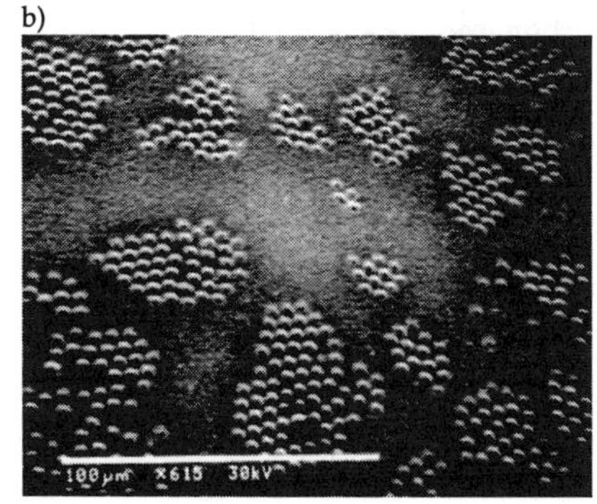

There are several potential mechanisms responsible for the microspheres ending up on the surface of the composite resin. For example, the polymerization process "squeeze" the microspheres out of the organic phase, or there may be an attraction between the microspheres and the aqueous phase or else between the microspheres and the stabilizing system that causes the microspheres to migrate to the resin surface. We believe that a combination of centrifugal forces generated as the monomer droplets tumble in the agitated suspension system and the relative chemistries of the organic phase, aqueous phase and the surface chemistry of the microsphere that establish the ultimate position of the microsphere. The motion of the monomer droplets imparts a radial acceleration on the microspheres forcing them to the surface of the droplet. If the microspheres have an affinity for the organic phase then they will remain constrained to the bead surface.

The ability to create unique hybrid resins with varying compositions opens up many potential applications. For example, a bulk polyacrylonitrile resin is different chemically than the styrenic microspheres assembled on its surface. The microspheres can be functionalized without affecting the nature of the support. Likewise, different reactions can be carried out on the acrylonitrile resin (formation of weak acid and weak base moieties) and on the microspheres (sulfonation, etc.) leading to hybrid resins. The gel phase of either resin can be manipulated to control the kinetics of each functional environment. Different acrylic- or styrenic-based monomers can be used. Functionalized microspheres with opposite charge or polarity can be assembled on the surfaces. Separate islands of specific and divergent chemical nature can then be built onto the surface of the resin. Functional groups can be introduced at various points along the polymer backbone through interactions between the polymer bead and reactive chemical intermediates.[1-3]

The ability to build divergent core/shell type polymers could be potentially advantageous in the catalysis area. For example, a metal loaded microsphere could be assembled onto a polymeric macrosphere. The microsphere would have kinetics favorable for facilitating a catalytic reaction while the proximity of the polymeric macrosphere could act to concentrate the reaction products for later recovery.

Carbonaceous Composites

Two different types of carbonaceous composite resins have been synthesized. The first type consists of carbonaceous microparticles assembled onto various copolymer supports. The second type consists of carbonaceous resins derived from the pyrolysis of copolymer composites. Figure 4a is an example of the first class of composite resins made by assembling carbonaceous microspheres onto the surface of gel styrene/DVB copolymer resins. The carbonaceous microspheres are more hydrophobic than polyacrylonitrile microspheres. As a result, the carbonaceous microbeads have a greater tendency to stay *in* the interface rather than migrate to the *top* of the surface.

The nature of the carbonaceous materials depends in large part on its polymeric precursor. It is known, for example that the pyrolysis of an acrylonitrile-divinylbenzene copolymer resin results in a relatively hydrophilic carbon due in part to residual nitrogen containing groups. Carbonaceous materials derived from styrene-divinylbenzene copolymer are considerably more hydrophobic. Thus starting with an acrylonitrile core and styrene/divinylbenzene derived carbonaceous microspheres, we were able to prepare carbonaceous resins with different hydrophobicities.

Alternatively a porous structure can be prepared by pyrolyzing the composite resins depicted in Figures 1-3 because the polystyrene microspheres degrade or burn away while the polyacrylonitrile support is converted into a carbonaceous material. The resultant carbonaceous resin material has holes on the surface that are one third the diameter of the microsphere (7.5μm and 5.3 μm microspheres resulted in holes of 2.5 and 1.8μm respectively) in the starting copolymer composite, suggesting that the size of the holes can be predicted accurately from the size of the starting microspheres.

Figure 4. SEM of carbonaceous microparticles assembled on a) gel polystyrene/DVB (8.5 wt. % of the monomer phase) b) pyrolysis of copolymer shown in Figure 1b.

a)

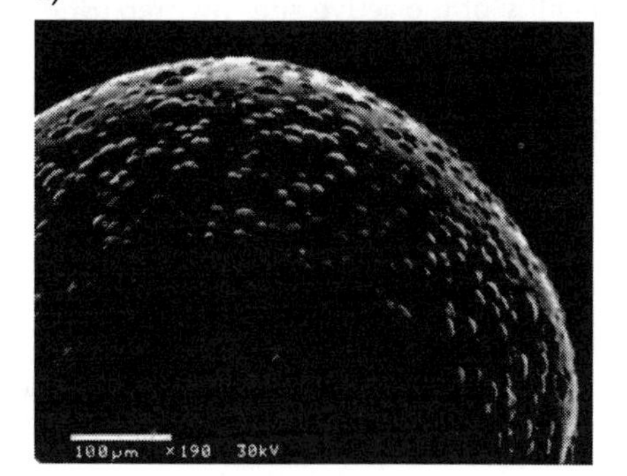

b)

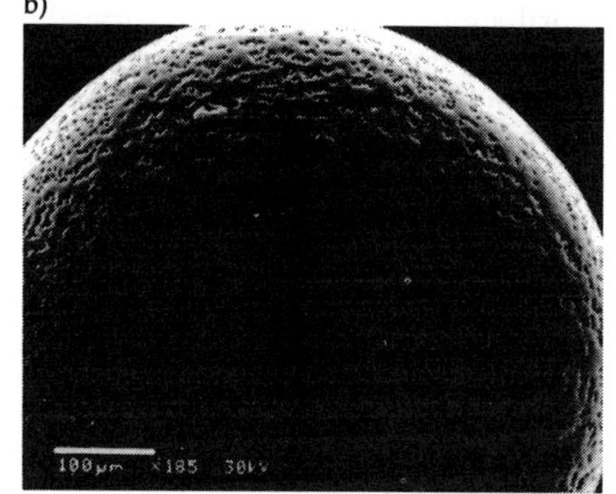

It is well known that activated carbon has the unique ability to act as a source of electrons and has a reduction potential of -0.14 V (vs. calomel electrode) indicating that it is capable of reducing gold from a gold chloride solution and similarly, silver from a silver nitrate solution.[12] We carried out these reactions with our porous carbonaceous materials to see if any site selectively for the holes existed. If anything, both gold and silver appear to reduce exclusively on the surface of the bead rather than in the holes (see Figure 5). The size of the crystals indicate that a large amount of free electrons are transported from the interior of the resin to the resin surface.

Figure 5. SEM of a carbonaceous material created with holes on the surface and then used to reduce a) gold from a gold chloride solution and b) silver from a silver nitrate solution.

a) b)

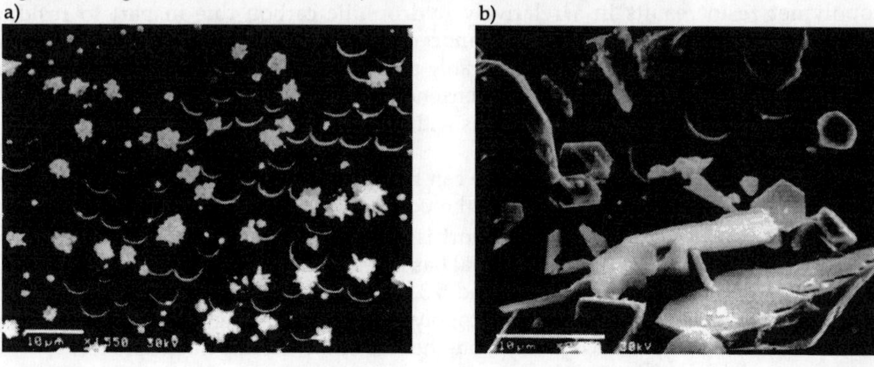

Core-shell resins are available through various other techniques, but the surface assembling technology allows for a much wider range of theoretical surface functionalized resins. Relatively expensive monomers can be used to create small microspheres; the microspheres are used in only one to two weight percent. Since the microspheres are on the surface, the diffusion pathway is greatly reduced. Furthermore, we have demonstrated the availability of the active sites by preparing a microsphere with amine-borane functionality. This functional group is capable of reducing gold, silver, platinum and palladium salts to the respective metal. After preparing a composite material, we added gold chloride solution and observed via optical microscopy that only the microspheres turned red, indicating the presence of colloidal gold in the matrix. (Figure 6.)

Figure 6. a) SEM of the surface of a composite of amine borane microspheres and a gel polystyrene (8% DVB) support. b) Optical micrograph of the resin after contacting with a gold chloride solution (dark areas represent reduced gold).

a) b)

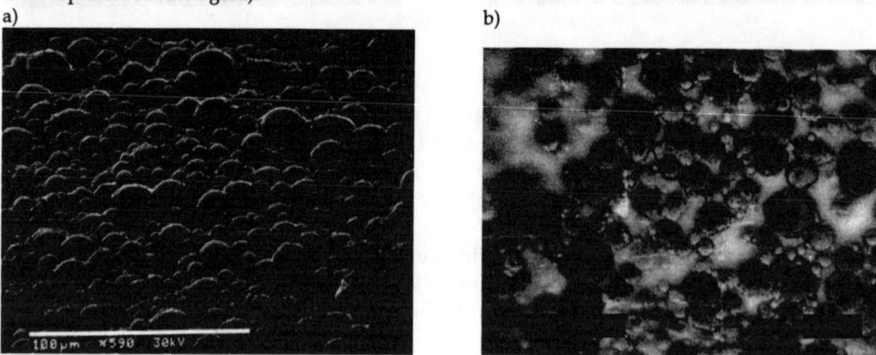

CONCLUSIONS

Novel composite resins have been developed wherein microstructures of varying shapes and sizes can be assembled onto the surface copolymer macroparticles. Placing the small reactive species on a large support resin offers several advantages: the small particles on the surface will have a short diffusion pathway for faster kinetics, the large support resins will deliver lower pressure drop in packed-bed applications, the relative low weight percent of microspheres, 1-2 percent, in the composite resin will allow for the use of more expensive, specialty monomers. Pyrolysis of these composites leads to hybrid carbonaceous materials with controlled hydrophobicities, pore size and functionality. The overall flexibility of the process will allow an experimenter to manipulate both the support and the assembled phases independently from one another.

BIBLIOGRAPHY

1. R. Kunin, Ion Exchange Resins,; Robert E. Krieger Publishing Co.: Huntington, New York, 1972.
2. F.G. Thorpe, in New Methods of Polymer Synthesis, J. R. Ebdon, Ed.; Chapman & Hall: N.Y., 1991.
3. J. March, Advanced Organic Chemistry, 4th Ed.. John Wiley & Sons: New York, 1992.
4. C.A. Pohl and S.C. Papanu, EP 0134099, 13 March 1985 and references therein.
5. T.S. Stevens and M.A. Langhorst, US 4,519,905, 28 May 1985 and references therein.
6. T.S. Stevens, M.A. Langhorst and O.W. Randall, 3rd, US 4,927,539, 22 May 1990.
7. J.S. Fritz, J.O. Naples, L.M. Warth, EP 0346037, 13 December 1989.
8. L.M. Warth and J.O. Naples, J. Chromatography , **462**, 165 (1989).
9. R.F. Strasburg, J.S. Fritz, J.O. Naples, J. Chromatography, , **547**, 11, (1991).
10. R.C. Ludwig, J. Chromatography , **592**, 101, (1992).
11. Y. Hanaoka, T. Murayama and S. Muramoto, US 4,447,559, 8 May 1984.
12. G.J. McDougall and C.A. Fleming, in Critical Reports on Appl. Chem., **19**, J. Wiley & Sons: N.Y., 1987.

EPR ANALYSIS OF COPPER EXCHANGED AND ENCAPSULATED COPPER (SALEN) COMPLEX IN ZEOLITE - Y

N.ULAGAPPAN AND V.KRISHNASAMY
Department of Chemistry, Anna University, Madras - 600 025, India.

ABSTRACT

Copper exchanged and copper-(salen) complex encapsulated in zeolite-Y were prepared. The encapsulated complex was analysed by TGA, DRS and IR, which confirms the presence of complex in zeolite. Both copper materials were examined by EPR spectra, and the calculated spin-Hamiltonian parameter indicates the copper complex is more covalent in character than NaCu-Y.

INTRODUCTION

Zeolites have molecular dimensional pores and ion-exchange capacity [1]. The metal exchanged zeolites have potential application in catalytic reactions, imaging and optical storage media, sensors, electrodes, batteries and membranes [2-4]. The molecular dimensional pore allows it to be used as a microvessel for carrying out shape selective catalysis and also to host for metal complexes in ship-in-a bottle manner [5]. These encapsulated metal complexes find place in catalytic applications [6]. Tools to characterize these materials are always limited due to complicated nature of the system. EPR can give some information about bonding nature of metal ions and metal complexes present in zeolites. EPR study of copper containing zeolites are of recent research interest [7,8]. Here we have investigated bonding nature of copper in zeolite encapsulated copper-(salen) complex and compared with other forms of copper species using spin-Hamiltonian parameters and molecular orbital coefficient calculations.

EXPERIMENTAL

Na-Y zeolite (Si/Al = 2.36) was used for preparation of Na,Cu-Y and Cu(Salen)-Y complex (salen = 1,6 bis (2-hydroxyphenyl)-2,5-diaza-1,5-hexadiene). The copper exchanged zeolite-Y was prepared by ion-exchange method. The copper (Salen)-Y composite was prepared by melting (140°C) Salen ligand with dry copper exchanged zeolite in nitrogen atmosphere for 4hr, followed by soxhlet extraction using chloroform untill the extract become colourless and by brine solution exchange to remove uncomplexed ligand and copper respectively. Copper-salen complex was prepared and recrystallised in ethanol.

Mat. Res. Soc. Symp. Proc. Vol. 368

ESR spectra of Na,Cu-Y, Cu-(salen)-Y and copper-salen complex were recorded at room temperature using quartz tubes on JEOL-FEIX ESR spectrometer operating at 9.2 GHz with 100 Khz modulation field. The Cu-salen-Y composite was analysed by TGA, IR and DRS. TGA was done in Mettler TA 3000 instrument at the rate of heating 20°C/min. in an ambient air flowing condition. IR was recorded (1200-1600 cm^{-1}) using KBr technique. DRS was recorded between 300-600 nm by Carry 2300 UV-VIS-NIR spectrophotometer.

RESULTS AND DISCUSSION

TGA, IR and DRS of Cu-Salen-Y

TGA analysis indicate losses at 118° (22.5%) and at 718° (9%). The first weight loss is due to combined water and complex removal and decomposition. The second loss is complex or ligand loss probably from sodalite cages. The losses account for existence of at least one complex per faujasite cage. The IR peak at 1420 cm^{-1} is due to Cu-salen ligand which is not present in Na-Y [9] and indicates the presence of a copper-salen complex in zeolite. The transition $^2B_{1g} \rightarrow {}^2B_{2g}$ is observed at 425nm in DRS spectra, which is characteristic for copper complexes.

EPR Spectra

EPR spectra of samples are shown in Fig.1 (Copper exchanged and zeolite encapsulated copper-salen complex). The spectra are typical of d^9 complex, possessing axial symmetry. The spectra show peaks in two different region ($g_{\parallel}$ and $g_{\perp}$ as reported in literature [10]. The observed spectra could be explained by the spin-Hamiltonian

$$H = \beta \bar{H} \bar{g} \bar{s} + \bar{S} \bar{A} \bar{I} \qquad (1)$$

where the hyperfine tensor A is measured in units of cm^{-1} and other terms have their usual meaning. The spin-Hamiltonian parameters determined are given in the Table 1.

Molecular orbital coefficients

The covalency parameter α^2 can be calculated using the following expression [11].

$$\alpha^2 = A_{\parallel}/0.036 + (g_{\parallel} - 2.0023) + 3/7\,(g_{\perp} - 2.0023) + 0.04 \qquad (2)$$

α^2 which describes the covalency of the σ bond between copper and its ligands has value of 1 if the bond is totally ionic and 0.5 if it is completely covalent. The covalent character of Cu-(Salen)-Y was found higher than NaCu-Y and it is comparable with that of free copper-salen complex.

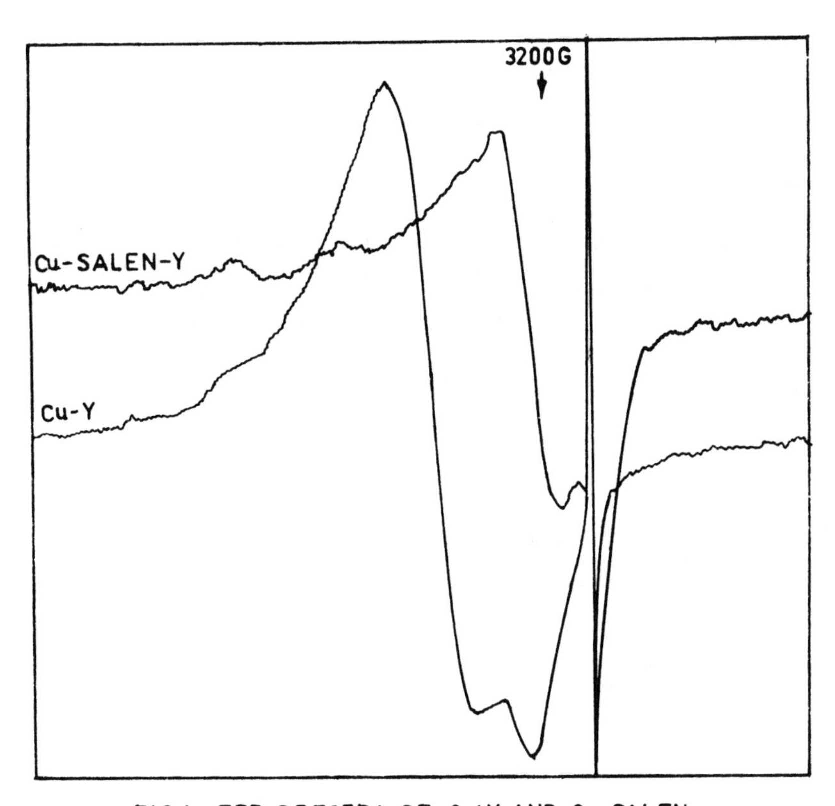

FIG.1 EPR SPECTRA OF Cu-Y AND Cu-SALEN

Table 1 Spin-Hamiltonian and Molecular orbital coefficients Nu,Cu-Y and Cu-Salen-Y

Sample	$g_{\parallel}$	$g_{\perp}$	$A_{\parallel}$ gauss	$A_{\perp}$ gauss	α^2
Na,Cu-Y	2.35	2.14	116.9	115.1	0.78
Dry Cu-Y [Ref.7]	2.38	2.07	136.4	-	0.85
Cu-Salen-Y	2.24	2.01	113.2	90.4	0.68
Cu-Salen	2.27	2.02	118.4	92.5	0.66

CONCLUSIONS

Encapsulated copper-salen complex in zeolite-Y was prepared and characterized. The EPR analysis of the copper exchanged, free copper-salen complex and copper-salen-Y were done. Spin-Hamiltonian and molecular orbital coefficient parameters of the copper speceies indicates that copper in complex form is more covalent than copper in exchanged one in zeolite.

REFERENCES

1. M.E.Davis, Ind. Eng. Chem. Res., **30**(8), 1675 (1991).
2. W.Holderich, M. Hesse and F.Naumann, Angew. Chem. Int. Ed. Engl., **27**, 226 (1988).
3. G.A.Ozin, A.K.Kuperman and A.Stein, Angew. Chem. Int. Ed. Engl., **28** (3), 359 (1989).
4. G.D. Stucky and J.E. MacDougall, Science, **247**, 669 (1990).
5. A.C. Huberechts, R.F.Parton, P.A.Jacobs and K.U.Leuven in Chemistry of Microporous Crystals, (Kodansha Ltd. Tokyo, 1991), p.225.
6. R.F.Parton, L. Uytterhoeven and P.A. Jacobs in Heterogeneous catalysis and Fine chemicals II, edited by M.Guisnet *et al.*, (Elsevier Science Publishers, Amsterdam, 1991), p. 395.
7. K.P.Wendlandt, F.Vogt, W.Morke and I.Achkar in Zeolite Chemistry and Catalysis edited by P.J.Jacobs *et al.*, (Elsevier Science Publishers, Amsterdam, 1991), p.223.
8. C.W.Lee, X.Chen, M.Zamadics and L.Kevan, J.Chem.Soc., Farad. Trans., **89** (22), 4137 (1993).
9. E.Paez-Mozo, N.Gabriunas, F.Lucaccioni, D.D.Acosta, P.Patrono, A.L.Ginestra, P.Ruiz and B.Delmon, J.Phy.Chem., **97**, 12819 (1993).
10. J.L.Rao, R.M.Krishna and S.V.J.Lakshman, Solid State Commun., **66**, 1185 (1988).
11. D.Kivelson and R.Neimann, J.Chem.Phys., **35**, 149 (1961).

SYNTHESIS AND CHARACTERIZATION OF SUPPORTED METAL CLUSTER CATALYSTS WITH WELL-DEFINED STRUCTURES

B. C. GATES
Department of Chemical Engineering and Materials Science
University of California, Davis, CA 95616

ABSTRACT

Nearly uniform supported metal clusters are the focus of this summary, which addresses their synthesis, characterization, and catalytic properties. Supported clusters represented as Ir_4 and Ir_6 were prepared on MgO powder by thermal decarbonylation of $[HIr_4(CO)_{11}]^-$ and $[Ir_6(CO)_{15}]^{2-}$, respectively. The most useful characterization technique for supported clusters is extended X-ray absorption fine structure spectroscopy, but the structure data are limited by the precision of the coordination numbers (about ±20%). Ir_4 and Ir_6 clusters on supports catalyze hydrogenation of cyclohexene and of toluene, and they are catalytically distinct from larger iridium particles, although the reactions are structure insensitive (proceeding at about the same rate per exposed metal atom) when catalyzed by metal particles larger than about 1 nm. Thus the concept of structure insensitivity does not extend to supported metal clusters, which are regarded as quasi molecular.

INTRODUCTION

Metals are among the most widely used catalytic materials, with applications including reforming of naphtha, abatement of automobile emissions, and hydrogenation of carbon monoxide. Because a catalytic metal is often expensive, it is usually applied in a highly dispersed form so that a large fraction of the metal atoms are exposed to reactant molecules and catalytically engaged. Thus the metal is usually a minor component in a composite material that is mostly an inexpensive porous support (carrier), typically a metal oxide or a zeolite.

Supported metals are classified on the basis of the metal particle size [1]: Metal particles larger than about 5 nm have surface structures resembling those of chunks of bulk metal and expose a distribution of crystal faces that is more or less independent of the particle size. Supported metal particles in the size range 1-5 nm are of great interest in catalysis because changes in the particle size lead to changes in catalytic properties for many catalytic reactions (classified as structure sensitive). Supported metal particles with diameters < 1 nm, referred to here as metal clusters to distinguish them from the larger particles, are the subject of this review. Supported metal clusters are gaining attention because a catalyst incorporating them (platinum clusters in the pores of LTL zeolite) is now applied industrially for selective reforming of naphtha to give aromatics.

Mat. Res. Soc. Symp. Proc. Vol. 368 © 1995 Materials Research Society

PREPARATION OF SUPPORTED METAL CLUSTERS

Synthesis of supported metal carbonyl clusters. The most common precursors of structurally simple supported metal clusters are metal carbonyl clusters. Methods for preparation of molecularly or ionically dispersed metal carbonyl clusters on metal oxide supports include reaction with the support surface and synthesis from mononuclear (single-metal-atom) precursors on support surface [2-6]. For example, $[Os_3(CO)_{12}]$ reacts by oxidative addition with OH groups of SiO_2 or of γ-Al_2O_3 to give predominantly $[(\mu\text{-}H)Os_3(CO)_{10}\{\mu\text{-}OM\}]$, where M = Si or Al, and the braces denote groups terminating the bulk metal oxide [7-11]. Deprotonation of a hydrido metal carbonyl cluster on a basic metal oxide support (e.g., $[H_4Os_4(CO)_{12}]$ on MgO or on γ-Al_2O_3) gives surface ion pairs $[H_3Os_4(CO)_{12}^-\{M^{n+}\}]$ [12], where M^{n+} is a cation exposed at the surface of the metal oxide support. Metal carbonyl clusters are also synthesized on supports from mononuclear metal carbonyl or metal salt precursors [6]. The surface chemistry is predicted from known solution chemistry [6]. For example, syntheses on basic MgO surfaces take place as in basic solutions in the presence of reducing agents, and syntheses on SiO_2 take place as in neutral solvents. $[Ir(CO)_2acac]$ reacts in the presence of CO on MgO to give $[HIr_4(CO)_{11}]^-$, which may be converted into $[Ir_8(CO)_{22}]^{2-}$, and ultimately into $[Ir_6(CO)_{15}]^{2-}$ [13]. Other clusters prepared by surface-mediated synthesis on MgO include $[Os_5C(CO)_{14}]^{2-}$ [14], $[Os_{10}C(CO)_{24}]^{2-}$ [14], $[Rh_5(CO)_{15}]^-$ [15], $[Pt_{15}(CO)_{30}]^{2-}$ [16], $[Pt_6(CO)_{12}]^{2-}$ [17], $[Pt_9(CO)_{18}]^{2-}$ [17], and $[Pt_{12}(CO)_{24}]^{2-}$ [16]. $[Ir_4(CO)_{12}]$ has been prepared on γ-Al_2O_3 [18] and $[H_4Os_4(CO)_{12}]$ on SiO_2 [19].

The syntheses referred to above take place in zeolite cages much as they do on surfaces of amorphous metal oxides. Syntheses in the supercages of the nearly neutral NaY zeolite are similar to those occurring on γ-Al_2O_3, e.g., those of $[Ir_4(CO)_{12}]$ [20] and $[Ir_6(CO)_{16}]$ [21] from $[Ir(CO)_2(acac)]$. Syntheses in the supercages of the more basic NaX zeolite are similar to those occurring on MgO, e.g., those of $[HIr_4(CO)_{11}]^-$ and $[Ir_6(CO)_{15}]^{2-}$ from $[Ir(CO)_2(acac)]$ [22]. These are examples of ship-in-a bottle syntheses, whereby the clusters formed in the zeolite cages are trapped (and stabilized) there because they are too large to fit through the zeolite apertures. Although $[Ir_8(CO)_{22}]^{2-}$ is synthesized on MgO from $[Ir(CO)_2(acac)]$, it does not form in NaX zeolite because it is too large to fit in the supercages.

Decarbonylation of supported metal carbonyl clusters. The preparation of supported metal clusters by decarbonylation of supported metal carbonyl clusters is illustrated by the decarbonylation of $[Ir_4(CO)_{12}]$ on γ-Al_2O_3 in helium at 473 K, which was shown by infrared and EXAFS spectroscopies to take place with little or no disruption of the tetrahedral metal frame [18]. The simple decarbonylation parallels that in the gas phase; electrospray mass spectrometry showed that CO ligands were removed one by one from a salt of $[HIr_4(CO)_{11}]^-$, giving the family of clusters $HIr_4(CO)_{11-x}$ (x = 0, 1, 2, ...11) [23]. Similarly, decarbonylation of MgO-supported $[HIr_4(CO)_{11}]^-$ [24], $[Ir_6(CO)_{15}]^{2-}$ [24], $[Pt_{15}(CO)_{30}]^{2-}$ [25], and $[Os_{10}C(CO)_{24}]^{2-}$ [26] also appears to take place without significant changes in the metal framework structures, as indicated by EXAFS spectroscopy.

Decarbonylation of $[Ir_4(CO)_{12}]$ and of $[Ir_6(CO)_{16}]$ takes place in NaY zeolite cages [21, 27]. These decarbonylations, in contrast to decarbonylations of the same clusters on MgO [28], are reversible [27]. Infrared spectra show that $[Ir_4(CO)_{12}]$ (or $[Ir_6(CO)_{16}]$) was decarbonylated by treatment in H_2 at 573 K [27]. When CO was adsorbed on the decarbonylated clusters at 77 K and the temperature raised with the sample in the presence of a few Torr of CO, mononuclear iridium carbonyls formed at about 240 K. These were converted at about 320 K into $[Ir_4(CO)_{12}]$ and at about 400 K into $[Ir_6(CO)_{16}]$, as shown by infrared spectra [27]. These results show that the reformation of the iridium carbonyl clusters involved chemistry more complicated than simple readsorption of CO onto the metal clusters; the clusters were broken up and then reconstructed from mononuclear fragments. Perhaps the confinement of the supercages helped to maintain the fragments in a position where the reconstruction could occur readily.

Deposition of size-selected gas-phase metal clusters. Beams of gas-phase metal clusters, size selected in a mass spectrometer and allowed to impinge on planar surfaces in an ultrahigh vacuum chamber, gave dispersed size-selected clusters on supports. The few such samples made in this way are Pt_n (where n = 1-6) on oxidized silicon wafers and on carbon films [29], and Pt_n and Pd_n (where n = 1-15) on Ag(110) single crystals [30].

STRUCTURAL CHARACTERIZATION OF SUPPORTED METAL CLUSTERS

Supported metal clusters are difficult to characterize because they are small, usually nonuniform, and present in only low loadings in solids that are usually amorphous. The most effective methods for identifying dispersed clusters have become available only recently, including, EXAFS spectroscopy, high-resolution electron microscopy, ^{129}Xe NMR spectroscopy, H_2 and CO chemisorption, and wide angle X-ray scattering. Size-selected platinum and palladium clusters on planar supports have been investigated with X-ray and ultraviolet photoemission spectroscopy [29, 30]. Each of the clusters Pt_n (n = 1, 2, ...6) on SiO_2 [29] has a distinct spectrum, but structural data for the supported clusters is still lacking. Results characterizing Pt_n (n = 1, 2, ...15) on Ag(110) indicate distinct core level binding energies for the different species; the line width was found to increase and then level off for clusters containing more than six platinum atoms [30].

The relatively new science of metal clusters on supports was almost without a structural foundation until the advent of EXAFS spectroscopy. This technique [31] provides quantitative structure data, which are most precise and accurate when the clusters contain only a few atoms and are nearly uniform [32]. The technique is limited because it provides only average structural information and relatively imprecise values of coordination numbers; the best data are obtained at low temperatures, typically 77 K.

EXAFS results representing some of the simplest and most thoroughly investigated supported metal clusters are shown in Table 1; these clusters were formed by decarbonylation of metal carbonyls. The structures formed from tetrairidium carbonyls (Table 1) are modeled as Ir_4 tetrahedra. These are the simplest and best defined supported metal clusters. The structures of supported metal clusters depend on the preparation conditions. For example, clusters may be oxidized by traces of O_2 impurities; increasing hydroxylation of the MgO support favors the formation of clusters larger than Ir_4, for example [35].

Table 1. Supported metal clusters formed by decarbonylation of supported metal carbonyl clusters: characterization by EXAFS spectroscopy.[a]

support	precursor	cluster modeled as	metal-metal first-shell coordination number	*R*, nm	$10^5 \cdot \Delta\sigma^2$, nm^2	E_0, eV	ref
γ-Al_2O_3	$[Ir_4(CO)_{12}]$	Ir_4 tetrahedra	2.9	0.269	3.1	0.83	18
MgO	$[HIr_4(CO)_{11}]^-$	Ir_4 tetrahedra	3.1	0.269	1.1	1.22	33
MgO	$[HIr_4(CO)_{11}]^-$	Ir_4 tetrahedra	3.1	0.270	2.3	0.50	34
MgO	$[HIr_4(CO)_{11}]^-$	Ir_4 tetrahedra	3.0	0.269	4.8	3.9	35
NaY zeolite	$[Ir_4(CO)_{12}]$ formed from $[Ir(CO)_2(acac)]$	Ir_4 tetrahedra	3.4	0.270	3.0	0.06	20
NaX zeolite	$[HIr_4(CO)_{11}]^-$ formed from $[Ir(CO)_2(acac)]$	Ir_4 tetrahedra	3.0	0.271	2.9	2.77	22

MgO	$[Ru_3(CO)_{12}]$	Ru_3	1.7	0.264	not stated	not stated	36
γ-Al_2O_3	$[Ru_3(CO)_{12}]$	Ru_6	3.1	0.264	not stated	not stated	36
γ-Al_2O_3	$[H_3Re_3(CO)_{12}]$	Re_3	2.0	0.267	0.86	-3.04	37
MgO	$[Ir_6(CO)_{16}]$ formed from $[Ir(CO)_2(acac)]$	Ir_6 in mixture with Ir rafts	2.7	0.272	4	-4.5	24
NaY zeolite	$[Ir_6(CO)_{16}]$ formed from $[Ir(CO)_2(acac)]$	Ir_6	3.6	0.271	3.5	-2.00	21
MgO	H_2OsCl_6	Os_{10}	3.8	0.267	0.37	1.2	38

[a]Notes: N is coordination number, R the average absorber-backscatterer distance, $\Delta\sigma^2$ the Debye-Waller factor, and ΔE_0 the inner potential correction. Typical experimental errors in N and R are approximately ±20% and ±2%, respectively. The structures depend on the pretreatment temperatures of the supports and the conditions of decarbonylation, which are stated in the cited references.

Table 2. EXAFS spectra of fresh and used supported metal cluster catalysts.

sample number	catalyst precursor	support	treatment/catalysis [a]	*N*,[b] fresh catalyst	*R*,[c] nm, fresh catalyst	*N*,[b] used catalyst	*R*,[c] nm, used catalyst	refs
1	$[HIr_4(CO)_{11}]^-$	MgO	none	3.2	0.271			23, 33, 41
2	$[Ir_4(CO)_{12}]$	γ-Al_2O_3	decarbonylation and catalysis of toluene hydrogenation	2.9	0.269	3.2	0.268	18, 23
3	$[HIr_4(CO)_{11}]^-$	MgO	decarbonylation and catalysis of toluene hydrogenation	3.2	0.271	2.8	0.269	23, 33, 41
4	$[Ir_6(CO)_{16}]$	NaY zeolite	decarbonylation and catalysis of toluene hydrogenation			3.9	0.270	23
5	$[Ir_6(CO)_{15}]^{2-}$	MgO	decarbonylation and catalysis of toluene hydrogenation			4.1	0.269	23
6	$[HIr_4(CO)_{11}]^-$	MgO	decarbonylation, treatment in H_2 at 573 K, and catalysis of toluene hydrogenation	3.1	0.270	3.1	0.270	23
7	$[Ir_6(CO)_{15}]^{2-}$	MgO	decarbonylation, treatment in H_2 at 573 K, and catalysis of toluene hydrogenation	3.8	0.269	3.7	0.270	23

8	$[Ir_4(CO)_{12}]$	γ-Al_2O_3	decarbonylation, treatment in H_2 at 573 K, and catalysis of toluene hydrogenation			5.4	0.267	23
9	$[Ir_6(CO)_{16}]$	NaY zeolite	decarbonylation, treatment in H_2 at 573 K, and catalysis of toluene hydrogenation	3.6	0.271	4.1	0.269	21, 23
10	$[HIr_4(CO)_{11}]^-$	MgO	decarbonylation, treatment in H_2 at 573 K, and catalysis of cyclohexene hydrogenation			3.3	0.271	23
11	$[HIr_4(CO)_{11}]^-$	MgO	decarbonylation, treatment in H_2 at 573 K, and catalysis of propane hydrogenolysis	3.35 ± 0.15	0.2730 ± 0.0005	3.26 ± 0.23	0.2735 ± 0.0006	34
12	$[Os_{10}C(CO)_{24}]^{2-}$	MgO	decarbonylation, treatment in H_2 at 573 K, and catalysis of *n*-butane hydrogenolysis	4.7	0.267	5.0	0.268	59

[a]Conditions of catalysis experiments stated in Table 3. [b]*N* is the first-shell metal-metal coordination number; [c]*R* is the metal-metal distance, both determined by EXAFS spectroscopy.

Notwithstanding the rapid progress in characterization of supported metal clusters, much remains to be learned. The structures and reactivities of supported metal clusters and the effects of supports on cluster structure and the nature of the cluster-support interface are less than well understood [24]. Powder metal oxide supports are structurally nonuniform, and it has been postulated that clusters reside preferentially at defect sites [39]; experiments with structurally well defined metal clusters on single-crystal metal oxides (or models of these consisting of thin layers of metal oxides on metals) are needed to determine how the support structure affects cluster structure and properties. Assessments of electronic properties of supported clusters are still not consistent with each other. Theoretical chemistry is just beginning to have an impact on the assessment of electronic properties, with calculations having been reported for Ir_4 and Ir_{10} [40], but it is still too early for theories to account reliably for the influence of the support.

CATALYSIS BY SUPPORTED METAL CLUSTERS

It is likely that many conventional supported metal catalysts incorporate metal clusters, but because of the difficulty of distinguishing the small clusters from larger metal particles, it has not been possible to identify or define the roles of clusters. Rather, evidence of catalysis by supported metal clusters has arisen only recently, made possible by syntheses that give catalysts containing metal almost exclusively in the form of clusters. The following section is a summary of catalytic results emphasizing supported metal clusters that have been characterized by EXAFS spectroscopy, with the data demonstrating the stability of the clusters and their presence in both the fresh and used catalysts.

Catalytic properties of supported clusters identified as primarily Ir_4 or Ir_6 were reported by Xu et al. [23], who investigated structure-insensitive reactions, cyclohexene hydrogenation and toluene hydrogenation. EXAFS spectra showed that the first-shell Ir-Ir coordination numbers characterizing both the fresh and used MgO-supported catalysts made by decarbonylation of supported $[Ir_4(CO)_{12}]$ or $[HIr_4(CO)_{11}]^-$ are indistinguishable from 3, the value for a tetrahedral metal frame, as in $[Ir_4(CO)_{12}]$ and $[HIr_4(CO)_{11}]^-$ (Table 2). The decarbonylated clusters retained this metal frame. EXAFS data show that the decarbonylated Ir_6 clusters had metal frames indistinguishable from the octahedra of the precursor hexairidium carbonyls, indicated by the Ir-Ir coordination number of approximately 4 (Table 2).

Catalytic activities (turnover frequencies) are summarized for these clusters and for conventional (structurally nonuniform) supported catalysts consisting of aggregates of iridium on supports. Turnover frequencies of the latter catalysts are based on the number of surface iridium atoms estimated from hydrogen chemisorption data. The decarbonylated iridium clusters are different in activity from each other and from the supported iridium particles on MgO (Table 3). The clusters represented as Ir_6 are several times less active than those represented as Ir_4, and both of these are markedly less active than the larger iridium particles.

Changing the support from MgO to γ-Al_2O_3 to zeolite NaY had little effect on the activities of the decarbonylated clusters (Table 3). The orders of reaction in toluene and in

Table 3. Catalytic activities of supported metal clusters and supported metallic particles [42].

sample number[a]	catalyst modeled as	support	catalytic reaction[b]	H_2 treatment temperature of catalyst, K	10^3 x TOF,[c] s^{-1}	ref
1	$[HIr_4(CO)_{11}]^-$	MgO	toluene hydrogenation	no treatment	0.00	23
2	Ir_4	γ-Al_2O_3	toluene hydrogenation	no H_2 treatment	0.94	23
3	Ir_4	MgO	toluene hydrogenation	no H_2 treatment	0.63	23
4	Ir_6	NaY zeolite	toluene hydrogenation	no H_2 treatment	0.25	23
5	Ir_6	MgO	toluene hydrogenation	no H_2 treatment	0.23	23
6	Ir_4	MgO	toluene hydrogenation	573	0.17	23
7	Ir_6	MgO	toluene hydrogenation	573	0.03	23
8	aggregates of about 20 atoms each, on average, formed from Ir_4	γ-Al_2O_3	toluene hydrogenation	573	9.9[d]	23
9	Ir_6	NaY zeolite	toluene hydrogenation	573	0.52	23

10	Ir_4	MgO	cyclohexene hydrogenation	573	18	23
11	Ir_4	MgO	propane hydrogenolysis	573	20	34
12	Ir crystallites	MgO	toluene hydrogenation	773	2.0	23

[a]Sample numbers match those of Table 2.
[b]Reaction conditions: Each reaction except cyclohexene hydrogenation carried out with vapor-phase reactants in a once-through plug flow reactor operated at atmospheric pressure. Cyclohexene hydrogenation carried out with liquid-phase cyclohexene in *n*-hexane solvent saturated with H_2 flowing through the stirred reactor, which was held at atmospheric pressure.
[c]TOF (turnover frequency) for toluene hydrogenation at 333 K; for cyclohexene hydrogenation at 298 K; for propane hydrogenolysis at 473 K with a H_2/hydrocarbon molar ratio of 2.5; for *n*-butane hydrogenolysis at 473 K with a *n*-butane/H_2/He molar ratio of 1/10/9.
[d]In contrast to the activities of the other catalysts, the activity of this catalyst is expressed per total Ir atom to facilitate the comparison with the results for sample 2. The comparison shows that the activity, even per total Ir atom, increased as a result of aggregation of the Ir caused by pretreatment of the sample in hydrogen.

H_2 were found to be approximately 0 and 1, respectively, both for the supported iridium cluster catalysts and the supported iridium particles.

The EXAFS data show that pretreatment in H_2 at 573 K did not lead to a significant change in the metal framework structure of MgO-supported Ir_4 or Ir_6, but it led to a decrease in their activities for toluene hydrogenation (Table 3). The results suggest that hydrogen bonded to the clusters and inhibited catalysis. However, the effect of hydrogen illustrated by these data is not the same as that observed for Ir_6 in NaY zeolite. The hydrogen pretreatment led to a doubling of the activity of this catalyst without a substantial change in the metal framework structure. The results imply that the different supports are associated with different effects of hydrogen treatment.

Evidence of an influence of the support on the stability of the supported metal clusters is shown by the data for a sample originally containing Ir_4 supported on γ-Al_2O_3. Following treatment in H_2 at 573 K, the catalytic activity (per total Ir atom) increased ten-fold (Table 3). This increase is explained, at least in part, by the EXAFS result for the used catalyst, which shows that the iridium had aggregated to give clusters with an average of about 20 atoms each (Table 2). Thus the data show that the support significantly affected the stability of the clusters but hardly affected their intrinsic catalytic activity. The results suggest that NaY zeolite helps to stabilize the clusters, perhaps by virtue of entrapment in the supercages.

Although the Ir_4 and Ir_6 clusters catalyze the same reactions as metallic iridium particles, their catalytic character is different, even for structure-insensitive hydrogenation reactions. It is inferred [23] that the clusters are metal-like but not metallic. Thus these data show the limit of the concept of structure insensitivity; this concept pertains only to catalysis by surfaces of structures that are present in three-dimensional particles about 1 nm in diameter or larger. The pattern of structure insensitivity that has been observed for numerous reactions (such as hydrocarbon hydrogenations) catalyzed by particles larger than 1 nm in diameter leads to the suggestion that these particles are metallic in character and the smaller clusters Ir_4 and Ir_6 are not metallic, but more nearly molecular in character from the point of view of catalysis [23].

If it is proper to regard supported metal clusters as quasi molecular [23], it follows that the support provides part of a ligand shell, which presumably helps to stabilize the cluster and affects the catalytic activity as ligands affect the catalytic activity of a molecular metal cluster. The lack of a significant support effect on the catalytic activities shown in Table 4 is consistent with the suggestion that all the supports mentioned in the table are metal oxides that are not much different from each other as ligands.

MgO-supported clusters formed by decarbonylation of $[HIr_4(CO)_{11}]^-$ were also investigated as catalysts for a structure-sensitive reaction, propane hydrogenolysis [34]. The decarbonylated cluster had an average Ir-Ir coordination number of 3.1, which implies that the tetrahedral metal frame was largely retained after decarbonylation. The catalyst, after use in a flow reactor at 473 K, was characterized by an Ir-Ir coordination number of 3.2, which implies that the cluster structure was largely retained. The lack of a significant change in the average cluster size indicates the stability of the supported catalyst, which is modeled as tetrairidium clusters. The catalyst performance data confirm the stability, indicated by a lack of change in the activity during operation in a flow reactor following a short break-in period.

The turnover frequency of the supported Ir_4 catalyst (Table 3) was found to be two orders of magnitude less than that of a conventionally prepared MgO-supported iridium catalyst containing iridium particles about 3 nm in average diameter.

The results are consistent with the identification of propane hydrogenolysis as a structure-sensitive reaction, but a reservation must be expressed about this inference: Because the supported crystallites are so much more active than the supported clusters, it is difficult to rule out the possibility that a small fraction of the clusters had aggregated on the support surface to give larger species and that these provided the surfaces for the observed catalytic activity. Such a concern for the toluene hydrogenation reaction should not be ruled out, but the likelihood of catalysis by undetected metal particles rather than metal clusters in this reaction is considered less likely than for propane hydrogenolysis because the catalyst preparation, catalysis, and EXAFS experiments were reproduced several times and because the reaction conditions were so mild (333 K, for example). Furthermore, data for cyclohexene hydrogenation are qualitatively in agreement with those for toluene hydrogenation, and the former reaction was carried out at only 298 K [23].

Data for *n*-butane hydrogenolysis catalyzed by MgO-supported clusters formed by decarbonylation of $[Os_{10}C(CO)_{24}]^{2-}$ and modeled as Os_{10} are also consistent with the structure sensitivity of hydrogenolysis [38].

SUMMARY AND ASSESSMENT [42]

Supported metal clusters are a new class of catalyst made possible by syntheses involving organometallic chemistry on surfaces, gas-phase cluster chemistry, and chemistry in zeolite cages. The synthetic chemistry would not have been possible without the guidance of new characterization science; EXAFS spectroscopy is the technique that has provided the most insight into structures of supported metal clusters. Clusters such as Ir_4, Ir_6, and Pt_n (where n is about 6) are small enough to be considered quasi molecular rather than metallic. Their catalytic properties are distinct from those of larger (metallic) particles, even for structure-insensitive reactions.

MgO-supported Ir_4 and Ir_6 clusters are stable catalysts, as shown both by EXAFS results indicating retention of the metal framework structures during catalysis (Table 4) and by the lack of significant changes in the catalytic activities during steady-state operation in flow reactors [23, 34]. Thus supported metal cluster catalysts seem to be robust enough for practical application, although questions about their possible regeneration remain to be answered. It may be fruitful to search for reactions for which supported metal clusters have catalytic properties superior to those of conventional supported metals. The catalytic activities of clusters modeled as Ir_4 and Ir_6 are less than those of supported iridium particles, at least for hydrocarbon hydrogenation and hydrogenolysis reactions, but the important opportunity in catalysis may be to find reactions for which the activity or selectivity of supported metal clusters is superior to those of conventional supported metals. The high selectivity of Pt/LTL zeolite catalysts for alkane dehydrocyclization appears to be the most persuasive indication of the value of supported metal cluster catalysts. The high selectivity of this catalyst for dehydrocyclization is related to its low selectivity for hydrogenolysis, which may be, at least in part, a consequence of the smallness of the platinum clusters.

ACKNOWLEDGMENT

This work was supported by the National Science Foundation (CTS-9300754).

REFERENCES

(1) Boudart, M. *J. Mol. Catal.* **1985**, *30*, 27.
(2) Gates, B. C.; Lamb, H. H., *J. Mol. Catal.* **1989**, *52*, 1.
(3) Gates, B. C.; Guczi, L.; Knözinger, H., Eds., *Metal Clusters in Catalysis*, Elsevier, Amsterdam, 1986.
(4) Iwasawa, Y. *Catal. Today* **1993**, *18*, 21.
(5) Ichikawa, M. *Adv. Catal.* **1992**, *38*, 283.
(6) Gates, B. C. *J. Mol. Catal.* **1994**, *86*, 95.
(7) Psaro, R.; Ugo, R.; Zanderighi, G. M.; Besson, B.; Smith, A. K.; Basset, J.-M. *J. Organomet. Chem.* **1981**, *213*, 215.
(8) Deeba, M.; Gates, B. C. *J. Catal.* **1981**, *67*, 303.
(9) Knözinger, H.; Zhao, Y. *J. Catal.* **1981**, *71*, 337.
(10) Psaro, R.; Dossi, C.; Fusi, A.; Ugo, R. *Res. Chem. Intermed.* **1991**, *15*, 31.
(11) Hsu, L.-Y.; Shore, S. G.; D'Ornelas, L.; Choplin, A.; Basset, J.-M. *J. Catal.* **1994**, *149*, 159.
(12) Krause, T. R.; Davies, M. E.; Lieto, J.; Gates, B. C. *J. Catal.* **1985**, *94*, 195.
(13) Kawi, S.; Gates, B. C., *Inorg. Chem.* **1992**, *31*, 2939.
(14) Lamb, H. H.; Fung, A. S.; Tooley, P. A.; Puga, J.; Krause, T. R.; Kelley, M. J.; Gates, B. C. *J. Am. Chem. Soc.* **1989**, *111*, 8367.

(15) Xu, Z.; Kawi, S.; Gates, B. C. *Inorg. Chem.* **1994**, *33*, 503.
(16) Puga, J.; Patrini, R.; Sanchez, K. M.; Gates, B. C. *Inorg. Chem.* **1991**, *30*, 2479.
(17) Xu, Z.; Rheingold, A.; Gates, B. C. *J. Phys. Chem.* **1993**, *97*, 9465.
(18) Kawi, S.; Chang, J.-R.; Gates, B. C., *J. Phys. Chem.* **1993**, *97*, 5375.
(19) Dossi, C.; Psaro, R.; Roberto, D.; Ugo, R.; Zanderighi, G. M. *Inorg. Chem.* **1990**, *29*, 4368.
(20) Kawi, S.; Chang, J.-R; Gates, B. C. *J. Phys. Chem.* **1993**, *97*, 10599.
(21) Kawi, S.; Chang, J.-R; Gates, B. C. *J. Am. Chem. Soc.* **1993**, *115*, 4830.
(22) Kawi, S.; Chang, J.-R; Gates, B. C. to be published.
(23) Xu, Z.; Xiao, F.-S.; Purnell, S. K.; Alexeev, O., Kawi, S.; Deutsch, S. E.; Gates, B. C. *Nature (London)*, **1994**, *372*, 346.
(24) van Zon, F. B. M.; Maloney, S. E.; Gates, B. C.; Koningsberger, D. C. *J. Am. Chem. Soc.* **1993**, *115*, 10317.
(25) Chang, J.-R.; Koningsberger, D. C.; Gates, B. C. *J. Am. Chem. Soc.*, **1992**, *114*, 6460.
(26) Lamb, H. H.; Wolfer, M.; Gates, B. C., *J. Chem. Soc., Chem. Commun.* **1990**, 1296.
(27) Beutel, T.; Kawi, S.; Purnell, S. K.; Knözinger, H.; Gates, B. C. *J. Phys. Chem.* **1993**, *97*, 7284.
(28) Xiao, F.-S.; Alexeev, O.; Gates, B. C., *J. Phys. Chem.*, in press.
(29) Eberhardt, W.; Fayet, P.; Cox, D.; Kaldor, A.; Sherwood, R.; Sondericker, D. *Physica Scripta*, **1990**, *41*, 892.
(30) Roy, H. V.; Fayet, P.; Patthey, F.; Schneider, W. D.; Delly, B.; Massobrio, C. *Phys. Rev. B*, **1994**, *49*, 5611.
(31) Koningsberger, D. C.; Prins, R., Eds. *X-ray Absorption: Principles, Applications, Techniques of EXAFS, SEXAFS, and XANES* Wiley, New York, 1988.
(32) Gates, B. C.; Koningsberger, D. C. *CHEMTECH* **1992**, *22*, 300.
(33) Maloney, S. D.; van Zon, F. B. M.; Koningsberger, D. C.; Gates, B. C. *Catal. Lett.* **1990**, *5*, 161.
(34) Kawi, S.; Chang, J.-R.; Gates, B. C. *J. Phys. Chem.*, in press, 1994.
(35) Triantafillou, N. D.; Gates, B. C. *J. Phys. Chem.* **1994**, *98*, 8431.
(36) Asakura, K.; Iwasawa, Y. *J. Chem. Soc. Faraday Trans.* **1990**, *86*, 2657.
(37) Fung, A. S.; Tooley, P. A.; Kelley, M. J.; Koningsberger, D. C.; Gates, B. C. *J. Phys. Chem.* **1991**, *95*, 225.
(38) Lamb, H. H.; Wolfer, M.; Gates, B. C. *J. Chem. Soc., Chem. Commun.* **1990**, 1296.
(39) Purnell, S. K.; Xu, X.; Goodman, D. W.; Gates, B. C. *J. Phys. Chem.* **1994**, *98*, 4076.
(40) Ravenek, W.; Jansen, A. P. J.; van Santen, R. A. *J. Phys. Chem.* **1989**, *93*, 6445.
(41) Maloney, S. D.; Kelley, M. J.; Koningsberger, D. C.; Gates, B. C. *J. Phys. Chem.* **1991**, *95*, 9406.
(42) Gates, B. C. *Chem. Rev.*, submitted.

INFRARED SPECTROSCOPIC CHARACTERIZATION OF SULFIDE CLUSTER-DERIVED ENSEMBLES

JAMES R. BRENNER* AND LEVI T. THOMPSON**
* Argonne National Laboratory, Department of Chemistry, 9700 S. Cass Ave., Bldg. 200, Room C-113, Argonne, IL 60439
** The University of Michigan, Department of Chemical Engineering, 2300 Hayward Ave., 3230 H.H. Dow Bldg., Ann Arbor, MI 48109-2136

ABSTRACT

The transition metal sulfide clusters $(MeCp)_2Mo_2(\mu\text{-}SH)_2(\mu\text{-}S)_2$, $(MeCp)_2Mo_2Co_2(\mu_3\text{-}S)_2(\mu_4\text{-}S)(CO)_4$ [MoCoS], and $(MeCp)_2Mo_2Fe_2(\mu_3\text{-}S)_2(CO)_8$, (MeCp = methylcyclopentadienyl) were used to prepare γ-Al_2O_3-supported catalysts. For comparison, a series of supported materials was also prepared using conventional incipient wetness impregnation.

Infrared spectroscopy of adsorbed species was used to characterize the sites in the cluster-derived and conventionally prepared catalysts. Nitric oxide chemisorbed onto the MoCoS/A catalyst was associated initially only with Co sites and then upon gentle heating shifted to the Mo sites, indicating that Co and Mo were in close proximity. In contrast, NO adsorbed onto both Co and Mo sites in the conventionally prepared materials and desorbed independently from these two types of sites. Infrared spectra of adsorbed thiophene and pyridine were similar for the cluster-derived and conventionally prepared catalysts. Thiophene reacted at 100 °C to produce both olefinic species. The most abundant products from thiophene HDS were 1-butene, cis-2-butene, and trans-2-butene. Displacement studies showed that thiophene, pyridine, and NO adsorbed to the same site. The most active sites for HDS and HDN contained both Mo and a late transition metal. The HDN product distributions suggested that Mo was selective for C=N bond cleavage while the late transition metals were more active for C=C hydrogenolysis.

I. EXPERIMENTAL

Synthesis. The clusters $(MeCp)_2Mo_2(\mu\text{-}SH)_2(\mu\text{-}S)_2$ [MoS], $(MeCp)_2Mo_2Fe_2(\mu_3\text{-}S)_2(CO)_8$ [MoFeS], $(MeCp)_2Mo_2Co_2(\mu_3\text{-}S)_2(\mu_4\text{-}S)(CO)_4$ [MoCoS], $Fe(\mu\text{-}S)_2(CO)_6$ [FeS], and $Cp_4Co_4(\mu_3\text{-}S)_2(\mu_3\text{-}S_2)_2$ [CoS] were synthesized according to procedures described in the literature (MeCp = methylcyclopentadienyl) [1-5]. Gamma-alumina (Catapal, 150 m^2/g) was calcined at 500 °C in dry air for 5 hours. A solution of the desired sulfide cluster was added to a slurry of the support using standard Schlenk techniques to produce a total metal loading of 1.0 wt. %. The resulting slurry was stirred under N_2 for one hour before the solvent was evaporated off under vacuum and the materials stored under N_2. The cluster-derived catalysts have been coded K/A, where K denotes the type of cluster and A denotes γ-Al_2O_3. Transmission electron microscopy has shown the cluster-derived materials to be molecularly dispersed [6].

A second series of catalysts was prepared using conventional synthesis methods to provide a basis for comparison with the cluster-derived materials. Ammonium heptamolybdate and Co, Ni, or Fe nitrate salts were impregnated onto γ-Al_2O_3 to the point of incipient wetness (IW) sequentially with Mo first to produce nominal loadings of 1.0 wt. % total metal and a 1:1 atomic ratio of Mo to promoter. After each impregnation step, the resulting material was calcined in dry air at 500 °C for 5 hours. A series of 1 wt. % Fe, Co, and Ni oxides supported on γ-Al_2O_3 were also prepared via the incipient wetness impregnation of the metal nitrates. The calcined γ-Al_2O_3-supported oxides were sulfided in flowing 2% H_2S/H_2 at 400 °C for at least 4 hours.

Infrared Spectroscopy. Thin wafers of the materials were pressed and infrared spectroscopy was performed inside a N_2-filled drybox. Details concerning the system are given elsewhere [6]. Because the cluster derived materials were air sensitive and the press could not be placed in the drybox, preparation of these materials involved pipetting a solution of the cluster in CH_2Cl_2 onto pressed wafers of calcined γ-Al_2O_3. Treatments were carried out by flowing gas

Mat. Res. Soc. Symp. Proc. Vol. 368 © 1995 Materials Research Society

through a treatment cell at 100 ccm and 10 psig. Following a purge with flowing He at 100 ccm for one hour, the wafers were transferred under N_2 to the spectroscopy cell. The temperatures used for the adsorption of NO, thiophene, and pyridine were 25, 100 and 60 °C, respectively.

II. RESULTS AND DISCUSSION

Catalytic Properties. The hydrodesulfurization (HDS) and hydrodenitrogenation (HDN) activities and product distributions of the sulfide cluster-derived catalysts have previously been shown to be similar to those of conventionally prepared, sulfided Mo-Co/A, Mo-Fe/A and Mo/A catalysts [6-9]. The HDS and HDN activities which are summarized in Table 1 suggest that the sulfide cluster-derived ensembles are adequate models of the active sites in commercial hydrotreatment catalysts.

Table 1 - Thiophene HDS and Pyridine HDN Activities

Catalyst	HDS Activity* (10^3 mol/mol O_2/s)	ΔE_{act} (kJ/mole)	HDN Activity* (10^6 mol/mol O_2/s)	ΔE_{act} (kJ/mole)
MoS/A	1.9	92	1.2	92
Mo/A IW	2.7	84	8.2	75
FeS/A	0.06	‡	1.3	142
MoFeS/A	4.3	79	5.2	79
Mo-Fe/A IW	3.8	71	13	172
CoS/A	0.07	‡	2.7	67
MoCoS/A	9.6	59	23	84
Mo-Co/A IW	7.3	92	48	105

*HDS and HDN activities at 365 °C and 1 atm
‡Too low to accurately measure

NO Adsorption. Evidence for NO adsorption on the γ-Al_2O_3 support was observed in the 1500-1700 cm^{-1} range when the NO dose was greater than 15 minutes. The "NO-on-γ-Al_2O_3" features were much more significant for the cluster-derived materials than for the conventionally prepared sulfide catalysts. This observation may be a consequence of the reaction of aluminum carbonates produced during the decarbonylation of the clusters with NO. Nitric oxide saturated the Mo or promoter sites after 2 minutes of exposure; therefore, the NO dose was restricted to 2 minutes in all cases.

The infrared spectra of NO adsorbed on the MoS/A and conventionally prepared Mo/A IW catalysts were remarkably similar (Fig. 1). The band positions suggested that NO adsorbed onto Mo as a dinitrosyl [10]. The infrared spectra of NO adsorbed onto the cluster-derived MoFeS/A have been compared with those for conventionally prepared Mo-Fe/A IW, Mo/A IW, and Fe/A IW sulfide catalysts in Fig. 2. There was a significant degree of overlap between the peaks corresponding to the symmetric stretches of NO on Fe and Mo (~1800-1810 cm^{-1}). The assignment of the asymmetric NO stretches (1745 cm^{-1} for Fe and 1700 cm^{-1} for Mo) was clear, however. Spectra for the conventionally prepared Mo-Fe/A catalysts indicated that NO adsorbed onto both Fe and Mo. In contrast, character of the MoFeS/A catalyst appeared be to dominated by Fe. In fact, the spectrum for MoFeS/A was similar to that for the Fe/A IW catalyst.

Nitric oxide adsorption on Co/A gave rise to two peaks, one at ≈1865 cm^{-1} and the other which overlapped the 1805 cm^{-1} peak for Mo (Fig. 3). Nitric oxide adsorbed onto the conventionally prepared Mo-Co/A catalysts produced three peaks that were approximately but not exactly the weighted sum of the spectra of NO adsorbed onto the Co and Mo sulfides. The strong similarities of the spectra for the MoCoS/A and Co/A IW catalysts suggested that NO adsorption sites on the MoCoS/A catalyst were predominantly Co. Nitric oxide desorbed from the Mo and Co sites independently for the conventionally prepared Mo-Co/A IW materials.

MOS/A
MO/A IW
AL2O3
2000
1900
1800
1700
1600
1500
Wavenumber (cm-1)

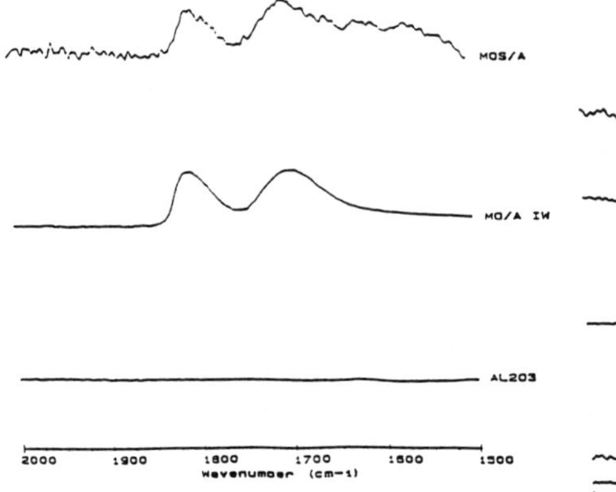

Fig. 1 - NO adsorption on Mo-containing catalysts and γ-Al_2O_3.

MoFeS/A
MoFe/A IW
Mo/A IW
Fe/A IW
Al2O3
2000
1900
1800
1700
1600
1500
Wavenumber (cm-1)

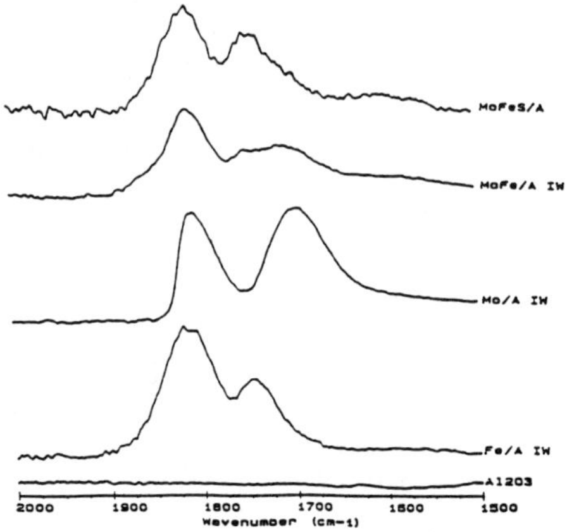

Fig. 2 - NO adsorption on Mo and Fe containing catalysts and γ-Al_2O_3.

MOCOS/A
MOCO/A IW
MO/A IW
CO/A IW
AL2O3
2000
1900
1800
1700
1600
1500
Wavenumber (cm-1)

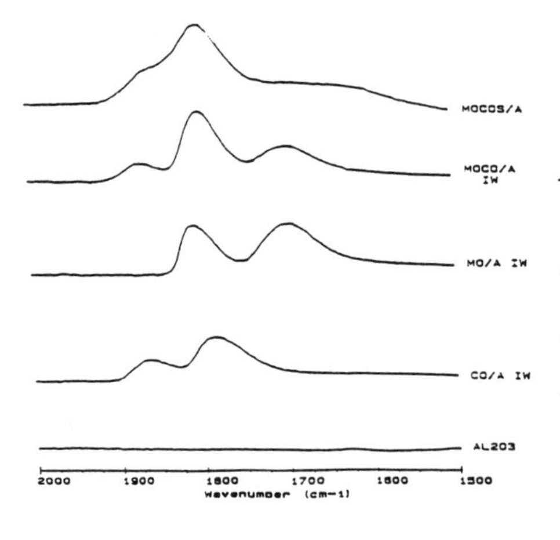

Fig. 3 - NO adsorption onto Mo- and/or Co-containing catalysts and γ-Al_2O_3.

NO @ 25 C Then He Purge for 0.5 hr @
25 C
100 C
200 C
300 C
400 C
2000
1900
1800
1700
1600
1500
Wavenumber (cm-1)

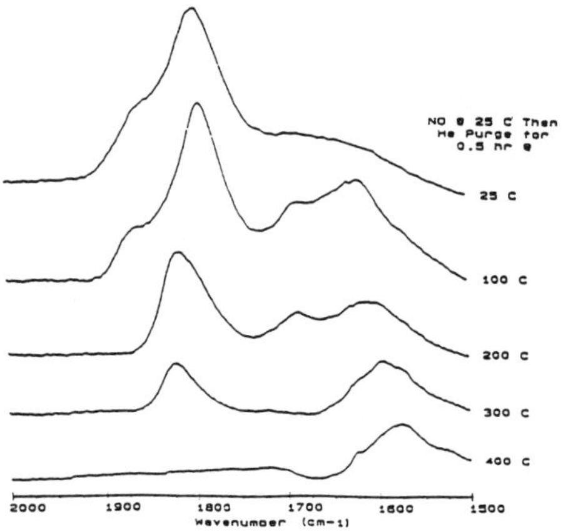

Fig. 4 - Desorption of NO from MoCoS/A catalyst.

For both the cluster-derived and conventionally prepared catalysts, NO desorbed from Co by 200 °C and from Mo between 200 and 300 °C. A series of spectra detailing the consequences of heating the MoCoS/A catalyst following NO adsorption is given in Fig. 4. Upon heating to 100 °C, the peak at 1865 cm^{-1} (NO on Co) decreased with respect to the peak at 1805 cm^{-1}. Concomitantly, the intensity of the peak at 1692 cm^{-1}, consistent with NO associated with unpromoted sites, increased. This result was reproduced several times. The ratio of the areas of the symmetric (shifted from 1805 cm^{-1} to 1820 cm^{-1}) and asymmetric (1692 cm^{-1}) NO-Mo stretches remained very high. After desorption at 300 °C, only the symmetric NO-Mo stretch was present. These results indicated that desorption of NO from Co had an affect on the bonding of NO to Mo and can be explained if either NO adsorption onto Co blocked NO adsorption onto Mo or if NO migrated from Co to Mo. Regardless of which explanation is correct, both imply that Mo and Co were in close enough proximity in MoCoS/A in that desorption of NO from Co had an effect on the bonding of NO to Mo.

Thiophene Adsorption. There were no C-H stretch features (3200-2800 cm^{-1}) attributable to thiophene on metal sulfides. The failure to detect these C-H stretch features was due both to the relatively low loadings (1 wt. % metal) employed and to the strong infrared absorption of γ-Al_2O_3 in this region. Spectra after thiophene adsorption for the cluster-derived catalysts are shown in Fig. 5. The strong similarities between the spectra of thiophene adsorbed onto cluster-derived and conventionally prepared catalysts again suggested that the sulfide cluster-derived ensembles modeled sites in conventionally prepared sulfide catalysts.

A sharp peak at 1253 cm^{-1} was observed only for the Mo-containing catalysts suggesting that thiophene adsorbed onto Mo. This feature is expected for the in-plane C-H bending of liquid and vapor-phase thiophene [11]. The C=C stretching frequencies at 1680 and 1470 cm^{-1} were absent in spectra of the cluster-derived catalysts. This observation is consistent with π-bound thiophene as opposed to S-bound thiophene. Features in the 1700-1850 cm^{-1} range were observed for all the catalysts. This range is too high to be attributable to aromatic C=C stretches and is more consistent with olefinic C=C stretches, suggesting that opening of the thiophene ring had occurred by 100 °C. The band at 1850 cm^{-1} indicated the presence of a terminal olefin; there are no C-H, C-S, S-H or other C=C vibrational modes between 1800 and 2000 cm^{-1}.

Much of the thiophene desorbed between 100 and 200 °C. The olefin desorption temperature was approximately 300 °C for the MoCoS/A catalyst. The temperature for olefin desorption from the MoS/A and MoFeS/A catalysts was between 300 and 400 °C. Reduction at 400 °C was sufficient to regenerate all the catalysts.

Pyridine Adsorption. The adsorption of pyridine onto the cluster-derived materials and blank γ-Al_2O_3 is shown in Fig. 6. The spectra for the Mo-containing materials contained a band at 1595 cm^{-1}. This band was not observed for the Co/A, Fe/A, and Ni/A materials. Pyridine populated the γ-Al_2O_3 first and then filled the Mo sites. This suggested that the Mo sites were less acidic than those on γ-Al_2O_3. Pyridine desorbed upon He purging at 100 °C.

Adsorbate Displacement Studies. Exposures to thiophene of up to 5 hours did not significantly influence the spectrum of an MoS/A catalyst that had been dosed with NO (Fig. 7). The NO; however, was completely removed upon H_2 reduction at 400 °C. A subsequent one-hour thiophene exposure at 100 °C produced both the thiophene-on-Mo vibration at 1253 cm^{-1} and the C=C stretch at 1800 cm^{-1}. The vibrational frequencies for NO and C=C bonds overlap, so it was not entirely clear whether or not NO displaced the olefinic species. Nonetheless, it was clear that NO displaced the thiophene that had adsorbed to Mo. Nitric oxide also completely displaced thiophene from Mo sites in the MoFeS/A and MoCoS/A catalysts. The NO and C=C stretching frequencies also overlapped for the MoFeS/A and MoCoS/A catalysts. Recall that most of the olefin adsorbed to the bimetallic catalysts was adsorbed to the late transition metal. Therefore, it was not clear whether NO displaced olefin from either Mo or the late transition metal sites.

For both the cluster-derived and conventionally prepared catalysts, pyridine adsorbed at 60 °C displaced all of the NO adsorbed at 25 °C (Fig. 8 for MoS/A). Much of this pyridine was displaced upon NO readsorption.

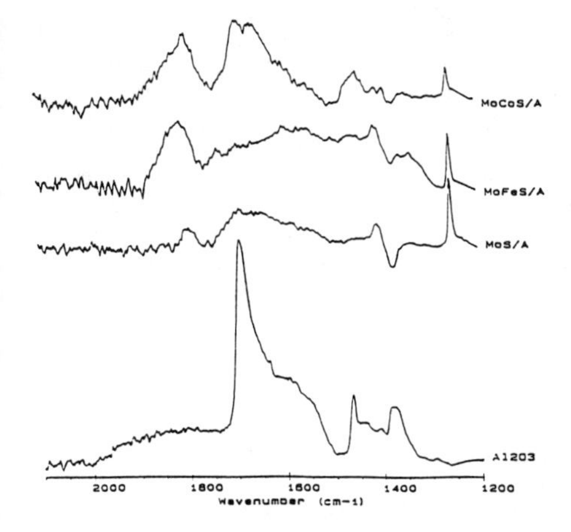

Fig. 5 - Thiophene adsorption onto cluster-derived catalysts and γ-Al_2O_3.

Fig. 6 - Pyridine adsorption onto cluster-derived catalysts and γ-Al_2O_3.

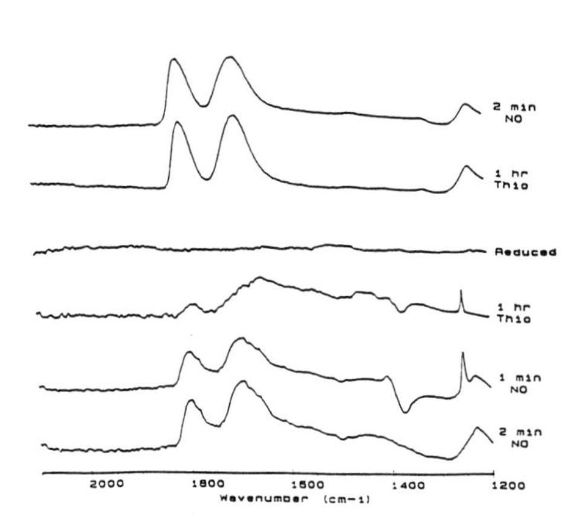

Fig. 7 - NO displacement by thiophene for MoS/A catalyst.

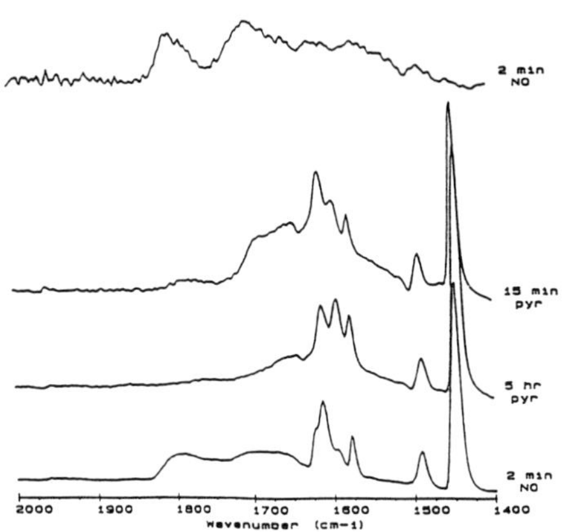

Fig. 8 - Pyridine displacement by NO and vice versa for MoS/A catalyst.

III. CONCLUSIONS

Infrared spectroscopic results indicated that the character of active sites derived from the sulfide clusters closely resembled that of sites present on the surfaces of conventionally prepared unpromoted and promoted Mo sulfide hydrotreatment catalysts. In particular, the positions and relative intensities of bands for NO adsorbed on the sulfide cluster derived ensembles were similar to those for the conventionally prepared HDS/HDN catalysts. This conclusion is consistent with our earlier findings that the catalytic properties of the cluster derived and conventionally prepared catalysts were similar. Evidence was also presented which suggested that Mo and the promoter element (Co or Fe) were in close proximity to each other. It is plausible, but not verified, that Mo and the promoter were bonded together forming highly dispersed cluster derived ensembles. Finally, our results indicated that the hydrogenolysis of C-S bonds in thiophene is facile while the hydrogenation of surface-bound sulfur is the rate limiting step during HDS over the cluster derived catalysts.

IV. REFERENCES

1. Rakowski DuBois, M., VanDerveer, M.C., Dubois, D.L., Haltiwanger, R.C., and Miller, W.K., J. Am. Chem. Soc. **102**, 7456 (1980).
2. Williams, P.D., Curtis, M.D., Duffy, D.N., and Butler, W.M., Organometallics **2**, 165 (1983).
3. Curtis, M.D., and Williams, P.D., Inorg. Chem. **22**, 2261 (1983).
4. Uchtman, V.A., and Dahl, L.F., J. Am. Chem. Soc. **91**, 3756 (1969).
5. Bogan, L.E., Lesch, D.A., and Rauchfuss, T.B., J. Organomet. Chem. **250**, 429 (1983).
6. Brenner, J.R., Ph.D. Dissertation, The University of Michigan, 1994.
7. Carvill, B.T. and Thompson, L.T., Appl. Catal. **75**, 249 (1991).
8. Brenner, J.R., Carvill, B.T., and Thompson, L.T., Appl. Organomet. Chem. **6**, 463 (1992).
9. Brenner, J.R., and Thompson, L.T., Catal. Today **21**, 101 (1994).
10. Rosen, R.P., Segawa, K., Millman, W.S., and Hall, W.K., J. Catal. **90**, 10 (1984).
11. Waddington, G., Knowlton, J.W., Scott, D.W., Oliver, G.D., Todd, S.S., Hubbard, W.N., Smith, J.C., and Huffman, H.M., J. Am. Chem. Soc. **71**, 797 (1949).
12. Topsøe, H., Clausen, B.S., Topsøe, N.Y., and Pedersen, E., Ind. Eng. Chem. Fundam. **25**, 25 (1986).

SELECTIVE HYDROGENATION OF CROTONALDEHYDE OVER COBALT BASED SELF-SUPPORTED CATALYSTS

A. N. PATIL*, M. A. BAÑARES*, X. LEI**, T. P. FEHLNER**, AND E. E. WOLF*

*Department of Chemical Engineering
**Department of Chemistry and Biochemistry,
University of Notre Dame, Notre Dame, Indiana-46556, U. S. A.

ABSTRACT

Complex cobalt-carbonyl ligand based clusters of clusters are used as molecular precursors for self-supported model catalysts. These precursors consist of two metal layers: an outer of the complex Co-carbonyl ligands, and a core of metal (e.g. Co or Zn) carboxylate groups. Partial thermolysis at low temperature (LT) of these materials under hydrogen results in almost completely decarbonylated material with a mainly unchanged carboxylate metal core. Complete pyrolysis at higher temperatures (HT) in hydrogen leads to mixed metal environment. These materials were used as a heterogenous catalyst in the gas phase hydrogenation of crotonaldehyde. The maximum yield of 27 % of desired product crotyl alcohol was observed when HT-CoCo was used as the catalyst at 423 K. The catalyst activity and the crotyl alcohol selectivity remained unchanged over 2 days of operation. The bimetallic ZnCo catalysts showed lower selectivity to crotyl alcohol than the CoCo catalysts.

INTRODUCTION

The selective hydrogenation of $\alpha-\beta$ unsaturated aldehydes to unsaturated alcohols is important in production of perfumes, flavorings and pharmaceuticals (1, 2). When carbonyl group is present in conjugation with C=C group, the most commonly used catalyst metals for hydrogenation such as Ni, Pt and Pd in the group VIII, preferentially reduce C=C bond yielding saturated aldehydes (compound III in Figure 1) and saturated alcohols (compound IV in Figure 1) (3, 4). This is possibly due to lower bond energy of the C=C bond than the C=O bond (5). Only Os and Co in the group VIII have been reported to be more selective to unsaturated alcohols (6, 7). For crotonaldehyde hydrogenation reaction, in order to improve selectivity to unsaturated alcohol (crotyl alcohol), the catalysts have been modified by alloying (8, 9), by adding promoters (10), incorporating strong interaction support (3) or by controlled poisoning (11, 12). The structure of catalyst is also found to be important (13). The crotyl alcohol yields up to 50 % have been reported in the literature (3-12). Most of these reactions are homogenous catalytic reactions carried out in presence of solvent at high pressures. The catalyst recovery and the separtion of the products from the solvents, the reactants and the catalysts are important considerations (14). Deactivation and time dependent selectivities over extended periods are also observed for these reactions (15, 16).

Previously, we have used the catalysts prepared by activating cobalt based carbonyl cluster carboxylate ligand complexes for hydrogenation of 1-3 butadiene (17, 18). These catalysts exhibit two different structures based on the activation (18). In this paper, our results of the gas phase hydrogenation of crotonaldehyde using the heterogeous self supported cobalt based clusters of clusters catalysts are presented.

EXPERIMENTAL

The molecular precursors $M_4O[(CO)_9Co_3CCO_2]_6$ (where M= Co or Zn) are used. The preparative methods for these molecular precursors have been described elsewhere (19,

Mat. Res. Soc. Symp. Proc. Vol. 368

20). These molecules have an O atom centered tetrahedral core of metal (M) atoms (see Figure 2). The six edges of the core metal tetrahedron are bridged by six cobalt carbonyl cluster carboxylate ligands.

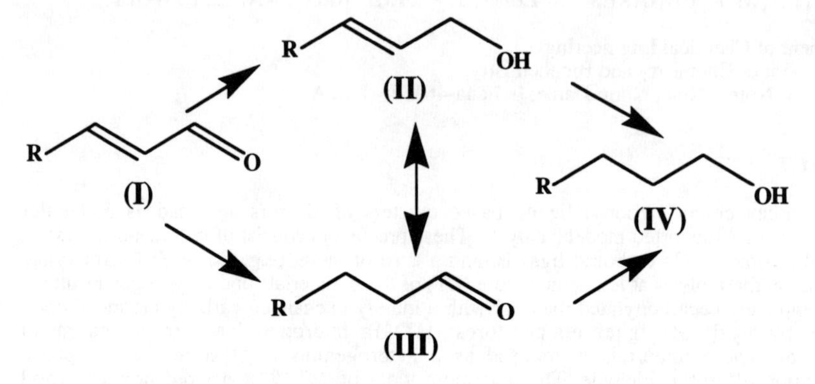

R = CH_3 for crotonaldehyde hydrogenation

Figure 1. Hydrogenation of α–β unsaturated aldehyde

The weight loss taking place for the different treatment procedures of the catalysts was measured with a Cahn RG Electrobalance. The sample temperature was increased linearly from room temperature to 673 K in 190 minutes in flowing helium or hydrogen (100 ml/min, NPT). The thermogravimetric analysis plot is shown in Figure 3.

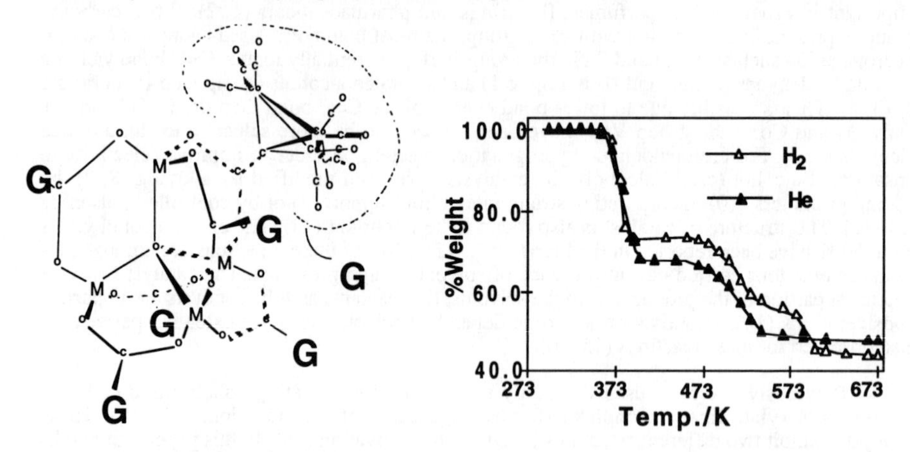

Figure 2. Schematic representation of catalyst precursor $M_4O[(CO)_9Co_3CCO_2]_6$

Figure 3. Thermogravimetric analysis plot for CoCo catalyst precursor

The hydrogenation of crotonaldehyde was studied in a quartz flow microreactor (4 mm i.d.) at atmospheric pressure in the temperature range 310-473 K. Typically 12 mg of the specific molecular precursor, confined with quartz wool, were activated under a hydrogen (99.999%, Mittler, South Bend, IN). In these measurements helium (99.999%, Mittler, South Bend, IN) was used to keep the total flow constant. Crotonaldehyde (99% +, Aldrich,

Milwaukee, WI) was introduced in the gas flow by bubbling reactant gas through crotonaldehyde at 273 K. At 273 K crotonaldehdye vapor pressure is 9.3 torr (21). This corresponds to 1.2 % of crotonaldehyde in the gas phase before the reaction. The reactor lines are at room temperature (298 K). The vapor pressures of the products of the reaction at 298 K is less than vapor pressure of crotonaldehyde at 273 K (21). Hence condensation of the products in the lines is not expected. Mass flows were controlled by electronic flow controllers (Brooks, mod. 5850-E, Brooks Instruments, Hatfield, PA) and the reaction system was automated and computer controlled. The reactor effluent was analyzed by gas chromatography on a Varian 3700 GC provided with a FID detector. A 10 % carbowax 20M on chromosorb 80/100 mesh packed column (Alltech, Deerfield, IL) was used to separate the effluents which were crotonaldehyde, crotyl alcohol, 1-butanol, 1-butanal and butane.

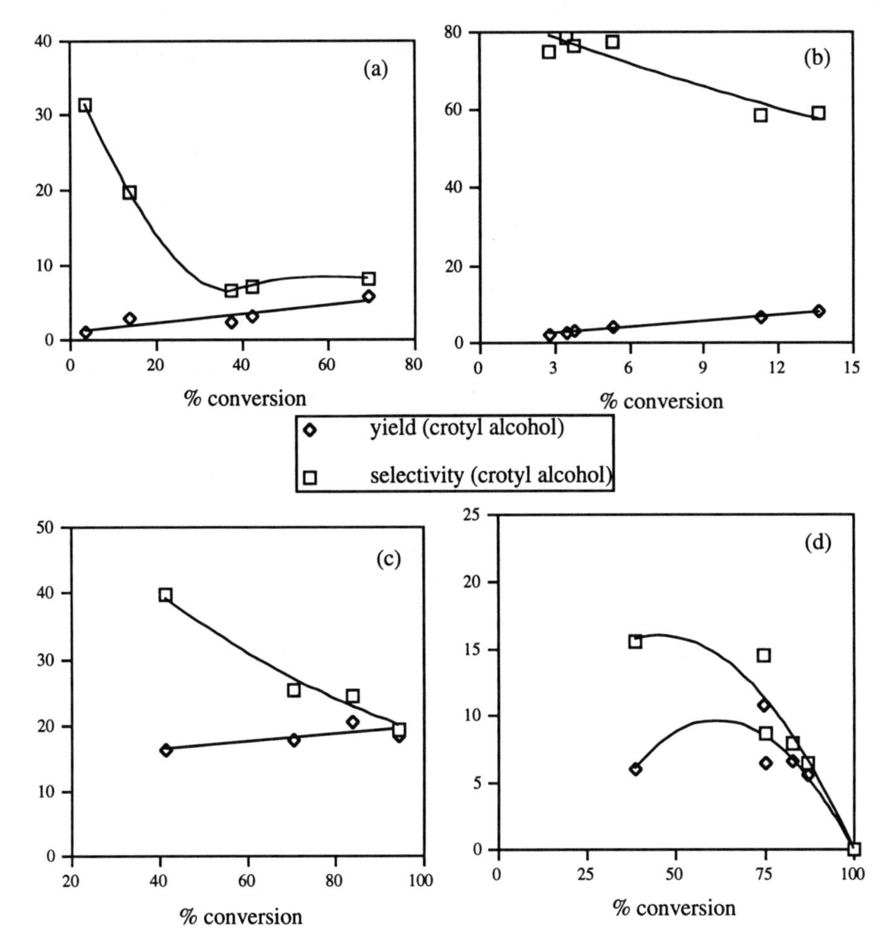

Figure 4. Plots of crotyl alcohol yield and crotyl alcohol selectivity versus % conversion for CoCo catalysts (a) LT-CoCo at 393 K (b) HT-CoCo at 393 K (c) HT-CoCo at 423 K (d) HT-CoCo at 453 K.

RESULTS AND DISCUSSION

The thermogravimetric analyses during the pyrolysis of the precursors exhibit two stages (Figure 3). For CoCo, the first stage of activation in helium corresponds to the loss of 37 carbonyl ligands, and the second stage, to the loss of the remaining carbonyl ligands and carbon dioxide molecules moieties from the carboxylate ligands. In hydrogen, the initial mass loss is less but the final mass loss is larger. The structure obtained in the first plateau under these treatments are designated as the low temperature (LT) catalysts. The catalysts resulting from treatments in the higher temperature regime are designated as the high temperature (HT) catalysts. The infrared spectroscopy and mass spectroscopy - temperature programmed decomposition experiments, reported elsewhere (18), show the carbonyls leaving first, followed by the carboxylates thereby leaving the core structure intact. The presence of the carboxylate ligands suggests that the core structure in the LT-species is similar to that observed for the initial precursors (see Figure 3).

Figures 4 a, b, c, d show crotyl alcohol yield and crotyl alcohol selectivity as a function of % conversion for the LT-CoCo catalyst at 393 K and for the HT-CoCo catalysts at 393 K, 423 K and 453 K respectively. By varying % of H_2 in the reactant stream, the crotyl alcohol conversion at given temperature and at given total flow was varied. It can be seen that for the LT-CoCo at 393 K, up to 70 % crotonaldehyde conversion was obtained (corresponding to 100 % hydrogen on crotonaldehyde free basis). At 70 % crotonaldehyde conversion, the rate of hydrogenation was 7.0×10^{-4} moles/hr g of catalyst. The corresponding turn over frequency was 1.2×10^{10} molecules/s cm^2 of the catalyst surface area. The catalyst surface area measurements are described elsewhere (18). The selectivity for crotyl alcohol drops rapidly from 30 % to 10 % at lower conversions and stabilizes around 10 %. The maximum yield of crotyl alcohol was 5 %. The HT-CoCo is not as active as LT-CoCo (for the same amount of precursor) at 393 K possibly due to lower surface areas (18) with maximum conversion of about 13 %. The crotyl alcohol selectivity is much higher than when the reaction was catalyzed by LT-CoCo at 393 K. The crotyl alcohol selectivity dropped from 80 % to 60 % with the increase in the conversion from 2 % to 13 %.

At 423 K, about 95 % crotonaldehyde conversion was achieved using the HT-CoCo catalyst. The crotyl alcohol yield was around 20 % and the highest crotyl alcohol yield we obtained in our experiments was 27 % (see Figure 6). The crotyl alcohol selectivity dropped by 50 % with the increase in conversion from 40 % to 100 %.

Table 1: Product distribution for the CoCo catalyzed hydrogenation of crotonaldehyde

	% conversion	butane	1-butanal	crotyl alcohol	1-butanol
LT-CoCo					
393 K	3.2	0	50	31.3	18.7
393 K	13.6	4.4	21.3	19.9	54.4
393 K	69.6	3.3	24.7	8.2	63.8
HT-CoCo					
393 K	13.6	0.7	27.9	58.9	12.5
423 K	41.3	0	33.4	39.7	26.9
423 K	70.8	1.8	25.4	25.4	47.4
423 K	94.4	2.1	19.5	19.5	58.9
453 K	38.4	8.6	48.4	15.6	27.4
453 K	74.9	4.2	33.5	8.7	53.6
453 K	100.0	4.0	0.0	0.0	96.0

At 453 K, HT-CoCo catalyst exhibited poor selectivity to crotyl alcohol. At 100 % conversion , no crotyl alcohol was detected in the product stream. The maximum yield of crotyl alcohol at 453 K was 8 %. At 100 % crotonaldehyde conversion, 96 % yield of 1-butanol was obtained. At 100 % crotonaldehyde conversion, the rate of hydrogenation was 6.8×10^{-4} moles/hr g of catalyst. The corresponding turn over frequency was 5.4×10^{11} molecules/s cm^2 of the catalyst surface area.

Table 1 lists the product selectivities at various conditions of crotonaldehyde hydrogenation with LT-CoCo and HT-CoCo catalysts. At 393 K, at low conversions crotyl alcohol was the most common product. At 423 K, more and more butanal and butanol was detected in the product stream. At 453 K, at higher conversions mostly butanol was formed. Significant amount of butane was also detected at 453 K. HT-CoCo catalyzed hydrogenation of crotonaldehyde at 423 K gives the highest yield of crotyl alcohol. At lower temperatures (393 K) the crotyl alcohol selectivity is higher, but the activity is low. At 453 K, the selectivities for crotyl alcohol are lower than the selectivities at 423 K. The lower yield of crotyl alcohol at 453 K is possibly due to consumption of crotyl alcohol formed via isomerization to butanal or via further hydrogenation to butanol. The experiments are in progress to investigate cause of lower yields of crotyl alcohol at 453 K

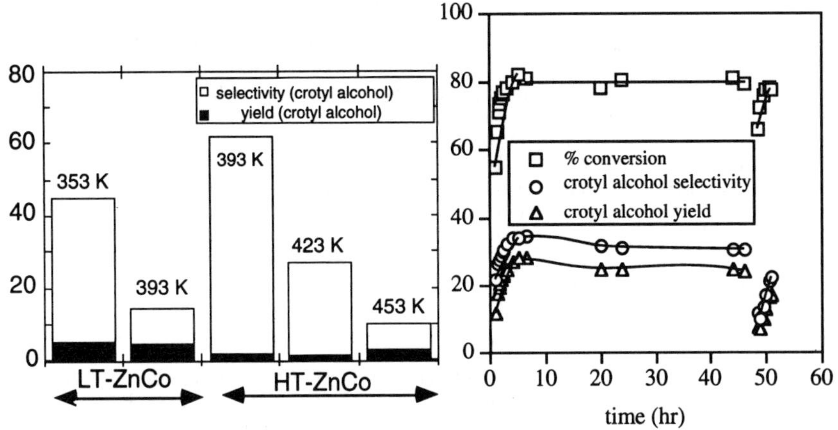

Figure 5 crotyl alcohol yields and selectivities for LT and HT ZnCo catalysts.

Figure 6 time dependence of crotonaldehyde conversion, crotyl alcohol yield and selectivity for HT-CoCo catalyst at 423 K

Figure 5 shows the highest crotyl alcohol yields obtained at different temperature conditions when LT-ZnCo and HT-ZnCo were used as catalysts. The corresponding values of selectivities are also shown in Figure 5. The LT-ZnCo at 353 K, yields maximum crotyl alcohol (about 5%) with 45 % selectivity. Overall, the ZnCo catalysts are much less selective to crotyl alcohol than the corresponding CoCo catalysts. Thus presence of Zn in place of Co in the core of cluster of clusters is not favorable for the higher yields of crotyl alcohol.

Figure 6 shows time dependence of crotonaldehyde conversion, crotyl alcohol yield and crotyl alcohol selectivity when the HT-CoCo was used as the catalyst. The reaction was carried out at 423 K with 80 % H_2 (crotonaldehyde free basis) in the reactant gas. The crotyl alcohol conversion increases sharply in the first few hours and levels off. The crotyl alcohol yield and the crotyl alcohol selectivity also stabilizes to about 27 % and 35 % respectively. The crotonaldehyde conversion, crotyl alcohol yield and crotyl alcohol selectivity remained constant when the reaction was continued over 2 days. The transitions occurring in the catalyst in the first few hours seem to be reversible. The similar transitional behavior was observed when crotonaldehyde flow was discontinued for about 2 hours (see Figure 6, between 45 - 48 hours) and was then restarted.

CONCLUSIONS

The molecular precursor clusters of clusters containing metal carbonyl cluster substituted carboxylate ligands upon partial thermolysis (LT catalysts) and complete thermolysis (HT catalysts), yield catalytically active materials. The CoCo catalysts exhibit high activity for hydrogenation of crotonaldehyde. The HT-CoCo at 423 K, gives the highest yield of desired product crotyl alcohol. The catalyst activity and selectivity are stable. The bimetallic ZnCo catalysts are less selective than one component CoCo catalysts.

ACKNOWLEDGMENTS

This work is funded by the National Science Foundation under grant CHE91-06933.

REFERENCES

[1] Arctander S., *Perfumes and Chemicals*, (Published by the author, Montaclair, NJ, 1969), p747.
[2] Weissermel K. and Arpe H. J., *Industrial Organic Chemistry*, (Verlag Chemie, Weinheim, New York, NY, 1978), p261.
[3] Vannice M.A. and Sen B., J. Cat., 115, 65 (1989).
[4] Chen Y. Z. and Wu J., Appl. Cat., 78, 1085 (1991).
[5] Pauling L., *The Chemical Bond*, (Cornell Univ. Press, Ithaca, NY, 1967).
[6] Sanchez-Delgado R. A., Andriollo A., Gonzalez E., Valencia N., Leon V. and Espidel J., J. Chem. Soc., Dalton Trans., (9), 1859 (1985).
[7] Nitta Y., Veno K. and Imanaka T., Appl. Cat., 56, 9 (1989).
[8] Raab C. G. and Lercher J. A., J. Mol. Cat., 75, 71 (1992).
[9] Bonnier J. M., Damori J. P. and Masson J., Appl. Cat., 42, 285 (1988).
[10] Marinelli J. B. L. W., Vleeming J. H. and Ponce V., Stud. Surf. Sci. Cat., 75, 1210 (1993).
[11] Hutchings G. J., King F., Okoye I. P. and Rochester C. H., Cat. Lett., 23, 127 (1994).
[12] Hutchings G. J., King F., Okoye I. P. and Rochester C. H., Appl. Cat. A, 83, L7 (1992).
[13] Waghare A., Oukaci R. and Blackmond D. G., Stud. Surf. Sci. Cat., 75, 2479 (1993).
[14] Lau C. P., Ren C. Y., Yeung C. H. and Chu M. T., Inorganica Chimica Acta, 191, 21 (1992).
[15] Lawrence S. S. and Schreifels J. A., J. Cart., 119, 272 (1989).
[16] Makouangou R., Dauscher A. and Touroude R., Stud. Surf. Sci. Cat., 75, 2475 (1993).
[17] Kalenik, Z., Ladna, B., Wolf, E. E., and Fehlner, T. P., Chem. Mater. 5, 1247 (1993).
[18] Banares M., Lei X., Fehlner T. P. and Wolf, E. E., J. Cat. (submitted).
[19] Cen, W., Haller, K. W., and Fehlner, T. P., Inorg. Chem. 31, 2072 (1992).
[20] Cen, W., Haller, K. W., and Fehlner, T. P., Inorg. Chem. 32, 995 (1993).
[21] Boublik T., Fried V. and Hala E., *The Vapor Pressures of Pure Substances*, (Elsevier, NY, 1984).

HETEROPOLY COMPOUNDS: EFFICIENT SUPERACID CATALYSTS

MAKOTO MISONO AND TOSHIO OKUHARA
Department of Applied Chemistry, Faculty of Engineering, The University of Tokyo, Bunkyo-ku, Tokyo 113, Japan

ABSTRACT

Our recent studies on the acidity and acid catalysis of heteropoly compounds are described, together with a brief introduction of the background. High catalytic activities were observed due to the combination of their superacidity, pseudoliquid behavior, acid-base bifunctionality and/or high surface concentration of proton. It was also shown how the acid strength and the number of acid sites as well as their tertiary structure were controlled by the constituent elements (hetero- and addenda-atoms and counter cations), and how those properties were correlated with the catalytic activities.

SUPERACID CATALYSIS OF HETEROPOLY COMPOUNDS - BACKGROUND

Heteropoly compounds ("metal oxide molecules"), are useful catalytic materials in fundamental studies as well as in practical processes [1-3]. One advantage is that their acid and redox properties which are most important chemical properties of catalysts can be controlled by proper choice of constituent elements, taking advantage of the molecular nature of heteropolyanions. To understand the catalysis of solid heteropoly compounds, it is necessary to distinguish between the primary (heteropolyanion), secondary (ionic crystal structure comprising cation and polyanion) and tertiary structure (particle size, pore structure, etc.). Accordingly, there are three prototypes for the catalysis of solid heteropoly compounds; (a) surface-, (b) bulk (I)[=pseudo-liquid]-, and (c) bulk (II)-types.

As solid acids, it has been fairly well established how to control the acidic properties by changing the counter cations. For example, in the case of pseudoliquid catalysis by the salts of small cations (e.g., Na^+), the acid amount and catalytic activity linearly decrease with the extent of neutralization (Fig. 1) [4]. As for large cations like Cs^+ and NH_4^+, the catalytic activity changes in a very unique way with the extent of neutralization as shown in Fig. 2 [5]. The acidic salts of $H_3PW_{12}O_{40}$(H3), particularly $Cs_{2.5}H_{0.5}PW_{12}O_{40}$ (Cs2.5), have high superacidity, and hence very high catalytic activity in such reactions as alkylations of aromatics (Reactions (1) and (2)) [6, 7] and alkanes (Reaction (3)) [8] and isomerization of n-butane (Reaction (4)) [9]. The combination of Pt and Pd with Cs2.5 enhanced much the efficiency for Reaction (4) [10].

Thus heteropoly compounds are promising alternatives to liquid acids like H_2SO_4 and $AlCl_3$ and therefore are catalysts for environmentally friendly (or benign) processes. Furthermore, careful preparation of the Cs salts can control the interparticle pore-size and reveals new shape selectivity in liquid-soild systems [11].

Mat. Res. Soc. Symp. Proc. Vol. 368 © 1995 Materials Research Society

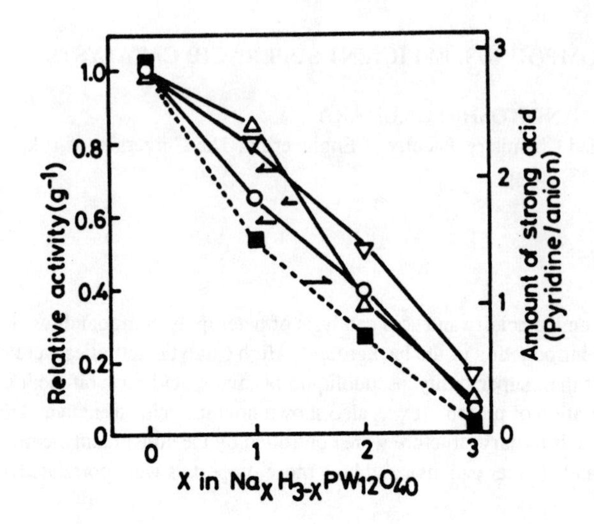

Fig. 1. Acidity and catalytic activities of $Na_xH_{3-x}PW_{12}O_{40}$. ■ : Number of acid sites (bulk) measured by pyridine absorption at 573K, o : Dehydration of 2-propanol (373 K), △ : Conversion of methanol to hydrocarbons (558 K), ▽ : Decomposition of formic acid (423 K).

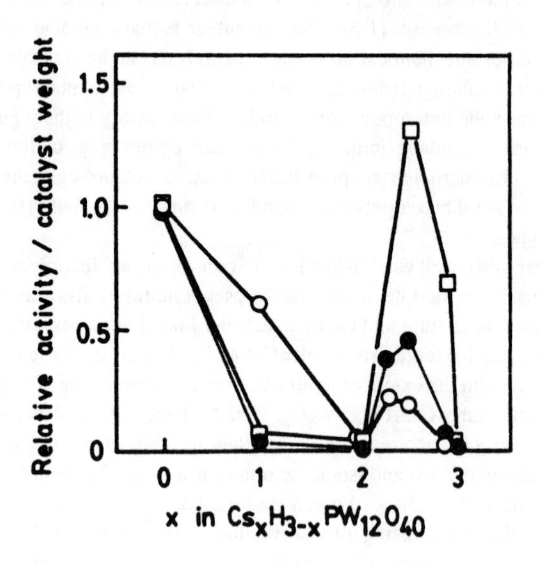

Fig. 2. Variation of catalytic activities as a function of the extent of neutralization by Cs (or the Cs content) of $Cs_xH_{3-x}PW_{12}O_{40}$. o : Dehydration of 2-propanol at 383 K after pretreatment at 383 K, ● : Dehydration of 2-propanol at 383 K after pretreatment at 573 K, ◻ : Conversion of dimethyl ether at 563 K after pretreatment at 573 K.

In this paper, we describe our recent work on the effects of heteroatom (central atom) and countercation on the acidic properties and their correlation with the acid catalysis in solution, gas-solid, and liquid-solid systems.

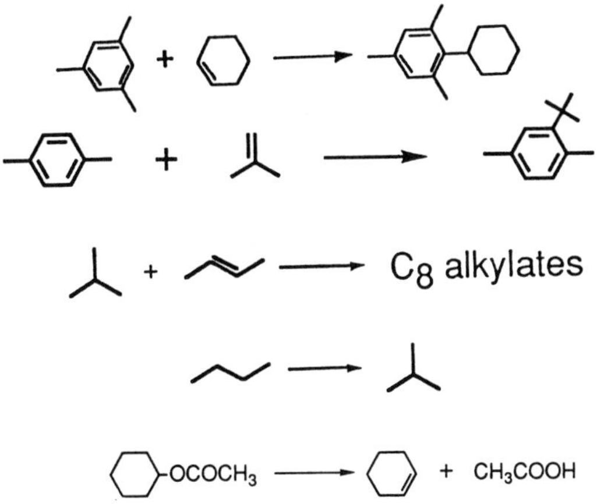

(1)

(2)

C_8 alkylates (3)

(4)

$OCOCH_3$ → + CH_3COOH (5)

ACID AMOUNT AND ACID STRENGTH

Hammett indicator test indicated that both H3 and Cs2.5 are strong acids ($H_0 < -8.2$) [12] and moreover superacids ($H_0 < -13.2$) [4]. Thermal desorption of ammonia shown in Fig. 3 demonstrates that H3 and Cs2.5 have similar strong acidity. This is reflected in the linear correlation between the number of surface acid sites and the catalytic activity (Fig. 4). The number of surface acid sites was estimated from the number of protons on the surface calculated from the surface area and the proton concentration in the solid bulk. Fig. 4 shows that the high activity of Cs2.5 is due to the very high concentration of superacidic protons on the surface, which was brought about by its very fine particles (or very high surface area). It is further noted in Fig. 4 that the catalytic activities per surface proton are much greater for heteropoly compounds than for other solid acids (zeolites, SO_4-ZrO_2, etc.). We tentatively attributed this to acid-base bifunctionality (combination of acidic proton and the basicity of polyanion) [6].

UV spectrophotometry of a Hammett indicator (pKa = -3.0) revealed that the acid strength in solution of the Keggin-type heteropolyacids having tungsten as addenda atom ($XW_{12}O_{40}$) increases with an increase in the oxidation number of heteroatom, X, (Co < B < Si, Ge < P), as shown in Table 1 [13]. As described below, the catalytic activity follows this order both for solution and solid heteropoly compounds, so that it may be reasonable to assume that the acid strength in the solid state is also in the order given in Table 1.

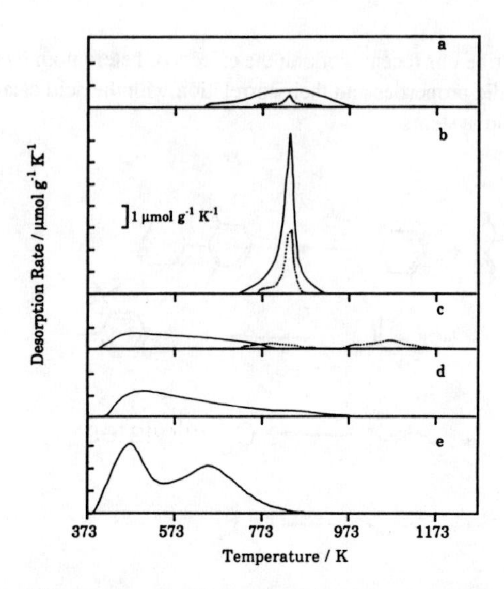

Fig. 3. NH_3 TPD profiles of several solid acids. (a) Cs2.5, (b) H3, (c) SO_4-ZrO_2, (d) SiO_2-Al_2O_3, and (e) H-ZSM-5. Solid lines: NH_3, dotted lines: N_2.

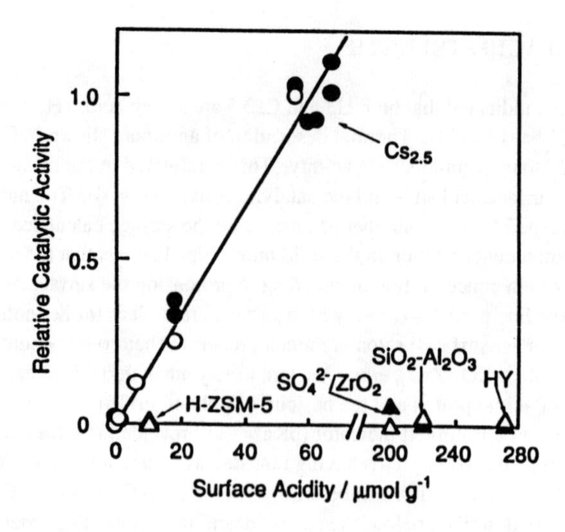

Fig. 4. Catalytic activity for alkylation of 1,3,5-trimethylbenzene with cyclohexene (open symbols; Reaction (1)) and decompsition of cyclohexyl acetate (solid symbols; Reaction (5)) as a function of surface acidity. (o, •): $Cs_xH_{3-x}PW_{12}O_{40}$, (△ , ▲): Other solid acids.

Table 1. H_o Values of Acetonitrile Solution of Heteropolyacids and PTS

Catalyst	H_o
$H_3PW_{12}O_{40}$	−2.14
$H_4SiW_{12}O_{40}$	−1.98
$H_4GeW_{12}O_{40}$	−1.69
$H_5BW_{12}O_{40}$	−1.55
$H_5FeW_{12}O_{40}$	−1.48
$H_6CoW_{12}O_{40}$	−0.88
PTS	−0.30

The proton concentration of the acid in acetonitrile was 4.86×10^{-3} mol dm^{-3}. The concentration of dicinnamylideneacetone (pK_a = −3.0) was 3.5×10^{-5} mol dm^{-3}.

Thus, the acidic properties of solid heteropoly compounds of Keggin-type are controlled as follows. The heteroatom greatly affects the acid strength and to a lesser extent the number of acid sites. The addenda atom has probably similar effects, reflecting those properties in solution [1], although firm experimental evidence has not been obtained yet. On the other hand, the counter cation becomes very important in the solid state and strongly affects the number of acid sites and tertiary structure, but not much the acid strength.

ACID CATALYZED REACTIONS OF ESTERS IN SOLUTION

Decomposition of isobutyl propionate (IBP) (Reaction (6), 401 K), ester exchange of IBP with acetic acid (Reaction (7), 343 K), ester exchange of IBP with n-propyl alcohol (Reaction (8), 343 K), and esterification of of propionic acid with isobutyl alcohol (Reaction (9), 343 K) have been studied in a homogeneous liquid phase using heteropolyacids; $H_mXW_{12}O_{40}$ (X = P, Si, Ge, B and Co) having the Keggin structure and $H_6P_2W_{18}O_{62}$ having the Dawson structure [14]. The catalytic behavior greatly depended on the kind of reaction, or the basicity of reactants.

In the case of Reaction (6), the catalytic activities were 60-100 times higher than H_2SO_4 and p-toluenesulfonic acid. Water present in the reaction system retarded the reaction. The activity order for Reaction (7) was probably similar. But the presence of water greatly accelerated Reaction (7), by creating a new reaction path via hydrolysis. In contrast, the differences in the activity of heteropolyacids and H_2SO_4 were not significant for Reactions (8) and (9), for which the effective acid strength of the catalysts probably leveled off as a consequence of the leveling effect of the reactant alcohols.

A good correlation was obtained between the rate constants of these reactions and the negative charges of the polyanions, as shown in Fig. 5 (A). The negative charge is closely related to the acid strength, as already shown in Table 1.

O-C(=O) → + C(=O)-OH (6)

O-C(=O) + CH_3-C(=O)-OH → O-C(=O) + C(=O)-OH (7)

O-C(=O) + OH →
O-C(=O) + OH (8)

OH + C(=O)-OH → O-C(=O) + H_2O (9)

ALKYLATION, AND DECOMPOSITION AND FORMATION OF ESTERS BY SOLID HETEROPOLYACIDS

Alkylation of 1,3,5-trimethylbenzene with cyclohexene at 343 K (Reaction (1)) and decomposition of cyclohexyl acetate at 373 K (Reaction (5)) were studied in liquid phase using solid heteropoly compounds [15]. The activities were measured for the heteropolyacids pretreated at 373 - 573 K, by which nearly constant rates were obtained for each heteropolyacid. The catalytic activities thus obtained and expressed by the initial rates exhibited a fair correlation with the negative charge of heteropolyanion (or the acid strength) (Fig. 5 (B)), similarly to the results in solution (Fig. 5 (A)).

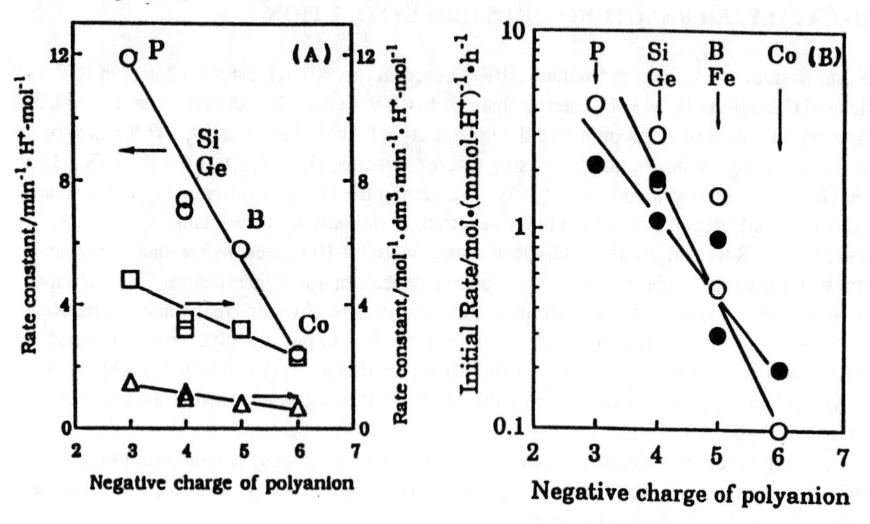

Fig. 5. Catalytic activity vs. negative charge of Keggin-type heteropolyanion. (A) Reactions in solution systems. ○ , △ , and □ ; Reactions (6), (8), and (9), respectively. (B) Reactions in liquid-solid systems. ○ and ● : Reactions (1) and (5), respectively.

Solid $H_3PW_{12}O_{40}$ catalyzed the alkylation of p-xylene with i-butene (Reaction (2)) in liquid phase at 303 K with 75% selectivity, while the selectivity of H_2SO_4 was only 7% [7]. The selectivity strongly depended on the acid strength when compared at a similar conversion level; $H_3PW_{12}O_{40} >> H_4SiW_{12}O_{40} >> H_5BW_{12}O_{40}$.

In the case of the synthesis of methyl t-butyl ether (MTBE) in gas-solid system catalyzed by the acid forms of heteropoly compounds, a large amount of reactant (or product) molecules were absorbed by the catalysts under the reaction conditions, and the pseudoliquid behavior strongly influenced the yield and selectivity of MTBE. Among heteropolyacids, $H_6P_2W_{18}O_{62}$ dispersed on silica showed the best performance, which was comparable or better than that of ion-exchange resin (Amberlyst 15); $H_6P_2W_{18}O_{62} >> H_3PW_{12}O_{40} > H_4SiW_{12}O_{40} \sim H_4GeW_{12}O_{40} > H_5CoW_{12}O_{40} \sim H_5BW_{12}O_{40}$ at 333 K[16].

References

1. Reviews; M. Misono, Catal. Rev. Sci. Eng., **29**, 269 (1987); **30**, 339 (1988); Y. Ono, in Perspectives in Catalysis, edited by J. M. Thomas and K. I. Zamaraev (Blackwell, Oxford, 1992) p 431; Y. Izumi, K. Urabe, and M. Onaka, Zeolite, Clay and Heteropoly Acid in Organic Reactions (Kodansha, Tokyo and VCH, Weinheim, 1992); I. V. Kozhevnikov and K. I. Matveev, Appl. Catal., **5**, 135 (1983); I. V. Kozhevnikov, in Acid-Base Catalysis II edited by H. Hattori, M. Misono, and Y. Ono (Stud. Surf. Sci. Catal., vol. 90, Kodansha, Tokyo and Elesevier, Amsterdam, 1994), p 21.
2. M. Misono, in Proc 10th Intern. Congr. Catal., Budapest, 1992, edited by L. Guczi, F. Solymosi, and P. Tetenyi (Elsevier, Amsterdam, 1993), p 69
3. M. Misono and T. Okuhara, CHEMTECH, November 1993, p 23.
4. T. Okuhara, A. Kasai, N. Hayakawa, Y. Yoneda, and M. Misono, J. Catal., **83**, 121 (1983).
5. S. Tatematsu, T. Hibi, T. Okuhara, and M. Misono, Chem. Lett., 1984, 865.
6. T. Okuhara, T. Nishimura, H. Watanabe, and M. Misono, J. Mol. Catal., **74**, 247 (1992).
7. H. Soeda, T. Okuhara, and M. Misono, Chem. Lett., 1994, 909.
8. T. Okuhara, M. Yamashita, K. Na, and M. Misono, Chem. Lett., 1994, 1450.
9. K. Na, T. Okuhara, and M. Misono, Chem. Lett., 1993, 1141.
10. K. Na, T. Okuhara, and M. Misono, J. Chem. Soc. Chem. Commun., 1993, 1422.
11. T. Nishimura, T. Okuhara, and M. Misono, presented at the 65th Annual Meeting of Chemical Society of Japan, Tokyo, March 1993.
12. M. Otake and T. Onoda, Shokubai, **17**, 13P (1975); H. Hayashi and J. B. Moffat, J. Catal., **77**, 473 (1982).
13. T. Okuhara, C. Hu, M. Hashimoto, and M. Misono, Bull. Chem. Soc. Japan, **67**, 1186 (1994).
14. C. Hu, M. Hashimoto, T. Okuhara, and M. Misono, J. Catal., **143**, 437 (1993).
15. C. Hu, T. Nishimura, T. Okuhara, and M. Misono, Sekiyu Gakkaishi, **36**, 386 (1993)
16. S. Shikata, T. Okuhara and M. Misono, Sekiyu Gakkaishi, **37**, 632 (1994)

SYNTHESIS AND PROPERTIES OF LANTHANIDE-EXCHANGED PREYSSLER'S HETEROPOLYANIONS

MARK R. ANTONIO, J. MALINSKY, AND L. SODERHOLM
Chemistry Division, Argonne National Laboratory, 9700 S. Cass Avenue, Argonne, IL 60439-4831, USA

ABSTRACT

Na^+ in the Preyssler heteropolytungstate anion $[NaP_5W_{30}O_{110}]^{14-}$ can be exchanged for a trivalent lanthanide ion. The potential significance of this new class of lanthanide heteropolyanions relates to their applications in catalysis science. This view follows from the fact that Keggin heteropolyanions and their free acids are used as heterogeneous solid catalysts and homogeneous solution catalysts. We describe synthetic conditions that lead to the incorporation of Ce^{3+} and Pr^{3+} within the Preyssler anion, and the coprecipitation of Ce^{3+} and the Preyssler anion. Initial studies indicate that the latter, coprecipitated, material deserves study for bifunctional catalytic activity.

INTRODUCTION

Heteropolyanions are discrete clusters, with sizes varying from 50-200 atoms, that can accept substantial reduction in their overall net charge without decomposition.[1-3] As such, these polynuclear anions serve to bridge the gap between the chemistry of small molecular clusters with discrete, atomic-like energy levels, on the one hand, and covalent solids with delocalized, bandlike energy levels, on the other. The prototypical heteropoly oxoanions have the 1:12 Keggin structure, $[PM_{12}O_{40}]^{3-}$ ($M \equiv Mo^{6+}$, W^{6+}). Upon electrochemical reduction with one or more electrons, the anion is transformed into an intensely colored, heteropoly blue $[PM_{12}O_{40}]^{n-}$, n=4, 5, 6. In view of their remarkable redox behavior, heteropolyanions have received considerable attention, both experimentally and theoretically, from both academic and industrial research groups. Principal interest stems from the potential technological spin-offs to catalysis science. Heteropolymolybdates and -tungstates, for example, show activity as acid as well as oxidation and ammoxidation catalysts.[3-8] Heteropolyanions and their free acids, i.e., heteropolyacids, are used as both heterogeneous solid catalysts and homogeneous solution catalysts.

Between 1968 and 1971, four new types of heteropolyanion clusters with rare earth (*RE*) elements were reported:[9-12] (1) decatugstometalates, $[REW_{10}O_{36}]^{n-}$; (2) dodecamolybdometalates, $[REMo_{12}O_{42}]^{n-}$; (3) bis(undecatungsto)metalates $[RE(PW_{11}O_{39})_2]^{n-}$; (4) bis(heptadecatungsto)metalates, $[RE(P_2W_{17}O_{61})_2]^{n-}$. Examples of the complexes prepared to date are summarized in Table I. More recently, the preparation of a number of bis(undecatungstocuprate)lanthanates, $[RE(CuW_{11}O_{39})_2]^{17-}$, as well as the germanium and boron derivatives, $[RE(GeW_{11}O_{39})_2]^{13-}$ and $[RE(BW_{11}O_{39})_2]^{15-}$, have been described,[13-14] Table I, Such heteropolyoxometalate clusters containing rare earth ions with dual valence states, such as Ce, Pr, Tb (3+ and 4+) or Sm, Eu, Tm, Yb (2+ and 3+), and multivalent (3+, 4+, 5+, 6+) transition metal ions of Group VIB are of particular interest. These can have unusual redox properties—facile, reversible reduction-oxidation—that may be exploited for catalysis. In fact, a recent study described the behavior of $[Ce^{4+}W_{10}O_{36}]^{8-}$, $[Nd^{3+}W_{10}O_{36}]^{9-}$, and $[Sm^{3+}W_{10}O_{36}]^{9-}$ for the homogenous, selective oxidation of cyclohexanol with H_2O_2 and for H_2O_2 decomposition.[15]

Mat. Res. Soc. Symp. Proc. Vol. 368 © 1995 Materials Research Society

Table I. Rare-earth (*RE*) element-containing heteropolyanions with the cluster charge (n) and the *RE* ions reported to form these structures.

Heteropolyanion	Charge (n)	*RE*
$[REW_{10}O_{36}]^{n-}$	9	3+ (Y, La-Yb)
	8	4+ (Ce, Th, U)
$[REMo_{12}O_{42}]^{n-}$	9	3+ (Ce)
	8	4+ (Ce, Th, U, Np)
	7	5+ (U)
$[RE(PW_{11}O_{39})_2]^{n-}$	11	3+ (La, Ce, Pr, Nd, Gd)
	10	4+ (Ce, U)
	9	5+ (U)
$[RE(SiW_{11}O_{39})_2]^{n-}$	13	3+ (Ce, Sm, Eu, Tb, Ho)
	12	4+ (Ce, U)
	11	5+ (U)
$[RE(CuW_{11}O_{39})_2]^{n-}$	17	3+ (La-Gd, Dy, Ho, Yb)
$[RE(P_2W_{17}O_{61})_2]^{n-}$	17	3+ (Ce, Tb, Dy, Ho)
	16	4+ (Ce, Tb, Th, U)
$[RESb_9W_{21}O_{86}]^{n-}$	16	3+ (La-Gd, Dy, Yb)
$[REP_5W_{30}O_{110}]^{n-}$	12	3+ (Y, Nd-Lu)
	11	4+ (U)

Among the large number and variety of heteropolyoxolanthanate clusters, it is most common that lacunary (i.e., unsaturated) polyanions act as ligands for the *RE* ions. In contrast, there are just two heteropolyanions of composition $[NaSb_9W_{21}O_{86}]^{18-}$ and $[NaP_5W_{30}O_{110}]^{14-}$ that encapsulate rare earth ions through Na^+ exchange,[16,17] the latter of which is known as the Preyssler anion, PA. A single-crystal X-ray structure study reveals that it contains Na^+ within a central cylindrical cavity formed by the cyclic arrangement of five $-PW_6O_{22}-$ groups assembled with D_{5h} symmetry.[18] Under relatively rigorous conditions, Na^+ can be exchanged with RE^{3+} ($RE \equiv$Y, Nd-Lu) to form heteropolyoxotungstate anions $[REP_5W_{30}O_{110}]^{12-}$.[17] The significance and interest of these *RE*-exchanged heteropolyanions relates to their possible utilization in separation and catalysis sciences. For example, the cavity through the Preyssler anion bears some resemblance to the tunnels connecting cages in microporous aluminophosphate molecular sieves and zeolites. In view of this, *RE*-ion exchanged Preyssler anions may have properties in common with *RE*-ion exchanged zeolites, which are in widespread use as hydrocarbon cracking catalysts and may have some impact on the catalytic reduction of NO.[19] The catalytic efficiency of the Preyssler anion for the selective oxidation of H_2S by molecular oxygen in aqueous solution was the subject of a recent investigation by Harrup and Hill.[20]

This report describes new preparative chemistry for the exchange of Na^+ in $[NaP_5W_{30}O_{110}]^{14-}$ with Ce^{3+} and Pr^{3+}. We have also performed initial studies on the coprecipitation of Ce^{3+} and the Preyssler anion. The motivation for study of the latter PA-supported Ce^{3+} complex stems from the preparation of rhodium and iridium polyoxometalates of $[SiW_{12}O_{40}]^{4-}$ and $[PMo_{12}O_{40}]^{3-}$. These have been described as bifunctional catalysts with diverse reaction chemistry, including olefin isomerization, hydrogenation, hydroformylation, and C-H activation processes.[21] Similarly, polyoxoanion-supported iridium and ruthenium (pre)catalysts have been prepared and characterized.[22,23] The cyclic voltammetry (CV) and *RE* L_3-edge XANES (X-ray absorption near edge structure) of $[REP_5W_{30}O_{110}]^{12-}$ and the Ce^{3+} salt of $[NaP_5W_{30}O_{110}]^{14-}$ are described herein.

EXPERIMENTS

The white, crystalline Preyssler salt, $K_{12.5}Na_{1.5}[NaP_5W_{30}O_{110}]•15H_2O$, was prepared according to the method of ref 17. The Ce^{3+} and Pr^{3+}-exchange reactions were based upon the following equation:

$$[NaP_5W_{30}O_{110}]^{14-} + nRE^{3+} \longrightarrow [REP_5W_{30}O_{110}]^{12-} + (n-1)RE^{3+} + Na^+$$

In a typical experiment for n=1, a colorless warm (60°C) solution of $Ce(NO_3)_3•6H_2O$ (53 mg, 0.121 mmole, dissolved in 3 cm^3 of H_2O) was added dropwise with stirring to a warm (60°C) solution of the Preyssler salt (1 g, 0.121 mmol, dissolved in 12 cm^3 H_2O). The resulting clear, colorless solution was sealed in a Parr 4746 Teflon-lined digestion vessel and heated at 165°C in a Lindberg/Blue M crucible furnace for 48 h. Upon cooling, 4 g of solid KCl was added to the clear yellow solution to precipitate a fine pale yellow powder, which was collected on No. 42 Whatman filter paper, rinsed with ca. 10 cm^3 of ice-cold water and dried in air. The same procedure was followed for n=2 in the equation above. Yields of 0.86–0.98 g (80–90% based upon $[NaP_5W_{30}O_{110}]^{14-}$ and an estimated empirical formula $K_{12}[CeP_5W_{30}O_{110}]•54H_2O$) were obtained. The Ce^{3+} exchange reaction occurs with both one and two equivalents of Ce^{3+} as either $Ce(NO_3)_3•6H_2O$ or $CeCl_3•7H_2O$. The exchange of Na^+ with Pr^{3+} also occurs in aqueous solutions of $Pr(NO_3)_3•6H_2O$ using the conditions described above. After heating at 165°C for 48 h, the clear colorless solution is worked up to produce a white powder.

The following method was used to coprecipitate Ce^{3+} with $[NaP_5W_{30}O_{110}]^{14-}$. A clear, colorless aqueous solution conatining $Ce(NO_3)_3•6H_2O$ (n=2) and the Preyssler salt (prepared exactly as above) was evaporated to dryness at 135°C for 2 h in a drying oven. This coprecipitation produced a pastel yellow solid.

Ce and Pr L_3-edge XANES data were collected at ambient temperature on beam line 4-3 (using a Si<220> double crystal monochromator) at SSRL and line X-23A2 (using a Si<311> double crystal monochromator) at NSLS. The vertical entrance slit width was set to 1 mm. The fluorescence signal, I_f, was detected by use of an ion chamber fluorescent detector (The EXAFS Co.) filled with argon and without a fluorescence filter. The powder samples were mounted at the conventional 45° incident—45° exit configuration in a helium purged sample box. Nitrogen was used to monitor the incident X-ray intensity, I_0. Due to the low concentrations of Ce and Pr, ca. 1.6 wt %, in the exchanged Preyssler anions, the fluorescence XANES, I_f/I_0, is not vitiated by self-absorption effects. The XANES data for the model compounds were obtained by the electron-yield, I_e/I_0, method. The normalization of the X-ray absorption data to a unit edge jump was performed according to conventional methods.[24]

Cyclic voltammetry data were obtained with a BAS 100B/W electrochemical analyzer, a BAS C-2 voltammetry cell stand, and BAS electrodes: 3.0 mm diameter glassy carbon working electrode (MF-2012); platinum wire auxiliary electrode (MW-1032); Ag/AgCl reference electrode with vycor tip (MF-2063). The electrolyte (1 M H_2SO_4) was prepared from 99.9999% H_2SO_4 (Alfa) and 18MΩ-cm water from a Millipore MILLI-Q™ water system. CV scans of the neat electrolyte, which was sparged and blanketed with nitrogen, did not reveal any evidence of electroactive impurities in the potential windows of interest, -0.60 to 0.20 V vs. Ag/AgCl. The scan rates were -100 mV/s and the anion concentrations were ca. 1 mM.

RESULTS AND DISCUSSION

The CVs of the Ce^{3+}- and Pr^{3+}-exchanged Preyssler anions are identical. As shown in Figure 1 for $[CeP_5W_{30}O_{110}]^{12-}$, the CV data exhibit 5 reversible redox waves attributable to W(6+/5+) and the formation of heteropoly blues. For comparison, the CV for the parent Preyssler anion is also shown in Figure 1. The differences in the CVs between the RE^{3+}-exchanged anions and the parent anion serve to confirm the exchange of Na^+ with Ce^{3+} and Pr^{3+}. In fact, the CVs for the Ce^{3+}- and Pr^{3+}-exchanged anions shown here are identical to the CVs for all the other RE^{3+}-exchanged anions, except Eu^{3+}.[17,25] Figure 1 also shows the CV for the yellow material obtained by coprecipitation of Ce^{3+} and $[NaP_5W_{30}O_{110}]^{14-}$. It has features that are similar to those in the CV of the Preyssler ion itself. Although additional studies are clearly required, the similarities of the CVs suggest that the Ce^{3+} cations are not interacting in a chemically significant way with the $[NaP_5W_{30}O_{110}]^{14-}$ anion framework. This situation of non-interacting cations and anions is similar to that exemplified by the bifunctional catalysts $[(Ph_3P)_2Rh(CO)]_4[SiW_{12}O_{40}]$ and $[(Ph_3P)_2IrH_2]_3[PMo_{12}O_{40}]$.[21]

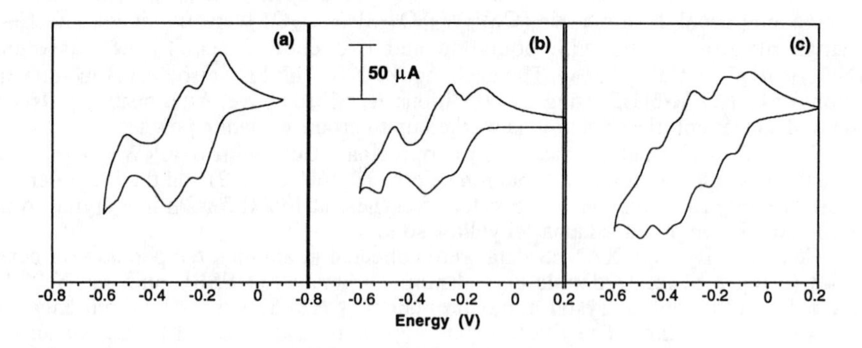

Figure 1. Cyclic voltammograms of (a) $K_{12.5}Na_{1.5}[NaP_5W_{30}O_{110}]\bullet 15H_2O$ (b) $K_{12.5}Na_{1.5}[NaP_5W_{30}O_{110}]\bullet 15H_2O$ coprecipitated with 2 equivalents of $Ce(NO_3)_3\bullet 6H_2O$ (c) $K_{12}[CeP_5W_{30}O_{110}]\bullet nH_2O$ in aqueous 1 M H_2SO_4 electrolytes.

The Ce and Pr L_3-edge XANES for the solid salts from the Ce and Pr-exchange reactions and the Ce coprecipitation are shown in Figure 2. The intense single edge resonances at 5721.6 eV (Ce L_3 XANES) and 5965.4 eV (Pr L_3 XANES) are due to $2p \rightarrow 5d$ electronic transitions. The edge profiles and positions found here are typical of those observed in L_3-edge XANES for trivalent Ce and Pr compounds,[26] such as for $Ce(NO_3)_3.6H_2O$ and $Pr(NO_3)_3.6H_2O$, see Figure 2. These spectra stand in sharp contrast to the L-edge XANES for compounds containing quadrivalent cerium and praseodymium, e.g., CeO_2 and PrO_2. For such materials, two well-resolved edge resonances are observed.[27] The XANES for the Ce^{3+}-exchanged-PA solid salt shown here is similar to that for the cerium-exchanged PA prepared according to the original method from aqueous Ce^{4+} [17]. In fact, the Ce XANES for the $K_{12}[CeP_5W_{30}O_{110}]{\bullet}nH_2O$ solid salt of Figure 2 is identical to that from the aqueous solution of $[CeP_5W_{30}O_{110}]^{12-}$ prepared from Ce^{4+}.[28] Whereas cerium is trivalent in the exchange products obtained with either Ce^{3+} or Ce^{4+} and in view of the fact that the corresponding CV data are identical, the differences between the XANES of the two solid state salts are surprising. They may suggest the presence of a Ce^{3+} impurity in the original preparations or, alternatively, different locations for Ce^{3+} within the cylindrical cavity of the Preyssler anion.[28] Additional studies to resolve this issue are in progress.

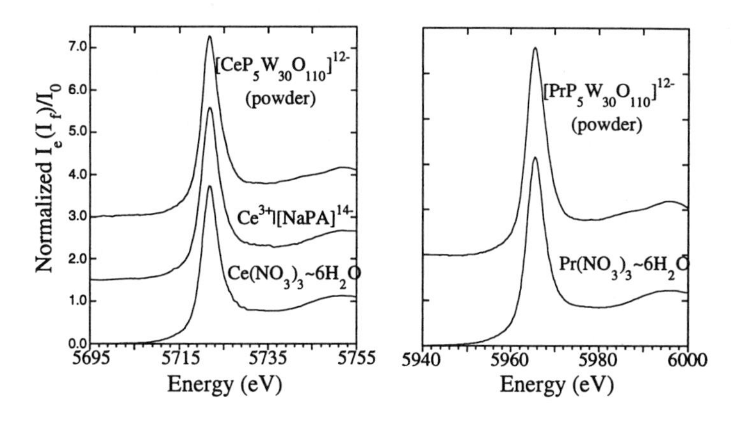

Figure 2. Solid-state L_3-edge XANES for (a) $K_{12}[CeP_5W_{30}O_{110}]{\bullet}nH_2O$, $K_{12.5}Na_{1.5}[NaP_5W_{30}O_{110}]{\bullet}15H_2O$ coprecipitated with 2 equivalents of $Ce(NO_3)_3{\bullet}6H_2O$; $Ce(NO_3)_3.6H_2O$; (b) $K_{12}[PrP_5W_{30}O_{110}]{\bullet}nH_2O$, $Pr(NO_3)_3.6H_2O$.

CONCLUSIONS

The exchange of Na^+ in the Preyssler anion, $[NaP_5W_{30}O_{110}]^{14-}$, with Ce^{3+} and Pr^{3+} was investigated. Under extreme conditions of temperature (165°C) and time (48 h), the direct exchange of Na^+ for Ce^{3+} and Pr^{3+} occurs from aqueous solutions of $Ce(NO_3)_3{\bullet}6H_2O$ (or $CeCl_3{\bullet}7H_2O$) and $Pr(NO_3)_3{\bullet}6H_2O$. The Preyssler anion was also precipitated with trivalent cerium. The cyclic voltammetry of the Ce and Pr exchange products confirm the Na^+ ion exchange, and the L_3-edge XANES show that Ce and Pr are trivalent. Although no catalysis screening studies were performed here, this work provides a new entry to the synthesis of heteropolyanions with potential catalytic activity.

ACKNOWLEDGMENTS

This work was supported by the U.S. DOE, BES—Chemical Sciences, under contract No. W-31-109-ENG-38.

REFERENCES

1. M.T. Pope, Heteropoly and Isopoly Oxometalates (Springer-Verlag, Berlin, 1983).
2. M.T. Pope, in Comprehensive Coordination Chemistry, Vol. 3, edited by G. Wilkinson, R.D.Gillard and J.A. McCleverty (Pergamon Press, New York, 1987), pp. 1023-1058.
3. M.T. Pope and A. Müller, Angew. Chem. Int. Ed. Engl. **30**, 34 (1991).
4. M. Misono, in Polyoxometalates: From Platonic Solids to Anti-Retroviral Activity, edited by M.T. Pope and A. Müller (Kluwer Academic, Dordrecht, The Netherlands, 1994) pp. 255-265; R.G. Finke, ibid., pp. 267-280; J.H. Grate, D.R. Hamm and S. Mahajan, ibid., pp. 281-305; R. Neumann, ibid., pp. 307-313; E. Cadot, C. Marchal, M. Fournier, A. Teze and G. Herve, ibid., pp. 315-326; E. Papaconstantinou, A. Ioannidis, A. Hiskia, P. Argitis, D. Dimotikali and S. Korres, ibid., pp. 327.335; C.L. Hill, G.-S. Kim, C.M. Prosser-Mccartha and D. Judd, ibid., pp. 359-371.
5. N. Mizuno and M. Misono, J. Molecular Catalysis **86**, 319 (1994).
6. M. Misono, Catal. Rev.—Sci. Eng. **29**, 269 (1987).
7. I.V. Kozhevnikov and K.I. Matveev, Russian Chemical Reviews **51**, 1075 (1982)
8. D. Sattari and C.L. Hill, J. Am. Chem. Soc. **115**, 469 (1993)
9. R.D. Peacock and T.J.R. Weakley, J. Chem. Soc. (A) 1971, 1836.
10. P. Baidala, V.S. Smurova, E.A. Torchenkova and V. I. Spitsyn, Doklady Chemistry **197**, 430 (1971), Engl. Trans.
11. E.A. Torchenkova, P. Baidala, V.S. Smurova and V.I. Spitsyn, Doklady Chemistry **199**, 568 (1971), Engl. Trans.
12. D.D. Dexter and J.V. Silverton, J. Am. Chem. Soc. **90**, 3589 (1968).
13. W. Qingyin, W. Enbo and L. Jingfu, Polyhedron **12**, 2563 (1993).
14. N. Haraguchi, Y. Okaue, Y. Isobe and Y. Matsuda, Inorg. Chem. **33**, 1015 (1994).
15. R. Shiozaki, H. Goto and Y. Kera, Bull. Chem. Soc. Jpn. **66**, 2790 (1993).
16. J. Liu, S. Liu, L. Qu, M.T. Pope and C. Rong, Trans. Met. Chem. **17**, 311 (1992).
17. I. Creaser, M.C. Heckel, R.J. Neitz and M.T. Pope, Inorg. Chem. **32**, 1573 (1993).
18. M.H. Alizadeh, S.P. Harmalker, Y. Jeannin, J. Martin-Frere and M.T. Pope, J. Am. Chem. Soc. **107**, 2662 (1985).
19. C. Yokoyama and M. Misono, Bull. Chem. Soc. Jpn. **67**, 557 (1994).
20. M.K. Harrup and C.L. Hill, Presented at the 208th National Meeting of the American Chemical Society, Washington, D.C., August 1994; paper INOR 567.
21. A.R. Siedle, R.A. Newmark, W.B. Gleason, R.P. Skarjune, K.O. Hodgson, A.L. Roe and V.W. Day, Solid State Ionics **26**, 109 (1988). A.R. Siedle, R.A. Newmark, K.A. Brown-Wensley, R.P. Skarjune, L.C. Haddad, K.O. Hodgson and A.L. Roe, Organometallics **7**, 2078 (1988). A.R. Siedle, C.G. Markell, P.A. Lyon, K.O. Hodgson and A.L. Roe, Inorg. Chem. **26**, 219 (1987)
22. W.G. Klemperer and B. Zhong, Inorg. Chem. **32**, 5821 (1993).
23. Y. Lin, K. Nomiya and R.G. Finke, Inorg. Chem. **32**, 6040 (1993)
24. M.R. Antonio, in Encyclopedia of Materials Characterization: Surfaces, Interfaces, Thin Films, edited by C. R. Brundle, C. A. Evans, Jr. and S. Wilson (Butterworth-Heinemann, Boston, 1992), p. 214.
25. M.R. Antonio, L. Soderholm, J. Muntean and G. Liu, in preparation (1994).
26. J.E. Sunstrom, IV, S.M. Kauzlarich and M.R. Antonio, Chem. Mater. **5**, 182 (1993).
27. Z. Hu, S. Bertram and G. Kaindl, Phys. Rev. B, **49**, 39 (1994).
28. M.R. Antonio and L. Soderholm, Inorg. Chem., in the press (1994).

CATALYTIC IMPLICATIONS FOR KEGGIN AND DAWSON IONS: A THEORETICAL STUDY OF STABILITY FACTORS OF HETEROPOLYOXOANIONS.

Shu-Hsien Wang[1] and Susan A. Jansen*
Chemistry Department, Temple University, Philadelphia, PA 19122
Shu-Li Wang
Department of Bioengineering, University of Pennsylvania, Philadelphia, PA 19104

ABSTRACT

Hetreropolyoxoanions(HPAs) have received wide spread attention due to their unique oxidation-reduction properties in catalysis. Two main structural motifs groups of HPAs have been described by Keggin and Dawson. The first is composed of twelve distorted, octahedrally-coordinated metal atoms forming trimeric and tetrameric subunits connected such that a closed shell results. The second is formed by fusing two Keggin units. The most common Keggin ions($XM_{12}O_{40}^{n-}$) with T_d symmetry and the Dawson ions($X_2M_{18}O_{62}^{n-}$) with D_{3h} symmetry are the α-isomers; while the β-isomers are generated by rotating one trimer 60^0 along its C_3 axis, thus producing the C_s symmetry for β-Keggin ions and D_{3d} symmetry for β-Dawson ions. Here, M is the framework metal atom, usually Mo, W, and X is the central heteroatom which can be the metal or nonmetal. From our previous studies, the instability of β-isomers caused by the "rotated" trimer can not be compensated by other coordination effects within the cluster and therefore defect states are retained between the HOMO-LUMO gap relative to the α-isomer. This enhances catalytic properties and causes thermal/photo lability. The catalytic function of α-Keggin ions are well studied; however, related research reports about Dawson ions are rare. The α-Dawson ion possesses stronger oxidizing ability and higher activities in oxidative dehydrogenation and oxygen addition reactions than the α-Keggin ions; the β-isomers have a greater tendency to be reduced than the α-isomers. Most of the characterization methods applied have not yielded the electronic or the structural requisites for the catalytic properties of the HPAs. In this work, the theoretical analysis is done through the application of crystal/tight binding theory coupled with the extended Hückel methodology. The stability trends and the redox behaviors provided by this work reliably trace the experimental data. In addition, the intrinsic instability and the redox effects in the β-isomers with respect to the α-isomers is assessed and a rationale for the observed catalytic and electronic develpoment of structrual function is also offered.

INTRODUCTION

The HPAs and their derivatives have been widely studied structurally and catalytically[1-9]. Most of these materials that are employed in oxidation-reduction reaction are the thermally stable α-isomers, typically based on the Keggin-type structures. Because of their stability, these materials function as acid catalysts and the acidities come from the stability of the anions in solution[1,4]. These properties can be modulated by their central oxoanions, countercations, and the framework metal atoms. Since there are numberous substitutional and geometrical isomers in this category, in this work, we only examine the factors of relative stability and acidity of the α- and β-Keggin ions and the α-Dawson ions as shown in Fig. 1 with substitution of the nonmetal central oxoanions, and the substitutional isomers of the framework metal atoms of the α-Keggin ions. The approach in this study is to decompose the HPA into its components, i.e. the trimer($M_3O_{13}^{8-}$) and the tetramer($M_4O_{18}^{12-}$), or the core ion(XO_4^{n-}) and the clathratic shells($M_{12}O_{36}$ for Keggin and $M_{18}O_{54}$ for Dawson)[10-12]. The models utilized in calculations are taken from the X-ray crystallographic data[3-15] and the computational methodology applied is the extended Hückel

----

[1]Sun Oil Fellow.

Mat. Res. Soc. Symp. Proc. Vol. 368 © 1995 Materials Research Society

scheme[16], which is a semiempirical quantum mechanics that has been successful in predictions of more accurate HOMO-LUMO energies in metal clusters, organometallics and surface interactions between the polymers and metal dopants. In this study, the iterated parameters are utilized; the analyses of electronic properties, the orbital frontiers and the density of states(DOS) provide insights for understanding the chemical behaviors of HPAs[10-12]. The limited study of the catalytic behavior[4,6] of the Dawson ions also restrict the comparisons between the theoretical study and the experimental results, however, the comparison among the isomers are addressed.

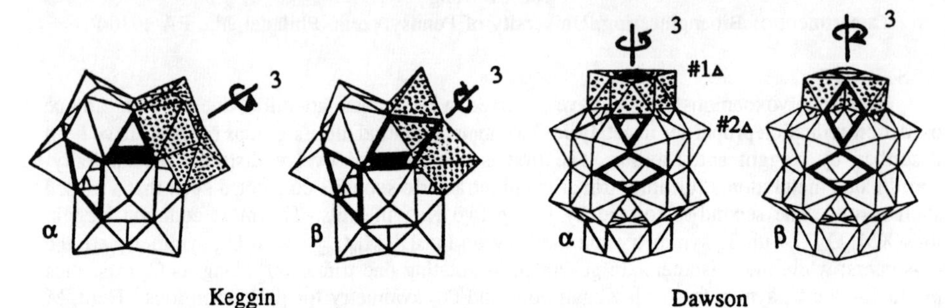

Fig 1. Structures of the heteropolyoxoanions.

RESULTS AND DISCUSSION

A. Geometrical Isomers: The most commonly observed or studied framework metal atoms of the HPAs are the Mo and W, the central oxoanions are the silicate, phosphate, and germanate, and the geometrical isomers are designated α and β[1-9]. The contributions of these constituents to stability effect can be interpreted by examinating the binding energies(B.E.), the overlap populations, the molecular orbital interaction diagrams and the DOS of the clusters[10-12]. The binding energy shown in Table I is the energy difference between the HPA cluster and its constituents: the oxoanion core and the clathratic shell. Thus, it can be refered to the stabilization energy. Moreover, the cluster assembly energy(C.A.E.) represents the stability by comparing the energy difference between the HPA and the sum of its atomic energies. In other words, the C.A.E. indicated the stability of the HPA shell that has been assembled and the B.E. tells how the central oxoanion helps to further stablize the shell by satisfying the coordination of the framework metal atoms, or by forming the chemical interactions.

Table I. Binding energy and cluster assembly energy of the Keggin and Dawson ions.

Species	B.E.(eV)	C.A.E.(eV)	Species	B.E.(eV)	C.A.E.(eV)
Keggin ions					
α-$SiW_{12}O_{40}^{4-}$	-5.02	-305	β-$SiW_{12}O_{40}^{4-}$	2.41	-326
α-$GeW_{12}O_{40}^{4-}$	-5.41	-332	β-$GeW_{12}O_{40}^{4-}$	-5.11	-334
α-$PW_{12}O_{40}^{3-}$	-5.45	-310	β-$PW_{12}O_{40}^{3-}$	-4.15	-335
Dawson ions					
α-$Si_2W_{18}O_{62}^{8-}$	-5.85	-514	α-$Si_2Mo_{18}O_{62}^{8-}$	-5.04	-500
α-$Ge_2W_{18}O_{62}^{8-}$	-6.67	-514	α-$Ge_2Mo_{18}O_{62}^{8-}$	-5.95	-500
α-$P_2W_{18}O_{62}^{6-}$	-6.73	-521	α-$P_2Mo_{18}O_{62}^{6-}$	-6.01	-507

The trends of stability shown in Table I reliably reproduce the experimental data that show Dawson ions to be more stable than the Keggin ions. The α-Keggin ions are generally more stable than the β-isomers which isomerize readily in solution[13]. This stability trend can also be examined by the orbital overlap population[10], indicating the strength of orbital interactions or, in short, the bond order. From our previous studies[10-12], the "trade-off" effect, based on the "core-shell model", between the X-O_p and the M-O_p defines the stability of the Keggin cluster. This analysis is also valid for the Dawson ions. Here, the O_p refers to the O shared by the central heteroatom X and the framework metal atom M. As shown in Table II, it is obvious that the greater the M-O_p, the more stable the anion as long as the bonding within the oxoanion is maintained. Therefore, this implies W-type Dawson ions tend to be more stable than the corresponding Mo-types.

Table II. The overlap population of the Keggin and Dawson ions where X= Si, Ge, P and M= W, Mo.

Species	X-O_p	M-O_p	Species	X-O_p	M-O_p
Keggin ions					
α-$SiW_{12}O_{40}^{4-}$	0.624	0.139	β-$SiW_{12}O_{40}^{4-}$	0.598	0.140
α-$GeW_{12}O_{40}^{4-}$	0.649	0.146	β-$GeW_{12}O_{40}^{4-}$	0.621	0.147
α-$PW_{12}O_{40}^{3-}$	0.645	0.148	β-$PW_{12}O_{40}^{3-}$	0.616	0.149
Dawson ions					
α-$Si_2W_{18}O_{62}^{8-}$	0.551	0.171	α-$Si_2Mo_{18}O_{62}^{8-}$	0.551	0.162
α-$Ge_2W_{18}O_{62}^{8-}$	0.415	0.183	α-$Ge_2Mo_{18}O_{62}^{8-}$	0.566	0.172
α-$P_2W_{18}O_{62}^{6-}$	0.562	0.183	α-$P_2Mo_{18}O_{62}^{6-}$	0.562	0.174

The HPAs are synthesized by acid condensation of metal oxides or their simple salts, possessing two dimensional, tetrameric lattice structures[1-9]. Upon condensation, trimeric subunits are formed by sharing the bridging oxygens. To acquire more understanding of the chemical or catalytic behaviors of the HPAs, it is necessary to decompose the structure into its building blocks: the trimer and the tetramer. Fig. 2 shows the DOS, a histogram of energy, of the Mo- and W- trimers and tetramers. There are numerous states, the so-called defect states, that intrude between -8.0 to -10.0 eV in the W-trimer. The effects noted in the DOS for the Dawson ions are similar. Table III shows the HOMO-LUMO of α-, β- Keggin and Dawson ions for comparison.

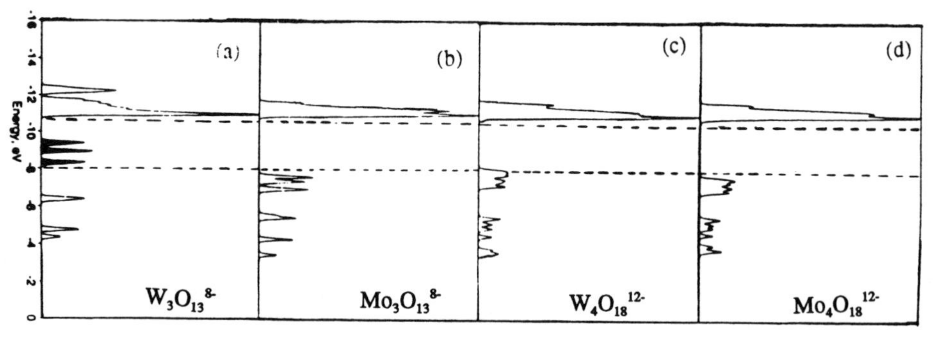

Fig 2. The DOS of the W-trimer(a) and tetramer(c), and the Mo-trimer(b) and tetramer(d). The defect states have been shaded.

Table III. The energies(eV) of HOMO-LUMO of the Keggin[10,11] and Dawson ions.

Species	HOMO	LUMO	Species	HOMO	LUMO
Keggin					
α-$SiW_{12}O_{40}^{4-}$	-10.86	-8.02	α-$SiMo_{12}O_{40}^{4-}$	-10.86	-7.65
β-$SiW_{12}O_{40}^{4-}$	-7.98	-7.89	β-$SiMo_{12}O_{40}^{4-}$	-7.66	-7.55
Dawson					
α-$Si_2W_{18}O_{62}^{8-}$	-10.73	-8.09	α-$Si_2Mo_{18}O_{62}^{8-}$	-10.73	-7.69

Generally speaking, the bigger the HOMO-LUMO gap, the greater stability of the species. However, the orbitals of the HOMO and the LUMO are dominated by O **2p** and W **5d** or Mo **4d** characters, and consequently show no differences upon substitution of the central oxoanions. However, from this information, it is possible to understand why the "rotated" β-Keggin ions are unstable and tend to be reduced more easily than the α-isomers. In fact, the HOMO-LUMO of the unsubstituted HPA is derived from the clathratic metal-oxide shell[10] since the valence state energy of X is either lower than the HOMO or higher than the LUMO of HPA for nonmetals and electropositive metals.

B. Substitutional Isomers: The substitutional effect of the framework metal atoms of the Keggin and Dawson have aroused much attention as the electronpositive transition metal substituents do not change the Keggin structure or stability limit but can enhance the catalytic properties by modification of the absorption-binding sites. It has been proven that the substituents tend to stay in the same trimer. In the Keggin ion, there are twelve chemically equivalent metal sites for the first substituent and leave four different metal sites for the second substituent as shown in Fig. 3 in an α-Keggin ion.

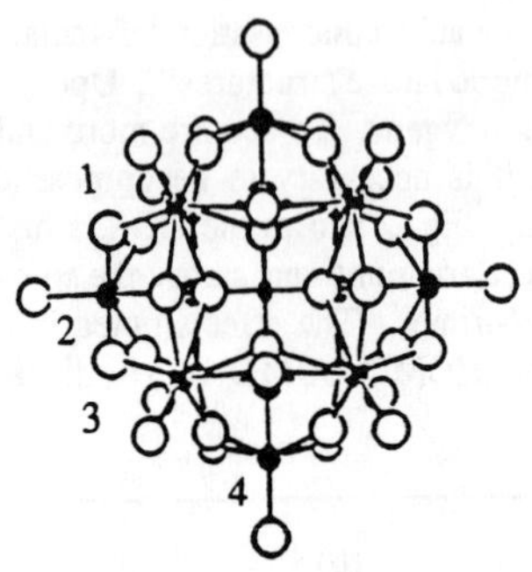

Fig 3. The substitutional sites of the framework metals. The numbers are the different sites for the second substituent.

Moreover, the possibility of observing a complete substituted trimer in an α-Keggin ion is greater than in an α-Dawson ion since there is only one type of trimer in Keggin and two types in Dawson as seen in Fig 1. In this work, the substituents, V and Cu, are selected to trace the periodic trend across the first row of the transition metals. The site-preference of the substituents in α-Keggin ions are examined by the binding energy or the core-shell model and the cluster assembly energy or the trimer model. Table IV. shows these energy effects of substitution.

The core-shell model does not predict a strong site preference for V up to three substituents, indicating that the first three electron reduction is chemically reversible as observed experimentally and these reduced electrons must populate the nonbonding states, typically the W

5d orbitals of the Keggin ion.

Table IV. Binding energy and cluster assembly energy(eV) of the Keggin ions.

Species	B. E.	C. A. E.	Species	B. E.	C. A. E.
α-$PW_{12}O_{40}^{3-}$	-5.45	-336.10			
α-$PVW_{11}O_{40}^{4-}$	-5.42	-336.39	α-$PCuW_{11}O_{40}^{7-}$	-5.03	-310.85
α-$PV_2W_{10}O_{40}^{5-}$	-5.40[a]	-336.66	α-$PCu_2W_{10}O_{40}^{11-}$	-4.62[a]	-285.60
α-$PV_3W_9O_{40}^{6-}$	-5.38[b]	-336.92	α-$PCu_3W_9O_{40}^{15-}$	-4.15[b]	-260.03

[a]There are no differences of the binding energies among the isomers with substituents in the same trimer or tetramer. [b]The substituents are in the same trimer.

As for Cu substituents, the "higher" reduction effect has to be considered. One V substitution is equivalent to one electron reduction whereas one Cu is equivalent to four electron reduction and thus the additional electrons will populate the antibonding states of the Keggin ion, producing greater instability. As the number of Cu substituent increases, this destabilization effect by populating the antibonding states is profound. This trend is also observed in the C.A.E, showing greater differences in substitution. Consequently, the more electropositive metal substituent, V, will not significantly diminish the stability of the Keggin ion, however, the more electronegative metal substituent, Cu, will destabilize the Keggin ion by the electron population effect between the HOMO-LUMO of the parent Keggin ion. The orbital interaction between V and O is more ionic, similar to that of W and O. The interaction between Cu and O is more covalent, giving significant changes in the frontier states. Many Cu-O antibonding states "intrude" and are populated. The substitutional effects of α-Dawson ions are similar to α-Keggin ions, however, the discussion is beyond this paper.

CATALYTIC IMPLICATIONS

The catalytic applications for the Keggin ions have been intensely studied whereas few studies of the Dawson ions are reported. As stated previously, the HPAs are widely used as redox and acid catalysts, the acidities or redox properties correlate with their stabilities. Therefore the analysis of their electronic properties will help to probe potential and realized catalytic properties. In oxidation-reduction, the frontier states or the HOMO-LUMO levels are implicated in the electron transfer and in the acidic catalysis. Both can be described in terms of the donor-acceptor concept. The acceptor state is the LUMO of the HPA and thus the W-type HPA will be a better oxidative catalyst than the Mo-type HPA since the acceptor state is mainly composed by valence, nonbonding **d** orbitals of the framework metals, of which the W is lower in energy ~ 0.8 eV than the Mo- Keggin ion. The acidity of HPAs can be divided into two types: the Lewis and the Brφnsted. The former is critical in oxidation processes in which reactants strongly interact with the surface. The latter is important in catalytic reactions that require hydrogen(hydronium or hydride) evolution. In this work, it has been observed that the bridging oxygens possess identical or very slightly more negative charge than the terminal oxygens. High negative charge creates the Brφnsted acid sites. Comparison of bond strength suggests that the bridging oxygen is more labile and thus implicated in acid catalyzed oxidative processes. The arguments based on electronic structural analysis are similar for α-Dawson ions.

The modification of catalytic properties by framework metal substitution is interpretated by two extremes. For electropositive substituent, e.g. V, its tendency of donating electrons to W or Mo based HPAs will promote the activity at the bridging oxygen, thus being better Brφnsted acid sites in acid catalysis and the "relative" deficiency of electrons at the substitutional sites will be able to localize the reduced electrons and tend to reduce oxygens. In addition, the substituent will also promote the Lewis acidity of the substituted HPAs. This localization effect is partially due to the higher energy state of the LUMO. In fact, competition between the Brϕnsted and Lewis acid character is observed in the substituted HPAs; in other words, the substituent(s) not only improve the Brϕnsted acidity but also the Lewis acidity. As for the electronegative substituent, e.g. Cu, the tendency to withdraw electrons from its neighboring atoms make the original framework metal atoms better Lewis acid centers, and localization of electron density is also observed. The new states at the frontier are generated by Cu-O covalent orbitals and thus activity will be modulated by Cu-O frontier character.

CONCLUSION

The HPA models utilized in this study illustrate the stability trends, acidities, and the redox properties observed in experiments. The computational methodology has given important insight into electronic requisites for catalytic activity. The clathratic shell model predicts all the chemical observations and the trimer model predicts the site preference for the framework metal substitution. From this study, the interpretations of the correlation among the acidity, stability, and substitutions of the central oxoanions and the framework metal atoms are supported by experimental results. The α-Dawson ions are generally more stable than and possess essentially the same energy of acceptor states as the α-Keggin ions. The tendency of isomerization from β → α is also observed. Future work will extend the studies to the β-Dawson ions and the organometallic species as the framework metal substituent of the Keggin ions.

APPENDIX

The parameters utilized in this study are as following: Si(3s: H_{ii} -16.3, ζ_1 1.383; 3p H_{ii} -8.5, ζ_1 1.383), Ge(4s: H_{ii} -16.0, ζ_1 2.16; 4p: H_{ii} -9.0, ζ_1 1.85), P(3s: H_{ii} -14.5, ζ_1 1.6; 3p: H_{ii} -9.7, ζ_1 1.6), O(2s: H_{ii} -27.16, ζ_1 2.28; 2p: H_{ii} -11.01, ζ_1 2.275), V(4s: H_{ii} -6.7, ζ_1 1.6; 4p: H_{ii} -3.4, ζ_1 1.6; 3d: H_{ii} -6.7, ζ_1 4.75, c_1 0.4558, ζ_2 1.5, c_1 0.7516), Cu(4s: H_{ii} -9.4, ζ_1 2.2; 4p: H_{ii} -4.06, ζ_1 2.2; 3d: H_{ii} -12.0, ζ_1 5.75, c_1 0.5722, ζ_2 2.1, c_1 0.6117), Mo(5s: H_{ii} -7.3, ζ_1 1.96; 5p: H_{ii} -3.6, ζ_1 1.96; 4d: H_{ii} -7.9, ζ_1 4.54, c_1 0.590, ζ_2 1.96, c_1 0.590), W(6s: H_{ii} -6.26, ζ_1 2.34; 6p: H_{ii} -3.17, ζ_1 2.34; 5d: H_{ii} -8.37, ζ_1 4.96, c_1 0.6685, ζ_2 2.16, c_1 0.5424).

REFERENCES

1. M. Misono and N. Mizuno, *J. Mol. Catal.*, 86, 309(1994).
2. R. I. Buckley and R. J. H. Clark, *Coord. Chem. Rev.*, 65, 167(1985).
3. M. T. Pope and A. Müller, *Angew. Chem. Int. Ed. Engl.*, 30, 34(1991).
4. A. K. Ghosh and J. B. Moffat, *J. Catal.*, 101, 238(1986).
5. R. A. Prados and M. T. Pope, *Inorg. Chem.*, 15, 2547(1976).
6. D. K. Lyon, W. K. Miller, T. Novet, P. J. Domaille, E. Evitt, D. C. Johnson, and R. G. Finke, *J. Am. Chem. Soc.*, 113, 7209 and 7222(1991).
7. R. Contant and R. Thouvenot, *Inorg. Chim. Acta*, 212, 41(1993).
8. R. G. Finke, M. W. Droege, and P. J. Domaille, *Inorg. Chem.*, 26, 3886(1987).
9. J. N. Barrows and M. T. Pope, *Inorg. Chim. Acta*, 213, 91(1993).
10. S.-H. Wang and S. A. Jansen, *Chem. Mater.* 6, 146 and 2130(1994).
11. S.-H. Wang, D. Singh, and S. A. Jansen, *J. Catal.*(submitted).
12. S.-H. Wang and S. A. Jansen, *Inorg. Chem.*(in preparation).
13. K. Y. Matsumoto, A. Kobayashi, and Y. Sasaki, *Bull. Chem. Soc. Jpn.*, 48, 3146(1975).
14. T. Yamase, T. Ozeki, and S. Motomura, *Bull. Chem. Soc. Jpn.*, 65, 1453(1992).
15. B. Dawson, *Acta Cryst.*, 6, 113, (1953).
16. S. A. Jansen and R. Hoffmann, *Surface Sci.*, 197, 474(1988).

PART V

Metal Oxides

SURFACE MODELS OF ALUMINA SUPPORTED COPPER OXIDE PREPARED BY LASER ABLATION: DISPERSION AND ELECTRON STRUCTURE CHANGES FOLLOWING REDUCTION AND OXIDATION CYCLES

B.HIRSCHAUER AND J.PAUL[§]
KTH/the Royal Institute of Technology, Physics III, 100 44 Stockholm, Sweden

§ For correspondence

ABSTRACT

We have explored and developed laser ablation as a tool to synthesize realistic planar models of high surface area catalysts as well as novel materials, not accessible with wet chemical methods. The process is illustrated by data for dispersion and electron structure modifications following thermal treatments of alumina supported copper oxide. Planar models of different wet and dry carrier/overlayer combinations were prepared and exposed to oxidizing and reducing atmosphere at 500°C. Characterization was made by XPS/ESCA. The data reveal the importance of surface hydroxyl groups in the two initial components to control coalescence and interdiffusion. The ablation route is part of our ongoing DENOX project and comparisons are made with other materials, tested for their materials properties and catalytic behavior in HC/NO reactions.

INTRODUCTION

The motivations for the present work are twofold: to document laser ablation as a synthesis method for model catalysts and to apply this method to create samples of alumina supported copper oxide. This effort is part of our ongoing project on alternative materials for NO reduction with hydrocarbons (HC) [1]. We concentrate on one catalyst, alumina/Cu_xO, with known DENOX capacity under reducing conditions and separate the influence of the carrier and of the impregnate on the different processes : (i) HC cracking, (ii) coke deposition, (iii) partial and complete oxidation of the HC, (iv) NO/coke reactions, (v) NO/HC and NO/cracking product interactions, and (vi) reactions between NO and partial oxidation products. Our feed contains nitric oxide but we also consider NO oxidation and NO disproportionation. Model studies with planar substrates as well as with high surface area powders are crucial in order to elucidate and separate the above mechanisms. Impregnation of the powders is done with a wet chemical speciation method, which allows us to control the dispersion and density of protons and metal ions on pure and acidified (Ti-doped) γ-alumina. Acid sites have a dual function for ion exchange and HC conversion. Single or polycrystalline copper samples serve to isolate the activity of the deposited phase but shortcomings in conventional thin film technologies motivated us to explore and develop techniques for laser ablation synthesis of planar models of catalyst/carrier combinations. Comparisons are made with our own materials as well as with conventionally prepared alumina supported copper oxides [2].

EXPERIMENTAL

The laser ablation unit is part of a multichamber ultra high vacuum (UHV) system for synthesis,

Mat. Res. Soc. Symp. Proc. Vol. 368 © 1995 Materials Research Society

surface analysis, and reaction studies [3]. We use a Lambda Physik Excimer laser (KrF, 248 nm) with a pulse duration of 20 ns (1.2 J/pulse) and a maximum repetition rate of 50 Hz. The image of the laser exit slit on the target has an area of 1.5 mm^2. Fig.1a represents characteristic yields and uniformities for the ablation of α-Al_2O_3 [4]. The target can be heated to 900°C, either during the ablation or as a post-treatment. Similar results have been reported, both for simple and complex oxides [5] but our experiments document that the dry oxides can be hydroxylated by ablation in moisture [6]. The cleanliness of our procedure was verified by XPS/ESCA (Fig.1b).

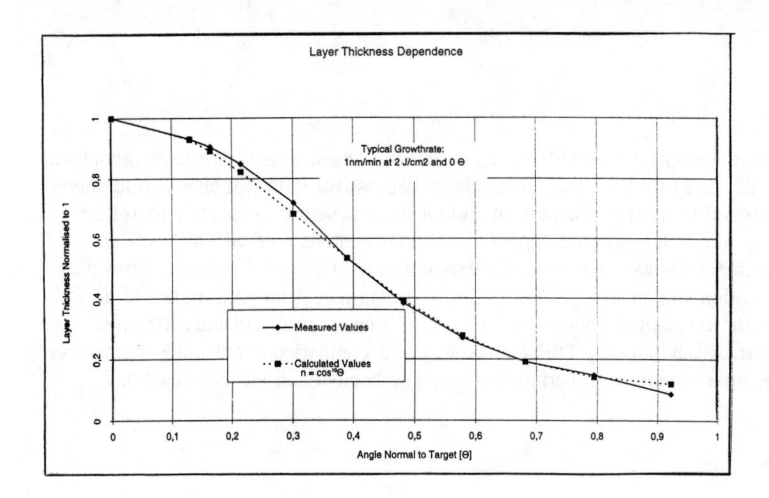

Figure 1a. Radial distribution of deposited α-alumina over a 3" target.

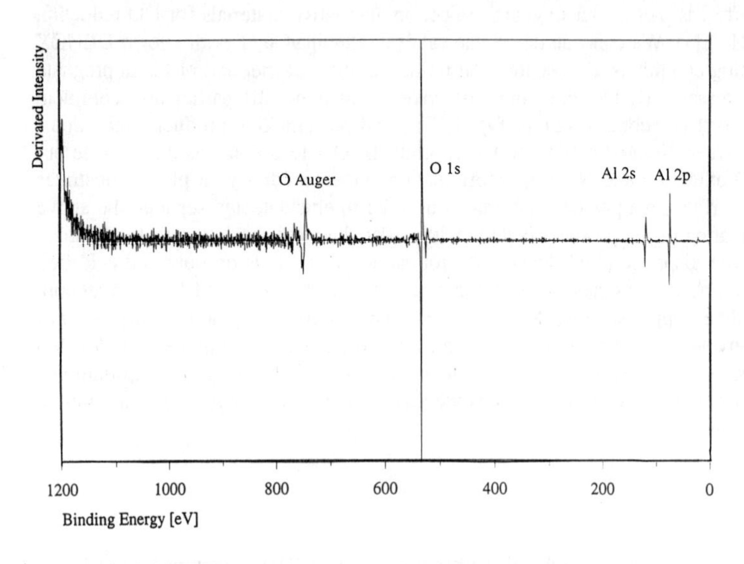

Figure 1b. XPS/ESCA spectrum, in its derivative form, of the same film.

The support discs were made of aluminum alloyed steel (3" diamater, 1mm thickness, Sandvik AB) of the same type used, as corrugated and rolled thin foil, in new types of catalytic converters. This steel contains 20% Cr, 5% Al, 0.018% C, 0.2% Si, 0.35% Mn, 0.02% Ce, and traces of S and P. The discs were sand blastered and annealed in air at 650°C for 1/2 h followed by 1h at 700°C and recooling over >1/2 h. Slow cooling is essential to equilibrate thermal stress. An aluminum oxide film forms as a results of Al diffusion from the bulk. The samples were stored in air at 120°C until introduced into the vacuum system. We choose this substrate material for its relevance as an industrial support for exhaust catalysts and because the surface segregated aluminum oxide guarantees a good adhesion to ablated alumina films.

This work reports ablation of thin overlayers (1 monolayer) of cuprous oxide, Cu_2O, on thick (μm) films of α-alumina. A thorough calibration of coverage is done by comparisons with bulk samples with known copper concentrations, titrated both during impregnation and *in situ* by CO adsorption [6] but equal amounts of copper oxide were ablated atop a 'bulk' alumina film in each case. The deposition was done at room temperature either in vacuum (10^{-10} torr) or in water vapor (1 torr), which yields four alumina/copper oxide combinations denoted dry/dry, dry/wet, wet/dry, and wet/wet.

Each sample was analyzed with X-ray photoelectron spectroscopy (XPS), as prepared, and after calcination (10^{-6} torr, O_2, 500°C), and reduction (10^{-6} torr H_2,500°C). These temperatures are typical for bulk catalysts [2]. Primary annealing was done simultaneously with the calcination step. This represents one choice. Annealing in vacuum reduces CuO and eventually Cu_2O, but an increased oxygen or water pressure will prevent this reduction. Materials prepared in aqueous solutions contain excess water, stable in vacuum at room temperature but not at elevated temperatures. This reaction can be followed with *in situ* techniques but we choose not to trace it for the present samples due to the presumed low water concentration in the overlayer [6]. Carbon removal, which is relevant for bulk samples exposed to air, is no issue for our planar models, prepared under UHV conditions (Fig.1b).

Photoemission (XPS/ESCA) spectra were obtained at normal emission with a VG Clam-2 analyzer (non-monochomatized Al Kα). We scanned three regions with 0.1 eV step-size and the following binding energies: 0-200eV, valence band, Al 2s/2p, and Cu 3s/3p; 500-600 eV, O1s; and 900-1000 eV, Cu2p. A survey scan, 0-1200 eV at 0.5 eV step-size revealed the presence of any contaminants. All spectra were obtained at room temperature. The use of several orbitals to trace the same element serves the purpose of internal intensity calibration and possibly concentration profile evaluation from the different penetration depths of electrons with different binding and thus kinetic energies.

RESULTS AND DISCUSSION

Figs. 2-5 shows ESCA spectra of equal amounts of copper oxide, ablated under dry or wet conditions, on a thick film of alumina, again prepared under dry or wet conditions. Three regions are displayed for each film and each figure shows intensities for the film as prepared and after calcination and reduction.

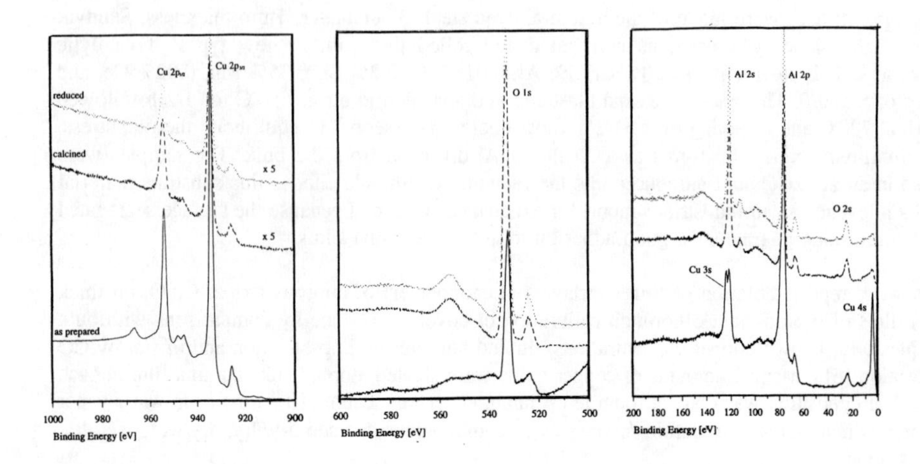

Figure 2. ESCA spectra of Cu_2O, ablated in vacuum, on a thick film of α-Al_2O_3, again ablated in vacuum (dry/dry). A full line represents 'as prepared' , a dashed 'calcined' , and a dotted 'reduced'.

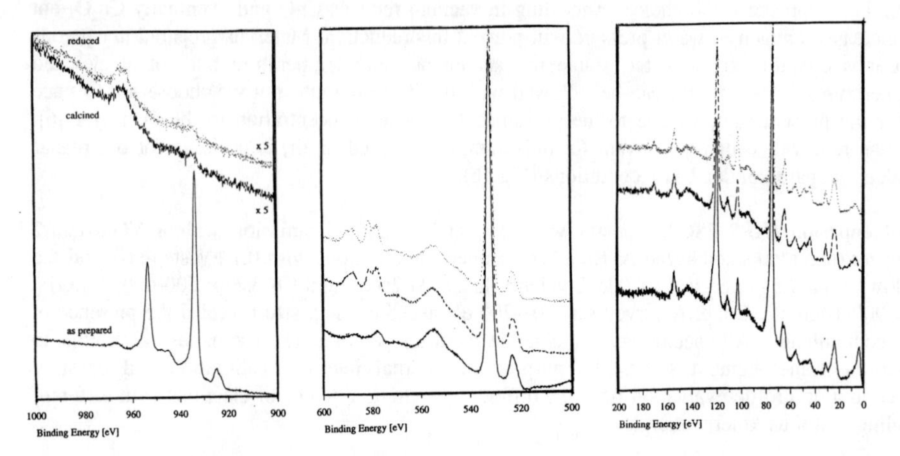

Figure 3. ESCA spectra of Cu_2O, ablated in vacuum, on a thick film of α-Al_2O_3, ablated in moisture (dry/wet). A full line represents 'as prepared' , a dashed 'calcined' , and a dotted 'reduced'.

A full analysis is presented in connection with our work on high surface area materials [6] but Figs.2-5 reveal a few distinctive characteristics:

- A relative insensitivity to calcination and reduction for the dry/dry and wet/wet samples. This agrees with intuition that materials with similar 'surfaces' give stable structures with high interfacial energies.

- Overall limited effects of reduction. This suggests a surprisingly stable dispersion of the copper overlayer even for a calcination temperature well below the 900°C required to form the bimetallic oxides [2].

- Significant changes of copper induced features after calcination, particularly at the Fermi level, for Cu_2O ablated in vacuum. Ablation in vacuum may cause an oxygen deficiency, well known for copper based high temperature superconductors [5].

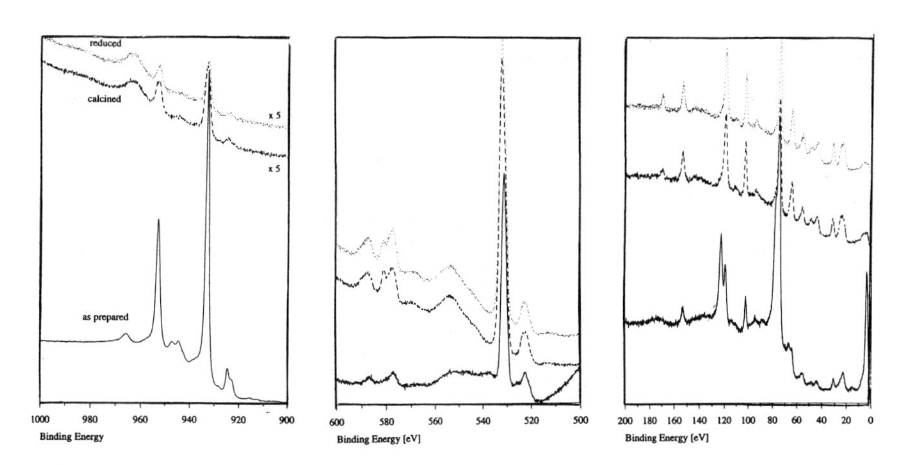

Figure 4. ESCA spectra of Cu_2O, ablated in moisture, on a thick film of α-Al_2O_3, ablated in vacuum (wet/dry). A full line represents 'as prepared' , a dashed 'calcined' , and a dotted 'reduced'.

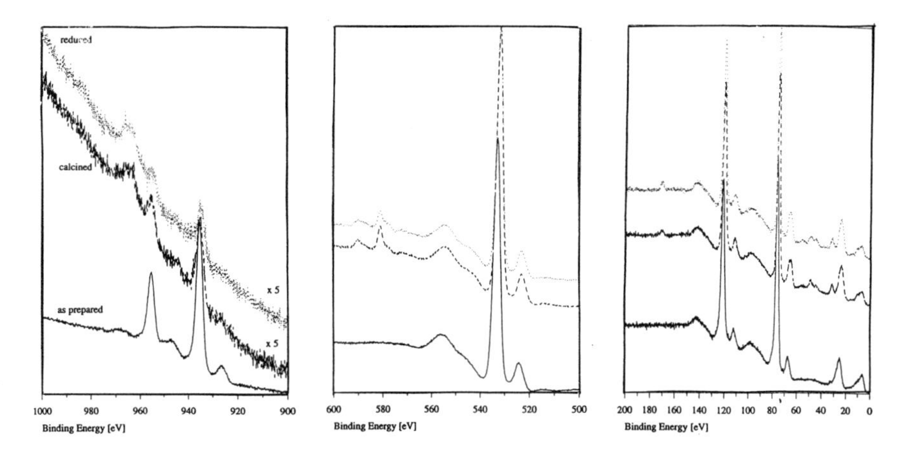

Figure 5. ESCA spectra of Cu_2O, ablated in moisture, on a thick film of α-Al_2O_3, again ablated in moisture (wet/wet). A full line represents 'as prepared' , a dashed 'calcined' , and a dotted 'reduced'.

ACKNOWLEDGEMENTS

Sandvik AB kindly provided us with samples of aluminum alloyed steel. We also acknowledge collaboration with researchers at Exxon CR.

REFERENCES

1. C.M Pradier and J. Paul, Proceedings Europcat-I, Montpellier, France; C.M. Pradier, H. Vikström, and J. Paul, Proceedings CAPOC 3, Brussels, Belgium; J. Paul, 1st World Conference on Environmental Catalysis, Pisa, Italy (accepted); J.Paul and L.O. Öhman, JECAT'95, Lyon, France (accepted); H. Lu, C.M. Pradier, and J.Paul, JECAT'95, Lyon, France (accepted); J.Paul , 10th World Clean Air Conference, Helsinki, Finland (accepted).

2. G. Ertl, R. Hierl, H. Knözinger, N. Thiele, and H.R. Urbach, Appl.Surface,Sci. 5 (1980) 49; G. Centi, C. Nigro, S. Perathoner, and G. Stella, in Environmental Catalysis, Ed. J. N. Armor (ACS Symposium Series 552, Washington, 1994); US-DOE Final Report Contract No.22-89PC89805 (Artur D. Little Co., Cambridge/MA, 1994).

3. J. Paul, C.M. Pradier, T. Levola, E. Supponen, and J. Vanhatalo, Rev.Sci.Instrum. (submitted).

4. B. Hirschauer, S. Söderholm, and J. Paul (manuscript).

5. Laser Ablation for Materials Synthesis, MRS Symposium Procedings 191, Eds. D.C Paine and J.C. Bravman (MRS, Pittsburgh, 1990); Laser Ablation in Materials Processing: Fundamentals and Applications, MRS Symposium Proceedings 285, Eds. B. Braren, J.J. Dubrowski, and D.P. Norton, (MRS, Pittsburgh,1992).

6. B. Hirschauer, L.O. Öhman, and J. Paul, (manuscript).

NEW NANOPHASE IRON-BASED CATALYSTS FOR HYDROCRACKING APPLICATIONS

DEAN W. MATSON, JOHN C. LINEHAN, JOHN G. DARAB, DONALD M. CAMAIONI, S. THOMAS AUTREY, AND EDDIE G. LUI
Pacific Northwest Laboratory [1], PO Box 999, Richland, WA 99352

ABSTRACT

Carbon-carbon bond cleavage catalysts produced *in situ* at reaction conditions from nano-crystalline hydrated iron oxides, show high activity and selectivity in model compound studies. Two highly active catalyst precursors, ferric oxyhydroxysulfate (OHS) and 6-line ferrihydrite, can be produced by a flow-through hydrothermal powder synthesis method, the Rapid Thermal Decomposition of precursors in Solution (RTDS) process. Model compound studies indicate that both catalyst precursors are active at a 400°C reaction temperature, but that there are significant differences in their catalytic characteristics. The activity of 6-line ferrihydrite is highly dependent on the particle (aggregate) size whereas the activity of the OHS is essentially independent of particle size. These differences are attributed to variations in the crystallite aggregation and particle surface characteristics of the two catalyst precursor materials. Catalytic activity is retained to lower reaction temperatures in tests using OHS than in similar tests using 6-line ferrihydrite.

INTRODUCTION

Essentially all processes in the petrochemical industry initially involve cleavage of the carbon-carbon bonds in complex organic molecules to produce simpler, lower molecular weight products. To be economically viable, these processes must be catalytically enhanced to speed up reaction rates, lower temperatures required for reaction, and to selectively cleave bonds to produce a desirable product distribution. Traditional hydrocracking catalysts, many of which are based on noble or other transition metals, tend to be both very expensive and highly toxic. Consequently, catalyst fabrication, recovery, recycling, and storage can all affect the economic viability of hydrocracking operations [2]. Development of highly efficient iron-based materials for processes involving carbon-carbon bond cleavage, including petroleum hydrocracking and coal liquefaction, offers the potential for decreasing catalyst costs as well as reducing the need for expensive catalyst recovery and recycling steps. Furthermore, if active iron-based catalysts can be produced containing ultrafine crystallites and high surface areas (i.e., having a large percentage of the iron on a catalytically active surface site), lower catalyst loading requirements may be achieved.

At the Pacific Northwest Laboratory (PNL) we are investigating the activities of iron-based powders as catalytic precursors in hydrocracking reactions and are developing new methods for producing iron-based nanocrystalline powders exhibiting the greatest catalyst potential. PNL's powder forming processes include the Rapid Thermal Decomposition of precursors in Solution (RTDS) process [3-5], a flow-through hydrothermal method, and the Modified Reverse Micelle (MRM) process [5], which uses microemulsion technology to precipitate ultrafine powders. Because these synthesis methods produce powders containing crystallite sizes in the nanometer size range, for certain precursor phases, conversion of these precursor powders to the active catalyst occurs rapidly under the conditions present in the reaction vessel. Using model compound studies we have previously demonstrated that iron-based powders generated by both the RTDS and MRM methods can produce highly active carbon-carbon bond scission catalysts under reaction conditions relevant to coal liquefaction processes [5]. In this paper we present recent results of model compound studies obtained using two of the more active catalyst precursor materials, 6-line ferrihydrite and ferric oxyhydroxysulfate, both of which have been produced by the RTDS method. We also attempt to relate powder characteristics to measured activity and use the results as a basis for mechanistic interpretation of the catalytic process.

Mat. Res. Soc. Symp. Proc. Vol. 368

CATALYST PRECURSOR SYNTHESIS AND EVALUATION

The RTDS continuous powder synthesis method involves formation of ultrafine solid particles by reaction of a solute species when pressurized homogeneous solutions are passed rapidly through a heated linear reactor (1-30 sec residence time) and then abruptly quenched by dropping the pressure. When using aqueous solvents for this process, well-known hydrothermal reactions leading to formation of oxides or hydroxides occur between metal-containing precursor solute species and the solvent itself [6]. Multiple-metal and/or doped oxide/oxyhydroxide powders can be easily produced by the appropriate choice and loadings of precursor solute species. RTDS products are collected as suspensions, and typically contain submicrometer-sized aggregates of nano-scale crystallites. The solid powders can be separated from the suspension by centrifugation, spray drying, or freeze drying. Details of the RTDS process have been presented elsewhere [3-5].

Six-line ferrihydrite is a poorly defined ferric oxyhydroxide phase named for the number of broad lines present in its powder X-ray diffraction pattern and is believed to be a hydrated substructure of hematite (α-Fe_2O_3) [7]. RTDS synthesis of this material was accomplished using feed solutions containing ferric nitrate (0.1 M) and urea (0.5 M). Under the hydrothermal conditions present in the reaction tube (~300°C, 4000-8000 psi pressure) the urea rapidly decomposes and reacts with the solvent to generate ammonia and carbon dioxide, creating a basic solution. The particulate product formed and collected in the RTDS suspension was similar in appearance, but crystallographically and catalytically distinct from material generated by adding ammonia directly to ferric nitrate solutions at ambient conditions. At higher processing temperatures or at the same RTDS processing conditions but without urea, hematite is the predominant iron oxide phase produced.

Ferric oxyhydroxysulfate (OHS) is a poorly crystallized variation of akaganeite (β-FeOOH), that contains tunnel structures in which chloride ions reside. In OHS some or all of the chloride ions are replaced by sulfate ions. OHS is formed naturally by bacterial oxidation of Fe(II) in acidic mine waters, or by a laboratory procedure involving low yields and a lengthy (30 day) dialysis step [7]. OHS was synthesized using the RTDS process by a low temperature reaction (100-150°C) of solutions containing the ferric and sulfate ions, followed by a brief period of dialysis (< 7 days).

Both the 6-line ferrihydrite and OHS phases can be produced using standard laboratory syntheses [7] and powders produced by those methods have been shown to be catalytically active for carbon-carbon bond cleavage reactions [8]. However, bench-top syntheses of these materials are batch processes that are not necessarily scalable to larger batch sizes, and both require lengthy dialysis steps that can be reduced or avoided by using the continuous RTDS method.

The RTDS-generated catalyst precursor powders discussed here were separated from the liquid phase by sedimentation/centrifugation. The solids were dried, first under flowing nitrogen, then in a vacuum drying oven at <100°C. The products were ground in an agate mortar where appropriate and sieved into three fractions: -325 mesh, +325/-230 mesh, and +230 mesh. Powders were characterized for phase and crystallite size by powder X-ray diffraction and for agglomerate size and shape by electron microscopy (SEM and TEM). Catalytic characteristics were evaluated by measuring the powder's ability to promote consumption of model organic compounds, including naphthyl bibenzylmethane (NBBM) and a series of diphenylmethane (DPM) derivatives (Fig. 1). Model compounds used for these studies were determined to be 99% pure or better by gas chromatography (GC) and NMR analysis. The catalyst test procedure consisted of loading fixed amounts of catalyst precursor, elemental sulfur, model compound, and 9,10-dihydrophenanthrene (a hydrogen donor solvent) in Pyrex tubes, sealing under vacuum, emersing in a fluidized sand bath for one hour, and analyzing the product distribution by GC [5].

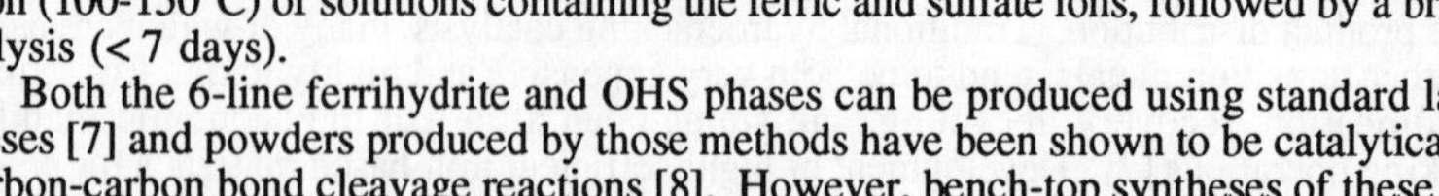

a b c d e
CH_2 CH_2 — CH_2
NBBM

a b c
CH_2 $(CH_3)_n$
DPM
n = 0,1,2,3

Figure 1. Model compounds used to evaluate catalytic activity of RTDS-derived powders.

RESULTS AND DISCUSSION

Electron micrographs of RTDS-generated OHS and 6-line ferrihydrite powders are presented in Figures 2 and 3. The OHS powders contain micrometer-sized agglomerates made up of submicrometer-sized spherical features (Fig. 2a). At higher resolutions obtained using TEM, the spherical features clearly do not have smooth surfaces, but rather are very irregular and consist of elongated crystallites extending outward from the sphere surface (Fig. 2b). SEM micrographs of sieved samples of RTDS-generated 6-line ferrihydrite (Fig. 3) suggest a much denser material than the OHS. Dark-field TEM micrographs of the 6-line ferrihydrite powders indicate that they consist of densely packed nanocrystallites. The higher microstructural density and strength of inter-crystalline agglomeration of the 6-line material was also reflected in its greater tendency to form hard masses on drying, requiring much greater effort to grind and sieve into the smaller fractions than the OHS. These obvious physical differences in particle morphology may play a significant role in determining the differences in catalytic capabilities of the two precursor materials, as discussed below.

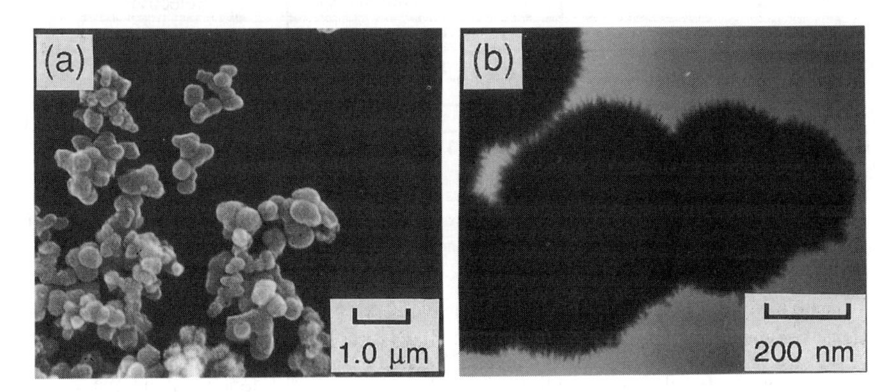

Figure 2. a) SEM micrograph of OHS powder produced by RTDS. b) TEM micrograph of the surfaces of the particles shown in Fig. 2a.

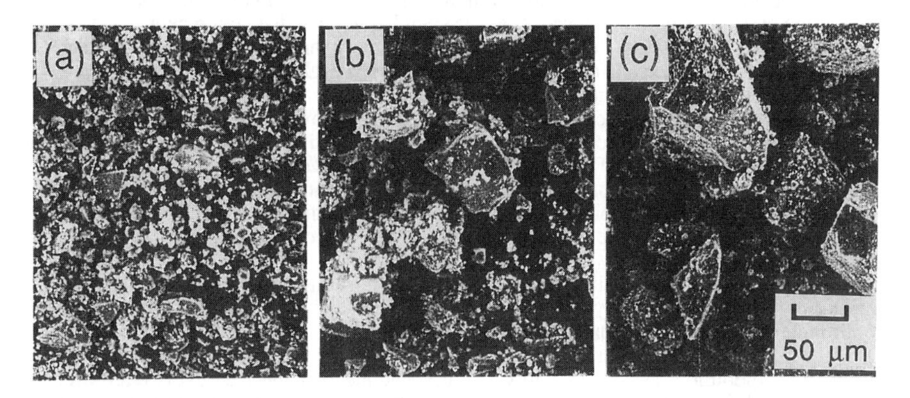

Figure 3. SEM micrographs of sieved 6-line ferrihydrite powder produced by RTDS. a) -325 mesh, b) -230/+325 mesh, c) +230 mesh.

The results of catalyst precursor screening of RTDS-generated powders using the model compound substrate naphthyl bibenzylmethane (NBBM) are shown in Table I. In addition to 6-line ferrihydrite and OHS, RTDS hematite (which is produced under the same RTDS conditions as 6-line ferrihydrite but in the absence of the urea in the feedstock solution) is included for reference. The Table also includes a thermal background value for consumption of the model compound under test conditions in the absence of added catalyst or elemental sulfur. In contrast to uncatalyzed thermal consumption of NBBM, in which essentially no selective bond cleavage was observed, NBBM consumption catalyzed by the iron-based materials preferentially cleaved bonds "a" and "b". Product distributions from catalyzed NBBM reactions yielded naphthalene, methylbibenzyl, methyl naphthalene, and bibenzyl almost exclusively.

Table I. Consumption of Naphthyl Bibenzylmethane in the Presence of Iron-Based Catalyst Precursors Produced by the RTDS Method[a]

Precursor Phase	NBBM Consumed (%)	Selectivity[b]
none	3 ± 2[c]	40-60
Ferric Oxyhydroxysulfate	>90[d]	97
6-line Ferrihydrite	>90	97
Hematite	39	85

[a]Test conditions: 3 mg catalyst precursor (-325 mesh), 3 mg elemental sulfur, 25 mg model compound, 100 mg 9,10-dihydrophenanthrene, 400°C, 1 hr.

[b]Selectivity defined as products of a and b bond cleavage as a percentage of products from total bonds cleaved.

[c]No sulfur added. Addition of sulfur without adding iron-based precursor yielded sulfur containing organic products.

[d]Model compound consumptions in excess of 90% were uncalibrated.

The results presented in Table I suggest that both OHS and 6-line ferrihydrite are excellent catalysts for the consumption of NBBM under the test conditions used. In previous work, however, we showed that the activity of RTDS-generated 6-line ferrihydrite as a dispersed catalyst precursor for the consumption of NBBM was strongly dependent on the particle (agglomerate) size of the precursor powder [9]. In fact, the majority of the catalytic activity observed for bulk ferrihydrite powders was attributed to a very small weight fraction of the starting material having the smallest particle sizes. These results suggested that mass transport limitations are an important consideration even in "nanocrystalline" materials, depending on the degree of the crystallite agglomeration. They also emphasized the distinction between crystallite size and the implied catalyst "dispersibility" in a reaction medium and the importance of comparing comparably sized and characterized powders when evaluating catalytic activities.

The effect of particle sieve size on activity of OHS as a catalyst precursor toward consumption of NBBM was also investigated. The results, shown in Figure 4, suggest little or no relationship between the observed catalytic activity and particle size for this material as was seen for the 6-line ferrihydrite. This lack of observed particle size/activity effect may result from the more open surface structure of the OHS particles (Fig. 2b) and/or weaker crystallite/crystallite interactions holding the agglomerated particles together. Evidence for the latter may be found in the fact that OHS required little or no grinding in order to accomplish sieving to the smaller particle sizes.

In addition to particle size/activity effects, dissimilarities were noted in the effect of reaction temperature on the catalytic activities of OHS and 6-line ferrihydrite. Although activities of both materials toward the consumption of NBBM dropped at reaction temperatures below 400°C, the 6-line ferrihydrite activity fell off more quickly. OHS retained measurable catalytic activity to temperatures as low as 250°C. We speculate that the free energy required for conversion of OHS to the active iron sulfide catalyst is lower than the corresponding energy for conversion of the 6-line ferrihydrite.

When OHS and 6-line ferrihydrite were evaluated as catalytic precursors for the consumption of the model compounds diphenylmethane and its methylated analogs (Fig. 1), the extent of reaction increased with degree of methyl substitution (Fig. 5). Products of the catalyzed reactions were almost exclusively the result of bond "b" cleavage in the methylated analogs. Toluene and methylated aromatic species derived from the substituted ring were the major products from the catalyzed reactions. The general increase in reaction rate of the model compounds having a systematically decreasing oxidation potential suggests that a redox mechanism may provide a rate determining step in the catalytic consumption of these species. Note that in the absence of catalyst, there was no corresponding correlation between consumption and degree of methyl substitution.

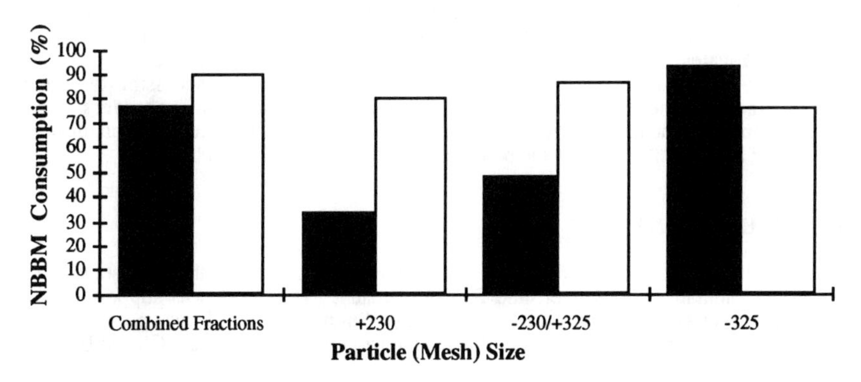

Figure 4. Comparison of the effect of particle size between 6-line ferrihydrite (white) and OHS (black) on the consumption of the NBBM model compound. Reaction conditions for 6-line ferrihydrite runs were the same as those used to obtain results shown in Table 1 except that reaction temperatures for the OHS runs were 360°C.

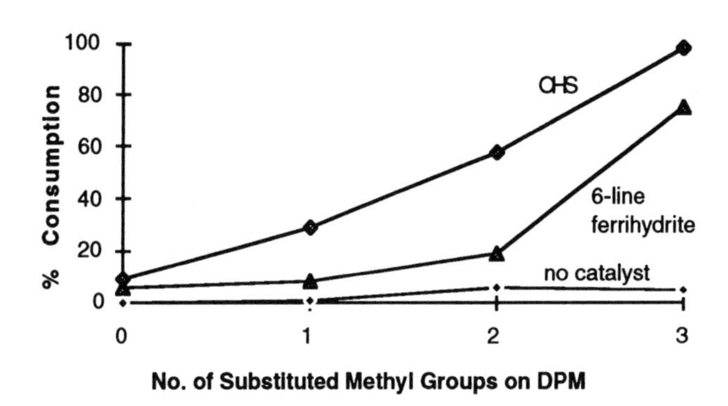

Figure 5. Consumption of diphenylmethane (DPM) and its methylated analogs (at 400°C, 1 hr reaction time, with elemental sulfur and 9,10-dihydrophenanthrene) in the presence of 6-line ferrihydrite, OHS, and no catalyst. Oxidation potentials of the model compounds decrease with increasing methyl substitution.

SUMMARY

The RTDS process can be used to produce highly active nanocrystalline iron-based catalyst precursor powders for hydrocracking applications. The two most active phases, ferric oxyhydroxysulfate and 6-line ferrihydrite, transform into active catalysts *in situ* under liquefaction conditions in the presence of elemental sulfur and a hydrogen donor. As dispersed heterogeneous catalyst precursors, the activity of the 6-line ferrihydrite phase is sensitive to the particle size whereas the oxyhydroxysulfate phase is not. This difference is attributed to major differences in the packing characteristics of individual crystallites in the particles making up the powders of the two materials. Individual crystallites in the oxyhydroxysulfate powders extend from the surfaces of spherical aggregates, with considerable surface area exposed to the solvent and model compound substrate molecules. Crystallites in the 6-line ferrihydrite powder particles are packed into dense aggregates. Differences between the effects of particle size and activity for these materials likely can be attributed to differences in the dispersibility and accessibility of active catalyst surfaces. Increasing activity of the iron-based precursor powders toward diphenylmethane with increasing methyl group substitution suggests that a redox mechanism may play the role of a rate limiting step in the bond cleavage reactions in these substrates.

ACKNOWLEDGMENTS

The authors gratefully acknowledge the following PNL staff for their assistance: G.E. Fryxell, J.E. Coleman, H.T. Schaef, and D.E. McCready. This work was supported by the U.S. Department of Energy, Office of Fossil Energy. Support for E.G.L was provided by the Associated Western Universities, Inc., Northwest Division under Grant DE-FG06-89ER-75522 with the U.S. Department of Energy.

REFERENCES

1. Pacific Northwest Laboratory is operated for the United States Department of Energy by the Battelle Memorial Institute under contract DE-AC06-76RLO 1830.
2. J.A. Rabo in Advanced Heterogeneous Catalysts for Energy Applications Vol. II, U.S. DOE report DOE/ER-30201-H1, p. 1.1 (1994)
3. D.W. Matson, J.C. Linehan, and R.M. Bean, Mater. Lett. **14**, 222 (1992).
4. J.G. Darab, M.F. Buehler, J.C. Linehan, and D.W. Matson in Better Ceramics Through Chemistry VI, edited by A.K. Cheetham, C.J. Brinker, M.L. Mecartney, and C. Sanchez (Mat. Res. Soc. Symp. Proc. **346**, Pittsburgh, PA, 1994) pp. 499-504.
5. D.W. Matson, J.C. Linehan, J.G. Darab, and M.F. Buehler, Energy and Fuels **8**, 10 (1994).
6. W.J. Dawson, Ceram. Bull. **67**, 1673 (1988).
7. U. Schwertmann and R.M. Cornell, Iron Oxides in the Laboratory: Preparation and Characterization (VCH Publishers, Inc., New York, 1991).
8. J.C. Linehan, D.W. Matson, and J.G. Darab, Energy and Fuels **8**, 56 (1994).
9. J.G. Darab, J.C. Linehan, and D.W. Matson, Energy and Fuels **8**, 1004 (1994).

INTERACTION OF Ni(II,III) AND SOL-GEL DERIVED ZrO_2 IN Ni/ZrO_2 CATALYST SYSTEM

H.C. Zeng*, J. Lin**, W.K. Teo*, F.C. Loh**, and K.L. Tan**
*Department of Chemical Engineering, Faculty of Engineering
**Department of Physics, Faculty of Science
National University of Singapore, 10 Kent Ridge Crescent, Singapore 0511

ABSTRACT

NiO/ZrO_2 catalyst has a selectivity for producing higher hydrocarbons, whereas NiO on classical supports gives rise to methanation of CO + H_2 mixture. In this study, tetragonal and monoclinic ZrO_2 gel carriers have been synthesized using sol-gel method. The interaction of nickel ions with a sol-gel derived catalyst support has been investigated using FTIR, DTA, and XPS. It is found that the nickel ions diffuses into the ZrO_2 continuously over 400 to 600 °C for both tetragonal and monoclinic ZrO_2, and forms a thermodynamically stable Ni/ZrO_2 solid solution at elevated calcination temperatures. Higher nickel surface contents are observed in tetragonal ZrO_2. In addition to Ni^{2+}, Ni^{3+} ions are also detected in Ni/ZrO_2 system. Cation ratio of Ni^{2+}/Ni^{3+} peaks at 700°C in the tetragonal ZrO_2. In the monoclinic case, the ratio remains constant at different elevated temperatures. The diffusion activation energy for Ni(II,III) on both tetragonal and monoclinic in ZrO_2 is 0.40 eV.

INTRODUCTION

ZrO_2-based ceramic materials have received increasing attention in recent years, owing to their prospect of industrial-scale applications. ZrO_2 matrices synthesized by the sol-gel process add new dimensions to catalysis research, magnetic and optoelectronic materials syntheses, corrosion prevention and many other important applications [1-16]. ZrO_2-supported transition metal oxide systems are also the topic of several current research projects [1-12]. Chemically and physically modified ZrO_2 is now widely regarded as a potential candidate in fabrication of a new generation of catalysts. The superiority of ZrO_2 in catalytic applications is ascribed to: (i) as a carrier material, ZrO_2 gives rise to an unique kind of the interaction between the active catalytic component and support, and (ii) ZrO_2 possesses four chemical properties, namely acidity or basicity as well as reducing or oxidizing ability [1-12].

In this paper, we report a systematic study of the interaction of Ni(II,III) with sol-gel derived tetragonal and monoclinic ZrO_2 supports by using the Fourier transform infrared spectroscopy (FTIR), differential thermal analysis (DTA), and X-ray photoelectron spectroscopy (XPS). Fundamental aspects of the metal ion diffusion, including those occur during the phase transformation, are also investigated.

EXPERIMENTAL

The ZrO_2 supports were prepared using a zirconium-n-propoxide-acetylacetone-water-isopropanol system described in the literature [17-19]. Briefly, Zr-n-propoxide was diluted in isopropanol solvent to give the sols a molar ratio of 1:45 Zr-n-propoxide to isopropanol.

Mat. Res. Soc. Symp. Proc. Vol. 368

Acetylacetone (acac) chelating agent was then added to the solution ([acac]/[Zr] = 1). Hydrolysis was started by adding deionized water drop-wise to the solution while stirring. The ratio of added water to Zr-n-propoxide ([H_2O]/[Zr]) was kept at a constant of 3. The prepared sol was dried at room temperature inside a fumehood for 10 days to generate a dry, yellowish and transparent gel. The gel was then heat-treated at 400 and 900 °C for 5 hours respectively to produce a tetragonal ZrO_2 catalyst carrier and a monoclinic ZrO_2. The Ni/ZrO_2 catalysts were prepared by impregnation method. This involved impregnating the ground gel powders in a 1.0 M solution of $Ni(NO_3)_2$ respectively for 4 hours, followed by filtrating and drying (100 °C, 2 hours), and finally calcining at various temperatures (T_c), 400, 500, 600, 700, 800, and 900°C, respectively for 5 hours in a static air furnace. The sample are identified in the text by their crystallographic structure and calcination temperature. For example a nickel sample supported on tetragonal ZrO_2 and calcined at 500°C is identified as t-500.

The crystallographic structures and polymorphic transition of the above Ni/ZrO_2 catalysts were examined with FTIR spectroscopy (Shimazu FTIR-8101) using the KBr pellet method. The phase transformation of the tetragonal ZrO_2 to monoclinic ZrO_2 was also studied with DTA (Shimazu DTA-50). To obtain information on their surface chemical composition, catalysts were studied using XPS (VG ESCALAB MKII, Mg K_∞, 120 W, constant analyzer pass energy of 20 eV). The carbon 1s peak at 284 eV was used for charge referencing in the peak assignment of XPS spectra.

RESULTS AND DISCUSSION

The IR vibration mode at 450 cm^{-1} in the as-prepared tetragonal gel and catalysts t-400 to t-700 is a typical tetragonal phase Zr-O vibration for the ZrO_2 matrix [19,20]. From the t-400 to t-700, the 450 cm^{-1} peak shifts to higher wavenumbers, revealing better crystallinity for the catalysts at higher calcining temperatures. In addition to the 450 cm^{-1} peak for the tetragonal phase, new absorption bands at 490 and 570 cm^{-1}, the monoclinic characteristic absorptions, emerge in the catalyst t-700, indicating that t-700 is a mixture consisting of the two polymorphic phases. For the catalysts t-800 and t-900, monoclinic ZrO_2 IR absorptions at 420, 500, 575, and 740 cm^{-1} are observed, which are fully developed in the spectrum for catalyst t-900 [19,20]. The characteristic IR absorptions for m-series are similar to those of t-900. In the DTA study for tetragonal gel from 400 to 1000°C, a broad endothermic peak located at around 800 °C is observed due to tetragonal to monoclinic phase transformation. This transformation parallels the development of the monoclinic IR absorptions.

XPS $Ni2p_{3/2}$ spectra for the t-series catalysts are displayed in Fig. 1. Two surface components, labelled as 1 and 2, can be clearly identified from these spectra. From the XPS data base and literature, peak 1 located at approximately 854 eV can be assigned to Ni^{2+} ions, while peak 2 at 856 eV is commonly attributed to either a Ni^{3+} surface species or $Ni(OH)_2$ [21-23]. It is important to note that since $Ni(OH)_2$ decomposes at 200°C and $NiCO_3$ at 365 °C [24-26], the formation of $Ni(OH)_2$ or $NiCO_3$ is highly unlikely. Furthermore, no O-H vibrational modes were observed in the FTIR study of the bulk catalysts. In view of these, the peak 2 component is therefore attributed to the Ni^{3+} ions [21,22]. For t-400 to t-900, the binding energies for both Ni^{2+} and Ni^{3+} progressively increase (about 1 eV), indicating a partial oxidation of the nickel ions, i.e. the formation of Ni-O-Zr surface species [27] (solid solution) at high T_c portion. As ZrO_2 is a surface abundant component, the chemical shift of Zr3d arising from the Ni-O-Zr formation is not as pronounced. From t-400 to t-900, the low-

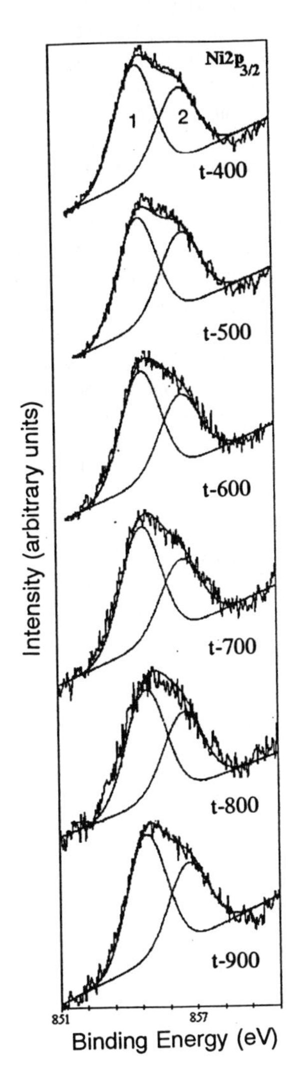

Fig. 1. XPS spectra of $Ni2p_{3/2}$ for the t-series catalysts at various calcination temperatures (400 to 900 °C).

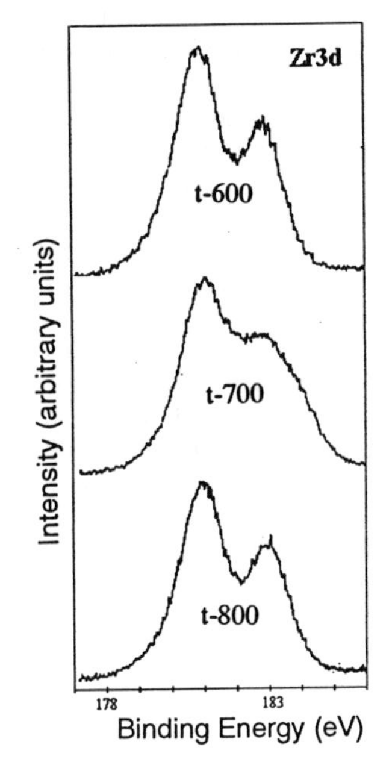

Fig. 2. XPS spectra of $Zr3d_{5/2,3/2}$ at T_c = 600, 700, and 800 °C.

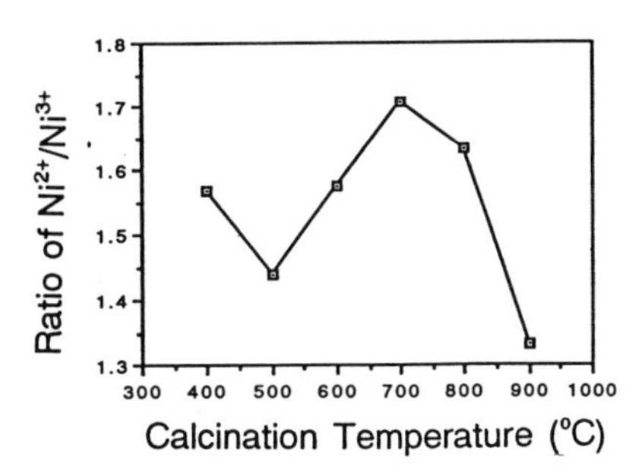

Fig. 3. Peak area ratio of Ni^{2+} to Ni^{3+} for the t-series catalysts versus T_c.

binding energy side ($Zr3d_{5/2}$) of Zr3d spectrum shifts from 181.4 eV to 181.6 eV. This positive chemical shift reveals that a small degree of the oxidation of Zr during the phase transformation also occurs concurrently. As shown in Fig. 2, the valley between $Zr3d_{5/2}$ and $Zr3d_{3/2}$ of the catalyst t-700 is filled with an oxidized component; also can be seen is that the base of the $Zr3d_{3/2}$ is broaden towards higher binding energy region by the new components. The positive components (t-700, Fig. 2) are interpreted as a disordered ZrO_2 matrix with a more ionic type of Zr-O bond, resulting in a higher oxidation state for Zr during the phase transition. In agreement, the amount of Ni^{2+} increases as the ratio Ni^{2+}/Ni^{3+} reported Fig. 3 reaches a maximum value 1.71 in the catalyst t-700, suggesting more electrons transferring through lattice and/or surface O^{2-} to Ni^{3+}.

Using XPS quantitative surface analysis, the absolute atomic concentration of an element can be calculated according a number of fairly crude assumptions [23]. Instead of measuring absolute concentration, a relative concentration of an element can be determined rather accurately and conveniently [23]. In the current study, a relative surface concentration ($R_{Ni/Zr}$) for Ni(II,III) is used, which is defined as the ratio of $Ni2p_{3/2}$ peak area to that of $Zr3d_{5/2,3/2}$. Fig. 4 reports the trend of the $R_{Ni/Zr}$ with T_c for the t-series catalysts. As can be seen in Fig. 4, the ratio decreases with T_c to 700 °C. Recalling that at this temperature, the tetragonal ZrO_2 starts to convert into the monoclinic, an abnormal sharp increase at 800 °C can be understood by reasoning that the phase transformation occurs significantly at this temperature. The surface area for tetragonal ZrO_2 is larger than that of the monoclinic [2], and more Ni(II,III) ions in the tetragonal phase are thus expected. The higher Ni(II,III) concentration at the point of 800 °C can be attributed to a re-establishment of the surface Ni(II,III) to ZrO_2 ratio, resulting from the expulsion (or segregation) of excess Ni(II,III) from tetragonal ZrO_2 matrix during the transformation to the monoclinic phase. Compared to the $R_{Ni/Zr}$ data for tetragonal samples, surface concentrations of Ni(II,III) for monoclinic samples are generally lower, noting that the surface area in the m-series is smaller. Since there is no major phase transformation occurring for the monoclinic samples during the heat treatments, a similar decrease from 400 to 600 °C is also observed. However, the $R_{Ni/Zr}$ becomes essentially a constant from 600 to 900 °C, which implies a thermodynamically stable Ni(II,III)/m-ZrO_2 mixture, or solid solution, is formed over this temperature range.

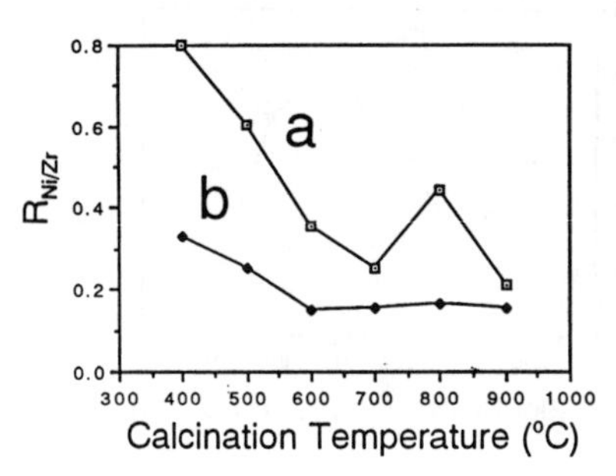

Fig. 4. $R_{Ni/Zr}$ versus T_c for: (a) the t-series catalysts, and (b) the m-series catalysts .

In a ZrO_2 matrix, the Ni(II,III) concentration $C_{Ni}(x,t)$ can be expressed as a function of diffusion distance (x), which starts from the surface ($x = 0$) and points towards the ZrO_2 substrate, and diffusion time (t). In the current study, the total Ni(II,III) ions (Q), including the surface Ni(II,III) and diffusing Ni(II,III), in each sample set is a constant. Using Fick's second law of diffusion [28]

$$D_{Ni} = Q^2/\pi t C_{Ni}^2(0,t) = a/R_{Ni/Zr}^2(0,t) \quad (1)$$

where D_{Ni} is the diffusion coefficient for a Ni(II,III) ion, a and t (constant calcining time) are constants, Arrhenius plots of the $lnD_{Ni} + C$ (C is a constant) versus $1/T$ (in K^{-1}) for the two catalyst series can be calculated and are displayed in Fig. 5. Both sets of the catalysts give a linear relationship between the lnD_{Ni} and $1/T$ (T_c = 400-600°C) and the slopes are very similar. The activation energy E_a for diffusion has the following relationship with the diffusion coefficient

$$D_{Ni} = D_o exp(-E_a/kT) \quad (2)$$

where D_o is a pre-exponential parameter and k is Boltzmann's constant [28]. The E_a calculated from these data according to Eq. (2) is 0.42 eV and 0.39 eV for t- and m-series, respectively, indicating no difference in the diffusion environment for Ni(II,III) ions in the two supports.

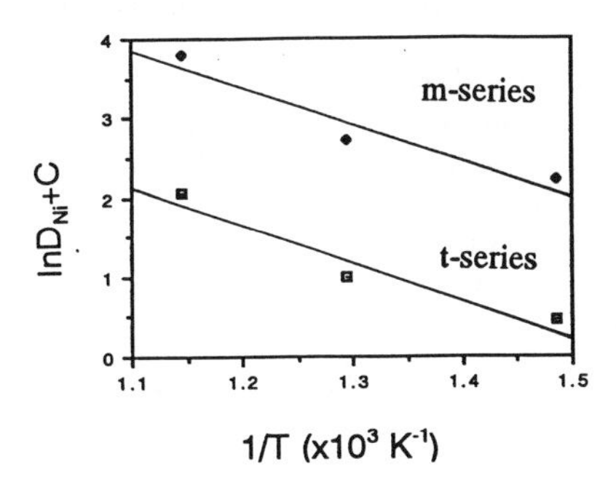

Fig. 5. Arrhenius plots of the $lnD_{Ni} + C$ versus $1/T$ for the t-series and m-series catalysts.

CONCLUSION

In summary, tetragonal and monoclinic ZrO_2 gel supports have been prepared by the sol-gel method. It has been found that the nickel ions diffuse into the ZrO_2 carriers progressively over 400 to 600 °C during the heat treatment for both tetragonal and monoclinic ZrO_2 matrices. A thermodynamically stable Ni(II,III)/ZrO_2 solid solution is formed at elevated temperatures. The content of Ni(II,III) in tetragonal ZrO_2 is higher than that in the monoclinic, confirming a larger surface area for the tetragonal carrier. Two oxidation states of $Ni2p_{3/2}$ (+2 and +3) are detected for the Ni/ZrO_2 catalysts. The population ratio of Ni^{2+}/Ni^{3+} peaks at 700 °C during the tetragonal to monoclinic phase transformation. However, in monoclinic case, the ratio remains constant throughout the course of experiments. The average diffusion activation energy determined for Ni(II) or Ni(III) ion in ZrO_2 supports is 0.40 eV at 400 to 600 °C.

ACKNOWLEDGMENTS

The authors gratefully acknowledge the research funding (RP3920644) for the experimental study of catalytic materials by the National University of Singapore and the technical assistance provided by Ms L.F. Ee.

REFERENCES

1. P.D.L. Mercera, J.G. van Ommen, E.B.M. Doesburg, A.J. Burggraaf, and J.R.H. Ross, Applied Catalysis **71**, 363 (1991).
2. P.D.L. Mercera, J.G. van Ommen, E.B.M. Doesburg, A.J. Burggraaf and J.R.H. Ross, Applied Catalysis **57**, 127 (1990).
3. H.C. Zeng and S. Shi, J. Non-Cryst. Solids (1995) in press.
4. H.C. Zeng, J. Lin, W.K. Teo, F.C. Loh, and K.L. Tan, J. Non-Cryst. Solids (1994) in press.
5. K. Tanabe, Mater. Chem. Phys. **13**, 347 (1985).
6. H.C. Zeng, J. Lin, W.K. Teo, J.C. Wu, and K.L. Tan, J. Mater. Res. **10**, (1995) in press.
7. P. Turlier, H. Praliaud, P. Moral, G.A. Martin and J.A. Dalmon, Appl. Catal. **19**, 287 (1985).
8. G.R. Gavalas, C. Phichitkul and G.E. Voecks, J. Catal. **88**, 54 (1984).
9. K.E. Smith, R. Kershaw, K. Dwight and A. Wold, Mater. Res. Bull. **22**, 1125 (1987).
10. S. Narayanan and G. Sreekanth, J. Chem. Soc., Faraday Trans. **1**,**85**, 3785 (1985).
11. P. Marginean and A. Olariu, J. Catal. **95**, 1 (1985).
12. L.A. Bruce, G.J. Hope and J.F. Mathews, Applied Catalysis **8**, 349 (1983).
13. M. Atik, J. Zarzycki and C. R'Kha, J. Mater. Sci. Lett. **13**, 266 (1994).
14. S. Sakka, J. Non-Cryst. Solids **73**, 651 (1985).
15. H. Dislich and E. Hussmann, Thin Solid Films **77**, 129 (1981).
16. K. Izumi, M. Murakami, T. Deguchi and A. Morita, J. Am. Ceram. Soc. **72**, 1465 (1989).
17. R. Guinebretiere, A. Dauger, A. Lecomte and H. Vesteghem, J. Non-Cryst. Solids **147&148** 542 (1992).
18. K. Yamada, T.Y. Chow, T. Horihata and M. Nagata, J. Non-Cryst. Solids **100**, 316 (1988).
19. J.C. Debsikdar, J. Non-Cryst. Solids **86**, 231 (1986).
20. C.M. Phillippi and K.S. Mazdiyasni, J. Am. Ceram. Soc. **54**, 254 (1971).
21. F. Arena, A. Licciardello and A. Parmalina, Catalysis Letters **6**, 139 (1990).
22. A. Torrisi, A. Cavallaro, A. Licciardello, A. Perniciaro and S. Pignataro, Surf. Interface Anal. **10**, 306 (1987).
23. A.B. Christie in Methods of Surface Analysis, edited by J.M. Walls (Cambridge University Press, New York, 1988), p. 127.
24. T.L. Webb and J.E. Kruger in Differential Thermal Analysis, Vol. 1 Fundamental Aspects, edited by R.C. Mackenzie (Academic Press, London, 1970) ch. 10, p. 330.
25. C.W. Beck, Am. Miner. **35**, 985 (1950).
26. R.W. Mallya and A.R.V. Murthy, J. Indian Inst. Sci. **43**, 87 (1961).
27. T.L. Barr, J. Vac. Sci. Technol. **A9**, 1793 (1991).
28. J.W. Mayer and S.S. Lau, Electronic Materials Science (Maxwell-Macmillan, New York, 1990), p. 183.

EXAFS, XPS AND TEM CHARACTERIZATION OF SnO_2-BASED EMISSION CONTROL CATALYSTS

PHILIP G. HARRISON*, WAYNE DANIEL, NICHOLAS C. LLOYD and WAN AZELEE
Department of Chemistry, University of Nottingham, University Park, Nottingham NG7 2RD, U.K.

ABSTRACT

Catalysts formed by sorption of Cr(VI) on to tin(IV) oxide have been investigated by EXAFS, XPS and TEM. The latter shows that uncalcined catalysts comprise uniform, homogeneous particles of dimensions 2nm which increase to 4nm on calcination at 400°C. Calcination at higher temperatures results in an increase in particle size to 20nm x 30nm at 600°C, *ca.* 60nm at 800°C, and 90-100nm at 1000°C. XPS data show the evolution of chromium(III) species upon calcination and both chromium(VI) and chromium(III) species exist at all calcination temperatures. Chromium *K*-edge XANES also shows the presence of tetrahedral chromium(VI) species at all calcination temperatures, but the relative abundance decreases with increase in calcination temperature. Corresponding EXAFS data indicate that the uncalcined catalyst contains sorbed four-coordinated $\{Cr_2O_7\}$ species whereas calcination produces six-coordinated $\{CrO_6\}$ species together with some residual tetrahedrally coordinated chromium. EXAFS data show that adsorbed $\{Cu(H_2O)_6\}$ is the copper(II) species present prior to calcination of Cu^{II}-SnO_2 catalysts.

INTRODUCTION

Catalysts based on platinum and rhodium represent the state-of-the-art in internal combustion engine emission technology. However, current usage of rhodium in these catalysts exceeds the Rh/Pt mine ratio, and hence it will become essential to reduce consumption in order to conserve the limited noble metal supply. Additionally, the price, strategic importance and general low availability of platinum group metals would, therefore, be very attractive to reduce dependence on noble metals and to seek viable alternative catalytic materials. Our studies have shown that catalysts based on tin(IV) oxide promoted with chromium and/or copper (Cr-SnO_2 and Cu-Cr-SnO_2 catalysts) exhibit excellent three-way catalytic activity - activity which is comparable to that shown by noble metals dispersed on alumina. This family of materials offers tremendous promise as cheap and efficient catalyst systems for the conversion of noxious motor vehicle emissions.

Knowledge of the solid-state chemistry of catalysts is invaluable to an understanding of their catalytic behaviour. However, the characterization of this type of composite oxide presents a major problem and is notoriously difficult. No single technique can yield anything but a very small amount of information, and a complete picture can only be gained by using a combination of methods. Here we report preliminary EXAFS, XPS, and TEM data for these systems.

EXPERIMENTAL

Catalyst samples were obtained by sorption onto tin(IV) oxide gel prepared by hydrolysis of tin(IV) chloride using aqueous ammonia. The resulting gel was washed exhaustively with triply distilled water until no residual chloride remained. The following

Mat. Res. Soc. Symp. Proc. Vol. 368 © 1995 Materials Research Society

procedure is typical. To a suspension of tin(IV) oxide gel dispersed in triply distilled water was added a solution of chromium(VI) oxide and the mixture stirred for 24h. The yellow mixture was filtered and the yellow solid air dried at 60°C for 24h. Copper(II)-doped tin(IV) oxide was prepared in a similar manner using a solution of copper(II) nitrate.

Electron microscopy studies were performed on a JEOL 2000FX operating at 200KeV. Samples were ground in a pestle and mortar and suspended in chloroform. A drop of the suspension was placed on a holey carbon formvar coated copper electron microscope grid and allow to dry in air.

All EXAFS data were collected at the Daresbury SRS on station 8.1 in fluorescent mode. Raw data were corrected for dark currents and converted to *k*-space using EXCALIB, and backgrounds subtracted usin EXBACK, to yield EXAFS functions $\gamma_{obs}(k)$. Model fitting was carried out using EXCURV92.

XPS data were obtained using a VG Ionex instrument.

THE NATURE OF Cr(VI)/SnO_2 CATALYSTS

The Sorption of Chromium(VI) Species Onto Tin(IV) Oxide Gel

Exposure of tin(IV) oxide gel to aqueous solutions of CrO_3 results in the sorption of chromium species onto the oxide giving orange powders after filtering and drying in air. The amount of chromium species sorbed onto the oxide is dependent on several variables including the concentration of the CrO_3 solution, the temperature, the time of exposure, and the separation and washing procedures adopted. For CrO_3 solution concentrations in excess of *ca.* 0.1M followed by filtering and drying but no washing, the loading achieved is linearly proportional to the concentration. Washing the catalyst dramatically decreases the loading indicating the weak nature of the adsorption.

Transmission Electron Microscopy - the Physical Nature of Cr(VI)/SnO_2 Catalysts

Samples for the Cr(VI)/SnO_2 (Cr:Sn 0.38) catalyst prior to calcination and also after calcination at tempertures up to 1000°C were examined by transmission electron microscopy. Representative micrographs are shown in Figure 1 (all scale bars represent 20nm). The uncalcined sample is homogeneous and highly crystalline (top left), and all areas analyse for tin and chromium in accordance with the synthesis ratio. Fundamental crystallites are approximately spherical and are of the order of 2nm in diameter with *ca.* 3 Å fringes being observed which correspond to the {110} lattice planes of tin oxide (*cf* 3.351Å for crystalline SnO_2). Pores of similar dimensions to the particles can also be observed. After calcination at 400°C the material has a more open texture and a range of particle sizes is observed. The micrograph shown (bottom left) corresponds to this structure and is built up from particles *ca.* 4nm in diameter. 3 Å Fringes are seen over the entire structure corresponding to rutile-structure tin(IV) oxide {110} lattice planes. The corresponding powder electron diffraction pattern is also shown in Figure 1. EDX microanalysis of such areas again gives elemental ratios consistent with those observed for the uncalcined sample. However, there are other particles of larger size *ca.* 10-15nm in diameter which tend towards hexagonal geometry and which are tin rich. For samples treated at 600°C two types of particle are observed (images not shown). The smaller, less regular shaped particles *ca.* 6-7nm in diameter have similar elemental composition (as measured by EDX analysis) to that of the uncalcined sample, but there are also a number of larger quasi hexagonal crystallites generally 20nm x 30nm in dimensions, but which may be up to 60nm in the longer dimension. By 800°C the analysed

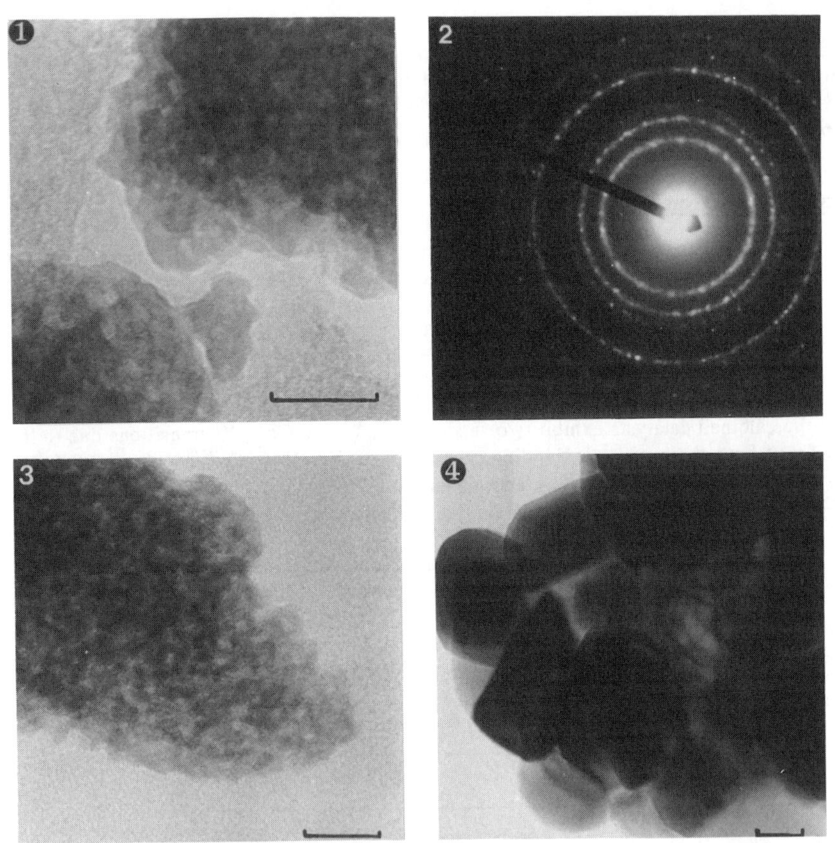

Figure 1. Transmission electron micrographs for the Cr(VI)/SnO_2 (Cr:Sn 0.38) catalyst. See text for identification.

samples comprise *ca.* 60nm distorted angular crystallites (bottom right). Calcination at 1000°C results in further increase in size of the particles which are now 90-100nm in size.

The powder XRD spectra of uncalcined samples of the Cr(VI)-SnO_2 catalysts exhibit only broad diffuse lines characteristic of nanoparticulate tin(IV) oxide. These peaks are seen to gradually sharpen on calcination, but are still relatively broad after calcination at 600°C. Only after calcination at high temperatures (*ca.* 1000°C) is a characteristically sharp spectrum due to crystalline SnO_2 obtained. When the concentration of chromium in the catalyst is low (*eg* for a Sn:Cr ratio of 0.015), no chromium-containing phase can be detected by XRD. Only at high ratios (*eg* 1:0.13) and at high calcination temperatures (1000°C) does a chromium-containing phase, Cr_2O_3, separate (XRD 2ϑ values at 24.5, 36.2, 41.56, 50.2, 63.5°). Average particle sizes calculated from XRD line broadening for Sn:Cr ratios in the range 0.011-0.132 indicate a uniform dimension of *ca.* 12nm for the uncalcined samples and samples calcined at 300°C. Calcination at 600°C produces particles of an average diameter of 34-39nm, whilst calcination at 1000°C gives an average particle size of >200nm.

Thus particle sizes derived from XRD data can be considered to be only semi-quantitative in nature, and a more rigorous assessment of particle size in these catalyst materials may only be obtained using electron microscopy. Em data also provides valuable evidence for the inherent crystallinity of uncalcined catalysts which cannot be detected by XRD, and also for the sintering effects observed for samples calcined at temperatures up to *ca.* 600°C.

X-ray Photoelectron Spectroscopy

Sn 3d, Cr 2p and O 1s XPS data for the Cr(VI)/SnO_2 (Cr:Sn 0.28) catalyst after calcination at various temperatures are presented in Table 1. The Sn 3d spectra comprise two peaks at binding energies of ca. 487eV and 495eV corresponding to the Sn $3d^{5/2}$ and Sn $3d^{3/2}$ transitions, respectively, and are essentially invariant with calcination conditions. The Cr 2p spectra for the uncalcined catalyst comprises two peaks at binding energies of 579.4eV and 588.7eV due to the Cr $2p^{3/2}$ and Cr $2p^{1/2}$ transitions of a chromium(VI) species. However, calcination produces quite dramatic changes to the form of the Cr 2p spectra, and the spectra of all the calcined catalysts exhibit two sets of Cr $2p^{3/2}$ and Cr $2p^{1/2}$ transitions due to the presence of both chromium(VI) (binding energies at *ca.* 579eV (Cr $2p^{3/2}$) and *ca.* 590eV (Cr $2p^{1/2}$)) and chromium(III) (binding energies at *ca.* 576.5eV (Cr $2p^{3/2}$) and *ca.* 587eV (Cr $2p^{1/2}$)) species (Figure 2). A semi-quantitative estimate of the relative ratios of Cr(VI) to

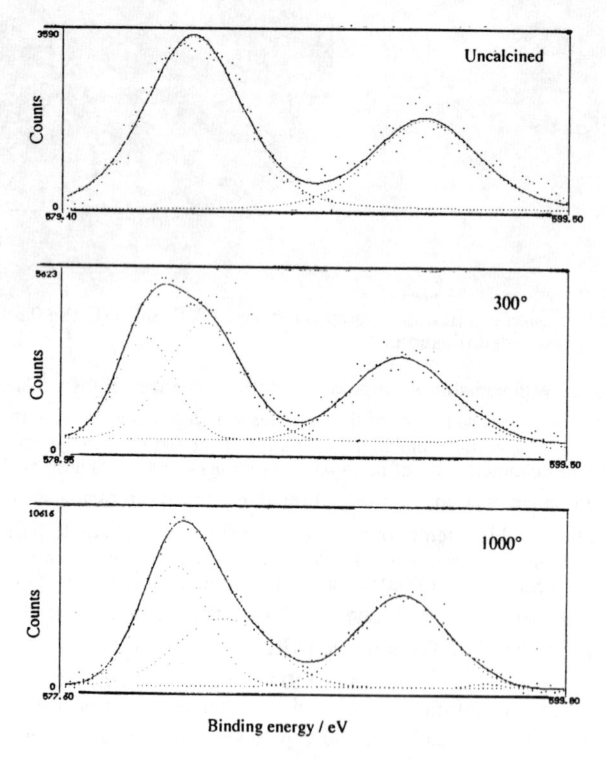

Figure 2. Chromium 2p XPS spectra for the Cr(VI)/SnO_2 (Cr:Sn 0.28) catalyst uncalcined, and calcined at 300∞C and 1000°C.

Cr(III) from the Cr $2p^{3/2}$ peak suggests that chromium(VI) species still predominate after calcination at 300°C (approximate Cr(VI):Cr(III) ratio 1.5), but approximately equal amounts exist after calcination in the temperature range 400-600°C. Chromium(III) species predominate at calcination temperatures ≥800°C. All oxygen 1s spectra exhibit a major peak at a binding energy of *ca.* 531eV with a much lower intensity shoulder on the higher binding energy side at *ca.* 532eV. The principal O 1s peak is assigned to lattice oxide anions. The shoulder to higher binding energy is less readily assigned, but has been observed previously for ceria and is most probably due to a surface oxygen-containing species.

Table 1. XPS binding energy data (eV) for the Cr(VI)/SnO_2 (Cr:Sn 0.28) catalyst calcined at various temperatures.

Calcination Temperature / °C	Sn $3d^{5/2}$ / Sn $3d^{3/2}$	Cr $2p^{3/2}$ / Cr $2p^{1/2}$	O 1s
Uncalcined	486.6 / 495.1	579.4 / 588.7	530.6 / 532.5
300	487.6 / 496.6	576.2 / 586.7 578.6 / 590.4	531.6 / 533.5
400	486.0 / 494.5	576.6 / 586.7 578.6 / 589.7	529.8 / 531.7
600	486.5 / 495	577.0 / 587.3 579.3 / 590.6	530.5 / 532.3
800	485.5 / 494.0	576.7 / 586.7 578.8 / 590.3	529.5 / 531.5
1000	487.0 / 495.5	576.4 / 586.5 578.6 / 590.3	530.4 / 532.5

Chromium *K*-Edge XANES and EXAFS

Chromium *K*-edge XANES and EXAFS data were obtained for the Cr(VI)/SnO_2 (Cr:Sn 0.12) catalyst sample. The prominent sharp pre-edge feature at +4.3eV in the XANES region is characteristic of the presence of tetrahedrally-coordinated chromium(VI) [1]. Figure 3 illustrates how the intensity of this features gradually decreases as the calcination temperature increases. Significant amounts of chromium(VI) are apparent even after calcination at 400°C, is still appreciable after calcination at 600°C, and can still be detected even after calcination at >800°C.

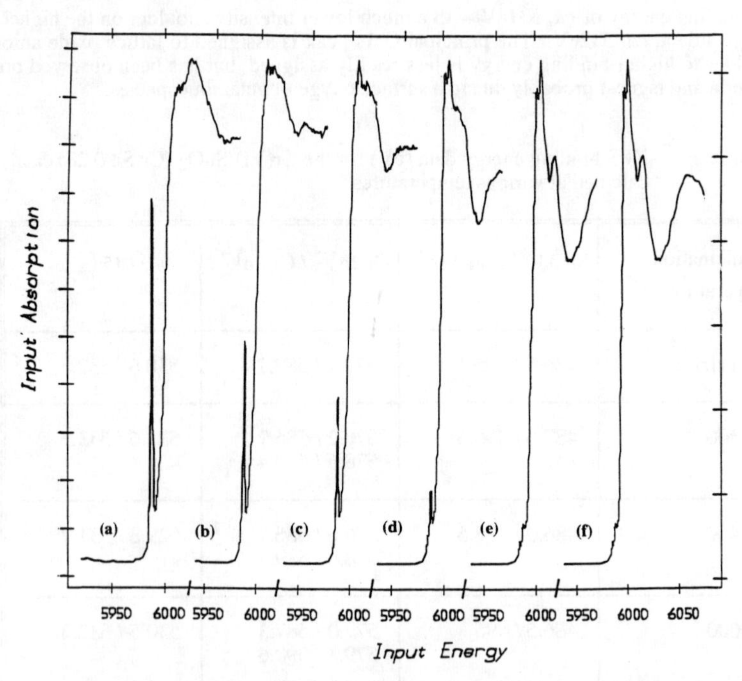

Figure 3. Chromium K-edge XANES region for the Cr(VI)/SnO_2 (Cr:Sn 0.12) catalyst (a) before calcination, and calcined at (b) 300°, (c) 400°, (d) 600°, (e) 800°, and (f) 1000°.

Figure 4. Observed (——) and calculated (-----) k^3-weighted Cr K-edge EXAFS spectra (upper) and observed (——) and calculated (-----) Fourier transform magnitudes (quasi-radial distribution function) of k^3-weighted Cr K-edge EXAFS spectra (lower) for the Cr(VI)/SnO_2 (Cr:Sn 0.12) catalyst before calcination (top left), and after calcination at 600° (top right) and 1000° (bottom left), and the Cu(II)/SnO_2 (Cu:Sn 0.39) catalyst before calcination (bottom right). Ordinate and abscissa axes are $k^3\gamma(k)$ and k/Å^{-1}, respectively, for the EXAFS spectra, and intensity and r/Å, respectively, for the Fourier transforms.

Figure 4

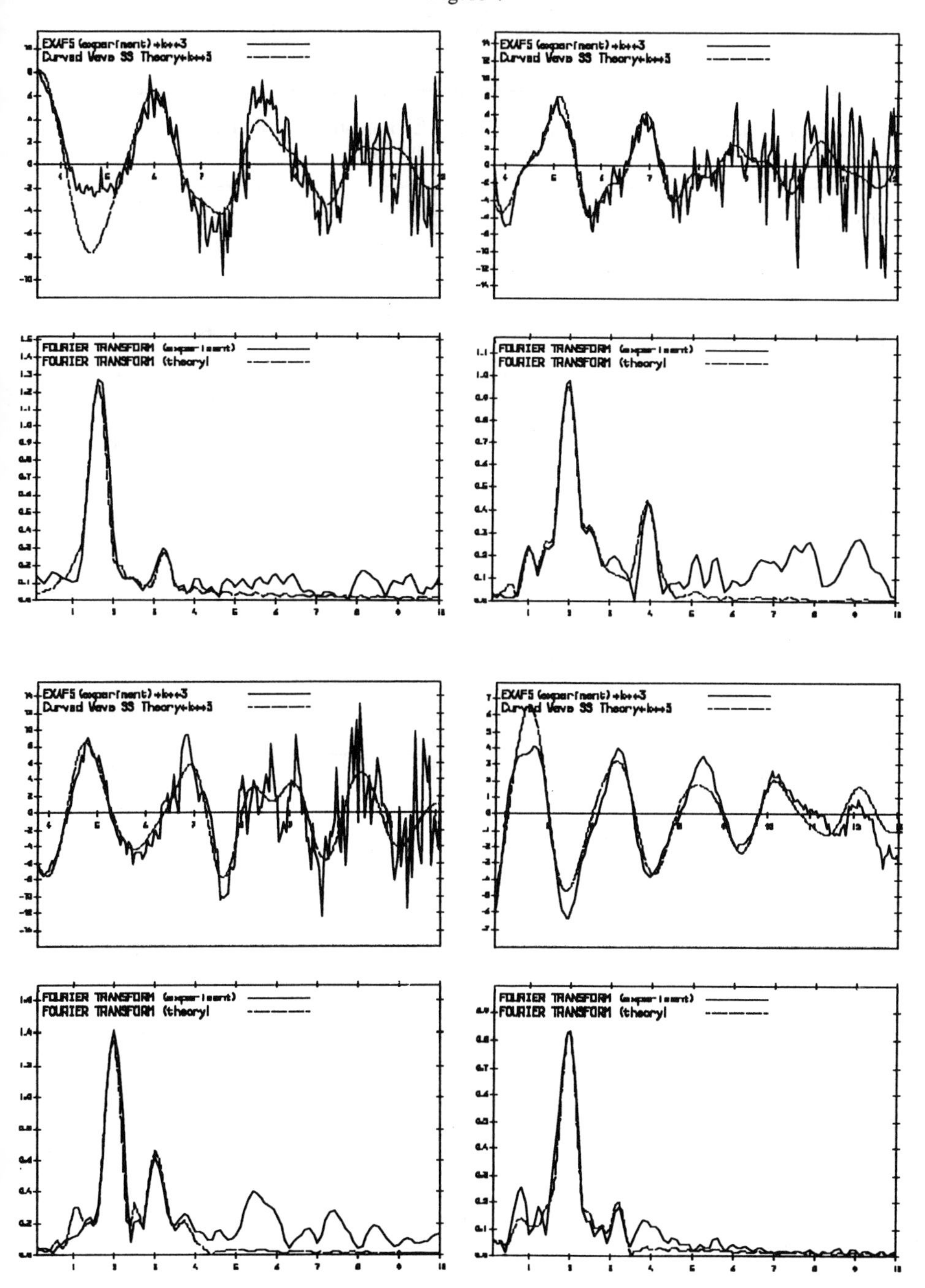

EXAFS (experiment)*k**3
Curved Wave SS Theory*k**3
EXAFS (experiment)*k**3
Curved Wave SS Theory*k**3
FOURIER TRANSFORM (experiment)
FOURIER TRANSFORM (theory)
FOURIER TRANSFORM (experiment)
FOURIER TRANSFORM (theory)
EXAFS (experiment)*k**3
Curved Wave SS Theory*k**3
EXAFS (experiment)*k**3
Curved Wave SS Theory*k**3
FOURIER TRANSFORM (experiment)
FOURIER TRANSFORM (theory)
FOURIER TRANSFORM (experiment)
FOURIER TRANSFORM (theory)

Table 2. Chromium *K*-edge EXAFS data for the Cr(VI)/SnO_2 (Cr:Sn 0.12) catalyst and copper *K*-edge EXAFS data for the Cu(II)/SnO_2 (Cu:Sn 0.38) catalyst

Catalyst sample	Atom type	Coordination number	Radial distance Å
Cr(VI)/SnO_2 - uncalcined	O	4.0	1.638
	Cr	1.0	3.284
Cr(VI)/SnO_2 - calcined at 600°C	O	0.3	1.760
	O	3.0	1.947
	O	3.0	1.966
	Cr	1.0	2.527
	Cr	3.0	2.986
	Cr	3.0	3.228
Cr(VI)/SnO_2 - calcined at 1000°C	O	0.4	1.655
	O	3.0	1.903
	O	3.0	2.001
	Cr	1.0	2.647
	Cr	3.0	2.889
	Cr	3.0	3.306
Cu(II)/SnO_2 - uncalcined	O	4.0	1.948
	O	2.0	1.964
	O	2.0	3.264

The chromium EXAFS data and the corresponding Fourier transforms for the Cr(VI)/SnO_2 (Cr:Sn 0.12) catalyst prior to calcination and after calcination at 600°C and 1000°C are illustrated in Figure 4. Coordination data are collected in Table 2. The data show that in the uncalcined catalyst the chromium is four-coordinated by oxygen atoms at a mean distance of 1.638Å with one chromium atom in the second coordination sphere at a distance of 3.284Å. The Cr-O distance compares with those determined by X-ray crystallography in $Na_2CrO_4.4H_2O$ [2], $Na_2Cr_2O_7.2H_2O$ [3], and CrO_3 [4] of 1.639Å, 1.663Å, and 1.673Å, respectively. The Cr...Cr contact of 3.284Å is comparable with that in crystalline $Na_2Cr_2O_7.2H_2O$ (3.160Å) [3]. We therefore conclude that the chromium(VI) species present in this catalyst is the $\{Cr_2O_7\}$ anion which is weakly bound probably via hydrogen-bonding contacts to the surface of the tin(IV) oxide particles.

The EXAFS data for the catalyst after calcination at 600°C and 1000°C are similar to each other, but quite different from that for the uncalcined catalyst. Now, although there is some residual oxygen at a Cr-O distance of 1.760Å (600°C) and 1.655Å (1000°C), the first coordination shell around chromium comprises six oxygen atoms at a distance of *ca.* 2.0Å. Subsequent coordination is by chromium atoms at distances >2.5Å. This mode of coordination is that found in Cr_2O_3 of the corundum structure [5], and hence we deduce that at high calcination temperatures crystallites of Cr_2O_3 are formed. However, both XRD and TEM data show that the composition of the catalyst is dominated by the sintering of tin(IV) oxide crystals.

THE NATURE OF Cu(II)/SnO_2 CATALYSTS

Also shown in Figure 4 and Table 2 are EXAFS data for the Cu(II)/SnO_2 (Cu:Sn 0.38) catalyst prepared by sorption from an aqueous solution of copper(II) nitrate on to tin(IV) oxide. These data show that the copper species present on the catalyst is the hexa-aquo $\{Cu(H_2O)_6\}$ ion which is presumably also hydrogen-bonded to the surface of the tin(IV) oxide particles. These data corroborate the conclusions formed from ESEEM studies [6].

CONCLUSIONS

Sorption on to particulate tin(IV) oxide produces a Cr(VI)-SnO_2 catalyst which comprises $\{Cr_2O_7\}$ species adsorbed on the surface of the 2nm oxide particles. Calcination of this catalyst results in an increase in particle size which is gradual up to a calcination temperature of 600°C. Calcination at temperatures in excess of this result in the formation of much larger particles. XPS shows that both chromium(III) and chromium(VI) exist even after calcination at 1000°C, although high temperature calcination produces Cr_2O_3.

Uncalcined Cu(II)/SnO_2 catalysts comprise hexa-aquo $\{Cu(H_2O)_6\}$ ions sorbed on to tin(IV) oxide.

Acknowledgements

We would like to thank the following for support: the Commission of the European Community, the Engineering and Physical Sciences Research Council, and the Government of Malaysia (for a scholarship to W.A.). We would also like to thank Dr. Carole Harrison for assistance with the electron microscopy.

REFERENCES

1. Kutzler, F.W., Natoli, C.R., Misemer, D.K., Doniach, and Hodgson, K.O., *J. Phys. Chem.*, **73**, 3274 (1980).
2. Ruben, H., Olovsson, I., Zalkin, A., and Templeton, D.H., *Acta Cryst.. B*, **29**, 2963 (1973).
3. Kharitonov, Y.A., Kuz'min, E.A., and Belov, N.V., *Kristallografiya*, **15**, 942 (1970).
4. Stephens, J.S., and Cruickshank, D.W.J., *Acta Cryst.. B*, **26**, 222 (1970).
5. Newnham, R.E., and de Haan, Y.M., *Z. Kristallogr.*, **117**, 235 (1962).
6. Harrison, P.G., Azelee, W., Goldfarb, D., Matar, K., and Zhao, D., *J. Chem. Phys.*, submitted.

CO OXIDATION BY Bi_2MoO_6 -γ(H) CATALYST

D. H. Galván, M. Avalos-Borja, S. Fuentes, L. Cota-Araiza,
J. Cruz-Reyes [a], F. F. Castillón [b] and M. B. Maple [c]

Instituto de Física-UNAM, A. Postal 2681, Ensenada, B. C. 22800 MEXICO
[a] Facultad de Química, UABC, Tijuana, B. C., MEXICO
[b] Departamento de Polímeros y Materiales, Universidad de Sonora, MEXICO
[c] Dept. of Physics and Inst. of Pure and Appl. Physical Sciences,
UC, La Jolla, CA 92093, USA

ABSTRACT

We present a study of a Bi-Mo catalyst, obtained by solid state reaction, for the oxidation of CO. Bismuth-molybdate catalysts form several phases (β-$Bi_2Mo_2O_9$, α-$Bi_2Mo_3O_{12}$ and γ-Bi_2MoO_6, among others), depending upon the Bi/Mo ratio and the preparation method [1]. The γ-Bi_2MoO_6 phase is the one that is stable at high temperatures, according to the phase diagram, and is the one we used as catalyst for the conversion of CO [2].

INTRODUCTION

Bismuth-molybdate catalysts are currently used for selective olefin oxidation and ammoxidation process [4]. The industrial importance of the catalysts is based on their unique ability to oxidize propylene and ammonia to acrylonitrile at high selectivity. These catalysts can from several phases (β-$Bi_2Mo_2O_9$, α-$Bi_2Mo_3O_{12}$ and γ-Bi_2MoO_6, among others) depending on the reaction temperature and the Bi/Mo ratio [5]. The γ-Bi_2MoO_6 is the phase that is stable at high temperatures, according to the phase diagram, and it is the one that we used as catalyst for the conversion of CO [3]. In this work we describe further studies on the structure characterization of the γ(H)- phase, by X-ray diffracton, Scanning Electron Microscopy (SEM), and Auger Electron Spectroscopy (AES), in order to establish a correlation with the catalytic properties.

EXPERIMENTAL PROCEDURE

Samples of Bi_2MoO_6 -γ(H) (the high temperature phase) were prepared by solid state reaction, starting from the stoichiometric proportions of bismuth and molybdenum [2]. We started with a 50% mol of Bi_2O_3 and Mo_3O_3 in order to obtain an atomic ratio of Bi/Mo =2. After obtaining a very uniform mixture, the sample was calcined in air in a covered Pt crucible at 1220 K for 24 hrs. The catalyst was allowed to cool to room temperature inside the oven. It is necessary to underline, that the sample melts at this temperature producing large, needle like crystals. The color of the crystals is light yellow, and it is uniform throughout the samples. X-ray diffractograms were obtained with a General Electric XRD6 diffractogram, using Cu Kα radiation at 40 Kv and 20 mA. Scanning Electron Microscopy (SEM) micrographs were obtained using a JEOL JSM-5300 microscope, in order to study the surface morphology.

Mat. Res. Soc. Symp. Proc. Vol. 368

The surface composition was determined using Auger Electron Spectroscopy (AES) with a scanning Auger microprobe (Perkin-Elmer PHI-595). Surface areas were determined by the BET method using nitrogen as adsorbate at 77 K. Samples were outgassed under high vacuum at 423 K during one hour, prior to area measurements. The carbon monoxide oxidation was used as a test reaction for these catalysts. The reaction was carried out at a temperature in the range of 483-873 K, in a conventional flow system under atmospheric pressure. The oxygen and carbon monoxide partial pressures were changed in order to obtain different CO/O ratios (0.2, 1.0, 5.0). The products were analyzed by thermal conductivity gas chromatography using a 3 mm stainless steel column filled with 3A molecular sieve at 343 K. The reaction rates were calculated at conversions of 5% (around 573 K). The catalysts were heated in air at 473 K.

RESULTS AND DISCUSSION

Fig. 1 shows an optical micrograph of the sample at the bottom of the crucible. The needle-like appearence is evident. Notice that the whole conglomerate resembles an interlacing straw arrangement.

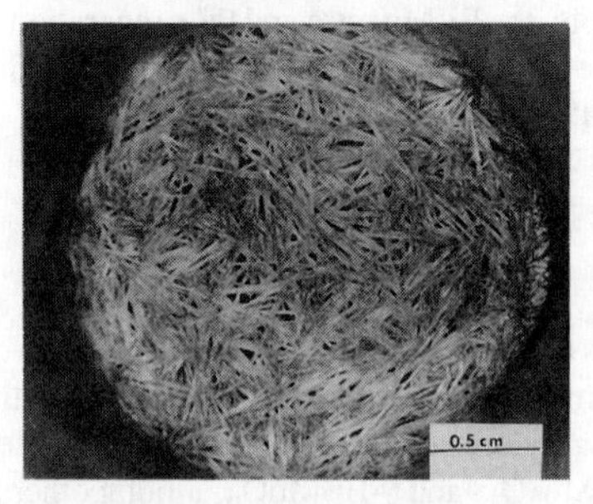

Fig.1 Optical micrograph of the sample at the bottom of the crucible

The surface morphology of the catalyst was studied using SEM, as it is shown in Fig. 2. Notice that the SEM image of these crystals about 60 mm in length and 10 mm wide is shown.

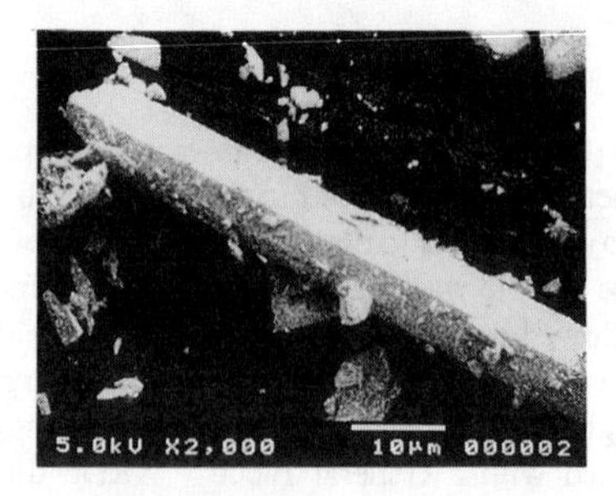

Fig. 2 SEM micrograph from the sample before the reaction test

X-ray diffraction was used to assess the phases involved in these catalysts. Fig. 3 shows an X-ray diffractogram pattern for this sample. The most prominent peaks ((-341), (341), (600), (080), etc.) agree with those reported for the γ(H)-Bi_2MoO_6 phase by Watanabe et al.[2]. The smaller peaks indicated by an arrow, correspond to koechlinite (the low temperature) phase. From theoretical calculations using the intensities reported in the JCPDF data files, it has been estimated that the catalyst is formed of a mixture of 90 % γ(H)-Bi_2MoO_6 phase and 10 % koechlinite γ(L) phase.

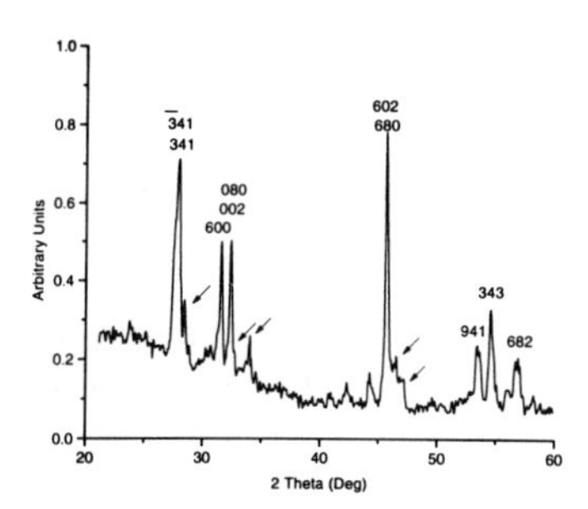

Fig. 3 X-ray diffraction from the sample calcined at 1220 K. Indexed peaks correspond to γ(H)-Bi_2MoO_6 phase and arrows point to koechlinite phase peaks.

The surface Bi/Mo ratio was determined by Auger analysis. The AES spectra of these needle-like platelets is shown in Fig. 4. From the peak heights, a Bi/Mo ratio of 1.15 was measured, which indicate that the sample was bismuth defficient, when compared with the stoichiometric ratio of 2.

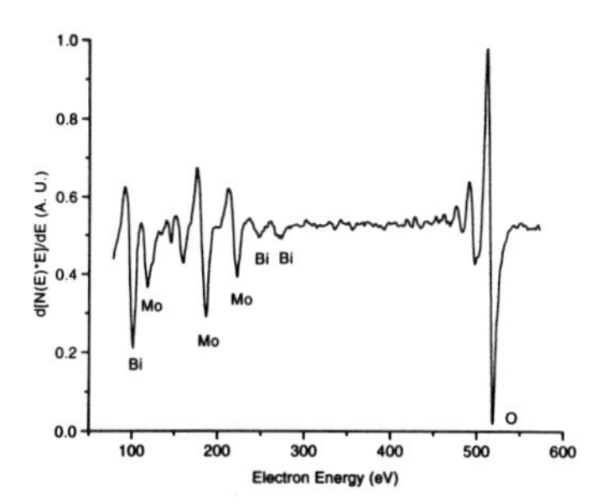

Fig 4 Auger spectrum showing bismuth, molybdenum and oxygen peaks.

The catalytic activity as a function of the reaction temperature was evaluated in the temperature range of 483-873 K. Fig. 5 shows the catalytic activity in terms of the percentage conversion of CO to CO_2, as a function of the reaction temperature, the full line corresponds to the sample reduced in H_2 and the dotted line corresponds to the sample calcined in O_2. The catalyst is very active for both cases, reaching 100 % conversion around 650 K.

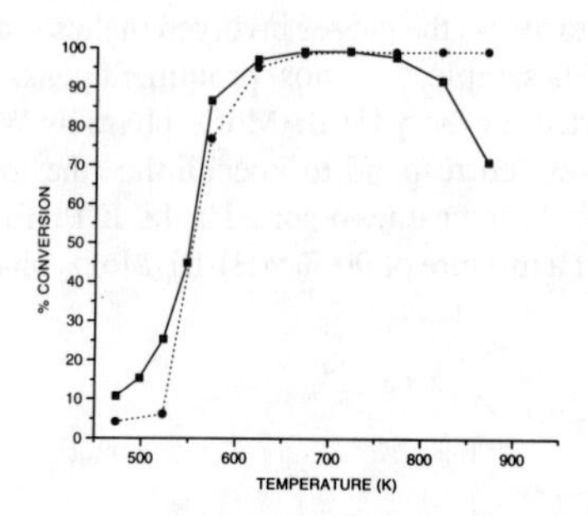

Fig. 5 Catalytic activity in terms of percentage conversion of CO to CO_2 as a function of the reaction temperature. Full line corresponds to samples reduced in H_2 and dotted line to samples calcined in O_2.

The surface area for the catalyst was about 1.89 m^2/g. This low surface area is mainly due to the high temperature used during preparation. Effort will be made to support these catalysts in order to increase their surface area and stability.

CONCLUSIONS

The structure of these catalysts, based on the solid state reaction, has been presented. XRD results shows that it is formed of 90 % γ(H)-Bi_2MoO_6 and 10 % koechlinite. The surface Bi/Mo ratio was determined by AES and found that it was bismuth defficient, as compared to the stoichiometric compound.
The surface area was very low, in the order of 1.89 m^2/g. This low surface area could be atributed to the high temperature used during preparation.
The catalytic activity, for the conversion of CO to CO_2, reaches 100 % at a temperature around 650 K, for both cases of oxidized as well as reduced.

ACKNOWLEDGMENTS

We thanks I. Gradilla, L. Rendón, G. Soto, E. Vilchis and A. Duarte for the help provided. M. B. Maple acknowledge support from the U. S. Dept. of Energy under Grant No. DE-FGO3-86ER 45230. D. H. G. acknowledges support from Conacyt.

REFERENCES

[1] T.Chen and G. S. Smith, J. Solid State Chem. **13** (1975) 288.
[2] A. Watanabe, S. Horiuchi and H. Kodama, J. Solid State Chem. **67** (1987) 333.
[3] D. H. Galván, S. Fuentes, M. Avalos-Borja, L. Cota-Araiza, J. Cruz-Reyes, E. A. Early and M. B. Maple, Cat. Letts. **18** (1983) 273.
[4] R. K. Graselli, in Heterogeneous Catalysis, Proc. of the IInd Symp. of the Industry. Cor porative Chemistry Programs of the Texas A & M University, April 1984, ed. B. L. Shapiro (Texas A & M Univ. Press, College Station) p. 182.
[5] Z. Bing, S. Pei, S. Shishan and G. Xiexiong, J. Chem. Soc. Faraday Trans., **66** (1990) 3145.

Mixed Metal Oxide Interfaces: An Experimental and Theoretical Assessment of Secondary Metal Influences in Metal Impregnated Aluminas

M. Malaty, D. Singh, R. Schaeffer, S. Jansen; Temple University, Phila. PA 19122; S. Lawrence; Saginaw Valley State University, University Center, MI 48710

Abstract:

Studies of the mixed-metal interface in metal impregnated alumina have indicated the possibility of much metal-metal and metal-substrate interaction. Studies were carried out on Ni,Cu/Al_2O_3 system which was evaluated to develop a better understanding of the forces that drive modification of the catalytic selectivity of Ni in the presence of Cu. Electron Paramagnetic Resonance (EPR), Powder X-ray Photoelectron Spectroscopy (XPS), X-ray Diffraction (XRD) and theoretical calculations were carried out on this bimetallic system, using Ni,Ag/Al_2O_3 as a reference as Ni shows negligible electron perturbation on co-adsorbance with Ag onto alumina. XRD results indicate that gross modification of the electronic fields of Ni and Cu are due to direct coupling and intercalation into the alumina matrix. As a result of this phenomena, these materials may form a good base for the development of novel ceramics based on mixed-metal interactions where the intermetallic perturbations are driven by the substrate effects.

Introduction:

A great deal of effort has focused on the analysis of single-component catalytic systems with the primary metal of interest being copper, nickel, platinum, and ruthenium impregnated into alumina, silica, or titania substrates. In this work we have investigated the effect of a secondary metal on the electronic and structural properties of the oxide. The influence of the secondary metal on the catalytic properties has been investigated for multiple systems though little direct structural evidence has supported the rationale behind the development.

Analysis of the effect of the secondary species in bimetallic catalysts has not focused on the evolving electronic changes but rather on the changes in catalytic activity and selectivity (1,2). The effect of various metals on the ability of Ni to catalyze many hydrogenation reactions has been extensively studied with Ni,Cu/Al_2O_3 being no exception (1-3). Cu acts as an inhibitor to the overall activity of Ni (1-3), but with accompanying enhancement of selectivity. Ag, on the other hand, has no observable effect on the catalytic activity of Ni.

Mat. Res. Soc. Symp. Proc. Vol. 368 © 1995 Materials Research Society

It is these two systems that will be studied in order to justify the secondary metal effect electronically, thus aiding in the development of a more inclusive and rigorous catalytic theory.

Experimental:

The supported precursors were prepared by the incipient wetness technique. After mixing the active component with alumina, each sample was dried overnight at 373-380K, crushed to a fine powder of uniform size, then calcined in compressed air flowing at 25 mL/min for 12 hours at 573K. Coverages were based on the theoretical limits based on the atomic radii and surface areas of the respective metals. For the single-component systems, a solution of the respective nitrate was mixed into the α-alumina. These catalysts were prepared to have the same metal coverage with the surface areas determined to be $284\pm3m^2/g$ and the metal-to-alumina ratio of 0.44 mol/mol. In the bimetallic systems the Ni coverage was the same as in the single component Ni/Al_2O_3. Solutions of the two metal nitrates were prepared, then added simultaneously to the alumina by incipient wetness technique. The second metal was in varying multiples of the mass of Ni used to produce a series of catalysts of varying weight ratios with surface areas of $291\pm20m^2/g$.

Results and Discussion:

<u>XPS</u>

An investigation of the precursor materials using XPS proved the difficulty in resolving the electronic states which were produced on adsorption of Ni^{2+} onto the substrate (4). In all instances, adsorption induced an increase in the spectral width of 40-50% (FWHM) when compared to the reference materials (Table 1). In addition, the binding energies for the $Ni(2P_{3/2})$ appear characteristic of at least two primary Ni phases which are independent of the secondary metal. Comparison of the FWHM and binding energies suggest that both the binary oxide and the metal aluminate contribute to the XPS signal. In addition, binding energy trends in the XPS as a function of metal concentration suggest that a greater percentage of binary oxide forms at higher metal concentration. It is assumed that the aluminate formation occurs initially at the surface of the alumina and as the surface becomes "saturated" the formation of the spinel may be limited. No effect was observed in the Ni XPS profile upon introduction of the secondary species. Binding energies for the 2P doublet were observed for copper oxides at 913eV and 933eV.

Table 1. Binding energies for Ni in materials

Sample:	$Ni(2P_{3/2})$ (eV)	FWHM (eV)	Sample:	$Ni(2P_{3/2})$ (eV)	FWHM (eV)
NiO	853.6	4.0	$Ni,Cu/Al_2O_3$ (II)	855.4	4.4
$NiAl_2O_4$	856.3	2.8	$Ni,Ag/Al_2O_3$ (I)	856.2	3.6
$Ni,Cu/Al_2O_3$ (I)	856.2	3.6	$Ni,Ag/Al_2O_3$ (II)	855.4	4.4

XRD

In an effort to understand the structural features of the mixed metal materials, several XRD analyses were performed. The data suggest a significant portion of the materials is amorphous while a fraction is crystalline. This is consistent with EPR and XPS analysis. A distinct diffraction pattern consistent with the existence of the boehmite phase of alumina is observed for all of the materials (5). This is unusual given the temperature at which the materials were calcined and may represent an unpredicted phase integration between the alumina lattice and the metal ion. The metal aluminate structure is a spinel derivative of boehmite. The formation of aluminates of the corresponding metal ions is reasonable, as beohmite contains periodic vacancies in its lattice normally occupied by M^{2+} ion. These are generally tetrahedral or octahedral sites in which Ni and Cu ions are known to fit readily. Provided that these sites are randomly occupied or incompletely satisfied by the metal ions, they will not produce a characteristic XRD pattern with the exception of some line broadening in the alumina pattern. Random filling of lattice vacancies would provide a direct structural precursor to the crystalline metal aluminate. In fact, processing the calcined materials at 1173K provides the crystalline spinel aluminates for such materials.

EPR

The EPR spectra of the Ni/Al_2O_3, Cu/Al_2O_3, $Ni,Cu/Al_2O_3$ and $Ni,Ag/Al_2O_3$ at 298K are shown in Figure 1. Ag/Al_2O_3 is not shown because Ag which is a d^{10} system, shows no signal when adsorbed on alumina. Under oxidizing conditions Ag^{2+} forms, however, a disproportionation occurs giving Ag^{+} and Ag^{3+} (d^{10} and d^{8}) instead (6).

This $Ni,Cu/Al_2O_3$ sprectrum consists of two predominate features; a broad cuprate signal with a resolved cupric signal superimposed on the broad signal. The Spin-Hamiltonian parameters produced by the $Ni,Cu/Al_2O_3$ are identical to those produced by Cu/Al_2O_3 and Ni/Al_2O_3. The broadresonanceisattributed to the surface oxides.

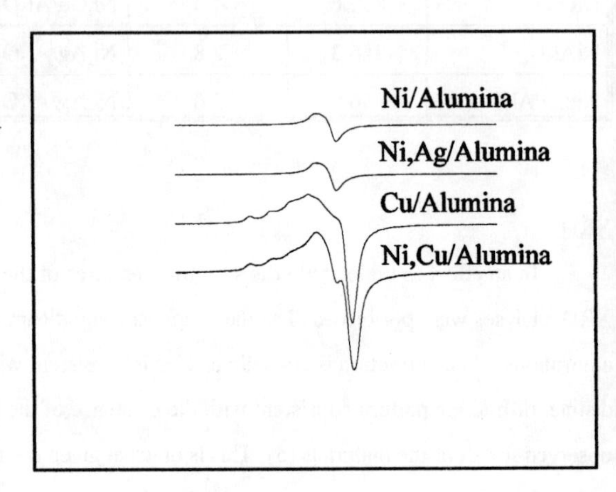

Figure 1. EPR Spectra of Ni/Al_2O_3, $Ni,Ag/Al_2O_3$, Cu/Al_2O_3 and $Ni,Cu/Al_2O_3$. Center field is at 3300G and sweep width is 2000G

From a comparisons of the spin densities for the bimetallic materials (Table 2), it becomes obvious that modification of the EPR active copper state is effected by the presence of nickel. No such effect was observed in the case of the Ni/Ag species. The origin of this phenomenon is not completely understood, however, it is believed to be due to differences in limits of metal ion segregation by the surface and bulk oxide lattices.

Table 2. Spin densities of the precursors

sample	Concentration: mol/mol (10^{-1})	spin density: mol/mol (10^{-4})
Ni,Cu	8.6	6.7
Ni,Ag	9.3	1.1
Ni	4.4	1.8
Cu	4.4	110
Ag	4.4	0.0017

Behavior:

Table 3 shows the catalytic data for one probe reaction of Ni,Cu species with Table 4 indicating the composition of the complex mixture produced. No significant modification in catalytic behavior was observed in the case of Ni,Ag oxides except for dilution of active species and thus the total reaction yield. Therefore the discussion of catalytic properties will be limited to the Ni,Cu bimetallic. The probe reaction considered for this work was the hydrogenation of crotonaldehyde. Typical products include the aliphatic aldehyde, alcohols, ether or a mixture of short chain hydrocarbons. In the case of the pure Ni/Al_2O_3, two major fractions are observed. The first is butanal in 81% and the remainder is a mixture of simple hydrocarbons. No other products were observed in detectable quantities. At a temperature of 473K and the addition of copper, butanol became the predominate reaction product. A smaller contribution of butanal was observed upon further addition of copper. Similarly the concentration of small hydrocarbon products was reduced. At lower temperatures the selectivity for butanal/butanol is reversed and the amount of simple hydrocarbon product is reduced.

Table 3. Catalytic results from Ni,Cu species

REDUCTION TEMPERATURE = 723K Hydrogenation Temperature = 473K					
WEIGHT RATIO CU:NI	BUTANAL	BUTANOL	CROTYL ALCOHOL	BUTYL ETHER	COMPLEX
0.0	81	--	--	--	19
0.5	19	64	--	5	12
1.0	25	61	--	5	9
2.0	35	54	--	3	8
HYDROGENATION TEMPERATURE = 423K					
0.0	84	--	--	--	16
0.5	83	13	1	1	2
1.0	81	17	1	--	1
2.0	78	19	2	--	--

Table 4. Components of the Complex

Compound	Percent
Methane	68
Ethane	6
Propane	25
Butane	0.3
CO, CO_2	0.7

Conclusions:

In this work we have observed that the resolved Ni,Cu/Al_2O_3 spectrum is a derivative of the two single-component systems, in which the copper species are octahedrally coordinated in the substrate spinel/oxide. Spin density data is consistent with this analysis. The observation of Ag, Ag_2O and the Ni-aluminate spinel suggest that no significant Ni-Ag interaction occurs. The broad unresolved peak indicative of CuO remaining at the surface which likely alloys with surface NiO. This "alloy" appears to be the chief component in influencing the catalytic activity of Cu.

References:

1. J.K.A. Clarke, Chem. Rev. 1975, **75(3)**, 291
2. J.H. Sinfelt, Accts. Chem. Res. 1983, **10**, 15
3. T.S. Cale, J.T. Richardson, J. Catal. 1983, **79**, 378
4. S.S Lawrence, Ph.D. Dissertation, Univ. of Missouri-St. Louis, 1989, (DA8922404)
5. H.F. McMardee, M.C. Morris, E.H. Evans, B. Paretzkin, W. Wong-Ng, "Methods and Practices in X-ray Diffraction," R. Jenkins (ed.), International Center for Diffraction Data, Swarthmore, PA, 1990, sec. 5.5.1.
6. W. Levason, M. Spicer, Coord. Chem. Rev. 1987, **76**, 45. M. Melnick, R.V. Parish, ibid, 156. H. Shmidbaur, K.C. Dash, Adv. Inorg. Chem. Radiochem. 1982, **25**, 236

OXIDATIVE CATALYTIC DECOMPOSITION OF TOXIC GASES USING HYDROXYAPATITE AND FLUORHYDROXYAPATITE

Timothy P. Palucka, Nicholas G. Eror, and Thomas A. McNamara
University of Pittsburgh, Department of Materials Science and Engineering, 848 Benedum Engineering Hall, Pittsburgh, PA 15261

ABSTRACT

An oxidative catalytic route for the decomposition of nerve gases was investigated using hydroxyapatite (HA, chemical composition $Ca_{10}(PO_4)_6(OH)_2$) and its partially fluorinated analog fluorhydroxyapatite (FHA, $Ca_{10}(PO_4)_6F_x(OH)_{2-x}$). Samples were prepared with surface areas ranging from 34 to 238 m^2/g to study surface area effects; 1.2 wt. % platinum was deposited on one substrate to investigate the effect of a transition metal on activity and selectivity. Reaction studies were performed using dimethyl methylphosphonate (DMMP), a nerve gas simulant, in a stream of 80 percent nitrogen and 20 percent oxygen at 573 K and atmospheric pressure. High surface area FHA samples showed an increase in the "protection period" (period of 100% conversion) with increasing fluorine substitution; such an increase was not seen for low surface area FHA samples. In the absence of platinum, the reaction products were methanol and dimethyl ether; with platinum, CO_2 was also obtained.

INTRODUCTION

Since the discovery of nerve gases in 1936 there has been a need to devise methods to protect soldiers and civilians from their deadly effects. The use of chemical reactants and absorbents for this purpose has been investigated, but both suffer from the limitation of being consumed during the process. A catalytic decomposition route is preferable since small amounts of catalyst could theoretically decompose large amounts of nerve gas without being consumed. In reality, phosphorous-based poisons tend to remain on the surface and deactivate the catalyst. Lee et. al.[1] and McNamara[2] sought to minimize this poisoning effect by using HA, a material with a phosphate-based lattice, as a catalyst. Continuing that line of inquiry, this study seeks to determine the effect of fluorine, a common component of nerve gases, on the decomposition reaction. Since F^- might readily react with HA, replacing the OH^- group to form FHA, it is important to know whether such substitution enhances or diminishes the activity and selectivity of the catalyst.

BACKGROUND

Hydroxyapatite is the mineral component of bones and teeth, having the composition $Ca_{10}(PO_4)_6(OH)_2$. The HA unit cell is hexagonal with a symmetry group of $P6_3/m$, and lattice parameters of $a = 9.43$ Å and $c = 6.88$ Å.[3] HA is easily susceptible to both cationic and anionic substitutions, with Sr^{2+}, Ba^{2+}, Pb^{2+} and Cu^{2+} being commonly substituted cations and F^- and CO_3^- commonly substituted anions. In addition, a range of calcium deficient HA samples can be prepared with Ca/P ratios varying from 1.67 (stoichiometric) to 1.50; the stoichiometric form has basic properties[4] while the calcium deficient form is more acidic.[5] These properties suggest a rich catalyst chemistry for HA. It has been used as a catalyst in the dehydration and

Mat. Res. Soc. Symp. Proc. Vol. 368 © 1995 Materials Research Society

dehydrogenation of alcohols,[5,6] and for the isomerization of 1-butene to cis-2-butene.[6] A Pb doped HA catalyst was shown to be effective in the selective oxidative coupling of methane to form ethane and ethylene.[7]

In fluorhydroxyapatite (FHA), fluorine substitution for OH^- is incomplete; a sample with 100% fluorine substitution is called fluorapatite (FA). The hexagonal FHA crystal experiences a shortening of the a axis with increasing fluorine substitution; one study reported an a axis length of 9.399 Å for 95% fluorine substitution.[8] To the best of our knowledge, FHA has not been investigated as a catalyst for any system prior to this study.

A typical nerve gas is a fluorine-containing phosphonate ester having the general composition R1R2FP=O, where R1 and R2 are alkyl or alkoxy groups, and R1, R2, F, and O are all bonded to the central P atom. For safety purposes, the less toxic nerve gas simulant dimethyl methylphosphonate (DMMP) $(CH_3O)_2CH_3P{=}O$ is often used for reaction studies. The reaction of interest is:

$$(CH_3O)_2CH_3PO + (x+15/2)\,O_2 \Longrightarrow 3CO_2 + 9/2\,H_2O + PO_x$$

Complete oxidation to CO_2 and H_2O is desired for protection of human beings; less desirable products include carbon monoxide, methanol, acetic acid, and dimethyl ether. The PO_x species tends to remain on the surface and acts as a poison; it can react with water to form H_3PO_4.

Previous research in this area has included fundamental adsorption studies of nerve gases and nerve gas simulants, as well as reaction studies. Single crystal studies of DMMP adsorption on Pt(111)[9] and Rh(100)[10] have been conducted, along with the oxidation of DMMP on Mo(110).[11] The adsorption of DMMP on α-Fe_2O_3,[12] SiO_2,[12] and Al_2O_3[13] has also been investigated.

Reaction studies include the oxidative decomposition of DMMP over a monolithic Pt/TiO_2/cordierite catalyst,[14] oxidation of the nerve gas isopropyl methyl fluorophosphonate (Sarin) over Pt/Al_2O_3,[15] and the hydrolysis of Sarin over γ-Al_2O_3.[16]

Graven and Weller[17] investigated the catalytic conversion of DMMP on a Pt/Al_2O_3 catalyst and found that in air the major decomposition products were methanol and CO_2, with dimethyl ether as a minor product; they postulated that methanol formation was by a hydrolytic reaction of DMMP with residual surface water on the catalyst or with water obtained as an oxidation product from DMMP.

The first use of HA as an oxidation catalyst for the decomposition of DMMP was by Lee et. al.,[1] who studied the reaction over Cu substituted hydroxyapatite of the general formula $Cu_xCa_{10-x}(PO_4)_6(OH)_2$. They reasoned that since poisoning of the catalyst surface by phosphorous species was a general problem encountered in the previous studies, the use of a catalyst with a phosphate-based lattice might decrease the poisoning effect; perhaps the phosphate poison would incorporate into or extend the lattice. They obtained protection periods of approximately 1.3 hours at 573 K during which the only products were CO_2 and H_2O; as the catalyst deactivated, methanol was also detected. As before, phosphorous-based compounds remained on the surface and led to deactivation.

McNamara[2] deposited Pt, Pd, Ni, and Co on HA samples to investigate the effect of these metals on the decomposition reaction at 573 K. He concluded that a 1.2 wt. % Pt/HA catalyst was most effective, with a protection period of approximately 8 hours; for comparison, a 1.2 wt. % Pt/γ-Al_2O_3 sample had a protection period of 15 hours.

EXPERIMENTAL

Commercially available HA (Aldrich) was used by itself and as the precursor for FHA synthesis. FHA was synthesized via the solid state route described by Maiti[18] which involved pestle-milling HA along with sufficient ammonium fluoride to give the desired fluorine substitution, followed by heating to 873 K for 20 hours. Percent fluorine substitution was determined using a fluoride-ion sensitive electrode (Orion 9609BN) and the method of Duff and Stuart.[19]

High surface area HA and FHA samples were synthesized according to the precipitation method of Aoki and Sakka.[20] This involved the slow titration of diammonium hydrogen phosphate solution (together with an ammonium fluoride solution for FHA synthesis) into a calcium acetate solution at room temperature with vigorous stirring; the pH was maintained at 10.6 throughout by the addition of concentrated ammonium hydroxide.

Reaction studies were carried out in a conventional fixed-bed reactor system operated at 573 K and atmospheric pressure with 300 milligrams of catalyst in a 1/4 inch quartz u-tube. The sample was pretreated at 573 K for 1 hour under nitrogen flow prior to the start of the reaction. Flow rates of 80 cc/min. of nitrogen and 20 cc/min. of oxygen were used, with a DMMP flow rate of 2.885 x 10^{-6} moles/min. Reaction products were analyzed using an on-line gas chromatograph (Gow Mac 750) equipped with a thermal conductivity detector (TCD) and a flame ionization detector (FID).

Samples were characterized using BET, x-ray diffraction (XRD), and transmission electron microscopy (TEM).

RESULTS AND DISCUSSION

Three HA samples were prepared with BET surface areas of 41(sample A), 139 (B), and 217 m^2/g (C). Reaction studies showed a significant increase in the protection period (period of 100% conversion) from 10 min. to 35 min. on going from A to B , but no change in going from B to C (Figure 1). However, the slope of the deactivation portion of the curve decreased markedly for sample C, due to the ability of the higher surface area sample to accomodate the PO_x surface poisons more readily. Indeed, at 7 hours sample C was still converting at 20%, while sample A dropped rapidly to 0% conversion within one hour. The major reaction product was methanol, with dimethyl ether as a secondary product. No CO_2 was detected in the product stream. As suggested by Graven and Weller,[17] it is possible that methanol formation is caused by a hydrolytic reaction of DMMP with water obtained as an oxidation product.

A series of FHA catalysts with percent fluorine substitutions of 26, 52, 79 and 99% were synthesized starting with the Aldrich HA precursor. BET surface areas ranged from 34 to 41 m^2/g. Figure 2 is a TEM micrograph of the 26% FHA sample, showing the characteristic acicular FHA crystals with a length of aproximately 150 nm and a width of 40 nm. Reaction studies across the series (including an HA sample) showed a uniform protection period of about

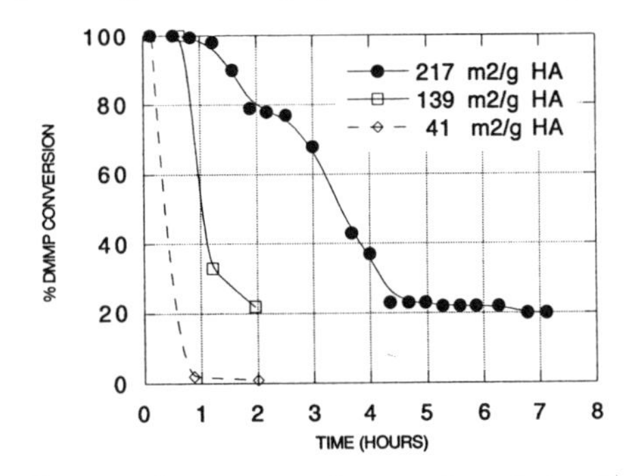

Figure 1. Time courses of conversion of DMMP over a series of HA catalysts with surface areas of 41 (sample A), 139 (B) and 217 m^2/g (C).

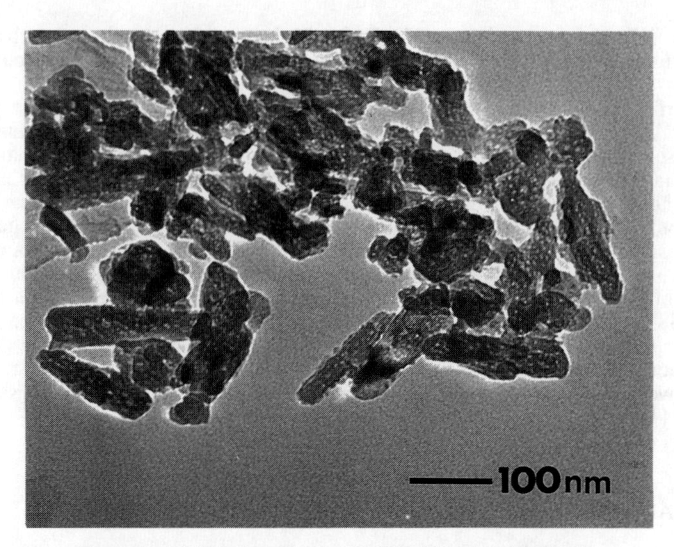

Figure 2. TEM micrograph of 26% FHA sample showing characteristic acicular shape and particle dimensions of approximately 150 nm in length by 40 nm in width.

10 minutes (Figure 3); differences in slope of the deactivation portion of the curve are artifacts based on slight variations in the sampling interval. The reaction products were the same as for the HA series described above. Figure 3 seems to indicate that no modification of surface properties was obtained by fluorine substitution; however, Figure 4 shows a significant increase in the generation of methanol during the protection period with increasing fluorine content, indicating some surface modification, probably increasing acidity, across the series. Since methanol is an undesired product, increasing fluorine substitution would appear to be detrimental to the selectivity of the decomposition reaction. Figure 5a shows XRD patterns for the complete FHA series; the peak shift in the XRD patterns in Figure 5b is indicative of changing lattice parameters due to fluorine substitution.

High surface area FHA samples showed an increase in protection period from 45 minutes to 90 minutes as the fluorine content was increased from 40% (238 m^2/g) to 77% (236 m^2/g) (Figure 6).

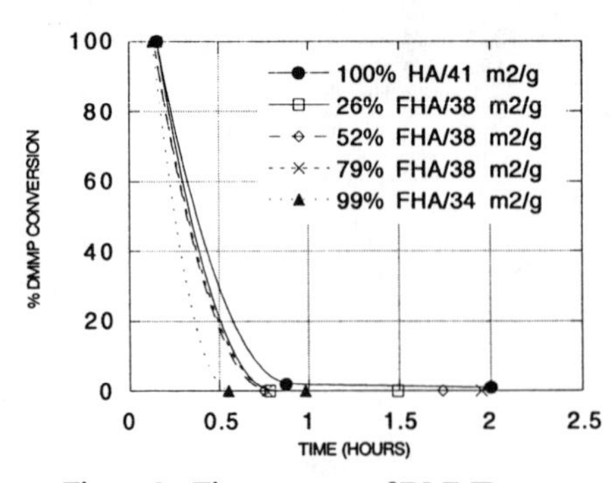

Figure 3. Time courses of DMMP conversion for complete FHA series.

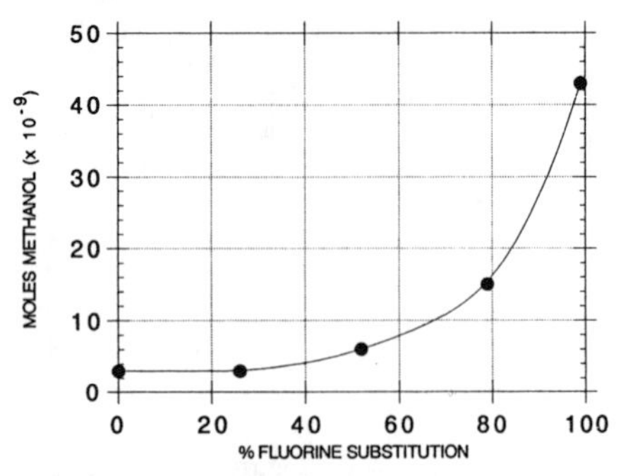

Figure 4. Methanol formation versus percent fluorine substitution.

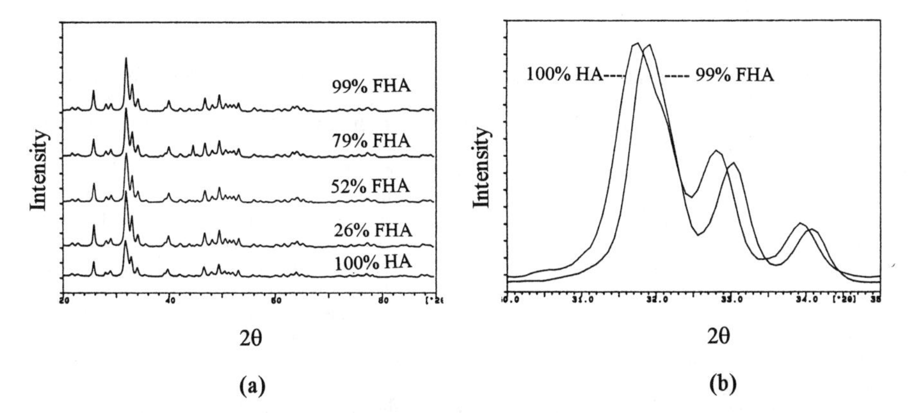

Figure 5. a) XRD patterns for complete FHA series; b) comparison of XRD patterns of HA and 99% FHA samples showing peak shift due to changing lattice parameters caused by fluorine substitution.

Perhaps any enhancement of the protection period due to fluorine substitution was obscured in the previous series due to the low surface areas of the samples and the resolution dictated by the sampling interval. Again, the reaction products were methanol and dimethyl ether, with no CO_2.

Figure 7 shows the effect of the addition of 1.2 wt. % Pt to the surface of the 77% FHA sample. CO_2 was present as a reaction product along with methanol and dimethyl ether for this reaction. The protection period of about 1.5 hours showed no increase over the pure substrate, but the rate of deactivation was markedly reduced by the addition of Pt. Indeed, at 20 hours the sample was converting at approximately 70%; at 49 hours a conversion of 30% was still being obtained. For comparison purposes, the data for the best previous catalyst synthesized by McNamara,[2] a 1.2 wt. % Pt on γ-Al_2O_3 sample with a surface area of 213 m^2/g is included in Figure 7. This sample had a protection period of approximately 15 hours, but had deactivated to <20% conversion after 29 hours.

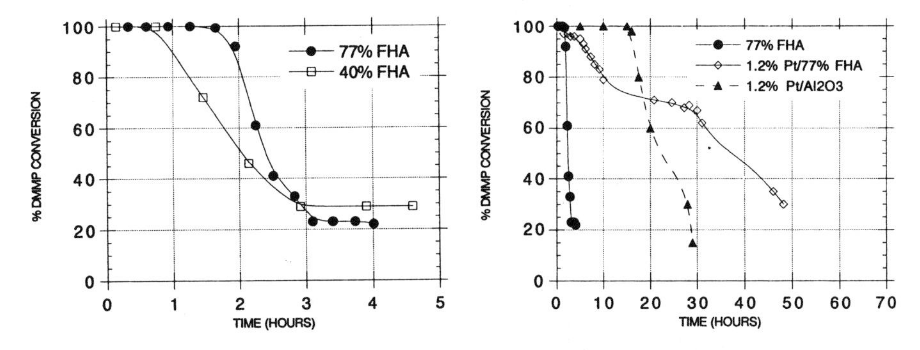

Figure 6. Time courses of conversion of DMMP for high surface area FHA samples showing increase in protection period with increasing fluorine content.

Figure 7. Time courses of conversion of DMMP comparing 1.2%Pt/77% FHA sample with 1.2% Pt/Al_2O_3.

CONCLUSIONS

1. Low surface area (approximately 40 m^2/g) HA and FHA samples showed protection periods of about 10 minutes; when the surface area was increased to > 200 m^2/g, the protection period increased to 90 minutes. The major reaction product was methanol, with dimethyl ether as a minor product. No CO_2 was formed.

2. The surface properties of HA (probably surface acidity) can be modified by substitution of fluorine to form FHA, as seen by the increase in methanol generated with increasing fluorine content. Since methanol is an undesired product for this reaction, such substitution seems detrimental to the selectivity of the catalyst. However, the high surface area FHA samples showed an increase in protection period with increasing fluorine content, indicating enhanced activity.

3. For 1.2 wt. % Pt on 77% FHA, no increase in the protection period was observed; however, the rate of deactivation was dramatically reduced, extending catalyst lifetime with low conversions to beyond 50 hours. CO_2 was formed as a minor reaction product along with methanol and dimethyl ether.

4. Additional studies involving a wider range of fluorine substitutions and variations in the percentage of Pt deposited are needed to further confirm these observations.

ACKNOWLEDGEMENT

This research was made possible by funds obtained from the AFOSR through the Materials Research Center (MRC) of the University of Pittsburgh.

REFERENCES

1. K.Y. Lee, M. Houalla, D.M. Hercules, and W.K. Hall, J. Catal. **145**, 223 (1994).
2. T.A. McNamara, MS Thesis, University of Pittsburgh, 1993.
3. A.S. Posner, A. Perloff, and A.F. Diorio, Acta. Cryst. **11**, 308 (1958).
4. Y. Imizu, M. Kadoya, H. Abe, H. Itoh, and A. Tada, Chem. Lett., 415 (1982).
5. C.L. Kibby and W.K. Hall, J. Catal. **29**, 144 (1973).
6. C.L. Kibby, S.S. Lande, and W.K. Hall, J. Am. Chem. Soc. **94,** 214 (1972).
7. Y. Matsumura, J.B. Moffat, S. Sugiyama, H. Hayashi, N. Shigemoto, and K.Saitoh, J. Chem. Soc. Faraday Trans. **90**, 2133 (1994).
8. H.G. Schaeken, R.M.H. Verbeeck, F.C.M. Driessens, and H.P. Thun, Bull. Soc. Chim. Belg. **84**, 881 (1975).
9. M.A. Henderson and J.M. White, J. Am. Chem. Soc. **110**, 6939 (1988).
10. R.I. Hedge, C.M. Greenlief, and J.M. White, J. Phys. Chem. **89**, 2886 (1985).
11. V.S. Smentkowski, P. Hagans, and J.T. Yates, Jr., J. Phys. Chem. **92**, 6351 (1988).
12. M.A. Henderson , T. Jin, and J.M. White, J. Phys. Chem. **90,** 4607 (1986).
13. M.K. Templeton and W.H. Weinberg, J. Am. Chem. Soc. **107**, 97 (1985).
14. C.C. Hsu, C.S. Dulcey, J.S. Horwitz, and M.C. Lin, J. Mol. Catal. **60**, 389 (1990).
15. R.W. Baier and S.W. Weller, I&EC Proc. Des. and Dev. **6**, 380 (1967).
16. A.E.T. Kuiper, J.J.G.M. van Bokhoven, and J. Medema, J. Catal. **43**, 154 (1976).
17. W.M. Graven, S.W. Weller, and D.L. Peters, I&EC Proc. Des. and Dev. **5**, 183 (1966).
18. G.C. Maiti, Indian J. Chem. **29A**, 402 (1990).
19. E.J. Duff and J.L. Stuart, Anal. Chim. Acta. **52**, 155 (1970).
20. A. Aoki and Y. Sakka, J. Mining.and Mat. Inst. Japan **106**, 861 (1990).

PART VI

Monoliths and Foams

CELLULAR CERAMIC SUBSTRATES

Jimmie L. Williams, Irwin M. Lachman, M.D. Patil and Donald L. Guile,
Corning Inc., Corning, NY 14831

ABSTRACT

Ceramic materials in monolithic honeycomb form offer the advantage of low pressure drop for automotive & stationary emissions control reactors, catalytic oxidation, adsorption, separation, and other applications. The ceramic materials can be extruded into cellular monoliths or coated onto a cellular ceramic substrate using conventional "washcoat" processes. Both extruded and coated ceramic substrates are prepared with a binder to promote interparticle adhesion and strength. For many applications it is advantageous to control pore structure, density, strength and surface area.

In this paper composition, fabrication and physical properties of cellular monolithic substrates will be reviewed. Emphasis will be on extruded materials used as catalytic supports for automotive, DeNOx and oxidation reactions.

INTRODUCTION

Extruded ceramic materials in cellular form offer the advantage of low pressure drop for automotive and stationary emissions control, catalytic oxidation, adsorption, separations, and filtration devices.[1-8] For example, thin membrane coatings on cordierite substrates are used to remove particulates from flue gases.[6] Other applications include the chemical process industries where high thermal shock resistance microcracked ceramics are used.[5]

Table 1 lists materials that have been fabricated in cellular form. The materials can be extruded into monolithic cellular bodies, or coated onto cellular substrates of different material using conventional "washcoating" processes. Extruded bodies may have various cellular shapes (e.g. square, triangular, hexagonal), cell densities (9 to 600 cells/in^2) and wall thicknesses (0.005" to 0.054"). Coating technology is widely known in the literature and will not be discussed. The first section of this paper will discuss fabrication of cordierite and other high temperature ceramics in cellular form. The second section will emphasize extrusion of high surface area materials and physical properties of the substrates.

Mat. Res. Soc. Symp. Proc. Vol. 368 © 1995 Materials Research Society

The most widely used material for automotive emission control is cordierite.[3] The unique properties of cordierite cellular monoliths make them ideally suited for automotive application requiring low thermal expansion, high thermal stability and mechanical strength. Extruded cordierite has low surface area and therefore is coated with a high surface area oxide upon which the precious metal catalyst is deposited.

New emissions standards in the U.S. and internationally will require the development of alternative technology to further reduce automotive, bus and truck emissions.[9] Copper exchanged ZSM-5 as well as other types of zeolites are being evaluated for lean burn engines.[10,11] Electrically heated metal monoliths are being developed to reduce cold start emissions.[12,13] The electrically heated catalyst (EHC) is capable of converting exhaust emission within five seconds after engine ignition. Adsorber systems have also been evaluated for cold start.[14-16] Application of EHCs for cold start will be discussed in a later section.

Extruded cellular ceramics are used for stationary emissions control of NOx by Selective Catalytic Reduction (SCR) with NH_3.[2] In SCR, NH_3 is injected into flue gas streams of utility boilers, gas turbines and cogeneration units to convert NOx into nitrogen and water vapor. These extruded monoliths are comprised of a uniformly distributed catalyst within the high surface area support. The catalyst may be part of the initial batch mixture, or, alternatively, may be impregnated into the extruded monolith after heat treatment. In any case, high surface area materials make up most, if not all, of the monolith instead of only being applied as a washcoat. Extruded vanadia/titania catalyst developed in Japan is the most widely used catalyst for SCR.[17-20] It is used in Europe and is gaining wide spread use in the U.S. as a result of the 1990 Clean Air Act. Various types of SCR catalytic substrates will be covered.

Table 1. Materials

Alpha and Gamma Alumina (Al_2O_3)
Cordierite ($2MgO \cdot 2Al_2O_3 \cdot 5SiO_2$)
Cordierite-Mullite ($2MgO \cdot 2Al_2O_3 \cdot 5SiO_2$ - $3Al_2O_3 \cdot 2SiO_2$)
Mullite ($3Al_2O_3 \cdot 2SiO_2$)
Mullite-Aluminum Titanate ($3Al_2O_3 \cdot 2SiO_2$ - $Al_2O_3 \cdot TiO_2$)
Silica (SiO_2)
Silicon Carbide (SiC)
Magnesium Aluminate Spinel ($MgO.Al_2O_3$)
Zirconia-Spinel (ZrO_2 - $MgO \cdot Al_2O_3$)
Titania
Zeolites
Carbon
Metal and Metal alloy powders

FABRICATION OF CORDIERITE AND HIGH TEMPERATURE MONOLITHS

The most important material used world-wide[3] for extruded monolithic substrates in automotive emissions control is cordierite, $2MgO \cdot 2Al_2O_3 \cdot 5SiO_2$. Compositions based on cordierite possess a unique combination of several critical characteristics: a) thermal shock resistance due to a low thermal expansion coefficient; b) porosity and pore size distribution suitable for washcoat application and good washcoat adherence; c) sufficient refractoriness because the melting point exceeds 1450°C; d) sufficient strength for survival in an exhaust environment; and e) compatibility with washcoats and catalysts.

Figure 1 shows a flow chart of the procedures used to fabricate batch materials into cellular monolithic form. Cordierite is a natural mineral but is not abundant enough to meet supply demand for automotive substrates. It is synthesized from readily available raw materials of high purity. Three typical batch compositions are shown in Table 2. The compositions used are close to the stoichiometric formula. Porosity and pore size distribution are effected by particle size of the raw materials. Particle size of the talc has also been shown to be a factor in controlling thermal expansion. Details of physical properties for cordierite are in reference (3).

Table 2. Extruded cordierite monolithic substrate compositions (wt%)

Raw Material	A	B	C
Kaolin	21.74	40.20	43.00
Talc, raw	39.24	19.40	38.00
Talc, calcined	----	19.80	-----
Alumina	11.23	3.68	-----
Hydrated alumina	17.80	16.90	19.00
Silica	9.99	-----	-----

The fabrication of cordierite follows the steps outlined in Figure 1; a) mixing of raw materials with an organic binder, b) kneading the material with addition of water to plasticize, and c) extrusion through proprietary dies developed for this process. The fourth step is uniform drying to insure removal of moisture to prevent cracking and to provide strength. The final step is firing at 1400°C to transform the extruded material into a strong body, with alignment of the Z-axes of the cordierite crystals within the plane of the extruded cell walls. The orientation of cordierite leads to overall low thermal expansion of the substrate along its axial dimension, resulting in high thermal shock resistance.[21,22]

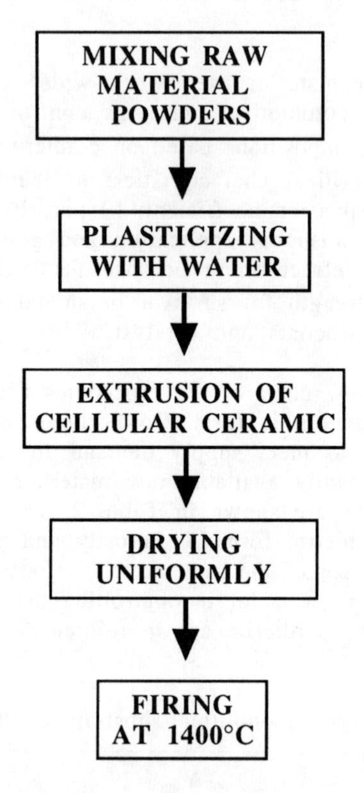

Figure 1. Fabrication and extrusion of cordierite.

Mullite-based compositions listed in Table 1 are fabricated similarly using the process outlined in Figure 1. Highly thermal shock resistant microcracked compositions have also been formulated based on mullite, mullite+aluminum titanate and zirconia+spinel. Properties of these materials are tabulated in reference (3).

EXTRUSION OF HIGH SURFACE AREA MONOLITHS

Extrusion of high surface area materials follows the procedure outlined in Figure 2. In the first step high surface area raw materials are mixed with inorganic & organic additives and binders. Inorganic additives used include glass fibers, glass flakes, and inorganic precursors in the form of sols or gels.[19,20] These additives will reduce cracking and insure adequate strength of the fired monolith. Organic additives include polyvinyl alcohol, cellulose, polymers and waxes. The catalyst or its precursor form can be added here or, in the case of zeolites, ion-exchanged.

In the second step water is added to plasticize the batch materials followed by extrusion. Drying to 100°C must be carried out slowly so that the water is driven off gradually in a very uniform manner to prevent cracking.

In the fifth step the firing temperature depends on the presence of catalyst in the initial batch. If catalyst is present the firing temperature must be kept low, 400-650°C, to prevent its inactivation by sintering. If catalyst are not present the temperature can be higher to develop optimum strength.

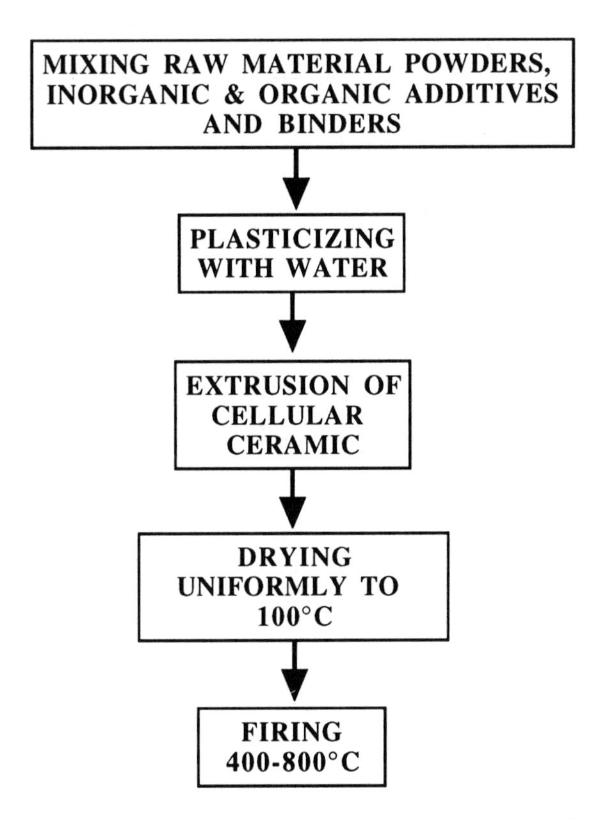

Figure 2. Fabrication of high surface areas cellular ceramic substrates.

Tabulated in Table 3 is the effect of addition of silica on the physical properties of alumina and titania with heat treatment temperature. The addition of silica retards the crystalline transformation of alumina and titania to their low surface area phases. The pore size contribution becomes bimodal and becomes courser with heat treatment temperature.

Table 3: Physical Properties of High Surface Area Monolithic Catalyst Structures.

Heat treat Temperature °C	Substrate Composition	Surface Area m2/g	%Open Porosity (by boiling H2O)	Median pore Size A (by Pore Volume Dist.)	Bulk Density g/cc
500	Alumina	202	63	75	1.31
1000	Alumina	80	59.1	145	1.49
500	93%Alumina + 7 % Silica	260	62.1	55,140 (bimodal)	1.04
1000	93%Alumina + 7 % Silica	155	64	82.5,190 (bimodal)	1.24
1250	93%Alumina + 7 % Silica	25.1	37.5	120,2300* (bimodal)	2.2
500	Silica	47	47	62	1.17
500	Titania	43	57.6	20,115,325	1.6
500	91% Titania + 9% Silica	98.3	55.9	40,325 (bimodal)	1.59
1000	91% Titania + 9% Silica	30		40,325 (bimodal)	1.64

*Measured by Hg Porosity

STATIONARY EMISSIONS CONTROL (DENOX)

Extruded cellular ceramics are being used for stationary emissions control of NO_x by Selective Catalytic Reduction (SCR) with NH_3.[17-20] In SCR, NH_3 is injected into flue gas streams of industrial & utility boilers, gas turbines, cogeneration units, and incinerators to react with NOx to form innocuous nitrogen and water vapor.

The extruded monoliths are formed from high surface area materials and catalyst formulations, both noble and base metals. The catalyzed substrates are uniformly made up of these materials. In some cases the monolith is extruded without catalyst in the initial mix, in which case the catalyst is later impregnated or ion-exchanged (zeolites) in the extruded monolith after heat treatment. The selection of catalyst is dependent on the desired operating temperature use window (200-600°C). Catalyst systems are combined in some cases to produce a wider DeNOx operating window.

Three types of commercial SCR catalyst are available; noble metal-based, vanadia/titania and zeolites, both natural and synthetic. Noble metal catalysts operate at low temperatures from 200°C to 280°C. vanadia/titania catalysts are usually used at temperatures ranging from 250°C to 400°C. Zeolite-based catalysts are suitable for higher temperature operation from 400°C to 600°C. The most widely used catalyst is vanadia/titania developed in Japan and used extensively in Europe, the U.S. and other countries. In recent developments interest in metallic monoliths for flue gas treatment has increased due to their mechanical durability.[24,25]

Zeolites can be combined with vanadia/titania to produce a catalyst with a wide DeNOx operating window that is more resistant to deactivation. Mordenite and ZSM-5 are the most widely used zeolites for SCR; however, others are also suitable for this application.[17,26] Agglomerates of zeolites and vanadia/titania type catalysts can also be incorporated into extruded monolith as separate discrete particles.[27] The Catalyst/Washcoat-in-the-Wall composite allows the combination of multiple catalysts in a monolith where they are separate to prevent interaction. Zeolites have also been incorporated into monolithic cellular form by in-situ growth[28,29]

Extruded alumina monolithic supports have been used to produce a more sulfur-tolerant catalyst.[30] The impregnation of alumina with rhodium in the presence of an acid produces a catalyst with noble metal driven deep within the walls of the support. The outer layer of alumina is capable of adsorbing sulfur dioxide reversibly.

Global demands on reduction of NOx emission limits have been placed on stationary combustion sources over the last decade.[24]
In the U.S. some state, regional and local governments have enacted more stringent legislation to drive down NOx emissions from new sources.
The provisions of the 1990 Clean Air Act, especially Titles IV and V, will impact future rulings on currently unregulated sources and existing facilities such as gas turbines, utility boilers, commercial heaters, and incinerators which will need to be retrofitted to meet these restrictions.

COLD START EMISSIONS

Most automotive emissions occur during the first two minutes after engine start-up. These emissions, referred to as "Cold Start" emissions, remain unchecked since the cellular ceramic catalytic substrate has not reached operation temperature. Electrically heated catalysts (EHCs) have been developed and evaluated for removal of cold start emissions. Adsorber systems have also been evaluated for this application.

The metal substrates used for EHCs are extruded and foil type monoliths of FeCrAl compositions.[12,13] These compositions yield high electrical resistance and oxidation resistance as well as durability. The extruded metal monolith is fabricated from metal powders (-325 mesh) following the general procedure outlined in Figure 1. EHC systems consist of a metal substrate (with electrodes) close coupled with a catalyzed ceramic substrate located immediately behind it. The EHC heats exhaust gases so that the ceramic substrate can begin conversion within five seconds after engine ignition. Engine test results for an extruded EHC system are listed below in Table 4.[12]

Table 4. EHC test results and California's LEV and ULEV emission standards.

Emissions, g/mi	EHC	LEV	ULEV
Non-Methane Organic Gases	0.017	0.075	0.04
CO	0.42	3.4	1.7
NOx	0.16	0.2	0.2

In 1997 California's stringent Low Emissions Vehicle (LEV) and Ultra Low Emissions Vehicle (ULEV) standards (Table 4) will require reduction in cold start emissions. California's LEV and ULEV regulations are more restrictive than the U.S. Federal standard and are being adopted by other states. It is expected that 25% of 1997 model automobiles sold in California will be Low Emissions Vehicles and 2% Ultra Low Emissions Vehicles if manufacturers are to meet California's designated fleet average. EHC systems are capable of meeting and surpassing these requirements.

HIGH TEMPERATURE CELLULAR CERAMIC SUBSTRATES

High temperature cellular ceramic compositions have been developed for catalytic combustion, as catalytic supports for oxidation reactions in the chemical process industries, and as molten metal filters.[5] Varying degrees of thermal shock resistance are needed so that mullite and alumina containing refractory compositions have been used as well as designed to utilize microcracking for enhanced thermal shock resistance.

Pt-Rh deposited on alumina supported on cordierite has been studied for ammoxidation for the production of nitric acid.[31] Laboratory studies have shown excellent yield with reduction in precious metal use. In the current industrial process, a mixture of ammonia and air is passed downward through a layered pad of Pt-Rh alloy gauze. Extra gauze layers are used to produce a pressure drop across the catalyst which gives a more even flow distribution. However the extra gauze results in greater precious metal loss.

Cordierite and mullite compositions have been used as supports for catalytic combustors for woodburning stoves, preventing creosote deposits, chimney fires and to improve combustion efficiency.[3]

Ceramic monoliths are used as supports for the manufacture of HCN using the Andrussow process.[3] In this process, Pt alloy gauze is supported on alumina-based cellular monoliths. A reaction mixture of methane, ammonia and air is passed through the catalyst bed at 1100°C to 1150°C to produce HCN.

Cellular ceramics in the future may be applied to catalytic oxidation reactions where refractoriness, pressure drop and contact times are important for yield, selects and elimination of pollutants. In most cases, ceramic monoliths will continue to be used as structural supports for carrying the catalyst/high-surface-area coating.

REFERENCES

1. CAPoC 2, 2nd International Congress on Catalysis and Automotive Pollution Control, Universite Libre de Bruxelles, Belgium, 10-13 Sept. 1990.
2. Proceedings of the 87th Annual Meeting of the A&WMA, Cincinnati, Ohio, 19 June 1994.
3. I.M. Lachman and J.L. Williams, Catalysis Today, **14**, 317 (1992).
4. S. Irandoust and B. Andersson, Catal. Rev.-Sci. Eng., **30** (3) (1988).
5. I.M. Lachman, Spechsaal, **119** (12) 116-19 (1986).
6. R.F. Abrams and R.L. Goldsmith, Paper No. 93-RP-138.04, Proceedings of the 86th Annual A&WMA, Denver, CO, 1993.
7. D. Crompton and A. Gupta, Paper No. 93-TP-31B.06, Proceedings of the 86th Annual A&WMA, Denver, CO, (1993).
8. M.D. Patil and I.M. Lachman, in Perpectives in Molecular Sieve Science, ACS Symposium Series 386, edited by W.H. Flank and T.E. Whyte, Jr., (American Chemical Society, Washington, D.C., 1988) pp 492-499.
9. K. Kollmann, J. Abthoff, and W. Zahn, SAE Paper No. 940469.
10. M.J. Heimrich and M.L. Deviney, SAE Paper No. 9307346.

11. D.R. Monroe, C.L. DiMaggio, and D.D. Beck, SAE Paper No. 930737.
12. L.S. Socha, Jr. and D.F. Thompson, SAE Paper No. 920093.
13. H. Mizuno, F. Abe, S. Hashimoto, and T. Kondo, SAE Paper No. 940466.
14. M.J. Heimrich, L.R. Smith, and J. Kitowski, SAE Paper No. 920847.
15. B.H. Engler, D. Lindner, E.S. Lox, K. Ostgathe, A. Sindlinger, and W. Muller, SAE Paper No. 930738.
16. J.K. Hochmuth, P.L. Burk, C. Tolentino, and M.J. Mignano, SAE Paper No. 930739.
17. R.M. Heck, J.M. Chen, and B.K. Speronello, Paper No. 93-MP-7.01, Proceedings of the 86th Annual A&WMA, Denver, CO, (1993).
18. K. Matsushita and T. Kenshi, U.S. Patent No. 4 010 238 (1988).
19. F. Nakajima et al., U.S. Patent No. 4 085 193 (1978).
20. L. Abe and T. Nakatsuji, U.S. Patent No. 4 520 124 (1985).
21. I.M. Lachman, R.D. Bagley, and R.M. Lewis, bulletin of the Amer. Cer. Soc., **69** (2), 202-205 (1981).
22. K. Sugiura and Y. Kuroda, J. Ceramic Assoc. of Japan, **63**, 579 (1955).
23. R.M. Heck, J.C. Bonacci, and J.M. Chen, Paper No. 87-52.3, Proceedings of the 80th Annual A&WMA, New York, NY, (1993).
24. L.J. Czarnecki, C. Libanai, and J.S. Rieck, Paper No. 94-RP131.06, Proceedings of the 87th Annual A&WMA, Cincinnati, Ohio (1994).
25. I.M. Lachman, M.D. Patil, J.L. Williams, and R.R. Wusirika, U.S. Patent No. 4 912 077 (1990).
26. M.J. Kiovsky, P.B. Koradia, and C.T. Lim, Ind. Eng. Chem. Prod. Res. Dev., **19**, 218 (1980).
27. J.L. Williams and I.M. Lachman, Man and His Ecosystem, Proceeding of the 8th World Congress, Vol. 4, pp 351-356, Elsevier, Amsterdam (1989).
28. S.M. Brown and G.M. Woltermann, U.S. Patent No. 4 157 375 (1979).
29. I.M. Lachman and M.D. Patil, U.S. Patent No. 4 800 187 (1989).
30. H.G. Stenger, Jr., E.C. Meyer, J.S. Hepburn and C.E. Lyman, Chem. Eng. Sci., **43** (8), 2067 (1988).
31. W.T. Elkington, M.S. thesis, Brigham Young University, 1989.

PEROVSKITE CATALYSTS: HIGH-SURFACE AREA POWDERS SYNTHESIS, MONOLITHS SHAPING AND HIGH-TEMPERATURE APPLICATIONS

VLADISLAV A. SADYKOV, L.A. ISUPOVA, S.F. TIKHOV AND O.N. KIMKHAI.
Boreskov Institute of Catalysis ST RAN, pr. Lavrentieva, 5, Novosibirsk, 630090, Russia.

ABSTRACT

Monolith perovskite catalysts of honeycomb structure for high-temperature applications are elaborated. Problems of powdered perovskites synthesis by efficient plasmochemical and mechanochemical methods as related to real structure and reactivity of these compounds are discussed. Pilot testing in reactions of fuels combustion, ammonia oxidation, methane conversion and sulphur dioxide reduction proved high activity and stability of these catalysts.

INTRODUCTION

Catalytic combustion is now considered as one of the promising ways to abate NO_X in flue gases of power plants, turbines and other sources [1]. Mixed oxides of transition and rare-earth metals possessing perovskite structure appear to be suitable for catalytic combustion due to their well-known stability at high temperatures in atmospheres containing steam and oxygen [2]. They also seem promising for other high-temperature processes such as methane conversion, ammonia oxidation, sulphur dioxide reduction etc. However, up to now any indications on the industrial production and large-scale application of these catalysts are absent.

To elaborate the technology of perovskite monolith honeycomb catalysts production, the methods of preparation of highly dispersed, chemically active powders with uniform phase composition and narrow particle size distribution should be invented. Moreover, any large-scale production of chemicals should be cheap and wasteless. These requirements are met by two methods used by us: 1. Mechanochemical activation (MAC) of solid starting compounds in high-powered ball mills with a subsequent thermal treatment [3]; 2. Arc plasma thermolysis of the mixed solutions of rare-earth and transition metals [4].

The work presented was undertaken to develop new methods of active powdered perovskites preparation, to shape monoliths by extrusion and to investigate their catalytic properties.

METHODS OF THE SAMPLES PREPARATION AND INVESTIGATION

As starting compounds, oxides, carbonates and nitrates of lanthanum and transition metals of "chem. pure"grade were used.

Such properties of the samples as specific surface, dispersion, crushing strength, real (defect) structure, surface composition were investigated by using conventional methods and equipment. Catalytic activities in the reactions of CO and butane oxidation were determined in the batch-flow systems, while activities in the reactions of methane conversion, sulphur dioxide reduction and ammonia oxidation were determined in fixed-bed flow systems [5].

Mechanical activation of the starting compounds was carried out using high-powered EI or AGO planetary ball mills [3]. Plasmochemical synthesis of perovskites was carried out using reactor of the type described in [4]. A flux of air arc plasma with an initial temperature in the

Mat. Res. Soc. Symp. Proc. Vol. 368

range of 4-6 thousands of K was generated by 1-3 plasmatrons operating at 180-220 V. Mixed nitrate solutions were injected into the reaction zone via spray nozzles. Powders were separated from a gas flow by using cloth bag filters.

Extrusion. The catalysts of simple (cylinders) or complex (rings, honeycomb monoliths) forms were prepared by extrusion of the plastic pastes composed of perovskite powders with addition of the acid peptizers and some surfactants. After optimizing rheological properties of the pastes and the regimes of drying and calcination for rings and micromonoliths, monoliths of various forms (rectangular, cylindrical, hexagonal prisms etc) with cross-section ca 70-90 mm and wall thickness ca 2 mm were formed using die (spinneret) designed and fabricated at the Institute of Catalysis and a plunger forming machine. After drying monoliths were calcined at temperatures in the range of 673-1373K for 2-4 hours.

RESULTS AND DISCUSSIONS

Plasmochemical route

To obtain highly dispersed perovskites with the uniform phase compositions, a search of the optimum conditions of synthesis was carried out by varying the ratio of the plasma flux to solution feed as well as the metals content in solution. The most efficient regimes were found to have current densities ca 40-60 A and the total supply of solution ranging from 1.5 to 3.0 g/sec , concentrations being ca 10-15 wt.% as calculated on MeO_X. At such parameters a cost efficiency of this method is comparable with that of the traditional wet methods.

The particles of perovskite powders thus obtained have nearly spherical form and rather uniform size distribution (mean diameter ca 0.1-0.3 mkm [6]). Such shape is determined by evaporation and shrinkage of a droplet in high-temperature zone with simultaneous decomposition of nitrates of the transition metals and lanthanum. According to TEM and XPD data, particles are composed from slightly misoriented blocks stacked in the (111) direction . As a result, high density of extended defects and surface steps is generated, which are not annealed due to a rapid quenching after flying off the high-temperature zone. As revealed by thermal analysis and IR data, on the surface of these particles substantial amounts of nitrates and hydroxiles are retained that is explained by a rather prolonged contact of the powders trapped in filters with flue gases containing water and NO_X.

Mechanochemical activation

A rather detailed investigation of the influence of the nature of starting compounds, type of mill, time of activation and temperatures of the activated mixtures sintering on the yield of perovskites has been carried out [5,7]. Higher oxides of Co and Mn were found to react more readily than suboxides that is of fundamental interest. It seems primarily to be connected with facile removal of oxygen from higher oxides in the course of activation [3], thus disordering the oxygen sublattice and enabling incorporation of lanthanum to form mixed La-O layers.

From the practical point of view, mechanical activation allowed to decrease temperatures of the effective solid state interactions from 1400 to 800-1000 K, while duration of synthesis was reduced from hundreds of hours typical to ceramic technology to several hours [5,7]. The optimum conditions of mechanochemical synthesis of pure perovskites were found to be 3 min activation in AGO with the subsequent annealing at 973K for 2 h. Particle size distribution for MCA samples is of continuous or bimodal type ranging from 10 to 0.1 mkm.

Reactivity of powdered perovskites

In general, for perovskites samples prepared by various methods, specific catalytic activity differs by a more than one order of magnitude (Table) thus evidencing structure sensitivity.

Table. Specific activities in butane oxidation at 673 K

Sample	La-Mn plasm.	La-Mn MCA	La-Co plasm.	La-Co ceram.	La-Cu plasm.	La-Ni plasm
W	1.0	0.07	0.25	0.1	0.03	0.016
Sample	Mn_2O_3	Co_3O_4	CuO	NiO		
W	0.4	1.6	0.1	0.35		

[W] = $\times 10^{2}$ ml Butane/$m^2 \times$ sec. Simple oxides are prepared by plasmochemical method.

When analyzing catalytic properties of perovskites, attention is usually focused on the point defects generated, i.e., by substitution in lanthanum sublattice [2]. However, for strontium-substituted cobaltites $La_{1-x}Sr_xCoO_{4+y}$ prepared via ceramic route, , we have found recently [8] that unsteady-state catalytic activity in CO oxidation for oxidized surface follows only surface cobalt concentration differing from the bulk one, any role of point defects being not observed. Steady-state activities (surface is partially reduced by the reaction mixture) rather well correlate with the density of extended defects being at maximum at x = 0.3-0.4. According to XPD and TEM data, at these values of substitution transition from hexagonal to cubic structure takes place accompanied by development of a disordered microdomain pattern (Fig.1). Decisive role of the extended defects was also demonstrated for lanthanum manganites [6].

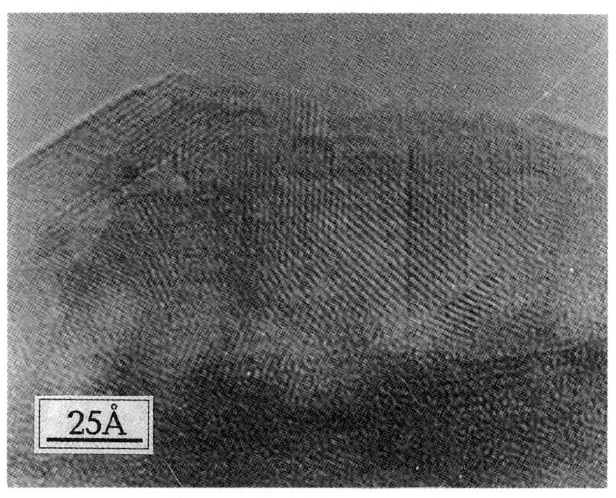

Fig. 1 Typical image of the particles of $La_{0.6}Sr_{0.4}CoO_3$ perovskite. JEM-400C machine

From the point of view of the atomic structure, this phenomenon appears to be determined by the surface arrangement of the most developed faces of (111) and (110) types. These planes are covered by tightly bound (and thus unreactive) bridged and triple-coordinated oxygen. Terminal oxygen of moderate reactivity bonded with transition metal cations is located only on the scarcely exposed faces of the (100) type. Assuming an analogy with the simple transition metal oxides [9], the most active weakly bound forms of oxygen can be assigned to clusters of TM ions located at surface steps including those emerging at outlets of extended defects. Since both factors -exposition of various faces and density of extended defects are known to depend strongly upon preparation conditions, it explains rather well the observed structure sensitivity. In this respect, two points deserve special comments:

1. High activity of plasmochemical perovskites appear to be explained by a spherical form of particles as well as by high density of extended defects, both factors creating surface steps.
2. Specific activities of perovskites but manganites are lower than those of TMO. As lanthanum ions are of low (if any) reactivity at moderate temperatures, these findings are quite predictable. Specificity of manganites seems to be associated with a change of the Mn oxidation state from 3+ to 4+ while entering perovskite lattice due to a well-known excess of oxygen in this compound. Indeed, highly charged manganese ions are known for their reactivity. In contrary, sharp decline of activity while going from NiO to nikelate could indicate a decreased ability of Ni(3+) to activate hydrocarbons.

Forming of monoliths

To form perovskites as monoliths, optimization of the rheological properties of pastes, regimes of drying and calcination was carried out [5] aimed also at obtaining catalysts possessing high activity and good thermal shock resistance. As binder, pseudoboehmite prepared by thermochemical activation of gibbsite was found to be the most suitable after peptizing by acetic or nitric acids. Such surfactants as glycerol, ethylene glycol, carboxymethyl hydro cellulose were used to facilitate extrusion and to suppress generation of cracks at drying stage. The catalysts obtained by extrusion of the optimized pastes with subsequent drying in controlled humidity conditions and calcination at 1173 K have specific surface values in the range of 30-50 m^2/g, total pore volume up to 0.3 ml/g, (mean pore radius in the range of 0.03-0.2 mkm) and mechanical strength up to 10-20 MPa/cm^2. Addition of alumina substantially enhances thermal stability and specific surface values though somewhat decreases catalytic activity at moderate temperatures. It could be explained by formation of the surface microphases of hexa-aluminate type having high melting point and decreased activities. The typical shapes of monoliths are given in Fig. 2 .

Pilot testing

Methane, propane -butane and gasoline combustion on monolith perovskites both in catalyst-supported and flameless regimes at GHSV ca 10,000/h was proved to decrease the level of nitrogen oxides emission to less than 50-100 ppm. The catalysts worked for two months at 1173-1373 K without loss of activity and monolith integrity withstanding daily start-up and shut-off [10].

Resistance to catalytic poisons (HF,HCl, SO_2). As far as the influence of HF is concerned, experiments modelling exhausts of alumina plants were carried out [10]. In this case (Fig.3) only manganites with a minimum content of alumina binder were found to retain a sufficiently high strength to work in such environments.

Two months work in the furnace for incineration of photomaterials revealed that monolith perovskites are more stable in respect of both activity and mechanical strength as compared with supported catalysts (Fig.4).

Methane conversion, sulphur dioxide reduction. Catalysts based upon nikelates of lanthanum at 1023-1073 K were shown to have more than order of magnitude higher activities in the identical conditions as compared with traditional nickel-alumina systems.

Ammonia oxidation into NO_X. Monoliths of lanthanum manganites when tested at 800-900 C demonstrated high activity, selectivity and thermal stability in this reaction that allows to use them in the second layer after Pt-Rh gauzes to ensure complete conversion of ammonia.

Fig. 2.The typical shapes of monoliths.

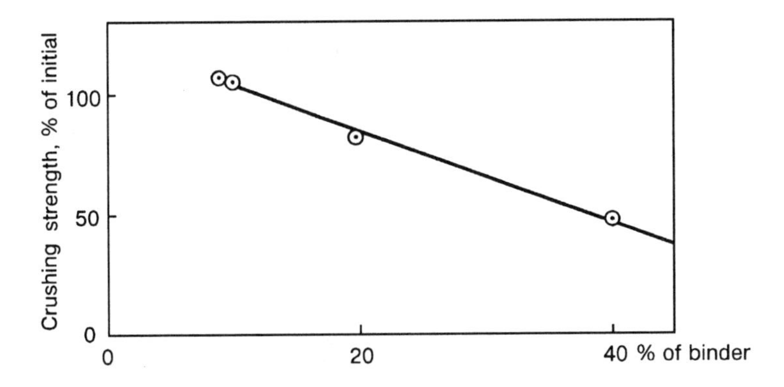

Fig. 3. Mechanical strength of manganite monoliths after HF action.vs binder content in wt.%

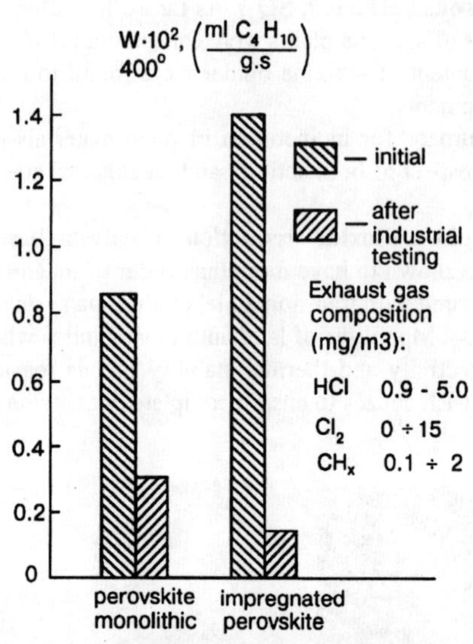

Fig. 4. Butane oxidation activities before and after industrial testing.

References

1. D.L. Trimm, Appl. Catal. **7**,249 (1983)
2. L.G. Tejuca, J.L.G. Fierro, J.M.D. Tascon, Adv. Catal. **36**,237 (1989).
3. E.G.Avvakumov, Mechanical Methods of the Chemical Processes Activation , 2nd ed.(Nauka Publishers, Novosibirsk, 1986),305 p.
4. V.D. Parkhomenko, P.N. Tsybulev, Yu.I. Krasnokutskii, Technology of the Plasmochemical Processes, (Vyssha Shkhola Publishers, Kiev, 1992), 350 p.
5. L.A. Isupova, V.A. Sadykov, L.P.Solovyova, M.P. Andrianova ,V.P.Ivanov,, G.N.Kryukova, V.N.Kolomiichuk, .,E.G.Avvakumov, I.A.Pauli, V.A. Poluboyarov, A.Ya. Rozovskii and V.F. Tretyakov, in Scientific Bases for the Preparation of Heterogeneous Catalysts,VI, preprints. (Louvain-la-Neuve, Belgium,1994),V.2, pp. 231-239.
6. S.F. Tikhov, V.A.Sadykov, E.A.Pack, O.N. Kimkhai, E.M. Moroz, V.P. Ivanov, G.N.Kustova and G.M.Alikina in Heterogeneous Catalysis, edited by L. Petrov and A.Andreev (Proc. VII Int. Symp., Bourgas, Bulgaria, 1991, v.2), pp. 423-428.
7. I.A.Pauli, E.G. Avvakumov, L.I. Isupova, V.A.Poluboyarov, V.A.Sadykov, Sib. Khim. Zhurn., **3**, 133 (1992).
8. L.A.Isupova, V.A.Sadykov, V.P.Ivanov, A.A.Rar, S.V. Tsybulya, M.P. Andrianova, V.N.Kolomiichuk, A.N.Petrov, and O.F.Kononchuk, React. Kinet. Catal. Lett. **53**, 223 (1994).
9. V.A. Razdobarov, PhD thesis, Boreskov Institute of Catalysis SD RAN, 1992.
10. V.A. Sadykov, L.A.Isupova, S.F.Tikhov, O.N.Kimkhai, A.Ya.Rosovskii, V.F.Tretyakov, V.V.Lunin, A.N.Petrov, P.N.Tsybulev, P.N.Voronin in Proc. 4th Europ. East-West. Conference & Exibition on Materials and Processes (CRISH "Prometey", St.-Petersburg, Russia, 1994), V.1.

MONOLITHS FOR PARTIAL OXIDATION CATALYSIS*

L. D. SCHMIDT AND A. DIETZ III
Department of Chemical Engineering and Material Science, University of Minnesota, Minneapolis, MN 55455

ABSTRACT

Monolith catalysts are effective for partial oxidation reactions at high temperatures at contact times between 10^{-4} and 10^{-2} sec. We summarize results for three reactions: (1) syngas by direct oxidation of CH_4, (2) olefins by oxidative dehydrogenation of alkanes, and (3) HCN by ammoxidation of CH_4, and we consider what features of monoliths create optimal selectivity with high conversions. Monoliths used are noble metal films coated on ceramic foams, extruded ceramics, ceramic fibers, and woven Pt-10%Rh gauze catalysts. Effects of metals and geometries of the ceramics are compared to show how these factors influence activity and selectivity.

INTRODUCTION

Gauze monoliths have been used for many years for nitric acid synthesis by the catalytic oxidation of NH_3 to NO.[1-7] The same catalyst is used for synthesis of HCN,[8]

$$CH_4 + NH_3 + \frac{3}{2}O_2 \rightarrow HCN + 3H_2O, \qquad (1)$$

which is essential in production of Nylon and methyl methacrylate polymers. These processes take place near atmospheric pressure at 800 and 1100°C, respectively, at contact times over the catalyst of ~10^{-4} sec. Commercial processes for HCN synthesis typically give 80% selectivity to HCN at 80% conversion of NH_3 and CH_4 using a Pt–10% Rh gauze catalyst.[9] We have studied these reactions extensively both experimentally[10-13] and by modeling[14] and have shown that they can be explained as occurring by the dehydrogenation of CH_4 and NH_3, with O_2 ideally oxidizing the hydrogen rather than forming CO.

We have recently shown that syngas (a mixture of CO and H_2) can be made by direct oxidation of CH_4,[15,16]

$$CH_4 + \frac{1}{2}O_2 \rightarrow CO + 2H_2, \qquad (2)$$

and this reaction has promise in efficient production of methanol and synthetic petroleum. The best catalyst for this process was found to be Rh coated on ceramic foam monoliths, with Pt

* This research is partially supported by DOE under grant No. DE-FG02-88ER13878 and by NSF under grant No. CTS-9311295

giving significantly lower H_2 selectivity.[17,18] Contact times in this process are also ~10^{-3} sec with the reactor operating nearly adiabatically at 1000°C.

More recently we have shown that olefins can be produced with high selectivity by oxidation of higher alkanes such as C_2H_6, C_3H_8, and C_4H_{10} with millisecond contact times.[19-21] The best catalyst for these reactions was found to be Pt on porous ceramic foams. Up to 70% selectivities to olefins could be obtained at 80% hydrocarbon conversion with adiabatic reaction temperatures of 900-1000°C.

It is the purpose of this paper to discuss what features of monolith catalysts provide high selectivities, and more generally, how monolith catalysts should be engineered for partial oxidation processes.

ADIABATIC REACTORS AT SHORT CONTACT TIMES

The key to these processes is that they can be operated autothermally (reactor ignites because reaction heat maintains reaction temperature without externally supplied process heat) and adiabatically (no heat losses by conduction or radiation) in lab reactors. These are conditions which would be used in industrial scale processes, and therefore such laboratory experiments can be directly translated into industrial conditions.

Only autothermal and adiabatic operation allows contact times much less than one second because any configuration requiring heat addition or removal to the reactor requires a sufficient size to provide heat transfer wall area. For typical heat exchangers, the time mandated is ~1 sec.

Short contact times also provide the possibility to avoid undesired reactions which will be significant at longer times. At sufficiently long times, reactions will go to thermodynamic equilibrium. One use of catalysts is simply to attain this equilibrium composition rapidly. Methane reaction with water or O_2 to form syngas are examples of equilibrium product distributions.

More important are the uses of catalysts to obtain products which are not predicted by equilibrium. According to thermodynamics, HCN and olefins produced by oxidation of alkanes are not predicted to form significantly at these temperatures. Of course, thermodynamic equilibrium allows that a catalyst can promote one reaction over another but can never take a single reaction from one side of equilibrium to the other. The First Law describes the final temperature in an adiabatic reactor from the enthalpies of reactions and their conversions, which for these examples is ~1000°C at typical feed conditions.

Experiments were carried out in quartz tubes with premixed gases flowing over monoliths sandwiched between coarse extruded monoliths acting as radiation shields. These are sealed into the reactor with alumina cloth gaskets for insulation as has been described in detail previously.[16-19] Inlet velocities were typically 1 m/sec, and gases accelerate upon entering the hot monolith by more than a factor of 10 by thermal expansion and the porosity of the monolith so that contact time in a catalyst 1 cm thick was ~1 msec. Over the 1mm thick gauze catalysts, the contact time was $\sim 10^{-4}$.

These lab reactors use approximately 1 cm^3 of catalyst (1.7 cm diameter and 1 cm thick for foam monoliths) and produce typically 10 lb/day of product. At these conditions a 1 ft diameter disk of monolith 1 cm thick would produce more than 1 ton/day of product. At high pressures and higher flow rates a catalyst of this size could produce amounts of products of typical world scale chemical plants.[9]

MONOLITH CATALYST STRUCTURE

Figure 1 shows 6 monolith catalysts used in our laboratories, and Table I lists some of the properties of these supports. The table shows property data consistent with those monoliths used in our laboratories. However in many cases some properties such as channel diameter and porosity could vary depending on the design specifications for a given monolith. In these experiments, for all types of monolith, the catalysts had similar sizes, contact times, surface areas, and porosities.

The catalysts we have used are primarily foams of α-Al_2O_3 which had been formed by impregnating porous organic foams with alumina slurry and firing to burn out the organic. We typically used 45 pores per inch foams which had surface areas much less than 1 m^2/gram. Ceramic fiber monoliths consisting of 10 μm diameter fibers were pressed or woven into a pack several millimeters thick. In some experiments we used extruded ceramic monoliths prepared from cordierite, millite, or α-Al_2O_3, but we found few advantages of extruded supports over foam supports, and conversions and selectivities were generally superior in foam monoliths.

Gauzes were formed from 80 μm diameter wires woven so that the spacing between wires was approximately 100 μm. Typically 5-20 of these gauzes were stacked into a pack which was approximately 1 mm thick. These are available primarily as Pt /10% Rh alloys which are used commercially in HNO_3 and HCN reactors.

Metals were deposited onto foam and fiber monoliths from concentrated metal salt solutions by adding enough of the solution to fill the catalyst to a desired loading. These were then heated and calcined to form metals. Typically a single impregnation produced 1 to 5% loading by weight, and these steps were repeated as necessary to attain the desired loading. This procedure produced a nearly uniform film of metal over the ceramic surface, which for heavy loadings was shiny and electrically conducting.

Figure 1: Typical monolith catalyst supports. Pictured clockwise from top left are 1) woven ceramic fibers, 2) a large channel etched Ni metal monolith, 3) a small channel etched Monel metal monolith, 4) Pt–Rh wire gauzes, 5) a Pt coated ceramic foam monolith (45 ppi), and 6) an extruded cordierite monolith.

After use for some time at 1000°C, the metal film broke up into large crystals of metal as noted by SEM. However for both syngas and olefins the catalyst activity and selectivity were completely insensitive to the microstructure, and performance was also insensitive to metal loading between 1 and 10%.

The surface areas of the catalysts were always much less than 1 m^2/gram. These are not typical supported catalysts but rather metal film catalysts which were deposited on the surfaces of the ceramic. We also examined the effects of adding a wash coat of γ-Al_2O_3 before adding metal and found no significant difference in activity or selectivity, a further argument that surface area is not important.

Table I: Monolith Properties

CATALYST	COMPOSITION	CHANNEL DIAMETER	POROSITY	THERMAL CONDUCT-IVITY (W / m K)	MAX. OPERATING TEMP. (°C)	MASS TRANSFER
Extruded Monolith	Al_2O_3	~600 μm	73–78%	9	2000	Low
	Cordierite				1500	Low
Foam Monolith	99.5 % Al_2O_3	444 μm	80%	8-11	2000	High
	Zirconia					High
Drawn and Etched Metal Monolith	Nickel	740 μm	60%	65-75	1100	Low
	Monel	300 μm	60%	100-120	1100	Moderate
Pt – Rh woven gauzes	Pt–10% Rh	80–100 μm	60–70%	78-81	1600	High
Ceramic Fibers (NEXTEL)	70% Al_2O_3 28% SiO_2 2% B_2O_3	30–50 μm	<60%	<10	1200–1500	High

We note several features of these catalysts which are different from conventional high area supported catalysts.

(1) These catalysts all have very low areas, and the surface area is nearly independent of the metal loading because the metal is either a film or very large particles.

(2) Since these catalysts operate at very high temperatures, only solids with very low volatility are suitable because the catalyst must have sufficiently low vapor pressure to not evaporate during the reaction. However, it is possible to overload the catalysts with metal to compensate for some evaporation because the surface area is not a strong function of the amount of metal. Metals such as Ni, Pd, and Ag were found to evaporate significantly in these experiments, while others such as Re formed volatile compounds.[16]

(3) Many conventional catalyst additives cannot be used because they quickly evaporate.

(4) Many typical impurities such as S or even Fe should not be significant catalyst poisons because they will quickly volatilize. On the other hand, we sometimes noted low selectivities from some catalysts which we associated with possible refractory impurities added in preparation of the foams or fibers.

(5) The morphology of the metals should rapidly approach equilibrium at the high temperatures of these experiments. Thus, crystal planes, surface defects, and alloy surface compositions should rapidly approach equilibrium structures and composition. On the other hand, reactions between metal and support (such as reaction of Ni with α-Al_2O_3 to form aluminates) may occur rapidly and unavoidably.

(6) Adsorption lifetimes are extremely short and reaction rates are extremely fast at these temperatures. We calculate that the adsorption lifetime of adsorbed H and C_2H_5 (heat of adsorption of 25 kcal/mole) to be 10^{-9} sec at 1000°C. Equilibrium coverages of most species are very low at these temperatures, even with gases at atmospheric pressure. Only adsorbed O and C (heats of adsorption greater than 50 kcal/mole) should build up significant coverages. With O_2 fed into the monolith, its coverage should be high in the early region of the monolith, but reactions to form H_2O and CO rapidly deplete adsorbed oxygen to very small coverages.

Thus we believe that these catalysts are considerably different than conventional supported catalysts. These metal surfaces should be nearly clean under these reaction conditions except for possible O and C adsorption. They are thus very simple catalysts whose properties are those of clean metal single crystal experiments. Therefore modeling of these experiments is possible using clean surface properties.[14,22] The choices of suitable catalysts are severely limited to those refractory metals or oxides which are stable under reaction conditions.

ROLE OF CATALYST GEOMETRY

Catalyst properties appear to be insensitive to catalyst area (all O_2 is consumed within approximately 1 mm of the catalyst under all conditions), but the selectivity and conversion of fuel are strong functions of the metal used. Rh yields syngas, Pt yields olefins, and Pd produces carbon which deactivates the catalyst.

However, the geometry of the catalyst appears to play a strong role in determining selectivity. For all reactions except HCN synthesis, the selectivity appears to attain steady state within a few minutes, and no deactivation occurs over many hours. For HCN synthesis, the selectivity to HCN rises from 25% to 70% over approximately 30 hours of operation, and the causes of this activation process will be discussed in a later publication.[23]

The heat transfer characteristics of the catalysts also appear to play a dominant role in performance. High thermal conductivity is very desirable to maintain autothermal operation, because it is essential for the hot surfaces in the early part of the monolith to preheat the reactants or the catalyst cannot be maintained ignited at high flow rates. Monoliths have very high thermal conductivities in spite of their high void fraction because they are continuous structures which do not rely on point contacts as does a pellet bed. Axial radiation in the catalyst is also important at these temperatures, and large channel catalysts should have higher radiative heat transfer than small channel catalysts.

We believe that high mass transfer within the monoliths plays an essential role in achieving high selectivities. While Reynolds numbers (typically 2 to 15 for most experiments) are too low for classical turbulence, the mass transfer rates appear to be much higher than that predicted in tubes or in packed beds. We suggest that surface roughness on a micrometer scale may be important in increasing mass transfer above that normally encountered in pellet beds. The pressure drop is found to increase more than linearly for these velocities, and this is also a characteristic of "turbulent" flow.

It is essential that these catalysts have open channels and that there be no "dead end pores" in the monoliths. For any series process where one desires an intermediate such as

$$C_2H_6 \rightarrow C_2H_4 \rightarrow C \rightarrow CO \tag{3}$$

where ethylene is desired it is essential to have a narrow residence time distribution with no regions where the intermediates may become trapped in the catalyst. Thus we want a catalyst with uniformly dispersed channels for gas flow but no pores in which intermediates can remain for long times because C or CO are the eventual products. We want a catalyst with channels which provide purely convective flow but no pores down which flow is diffusive. Again, this is in strong contrast to conventional catalysts in which most surface area is within the pores of porous pellets down which reactants and products must diffuse after they are convected to the external surfaces of the pellets.

SUMMARY

Low area, high temperature, short contact time reactors have great promise for partial oxidation processes as we have illustrated in three simple processes. The microstructure of the catalyst now plays a quite different role than in slow, high area catalysts where pore diffusion dominates and pressure drop is an important consideration.

Monolith catalysts appear to be essential to achieve these objectives, and their microstructures may have large effects on reactor properties. Characteristics such as channel sizes, size distribution, shapes, and connectedness may be very important. The absence of dead end pores may also be essential. The thermal properties of these monoliths may also be very significant.

Thus the geometry as well as the chemical composition of these catalysts need to be designed for optimum performance in various applications. Different reactions and different metals may present quite different requirements. Also the thermal stability of catalysts and their resistance to thermal stresses in ignition and extinction may need to be considered in detail.

REFERENCES

1. Handforth, S. L., and Tilley, J. N. , Ind. & Eng. Chem. **26** (12), 1287-1292 (1934).
2. Pignet, T., and Schmidt, L. D. , Chem. Eng. Sci. **29** 1123-1131 (1974).
3. Pignet, T., and Schmidt, L. D. , J. Catal. **40** 212-225 (1975).
4. Heck, R. M., Bonacci, J. C., Hatfield, R., and Hsiung, T. H. , Ind. Eng. Chem. Process. Des. Dev. **21** (1), 73-79 (1981).
5. Lee, H. C., and Farrauto, R. J. , Ind. Eng. Chem. Res. **28** 1-5 (1989).
6. Farrauto, R. J., and Lee, H. C. , Ind. Eng. Chem. Res. **29** (7), 1125-1129 (1990).
7. Horner, B. T. , Platinum Metals Rev. **35** (2), 58-64 (1991).
8. Pan, B. Y. K. , J. Catal. **21** (1), 27-38 (1971).
9. Satterfield, C. N. *Heterogeneous Catalysis in Industrial Practice;* 2 ed.; (McGraw-Hill, Inc., New York, 1991), pp. 267-337.
10. Hasenberg, D., and Schmidt, L. D. , J. Catal. **91** (1), 116-131 (1985).
11. Hasenberg, D., and Schmidt, L. D. , J. Catal. **97** (1), 156-168 (1986).
12. Hasenberg, D., and Schmidt, L. D. , J. Catal. **104** (2), 441-453 (1987).

13. Hickman, D. A., Huff, M., and Schmidt, L. D. , Ind. Eng. Chem. Res. **32** 809-817 (1993).

14. Waletzko, N., and Schmidt, L. D. , AIChE J. **34** (7), 1146-1156 (1987).

15. Hickman, D. A., and Schmidt, L. D. , Science **259** 343-346 (1993).

16. Torniainen, P. M., Chu, X., and Schmidt, L. D. , J. Catal. **146** (1), 1-10 (1994).

17. Hickman, D. A., and Schmidt, L. D. , J. Catal. **138** 267-282 (1992).

18. Hickman, D. A., Haupfear, E. A., and Schmidt, L. D. , Catal. Lett. **17** 223-237 (1993).

19. Huff, M., and Schmidt, L. D. , J. Phys. Chem. **97** (45), 11815-11822 (1993).

20. Huff, M., Torniainen, P. M., Hickman, D. A., and Schmidt, L. D. , (1993c).

21. Huff, M., and Schmidt, L. D. , J. Catal. **submitted** (1994).

22. Hickman, D. A., and Schmidt, L. D. , AIChE Journal **39** (7), 1164-1177 (1993).

23. Dietz III, A. G., and Schmidt, L. D. , (to be published).

RETICULATED CERAMICS FOR CATALYST SUPPORT APPLICATIONS

TRUETT B. SWEETING, DAVID A. NORRIS, LAURIE A. STROM AND JEFFREY R. MORRIS, Hi-Tech Ceramics Inc., Alfred, NY 14802

I. INTRODUCTION

The primary use of reticulated ceramics is in molten metal filtration. However, other applications have emerged which utilize the porous nature of these materials, including low mass kiln furniture, low NO_x infrared burners, gas diffusers, hot gas filters, sensors and catalyst supports. In this paper, the use of reticulated ceramics as catalyst supports will be reviewed with emphasis on the manufacture, structure and properties.

II. STRUCTURE AND TEXTURAL PROPERTIES

The process of converting an organic precursor into a ceramic replica has been described by Sherman et al[1] and Brown and Green[2]. Essentially the webs of the organic precursor are coated with a ceramic slurry, dried, then fired to remove the organic and sinter the ceramic. The resulting ceramics can range from 10 to about 30% of theoretical density. On the low end, density is limited by the formation of a structurally sound web, the upper limit results from blockage of the pore channels. Fig. 1 shows the reticulate structure illustrating the web and pore geometries.

The average pore size and pore size distribution are fixed by the precursor and any shrinkage that may take place during the sintering process. It is convention to classify foams by their pores per inch or ppi; however, this should not be taken as a literal interpretation. Pore size is typically measured by optical techniques in which pore diameter is measured directly from photographs. Fig. 2 shows typical pore size distributions for different classifications of 92% alumina foams.

Table I contains texture related properties of 92% alumina foams of various pore sizes. The internal or geometric surface areas shown were reported by Remue[3]. The internal surface area was calculated assuming uniform spherical pores. Surface area and pressure drop increase with a decrease in pore size. Fig. 3 shows internally generated data which shows the effect of flow rate on pressure drop through a one inch thick part over a range of pore sizes.

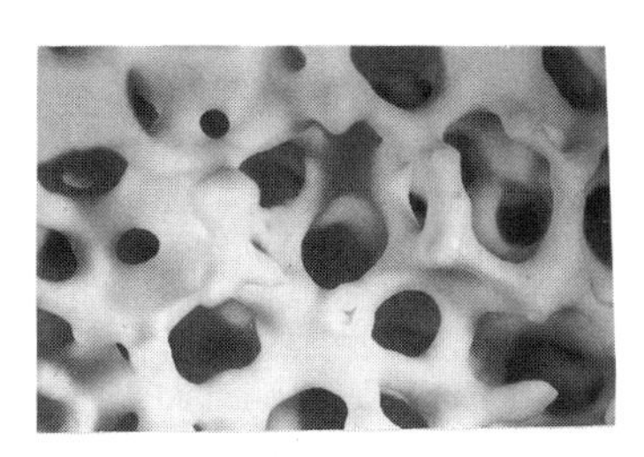

Figure 1
Structure of 10 ppi zirconia foam

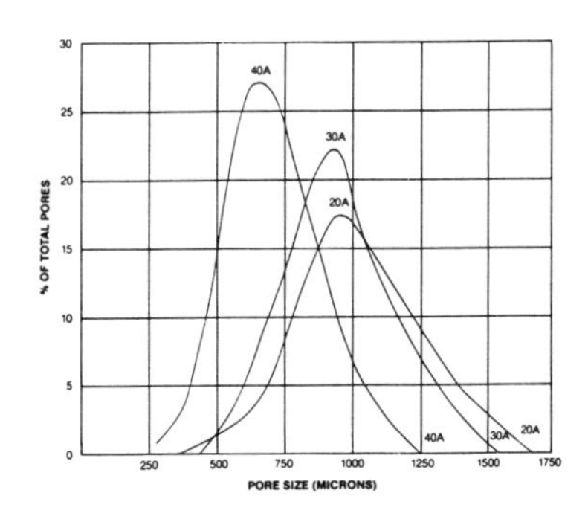

Figure 2
Pore size distribution of 92% alumina foams

Mat. Res. Soc. Symp. Proc. Vol. 368 © 1995 Materials Research Society

TABLE I

PROPERTIES OF CERAMIC FOAM AT VARIOUS PORE SIZES			
PORE SIZE	AVERAGE PORE DIAMETER(m)	CALCULATED SURFACE AREA(m^2/m^3)[1]	PRESSURE DROP* (inches WC)[1]
10	1521	2290	0.90
30	759	4370	2.28
45	420	8000	4.28
65	289	11200	5.1
80	208	16000	13.27

[1] D. Remue, MS Thesis, Univ. of Houston
*Pressure drop for 1"thick sample at 9.8 ft/sec

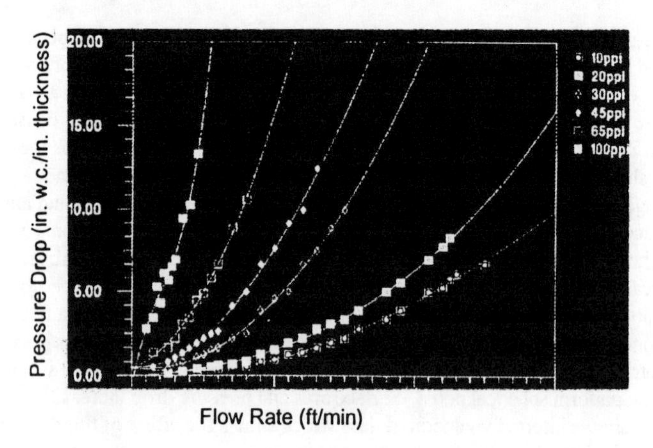

Figure 3
Pressure drop versus flow rate for various pore sizes

III. THERMAL AND MECHANICAL PROPERTIES

The material choice for a catalyst support depends on several factors including chemical and thermal stability in the process environment, compatibility with the catalyst, thermal shock resistance, and strength. Table II compares the properties of several reticulated ceramic materials. The low thermal expansion materials, Cordierite and Lithium Alumino Silicate (LAS), exhibit excellent thermal shock resistance and can be used to about 1300°C. PSZ(Mg) is a partially stabilized zirconia, stable in oxidizing atmospheres and usable to about 1700°C. Two grades of aluminum oxide are available, a 92% alumina and a high purity 99.5% grade. The 92% alumina contains an 8% silica addition which forms mullite during firing, reducing the thermal expansion and improving thermal shock resistance. The aluminas can be used to about 1600°C depending on the application. The OBSiC material is an oxide bonded silicon carbide which consists of SiC grains dispersed in a mullite matrix. The high thermal conductivity of the silicon carbide enhances the heat transfer capability of this material.

Reticulated ceramics can be fabricated from other compositions as well, including mixed oxides such as titinates and non oxides such as sialon. Functionally gradient materials which incorporate catalytically active materials into the surface region are also possible.

The thermal conductivity of reticulated ceramics is dominated by the pore structure. Fig. 4 shows the thermal conductivity of the two pore sizes of alumina foam. At the higher temperatures where radiation heat transfer is significant, the conductivity of the larger pore size material begins to increase rapidly, exceeding that of the fine pore material. In the low temperature regime, the conductivity of the finer pore sizes is higher as a result of the increased paths for heat conduction. It is in this lower temperature region where the contribution from the inherent material conductivity will be more significant. Since the structure of reticulate ceramics is similar in all directions, the axial and radial thermal conductivity will also be similar. For catalytic applications where radial heat transfer is desirable, this characteristic is advantageous.

TABLE II

PROPERTIES OF SELECTED RETICULATED CERAMICS*

	DENSITY (glcm^3)	MOR (psi)	COMPRESSIVE STRENGTH (psi)	THERMAL EXPANSION (10^{-6}/°C)	RELATIVE THERMAL SHOCK
LAS	0.36	190	150	1.2	EXCELLENT
Cordierite	0.45	192	215	2	EXCELLENT
OBSiC	0.60	240	155	5.5	V.GOOD
92 Al_2O_3	0.66	418	301	7.5	V. GOOD
99.5 Al_2O_3	0.64	324	292	8.7	GOOD
PSZ(Mg)	0.84	256	146	7.9	V. GOOD

*All samples 30 ppi

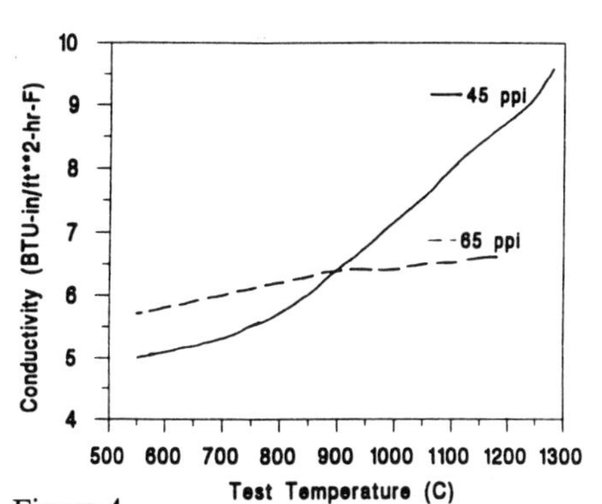

Figure 4
Thermal conductivity of 92% alumina foams

IV. WASHCOATING

In many catalyst applications, washcoats are applied to the surface of the supports to provide additional surface area for the catalyst to attach to. Fig. 5 shows the effect of various levels of washcoat on the surface area of a 20 ppi LAS substrate. The surface areas were determined by single point nitrogen absorption. The washcoat used in this study was "HPA" a blend of calcined hydrated alumina produced by Hi-Tech Ceramics. During a calcination step, the hydrated aluminas are converted into high surface area transition aluminas.

The amount of washcoat that can be successfully applied is limited by the thickness that will adhere without cracking and spalling of the coating. In the case of HPA washcoat, the limit is in the 5-6% range. The uniformity of the washcoat application is essential since it will determine the uniformity of the actual catalyst. Washcoat uniformity is process dependent as well as structure dependent. For a 10 ppi reticulate, uniform coatings can be obtained through a two inch cross section. For 65 ppi, cross sectional uniformity would be limited to about a half an inch.

Fig. 6 shows the HPA washcoat on an alumina reticulate. In this case the washcoat thickness is about 40 microns.

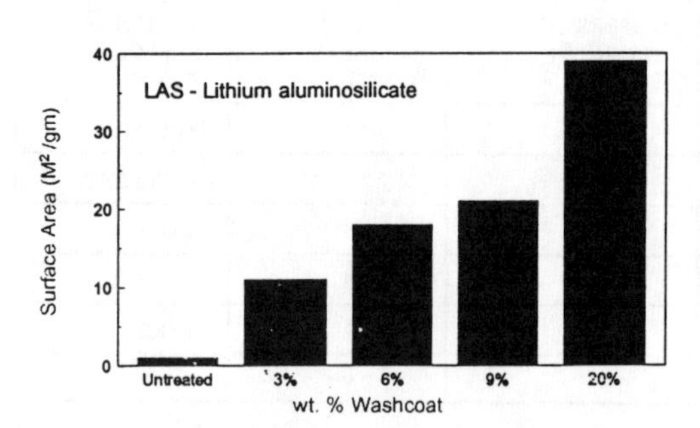

Figure 5
Surface area of 20 ppi LAS foam as a function of washcoat level

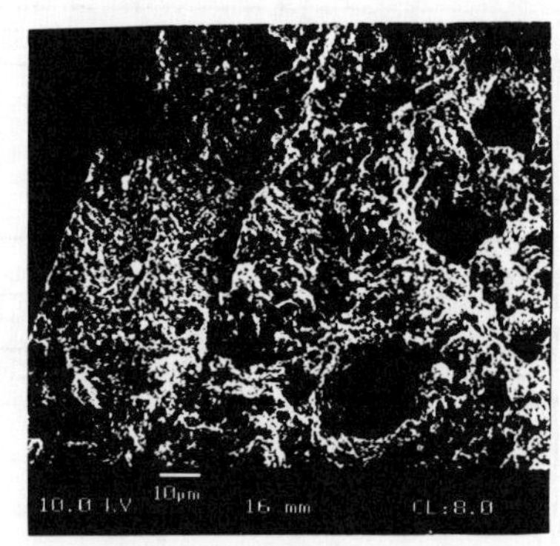

Figure 6
Cross section 30 ppi alumina strut showing HPA washcoat

V. SHAPE CAPABILITIES

The manufacturing process for reticulated ceramics is very flexible resulting in a wide range of shape capability. Table III summarizes the sizes available for various geometries in 45 ppi foam. For other pore sizes, some modifications may be required. For example, in 10 ppi foam, cross section less than 0.5" are not practical. For 80 ppi foam, cross sections are limited to about one inch. Typical tolerances are $^{+}$/- .060", although this value can vary depending on size and geometry.

This shape flexibility allows reticulated ceramics to be used in a variety of reactor designs. Process tubes can be easily loaded with cylindrical segments. Large diameter bed reactors can be fitted with multi-piece planar shapes. The tubular configuration allows for radial flow designs. In general, the cost per volume of reticulated ceramic will decrease as the part size increases. By taking advantage of the size capability, cost effective designs can be generated.

TABLE III

SIZE CAPABILITY FOR 45 PPI RETICULATE

SHAPE		DIMENSIONS (inches) MAXIMUM	MINIMUM
PLANAR	Length (or diameter)	24	0.5
	Width	24	0.5
	Thickness	2	0.125
CYLINDRICAL	OD	3	.25
	Length	18	.25
TUBULAR	OD	12	0.5
	Wall thickness	2	0.125
	Length	12	0.25

VI. APPLICATIONS

The choice of the appropriate catalyst support for an application depends on fluid flow requirements, conversion efficiency, durability and cost effectiveness. The tortuous path through reticulated ceramics provides turbulence within the structure and enhances the probability of interaction between the reactants and the catalyst. The net result is high mass transfer rates leading to high conversion efficiencies in those reactions limited by mass transfer. Fig. 7 shows the percent conversion of methane as a function of temperature for a 30 ppi cordierite foam with a platinum catalyst. For a standard 400 cell/in^2 honeycomb monolith with the same catalyst and tested in the same manner, conversion efficiencies were reported [4] to be in the 10% range. Although the pressure drop through the reticulate was about twice that of the honeycomb, these values need to be compared at equal conversion rates.

Evaluation of ceramic foams as catalyst support is underway in both the academic and industrial communities. On going work by Richardson[5] at the University of Houston is evaluating ceramic foams for steam reforming and other applications. In the case of steam reforming, a simulation comparison of a packed bed to a ceramic foam showed that tube length could be reduced by over a factor of two when using ceramic foam. Schmidt[6] at the University of Minnesota demonstrated foam supports were effective for the direct oxidation of methane to syngas and other short contact time reactions.

Campbell[7] compared the use of catalyzed ceramic foams to platinum gauze for the production of nitric acid by the oxidation of ammonia. His work showed that under controlled test conditions the ceramic foam provided similar or improved performance to the gauze while utilizing a factor of seven less platinum. Buck[8] described the use of ceramic foams in solar methane reforming. Following on the work of Magrini and Webb[9] the photocatalytic decomposition of aqueous organic compounds using anatase TiO_2 supported on reticulated ceramic has been studied by the Solar Energy Research Institute.

In the environmental area, reticulated ceramics are being used for both catalyzed and uncatalyzed reactions to enhance the destruction of volatile organic compounds in incinerators. These materials are also being investigated a catalytic converter substrates for small engines which are targeted for emission reductions in the late 1990's. Other proprietary programs are in progress in major U.S. companies as well as overseas.

Reticulated ceramics are also used in the chemical processing industry to support other catalyst media or as insulation media to retain heat within the reaction layers. The use of the reticulate in this manner also tends to improve flow distribution resulting in improved efficiencies.

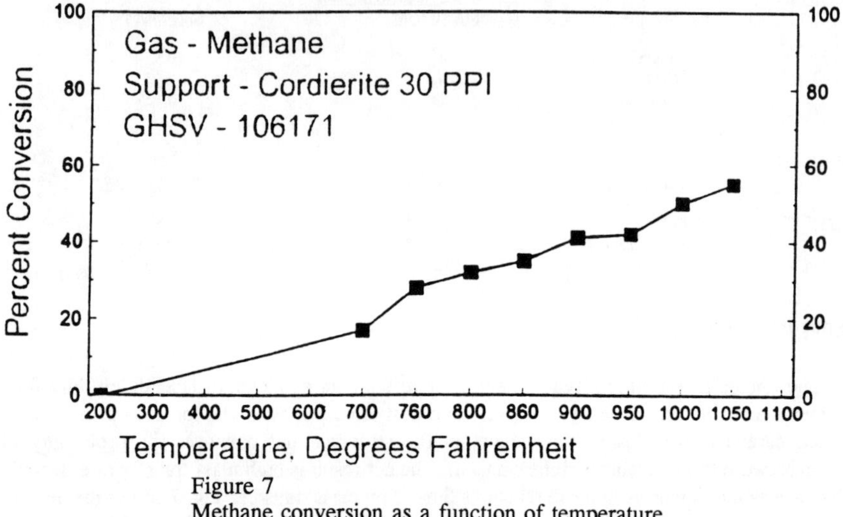

Figure 7
Methane conversion as a function of temperature for platinum catalyzed 30 ppi cordierite foam

VII. SUMMARY

Reticulated ceramics offer an interesting alternative to existing catalyst support media. The tortuosity and high surface area inherent in the structure of these materials results in high activity and conversion rates. Pressure drops are generally higher than honeycomb configurations and lower than packed beds. The range of materials available allows matching of properties to the application requirements. Shape flexibility will enable new design configurations to be developed or optimization of existing designs. As more information is generated in ongoing programs the fit of these materials in catalyst support applications will emerge.

REFERENCES

1. A. J. Sheeman, R. H. Tuffias, and R. B. Kaplan, Ceramic Bulletin, **70**, 1025 (1991).
2. D. Brown and D. Green, J. Amer. Ceram. Soc., **77**, 1467 (1994).
3. D. Remue, M.S. Thesis, University of Houston, (1993).
4. J. Robinson (private communication).
5. J. Richardson (private communication).
6. D. Hickman and L. Schmidt, J. Catal. **138**, 267 (1992).
7. L. Campbell, U.S. Patent No. 5336656 (9 Aug 1994).
8. R. Buck, Mabele, H. Bauer, A. Seitz, R. Tamme, ASME Solar Energy Conference, San Francisco, Ca 1994.
9 K. Magrini, J. Webb, ASME International Solar Energy Conference (1990).

CERAMIC FOAM CATALYST SUPPORTS PREPARATION AND PROPERTIES

JAMES T. RICHARDSON* AND MARTYN V. TWIGG**
*Department of Chemical Engineering, University of Houston, Houston, TX 77204-4792
**Johnson Matthey Catalytic Systems Division, Royston, Herts. SG8 5HE, United Kingdom

INTRODUCTION

Successful catalyst supports must fulfill different requirements simultaneously [1]. By serving as a base for catalytic components, they provide good activity and selectivity, but certain physical properties are equally important. For example, small catalyst particles are desirable in order to avoid external heat and mass transfer limitations and to provide high effectiveness for the active component dispersed throughout the particle. However, small particles pack with low bed porosities and unacceptable pressure drop results. Open-pore ceramic foams show promising properties that could overcome these difficulties. They have good high temperature resistance, low bulk density and tortuous flow patterns, together with open porosity as high as 85% formed from megapores 0.04 to 1.5 mm in diameter [2]. Characteristic parameters include cell size, window size and surface area, all correlated with the number of pores per inch [PPI]. Ceramic foams were first used as molten metal filters [3,4] and catalytic combustion devices [5-7]. Catalytic applications are beginning to appear and these have been reviewed by Twigg and Richardson [8].

Preparation starts with an organic precursor, such as polyurethane, having the same porosity as the desired final product [9-14]. The pores of the organic precursor are filled with an aqueous slurry of the ceramic together with appropriate amounts of wetting agents, dispersion stabilizers and viscosity modifiers. If a low viscosity slurry is used and excess slurry removed by blowing air through the foam, a coating of ceramic is left on the plastic after drying and calcining above 1000°C burns away the organic precursor and sinters the ceramic. The ceramic is a positive image of the plastic. Bulk densities are low, porosities high, but mechanical strength is relatively low. However, by using thixiotropic ceramic slurries with increased viscosity [13,14], a negative image of the plastic is formed. Excess slurry is removed only at the external surface of the structure, and the organic pores remain filled with ceramic, resulting in smaller pores, higher bulk densities, and lower porosities, but mechanical strength is higher. Ceramic foams are preshaped into a wide variety of forms and sizes (e.g. cylinders, rings, rods, or custom-designed configurations), by machining the plastic foam either before or after soaking in the ceramic slurry and drying.

For high surface areas, a washcoat may be added with the same procedures used for monoliths [15], attaining surface areas above 30 m^2g^{-1}. Washcoated foams have been loaded with metals and oxides [14,16], zeolites [15] and carbon [16]. Chemical vapor deposition methods have also been suggested [17,18].

Megaporosity provides a tortuous path for internal gas flow, and turbulence is much enhanced. Tracer experiments have confirmed that radial flow is much higher in foams than in equivalent beds of particles [19]. This results in forced convective flow within the structure, a feature known to produce better performance than conventional pellets that allow only diffusional transport through meso- and macro-pores [20-25]. High porosity also provides much lower pressure drop. Higher thermal conductivity should result from the continuous web-like structure of the foam, thereby providing improved heat transfer into and throughout the foam. Foam-supported catalysts are thus expected to display maximum advantages in reactions that are chemically fast but suffer significant diffusion limitations. Compared to conventional porous catalysts in such situations, selectivity is also likely to be markedly improved with foam catalysts that minimize reactions leading to by-products. These factors have been demonstrated in reported applications that include ammonia oxidation [26], catalytic combustion [27-31], partial oxidation [33-35], steam reforming [14], auto exhaust clean-up [10, 36-46], and solar processes [47-51].

Mat. Res. Soc. Symp. Proc. Vol. 368 © 1995 Materials Research Society

In this paper we expand upon and quantify the most advantageous properties of foam supports: (1) higher activity due to lower effective particle diameters, (2) lower pressure drop, (3) enhanced mass transfer, and (4) better heat transfer. The attractiveness of preformed structures is obvious, since this allows easier reactor loading and enables the catalyst to match the shape of specific devices.

EXPERIMENTAL

Ceramic foam pellets were obtained from Hi Tech Ceramics in the form of cylindrical pellets, 1.27 cm diameter and 2.54 cm length, 10 to 80 PPI, and with a washcoat of 6 wt% hydrated alumina. Foam properties are given in Table I. The most convenient characteristic length for ceramic foams is the average pore diameter, d_c, which is related to the hydraulic radius, r_h, by

$$d_c = 4r_h = 4\epsilon/a_v \qquad (1)$$

with a_v the surface area per unit volume of the foam, ϵ the pellet porosity, and assuming cylindrical geometry for the pores.

Pellets of the 30 PPI foam were soaked in a 0.03M solution of $RhCl_3 \cdot 3H_2O$ for one hour, shaken dry and calcined in hydrogen for 4 hours at 1000°C. The resulting catalyst contained 0.7 wt% Rh, had a total surface area (BET) of 4 m^2g^{-1} and a metal surface area (hydrogen chemisorption) of 7.2 m^2g^{-1}. Other pellets were loaded with Pt using the same procedure, except hexachloroplatinic acid solution was used to give a metal loading of about 9 wt%.

Two types of catalytic measurements were made: (1) carbon dioxide reforming of methane, and (2) oxidation of carbon monoxide. In both cases, the foam pellet was contained in a quartz tube in such a way that gas flow around the ceramic was avoided. Details of the procedures are given elsewhere [52, 53].

Pressure drop across the foam pellet was measured as a function of pore density and superficial velocity using air at ambient temperature [53]. Heat transfer studies were also made by measuring the inlet and outlet gas temperatures in a flow of nitrogen while the reactor temperature was maintained at a fixed value [54].

TABLE I

Typical Properties of Ceramic Foams[a]

92% α-Al_2O_3, 18% mullite

0.06 wt% hydrated alumina washcoat

PPI	PORE DIAMETER	BULK DENSITY	POROSITY	a_v
	mm	g cm^{-3}	%	$cm^2\ cm^{-3}$
10	1.51	0.54	84	22.3
30	0.74	0.70	80	43.1
45	0.42	0.69	80	76.2
65	0.29	0.74	79	108
80	0.21	0.71	79	151

[a] Courtesy Hi-Tech Ceramics, Inc.

RESULTS AND DISCUSSION

Previous studies identified rhodium as an effective catalyst for CH_4-CO_2 reforming without carbon formation [55]. Figure 1 shows a comparison between a Rh-loaded ceramic 30 PPI foam and a conventional 3 mm pellet catalyst, each with 0.5 wt% Rh, and with

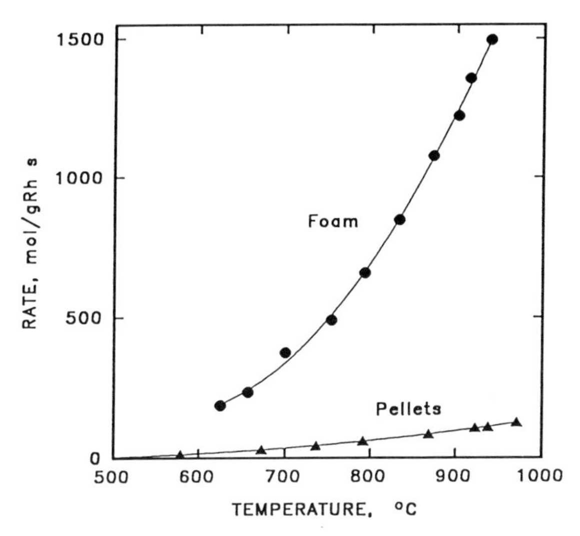

Figure 1. Comparison of CO_2-CH_4 reforming on Rh-loaded ceramic foams and pellets.

dispersions of 7.2% and 11.0% respectively. Figure 1 shows a large difference, for example, at 900°C the foam has a rate of 1250 mol $h^{-1}gRh^{-1}$, whereas the pelleted catalyst is 100 mol $h^{-1}gRh^{-1}$. The most logical explanation for this difference is that the foam has a very high effectiveness factor compared to the pellet, but there could also be an enhancement in rate due to better heat transfer through the ceramic solid structure. These features are still being explored.

Pressure drop-flow rate measurements were made for a large number of ceramic foams with varying pore size [53]. Typical results for the 30 PPI foam pellet are shown in Figure 2. The flow is non-Darcian and agrees with measurementsreported by Philipse and Schram, [56]. The friction factor has an exponential dependence of -0.131 on Re_h, the Reynolds number expressed in terms of the cell diameter d_c. This supports the conclusion the flow is turbulent.

The best fit to the data was found to be the popular Ergun equation [57], expressed in terms of the hydraulic parameter (e.g. d_c)

$$DP/L = [A(1-\epsilon)\nu U/d_c + Bd_g U^2]/\epsilon\, d_c \tag{2}$$

where DP/L is the pressure drop (Pa cm^{-1}), ϵ the porosity, ν the viscosity (poise) , d_g the gas density (g cm^{-3}) , U the gas velocity (cm s^{-1}), and A = 7.719 and B = 0.1183, which are close to the particle values of 6.667 and 0.1167 respectively.

However, particles with the same external surface area as the 30 PPI foam have a diameter of 1.4 mm and pack with a bed porosity of about 0.35, resulting in a much higher pressure drop than the foam. For example, ambient air flowing through a-30 PPI pellet at 650 cm s^{-1} has a measured pressure drop of 1.25 kPa cm^{-1}; the equivalent pellet gives an estimated 10 kPa cm^{-1}, a factor of eight higher.

Ceramic foams with their high tortuosities are expected to display enhanced mass transfer properties, yet no systematic investigation of these parameters has been reported. We adopted the procedure of measuring catalytic conversions under conditions deliberately selected to ensure mass transfer limitations, using carbon monoxide oxidation with platinum catalysts [53]. Tests confirmed the system was operating in an external diffusion-controlled regime at higher temperatures. Rate data were taken at 550°C for increasing velocities, and the mass transfer coefficient, k_c, calculated assuming first order dependence. Fluid properties were used to find the mass transfer factor, j_d, which was then

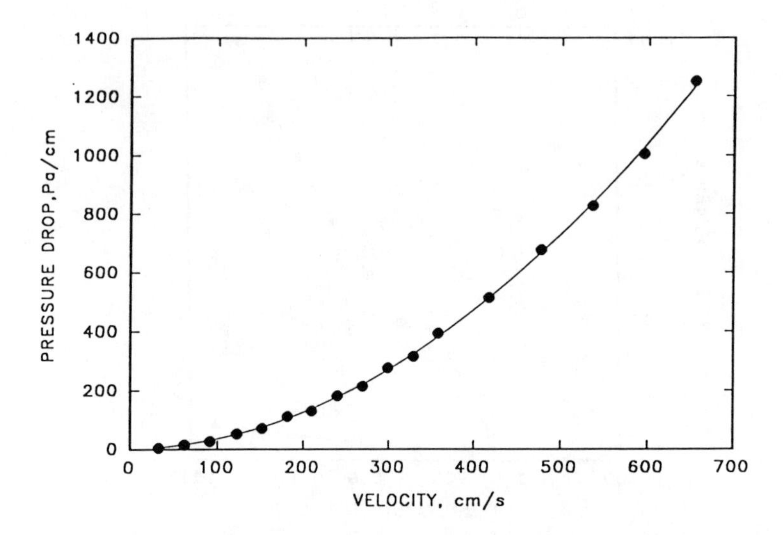

Figure 2. Pressure drop in the 30 PPI ceramic foam pellet.

correlated with the Reynolds number, Re_h. The results (Figure 3) correlated well with

$$\epsilon j_d = 0.326 Re_h^{-0.366} \tag{3}$$

which is within accepted range of precision for the hydraulic equivalent of the accepted Satterfield equation ($0.487 Re_h^{-0.36}$). This agreement indicates standard correlations for mass transfer coefficients are acceptable for ceramic foams. However, again the high surface area-high porosity feature of the foam gives an advantage. Comparisons between mass transfer coefficients under identical conditions show the foam is higher by a factor of 2-3 than equivalent pellets.

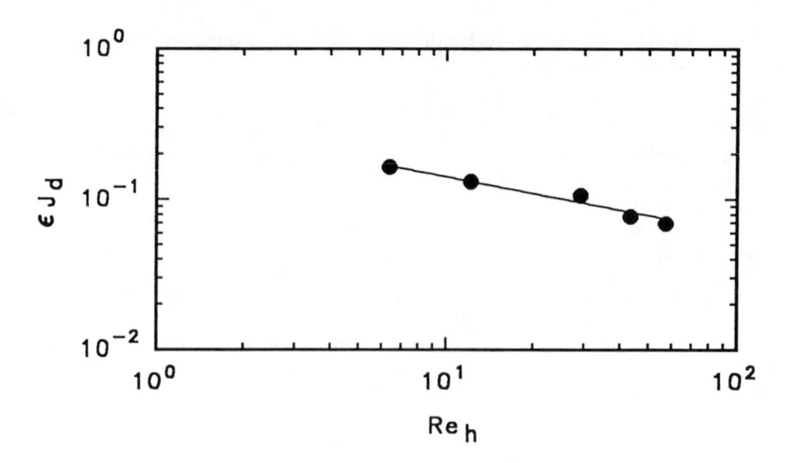

Figure 3. Mass transfer factor correlation for the 30 PPI foam.

Heat transfer into foams is expected to be higher than packed particles because of added conduction through the struts and forced convection into the pores due to their larger size [58, 59]. Experiments [54], in which inlet and outlet temperatures were measured at increasing flow rates, resulted in a correlation for the wall convective heat transfer coefficient, h_{cvw}, of the form

$$h_{cvw} = 0.755 k_g Re_h^{0.062} / d_c \quad (4)$$

where k_g is the thermal conductivity of the gas.

Precise comparisons between packed beds and ceramic foam structures are complex since many parameters are interdependent. We simulated the performance of a conventional steam reformer [60] and compared it to one containing a ceramic foam cartridge loaded to achieve equivalent intrinsic activity per gram of catalyst. The most dramatic result is a decrease in the required length of the reformer tube by about a factor of two, a consequence of the higher effectiveness factor and heat transfer. The higher porosity of the foam and the shorter tube decreased the pressure drop by a factor of ten.

These advantages of lower pressure drop, improved mixing, better heat transfer and prefabrication are ideal for highly exo- and endothermic reactions requiring long narrow reactor tubes and for selectivity control at low contact times with high space velocities.

ACKNOWLEDGMENTS

Research at the University of Houston reported in this paper was supported by Sandia National Laboratories, Albuquerque, N.M., U.S.A. under contract No: 55-4032 and by the Texas Higher Educational Coordinating Board ATP Program, Grant No: 003652121 ATP. We are grateful for the contributions of M. Garrait, D. Remue, and J-K Hung.

REFERENCES

1. J. T. Richardson, *Principles of Catalyst Development*, (Plenum Press, New York, 1989), p. 26.
2. L. J. Gibson and M. F. Ashby, *Cellular Solids, Structures and Properties*, (Pergamon Press, Oxford, 1988).
3. J. W. Brockmeyer and L. S. Aubrey, Ceram. Eng. Sci. Proc. **8**, 63 (1987).
4. P. K. Serville, R. Clift, C. J. Withers, and W. Keidel, Filtr. Sep. **26**, 265 (1989).
5. V. A. Maiorov, L. L. Vasil'ev and V. M. Polyaev, J. Eng. Phys. **47**, 1110 (1984).
6. R. Viskanta, in *Proceedings of the Third ASME/JSME Joint Thermal Engineering Conference*, edited by J. R. Lloyd and Y. Kurosaki (ASME/JSME, New York, 1991) p. 163.
7. F. Anderson, Prog. Energy Combust. Sci. **18**, 12 (1991).
8. M. V. Twigg and J. T. Richardson, presented at the Sixth International Symposium for the Scientific Bases for the Preparation of Catalysts, Louvain-la-Neuve, Belgium, 1994 (to be published).
9. K. Schwartzwalder and A. Somers, U.S. Patent No. 3,090,094 (1963).
10. F. Druche, Ger. Offen. DE 3,510,176 (1986).
11. H. Kondo, H. Yoshida, Y. Takeuchi, S. Nakagawa, JP 62 61,645 (1987).
12. F. F. Lange and K. T. Miller, Adv. Ceram. Mater. **2**, 827 (1987).
13. M. V. Twigg and W. M. Sengelow, U. S. Patent No. 4,810,685 (1989).
14. M. V. Twigg and W. M. Sengelow, U. S. Patent No. 4,863,712 (1989).
15. I. Satoyuki and S. Nonaka, Jpn. Kokai Tokkyo Koho JP 03 123,641 (1991).
16. I. Satoyuki and M. Inoe, Jpn. Kokai Tokkyo Koho JP 03 122,070 (1991).
17. R. A. Clyde, U.S. Patent No. 3,998,758 (1976).
18. R. A. Clyde, U.S. Patent No. 3,900,646 (1975).
19. T. Mizrah, J. P. Gabathuler, L. Gauckler, A. Baiker, L. Padeste, and H. P. Meyer, presented at the First International Symposium and Exposition on Ceramics for Environmental Protection, Koln, Germany, 1988 (unpublished).

20. A. Nir and L. Pismen, Chem. Eng. Sci. **32**, 35 (1977).
21. A. Rodriques, B. Ahn, and A. Zoulanian, J. AIChE **28**, 541 (1982).
22. D. Cresswell, Appl. Catal. **15**, 103 (1985).
23. A. Rodriques and R. Quinta Ferreira, AIChE Symp. Ser. **84**, 80 (1988).
24. A. Rodriques and R. Quinta Ferreira, Chem. Eng. Sci. **45**, 2653 (1990).
25. R. M. Quinta Ferreira, M. M. Marques, M. F. Babo, and A. E. Rodrigues, Chem. Eng. Sci. **47**, 2909 (1992).
26. L. E. Campbell, U.S. Patent No. 5,256,387 (1993); 5,217,939 (1993).
27. M. Haruta, Y. Souma, and H. Sano, J. Hydrogen Energy **7**, 729 (1982).
28. T. Inui, T. Kuroda, and T. Otowa, J. Fuel Soc. Jap. **64**, 270 (1985).
29. T. Inui, Y. Adach, T. Kuroda, M. Hanya and A. Miyamoto, Chem. Express **1**, 255 (1986).
30. K. Mangold, G. Foerster and W. Taetaner, Ger. Offen. DE 3,732,653 (1989).
31. K. Mangold, W. Taetzner, Ger. Offen. DE 3,731,888 (1989).
32. D. A. Hickman and L. D. Schmidt, Science **259**, 343 (1993).
33. M. Huff and L. D. Schmidt, J. Phys. Chem. **97**, 11815 (1993).
34. K. A. Vonkeman and L. V. Jacobs, Eur. Pat. Appl. EP 576,096 (1993).
35. P. M. Torniainen, X. Chu and L. D. Schmidt, J. Catal. **146**, 1 (1994).
36. G. Weldenbach, K. H. Koepernik and H. Brautigam, U.S. Patent No. 4,088,607 (1978).
37. T. Narumiya and S. Izuhara, U.S. Patent No. 4,308,233 (1981).
38. H. Hondo, H. Yoshida, Y. Miura, Y. Takeuchi and S. Nagagawa, JP 63 883,049 (1988).
39. A. Muramatsu and K. Yoshida, Jpn. Kokai Tokkyo Koho JP 04 04,237 (1992).
40. K.Tabata, I. Matsumoto, T. Matsumoto, J. Fukuda, Jpn. Kokai Tokkyo Koho JP 04 04,019 (1992).
41. Y. Watabe, K. Irako, T. Miyajima, T. Yoshimoto and Y. Murakami, SAE Technical Paper 830082 (1983).
42. J. J. Tutko, S. S. Lestz, J. W. Brokmeyer and J. E. Dore, SAE Technical Paper 840073 (1984).
43. T. Inui and T. Otowa, Appl. Catal. **14**, 83 (1985).
44. M. Kawabata, S. Matsumoto, K. Kito, H., Yoshida, JP 01 143,645 (1989).
45. T. Mizrah, A. Maurer, L Gauchler and J-P Gabathuler, SAE Technical Paper 890172 (1989).
46. M. Nitsuta and M. Ito, Jpn. Kokai Tokkyo Koho JP 02 173,310 (1990).
47. R. E. Hogan, Jr., R. D. Skocypec, R. B. Diver, J. D. Fish, M. Garrait, and J. T. Richardson, Chem. Eng. Sci. **45**, 2751 (1990).
48. R. Buck, J. F. Muir, R. E. Hogan, Jr., and R. D. Skocypec, *Solar Energy Materials, Proceedings of the 5th Symposium on Solar High-Temperature Technologies, Davos, Switzerland, August 1990* **24**, 449 (1991).
49. R. E. Hogan, Jr., and R. D. Skocypec, J. Solar Eng. Eng. **114**, 106 (1992).
50. R. D. Skocypec and R. E. Hogan, J. Solar Eng. Eng. **114**, 112 (1992).
51. J. F. Muir, R. E. Hogan, Jr., R. D. Skocypec and R. Buck, The CAESAR project, Sandia Report SAND92-2131 (1993).
52. M. Garrait, *A Ceramic Matrix Catalyst for Solar Reforming* M.S.ChE. Thesis, Department of Chemical Engineering, University of Houston, (1989).
53. D. Remue, *Properties of Ceramic Foam Catalyst Supports* M.S.ChE. Thesis, Department of Chemical Engineering, University of Houston, (1993).
54. J-K Hung (private communication).
55. J. T. Richardson and S. A. Paripatyadar, Appl. Catal. **61**, 293 (1990).
56. A. P. Philipse and H. L. Schram, J. Am. Ceram. Soc. **74**, 728 (1991).
57. S. Ergun, Chem. Eng. Prog. **48**(2), 89 (1952).
58. L. B. Younis and R. Viskanta, Int. J. Heat Mass Transfer **6**, 1425 (1993).
59. W. H. Meng, C. McCordic, J. P. Gore and K. E Herold, ASME/JSME Thermal Engineering Proceedings **5**, 181 (1991).
60. J. T. Richardson, S. A. Paripatyadar, and J. C. Shen, AICHE J. **34**, 743 (1988).

PREPARATION OF $La_{1-x}Sr_xMnO_3$ PEROVSKITE CATALYSTS SUPPORTED ON CERAMIC FOAM MATERIALS

Z.R.ISMAGILOV, O.Yu.PODYACHEVA, A.A.KETOV, A.BOS* AND H.J.VERINGA*
Institute of Catalysis, 630090, prosp. Ak.Lavrentieva, 5, Novosibirsk, Russia
*ECN, P.O.BOX 1, 1755 ZG Petten, Netherlands

ABSTRACT

The new method of preparation of $La_{1-x}Sr_xMnO_3$ catalysts supported on ceramic foam materials was developed. The synthesized supported perovskite catalysts were examined by X-ray diffraction analysis and BET methods. It was shown that in coated samples there is a pure perovskite phase on the support surface up to temperatures of 1000°C.

Unsupported $La_{1-x}Sr_xMnO_3$ catalysts were synthesized applying co-precipitation technique using La, Sr and Mn acetylacetonates and were studied by X-ray diffraction analysis. Temperature programmed reduction method revealed that catalysts contain 2 types of active centers: low temperature (α) and high temperature (β), the quantity of these centers and correlation between them depend on the value of x. It was shown that supported perovskite catalysts, similarly to the massive ones, maintain two types of active centers (α and β).

The activity of $La_{1-x}Sr_xMnO_3$ on mullite foam was characterized in the methane oxidation reaction.

INTRODUCTION

The catalytic combustion of organic fuels has the advantage that complete combustion without formation of thermal NO_x is possible [1]. A new approach for exhaust gas treatment of diesel engines is required . It is found that mixed oxides such as perovskites are very attractive catalysts at high temperatures [2,3]. Supported perovskite oxides are also very promising as catalysts in the processes of environmentally clean energy production.

For the application of such catalytic systems at high temperature, it is necessary to apply these oxide catalysts as an active layer on a suitable support material having high thermal stability. Both ceramic and metal substrates can be suitable for this purpose. Supports, which exhibit a low pressure drop and high thermal stability, are the ceramic monoliths and ceramic foams.

The present work is devoted to the preparation of $La_{1-x}Sr_xMnO_3$ catalysts supported on ceramic foam materials.

EXPERIMENTAL

Preparation of supported perovskite catalysts

Mullite foams were used as supports in this study. These materials were synthesized using procedures described for example in [4]. For the preparation of supported perovskite catalysts

Mat. Res. Soc. Symp. Proc. Vol. 368 © 1995 Materials Research Society

the technique of «polymer coating» was developed. The ceramic foam was covered by a thin epoxy film by dipping a sample in a solution of epoxy resin in acetone. Then the powder of the synthesized perovskite powder was passed through the support. The samples obtained were dried at 110° C for 1 hour and calcined at 800°C for 6 hours in air. The quantity of the active component in the synthesized supported catalysts ranged from 30 to 40 wt.%.

Preparation of the unsupported perovskite catalysts

Perovskite-type oxides were prepared by co-precipitation of La, Sr and Mn acetylacetonates. A stoichiometric amount of acetylacetone was added to the solutions of La, Sr and Mn nitrates. The complexes obtained were precipitated by addition of the necessary amount of ammonium. The precipitates were washed and calcined at 800° C for 6 hours. A series of perovskite samples $La_{1-x}Sr_{x}MnO_3$ with x value within 0.2-0.8 were synthesized. The perovskites with x=0 and x=1 were prepared using the citrate method. The temperature of calcination of $LaMnO_3$ was 850° C, whereas for $SrMnO_3$ - 1200° C.

Characterization of catalysts

The phases present in the products were identified by powder X-ray diffraction analysis (XRD). Specific surface area was measured by the BET method.

Thermoprogrammed reduction

Both supported and unsupported perovskite catalysts were examined by thermoprogrammed reduction (TPR).

The catalysts (M=15-75 mg, fraction 0.3-0.6 mm) were treated in a He flow at 200° C for 1 hour. Then the samples were cooled in a He flow to room temperature and TPR was performed within the temperature interval of 30-1000° C. The reduction mixture was 5% H_2 in Ar, the flow rate - 50 ml/min, the heating rate - 10° /min.

Measurement of catalytic activity in the methane oxidation reaction

The reaction was carried out in a conventional flow reactor by feeding a gas mixture of 2% CH_4 in air over the catalyst bed packed with 0.6 g of catalyst (fraction 0.3-0.6 mm) at a flow rate 71 ml/min. Before the experiments the catalysts were treated in air at 700° C for 1 hour.

RESULTS AND DISCUSSION

A number of perovskite samples $La_{1-x}Sr_{x}MnO_3$ with different x value (x=0.2 - 0.8) were prepared. The XRD patterns of the synthesized perovskites are given in Fig.1.

Till x=0.6 a single perovskite phase based on the $LaMnO_3$ structure exists. Starting from x=0.6 a complex mixture of phases appears, one can see the main phase based on $LaMnO_3$ structure, certain amount of perovskite phase based on $SrMnO_3$ and on traces of the $SrCO_3$ phase.

The specific surface area of the unsupported perovskite catalysts varied within the interval of

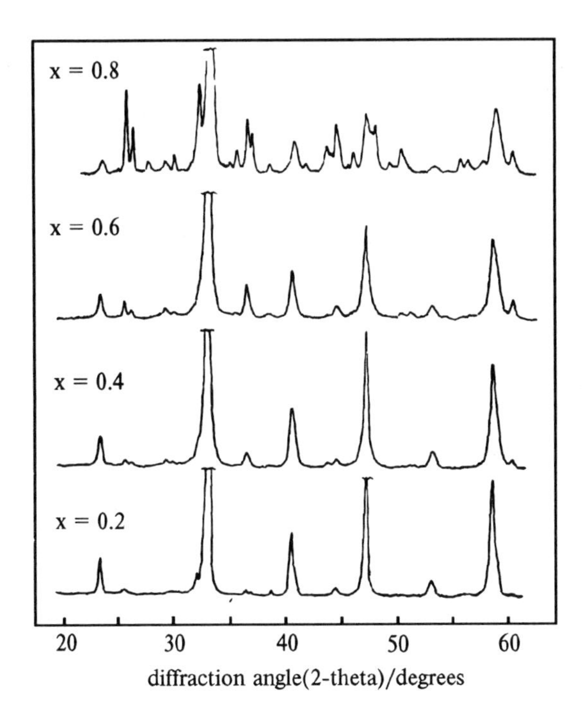

Fig. 1. XRD patterns of $La_{1-x}Sr_xMnO_3$

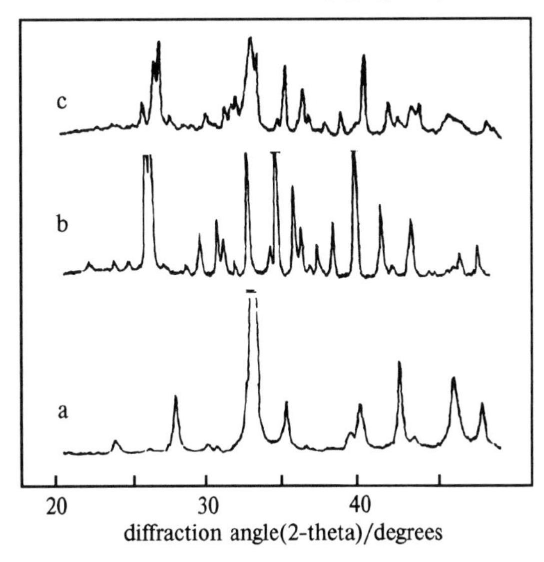

Fig.2. XRD patterns of: a) $La_{0.25}Sr_{0.75}MnO_3$; b) mullite; c) $La_{0.25}Sr_{0.75}MnO_3$ on mullite

2.5-14.1 m^2/g.

Using the technique of «polymer coating» a series of supported perovskite catalysts was synthesized. The $La_{1-x}Sr_xMnO_3$ powders with x=0.2; 0.25; 0.4; 0.6; 0.75; 0.8 were applied on mullite foams.

The synthesized supported perovskite catalysts were examined by X-ray diffraction analysis. The typical XRD pattern of the prepared catalysts (with $La_{0.25}Sr_{0.75}MnO_3$ on mullite foam taken as an example) is given in Fig.2. This diffractogram shows that only the perovskite phase is detected. The content of other phases that can be formed during the coating procedure is below detection limit. Analogous diffractograms were obtained for all synthesized supported catalysts.

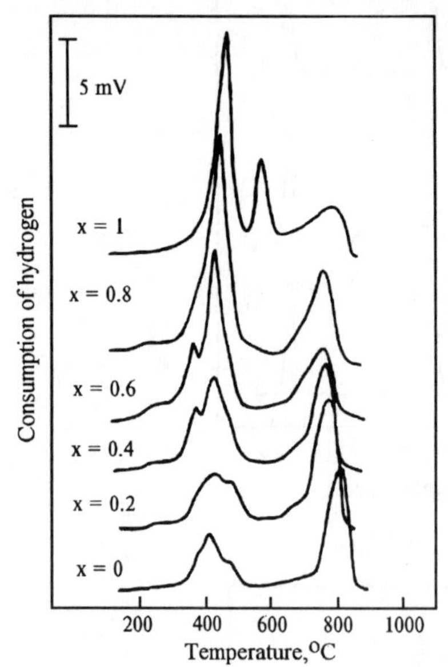

Fig. 3. TPR spectra of $La_{1-x}Sr_xMnO_3$

The thermal stability of the system under study was examined. The perovskite catalyst $La_{0.75}Sr_{0.25}MnO_3$ on mullite foam was calcined at 700, 800, 900, 1000° C for 10 hours and XRD patterns of the samples were obtained. The changes in the diffractogramms were also below the detection limit. Therefore, it can be concluded that the interaction between an active component and ceramic support does not alter the active component up to a temperature of 1000°C.

The value of the specific surface area of the supported perovskite catalysts was within the interval of 1-3 m^2/g.

TPR is used in catalysis to characterize the active centers of catalysts. The main aim of these experiments was to find the similarities and differences of the supported and unsupported perovskite catalysts.

The TPR spectra of unsupported $La_{1-x}Sr_xMnO_3$ (x=0-1) are given in Fig.3. The spectra consist of two peaks: a low temperature one (300-550°C) and a high temperature one (600-900°C) with maximum temperatures T_{max}= 415-460°C and T_{max}=760-810°C. Considering numerous literature data on thermoprogrammed desorption of oxygen from perovskites, it can be concluded that the low temperature form of absorption (α) correlates with weakly bonded, surface oxygen, whereas the high temperature form of absorption (β) represents strongly bonded (possibly lattice) oxygen. Correlation between the α and β forms of absorption depends on the value of x. The β form has a maximum value when x=0, whereas the α form has a maximum when x=0.8.

The value of x in $La_{1-x}Sr_xMnO_3$ influences the reducibility of perovskite catalysts. The higher x, the easier the reduction. The probable explanation is as follows: the substitution of La by Sr in $La_{1-x}Sr_xMnO_3$ leads to oxidation of the Mn cation (Mn^{3+} ->Mn^{4+}), in this case the substituted oxide becomes easier to reduce than $LaMnO_3$. A similar result was observed for the $LaCoO_3$ system [5].

The TPR spectra of $La_{1-x}Sr_xMnO_3$ supported on mullite foam are given in Fig.4. It is obvious that the supported catalysts contain analogous types of active centers compared with the unsupported ones; the TPR spectra consist of a low temperature (α) and high temperature (β)

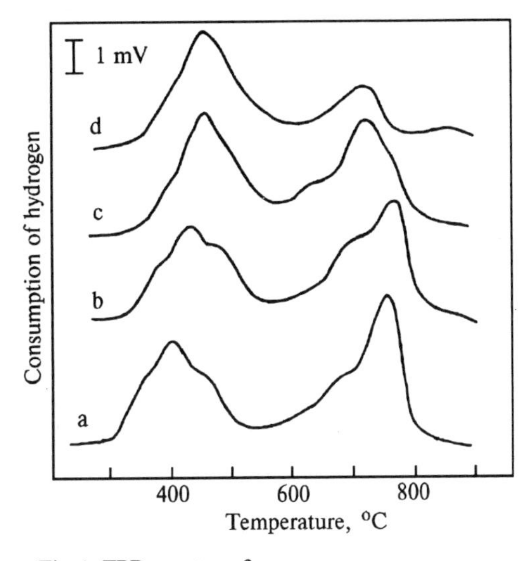

Fig.4. TPR spectra of :
a) $La_{0.8}Sr_{0.2}MnO_3$ on mullite
b) $La_{0.6}Sr_{0.4}MnO_3$ on mullite
c) $La_{0.4}Sr_{0.6}MnO_3$ on mullite
d) $La_{0.2}Sr_{0.8}MnO_3$ on mullite

form. Correlation between α and β forms depends on the value of x . Similarly to the unsupported perovskites, the α form has a maximum for x=0.8 (Fig.4).

The catalytic activity of supported perovskite catalysts was measured using the methane oxidation reaction.

The temperature dependencies of methane oxidation on $La_{1-x}Sr_xMnO_3$ on mullite foam are given in Fig.5. From this figure we notice that under the conditions studied there is no essential difference in the level of activity of supported perovskite catalysts with different x values.In the kinetic controlled region (low temperature) the activity of the supported perovskite catalysts is insufficient to exhibit the variations due to x. In the diffusion controlled region the reaction rate is limited by mass transfer to the surface of the catalyst, therefore the chemical composition of the catalyst does not significantly influence the reaction speed.

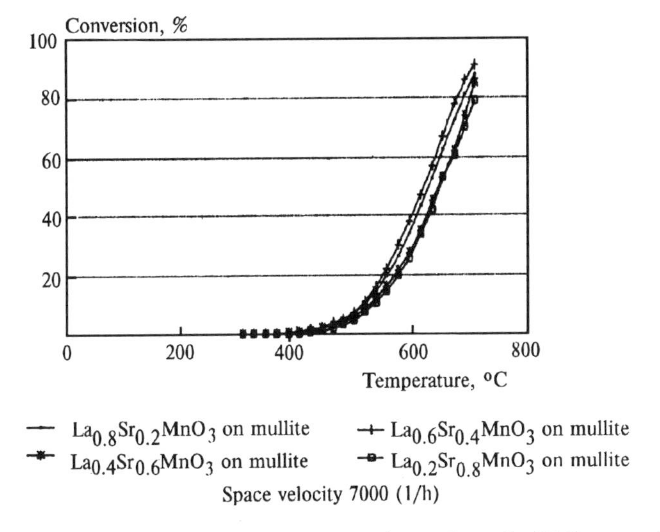

Fig. 5. Temperature dependence of methane conversion on $La_{1-x}Sr_xMnO_3$ on mullite

The catalytic properties of unsupported $La_{1-x}Sr_xMnO_3$ catalysts are well known [2, 6]. Our TPR results and measurements of the activity of supported $La_{1-x}Sr_xMnO_3$ catalysts show that the coating procedure developed here can be used for the synthesis of perovskite catalysts supported on a ceramic foam. The catalysts prepared combine the high catalytic activity of unsupported $La_{1-x}Sr_xMnO_3$ catalysts and unique properties of the support foam structure.

CONCLUSIONS

A new method for preparation of perovskite catalysts supported on ceramic foam materials was developed, called the technique of polymer coating. Synthesized perovskite powders were applied to ceramic mullite foams using this procedure. By XRD analysis of $La_{1-x}Sr_xMnO_3$ on ceramic foam only a pure perovskite phase is detected up to a temperature of 1000°C.

The technique of co-precipitation of La, Sr and Mn acetylacetonates was applied to prepare perovskite powders with relatively high specific area. The results obtained show that the coating procedure developed here could be used for the synthesis of supported perovskite catalysts.

ACKNOWLEDGMENT

The research was subsidized by the grant of Dutch Organisation for Scientific Research NWO in a scientific programme between the Netherlands and the Russian Federation.

The authors thank Dr. George Barannik for helpful discussions and suggestions in the preparation of this article.

REFERENCES

1. Z.R. Ismagilov and M.A. Kerzhentsev, Catal. Rev.- Sci. Eng., **32**, Nos 1&2, 51 (1990).
2. H. Arai, T. Yamada, K. Eguchi and T. Seiyama, Appl. Catal., **26**, 256 (1986).
3. L. Tejuca, J. Fierro, and J. Tascon, Adv. in Catalysis, **36**, 237 (1989).
4. GB Patent No 916784, C08J, 30.01.1963.
5. J. O. Petunchi and E. A. Lombardo, Catal. Today, **8**, 67 (1990).
6. J. G. McCarty and H. Wise, Catal. Today, **8**, 96 (1990).

PART VII

Porous Materials

SOL-GEL STRATEGIES FOR CONTROLLED POROSITY CERAMIC MATERIALS: Thin Film and Bulk

C. JEFFREY BRINKER*,**, RAKESH SEHGAL**, NARAYAN K. RAMAN**, SAI S. PRAKASH**, and LAURENT DELATTRE**
*Sandia National Laboratories, Advanced Materials Lab, 1001 University Blvd. SE Albuquerque, NM 87106
**The UNM/NSF Center for Micro-Engineered Ceramics, The University of New Mexico, Albuquerque, NM 87131

ABSTRACT

Using sol-gel processing techniques it is possible to vary the condensation pathway over wide ranges to form primary species ranging in structure from oligomers to polymers to particles. The porosity of the corresponding dry gels depends on the size and structure of the primary species, the organization of these structures, often by aggregation, to form a gel, and the collapse of the gel by drying. This paper reviews these ideas in the context of forming thin film or bulk specimens. Several strategies are introduced to control porosity on length scales of interest for catalysis and catalytic membrane reactors: 1) aggregation of fractals; 2) management of capillary pressure; 3) surface derivatization; 4) relative rates of condensation and evaporation; 5) the use of organic templates and 6) sintering. These strategies are contrasted with the more traditional particle packing approach to preparing controlled porosity materials.

INTRODUCTION

In the sol-gel process, colloidal dispersions of oligomers, polymers, or particles *sols* are transformed to liquid-filled solids *gels* and dried by evaporation to form *xerogels* or by supercritical fluid extraction to form *aerogels* [1]. The utility of sol-gel processed materials in the field of catalysis is that single and multi-component ceramics can be prepared in both thin film and bulk form with excellent control of microstructure, viz. surface area, pore volume, pore size and pore size distribution. In addition it is possible to modify or *derivatize* the pore surfaces by liquid or vapor phase techniques to "custom-tailor" pore size, pore surface chemistry, and catalytic activity for specific applications. This paper first contrasts the processing of bulk and thin film specimens. Then a brief review of several strategies to control dry gel microstructure is presented.

BULK VERSUS THIN FILM PROCESSING

Bulk specimens used for catalyst supports and adsorbents are typically prepared by allowing a rather concentrated sol to gel (through the formation of covalent bonds or electrostatic interactions) followed by aging (a process of strengthening through further condensation reactions and ripening) and evaporation of the pore fluid to produce a xerogel. During the drying stage, capillary tension P is developed in the pore fluid by the creation of liquid-vapor menisci. The magnitude of P is given by the Laplace equation [2]:

$$P = -2\gamma\cos(\theta)/r_p \tag{1}$$

where γ is the pore fluid/vapor surface tension, θ is the contact angle, and r_p is approximately the pore radius. The tension developed in the liquid is transferred to the solid network causing it to shrink. Shrinkage is resisted by the bulk modulus of the network K_p which increases with shrinkage or relative density as a power law [3]:

$$K_p = K_o(V_o/V)^m \tag{2}$$

where K_o is the bulk modulus of the initial gel, V_o is the initial gel volume, V is the shrunken volume, and m is an exponent which has been found experimentally to range between 2.5 and 4. Shrinkage stops at the *critical point* when the maximum capillary tension developed in the pore fluid

Mat. Res. Soc. Symp. Proc. Vol. 368 © 1995 Materials Research Society

P_c is balanced by the compressive stress developed in the network. At this point the volumetric strain ε_v attributable to drying is [4]:

$$\varepsilon_v = \frac{\sigma_y}{K_p} = \left(\frac{1 - \phi_s}{K_p}\right)\left(\frac{-\gamma\cos(\theta)}{r_p}\right) \tag{3}$$

where σ_y is the stress on the solid network on a face normal to the y direction and $\phi_s$is the volume fraction solids. Continued removal of solvent beyond the critical point normally occurs with no further change in volume, thus it is the extent of shrinkage preceeding the critical point that establishes the final pore volume, average pore size, and surface area.

Thin films are normally prepared by depositing a thin layer (1-10 μm-thick) of a more dilute sol on a substrate by dip- or spin-coating. This layer quickly thins by gravitational or centrifugal draining and evaporation, which also serves to concentrate the sol. Depending on the relative rates of covalent bond formation (condensation rate) or aggregation and pore fluid evaporation, gelation may or may not precede complete drying. In the former case, the gel film, once created, shrinks one-dimensionally in the direction z normal to the substrate surface. Shrinkage stops when the network stress is balanced by the capillary pressure as described above for bulk specimens. In the latter case we expect that a physical gel is formed in the vicinity of the drying line due to the strong dependence of viscosity K_G on drying shrinkage, for example [5]:

$$K_G = K_{Go}(V_o/V)^3 \tag{4}$$

where K_{Go} is the initial viscosity of the gel. Shrinkage stops when the capillary pressure is balanced by the viscous resistance. The total shrinkage that occurs by the critical point is calculated in a manner similar to Eq. 3 but with replacement of volumetric strain with strain rate and bulk modulus with network viscosity and allowing the properties to evolve over a characteristic time.

Scherer has shown that the stress distribution in a viscous gel-film is given as [6]:

$$P = -P_c = P_s + \left(\frac{K_G}{\beta}\right)\left(\frac{V_E}{h}\right)\left(\frac{\alpha \cosh(\alpha z/h)}{\sinh(\alpha)}\right) \tag{5}$$

where P_s is the syneresis stress in the liquid, V_E is the evaporation rate, h is the film thickness, z is the dimension along the film thickness,

the parameter $\alpha = \sqrt{\frac{\eta_L h^2 \beta}{D K_G}}$, the parameter $\beta = \frac{1+N}{3(1-N)}$,

N is Poisson's ratio, D is permeability, and η_L is the viscosity of the pore fluid. Assuming the stress due to syneresis to be constant (drying-related stresses are much larger), we have:

$$\frac{2\beta\,\gamma_{LV}\cos(\theta)}{r_{p\infty}} = \frac{K_{G\infty}V_{E\infty}\alpha_\infty \coth(\alpha_\infty)}{h_\infty} - \frac{K_{Go}V_{Eo}\alpha_0 \coth(\alpha_0)}{h_0} \tag{6}$$

The subscripts 0, ∞ denote values at gelation (initial) and at the critical point (final), respectively. Figure 1a shows the evolution of the network stress P and viscosity K_G as a function of time based on initial conditions listed in Table 1. The rising values of network stress and viscosity establish the pore size at the critical point Fig. 1b and the final values of film thickness, bulk viscosity, and permeability (Table 1).

Table 1. Model prediction of film microstructure during drying

Initial Values (constants)	
Porosity	80 %
Film Thickness	1 μm
Bulk Viscosity	0.1 GPa.s
Surface Area (constant)	600 m^2/g
Dip-coating Speed	20 cm/min.

Final values predicted by model	
Porosity	9.3 %
Film Thickness	2205 Å
Bulk Viscosity	9.3 GPa.s
Pore Diameter	3.46 Å
Liquid Permeability	1.36 $Å^2$
Shrinkage Time	~ 12 s

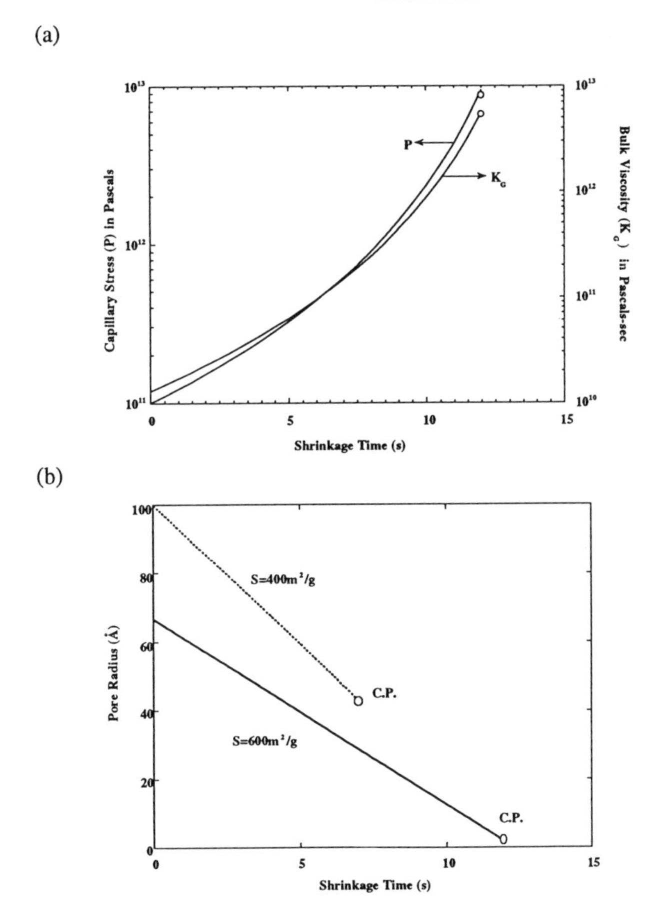

Figure 1. (a) Evolution of capillary stress and bulk viscosity with shrinkage time. Initial gel surface area = 600 m^2 /g. (b) Evolution of pore radius with shrinkage time. C.P. = critical point where shrinkage stops. S = initial gel surface area.

The primary processing feature that distinguishes thin film formation from bulk xerogel formation is the overlap of the aggregation, gelation, and drying stages. Since drying commences before gelation, thin film gels are likely to be more concentrated than their bulk counterparts, causing their initial pore size to be smaller. In addition, this overlap along with the inherent thinness of films has the effect of reducing the characteristic time scale of the process (seconds versus hours or days). So we expect films to be less highly condensed at the gel point and to undergo less condensation during drying. This causes the initial value of the modulus to be lower and may serve to reduce the value of the exponent m. Based on Eq. 3 (and related analyses for films), these combined effects cause gel films to experience greater drying shrinkage and xerogel films to exhibit smaller average pore size than corresponding bulk specimens.

STRATEGIES TO CONTROL MICROSTRUCTURE

Particle Packing

Virtually all particle packing concepts utilize particles that are packed together to create pores of a size related to the primary particle size [7]. Ideally if monosized particles could be assembled into a colloidal crystalline lattice, it would be possible to mimic the monodispersity of zeolite channel systems (but over a wider range of pore size). Unfortunately it is not yet possible to avoid some distribution of particle sizes, so attempts to form spatially extensive colloidal crystalline bulk or thin film specimens have been thwarted. The current state of the art is to prepare particles with quite narrow particle size distributions that are more or less randomly close-packed without aggregation (which would create a second class of larger pores). The advantage of this approach is that porosity is independent of the particle size. For example, random dense packing of monosized spherical particles always results in about 33% porosity. Particle packing is the basis of commercial γ-Al_2O_3 Knudsen separation membranes currently supplied by U.S. Filter and Golden Technologies with nominal pore diameters of 4.0 nm. In order to arrive at smaller pore sizes appropriate, for example, in gas separation applications, it is necessary to prepare smaller particles. Although the synthesis of appropriately small particles has been demonstrated [8], they have proven to be difficult to process into supported membranes due primarily to problems with cracking. Avoidance of cracking, which is essential for membrane performance, may be a fundamental limitation of the "particle approach" to the preparation of gas separation membranes. The problem is that the thickness of the electrostatic double layer erected around each particle to avoid aggregation does not decrease proportionally with particle size. Thus as the particle size is diminished, the tightly bound solvent layer comprises an ever increasing volume fraction f_s of the depositing film at the instant it gels. The removal of this liquid during subsequent drying creates a tensile stress (σ) within the plane of the membrane that results in cracking [9]:

$$\sigma = [E/(1 - \nu)][(f_s - f_r)/3] , \qquad (7)$$

where E is Young's modulus (Pa), ν is Poisson's ratio, f_s is the volume fraction solvent at the gel point, and f_r is the residual solvent (if any) in the fully dried film. From Eq. 12 we see that σ is directly proportional to f_s. The result is that cracking is more likely to occur in particulate membranes as the particle size is reduced [10].

Aggregation of Fractals

A strategy that is generally applicable to the wide range of polymeric sols characterized by a mass fractal dimension is aggregation. Although aggregation is generally avoided in the particulate approach, it may be exploited to control the porosity of films (and probably bulk specimens) prepared from polymeric sols. This strategy relies on the scaling relationship of size r_c and mass M of a fractal cluster:

$$M \propto r_c^D \qquad (8)$$

causing the porosity Π_f of a mass fractal cluster to increase with cluster size as:

$$\Pi_f \propto r_c^{(d-D)} \tag{9}$$

where the mass fractal dimension D is less than the embedding dimension d (for our purposes $d = 3$). When the individual fractal clusters comprising the dilute sol are concentrated by evaporation during dip- or spin-coating, this porosity is incorporated in the resulting film or membrane provided that: 1) the clusters do not completely interpenetrate and 2) there exists no monomer or small oligomeric species that are able to "fill-in" the gaps of the fractal clusters [11]. In practice the first criterion appears satisfied for D greater than about 2. Under these conditions the probability of cluster intersection is high (and that of cluster interpenetration low) during sol concentration. Figure 2 [12] shows the volume fraction porosity of a series of films prepared from multicomponent silica sols under conditions where D equaled 2.4 and there existed no detectable monomers or dimers (as determined by ^{29}Si NMR). We see that porosity increases uniformly with cluster radius of gyration (measured by dynamic light scattering of the dilute sol prior to film deposition). Corresponding pore sizes and surface areas listed in Table 2 [12] also show consistent increases with cluster size.

For D less than about 2, cluster interpenetration can completely mask the porosity of individual clusters. For example, Figure 3 shows the refractive index and volume fraction porosities for a series of silica films deposited from sols characterized by $D < 2$. We observe the film porosities to be approximately 10% regardless of the aging times employed prior to film deposition. Corresponding N_2 sorption studies (-196°C) revealed Type II isotherms for this series of films characteristic of adsorption on non-porous materials, while limited CO_2 sorption studies (0°C) indicated Type I isotherms characteristic of microporosity. The point is that for conditions that promote cluster interpenetration, rather dense films with small pore sizes are obtained regardless of cluster size. This situation is beneficial for the preparation of ultrathin membranes on porous supports, because aging can be employed to grow polymers large enough to be trapped on the support surface without suffering an increase in pore size [13].

Relative Rates of Condensation and Evaporation

The overall extent of condensation along with its distribution is influential in establishing the initial modulus of chemical gels and the initial viscosity of physical gels. Intuitively it should also affect the value of the exponent m. Thus we expect (Eqs. 3 and 6) the extent of condensation to be quite influential is determining the extent of shrinkage during drying and correspondingly the xerogel pore volume and pore size. Since the extent of condensation depends on the product of time and condensation rate, factors that control condensation kinetics and evaporation rates are operative in controlling microstructure. For silicates, the concentration of acid or base catalysts and the molar hydrolysis ratio r (H_2O/Si) are commonly used processing parameters to affect the condensation rate. For more electropositive metals, the condensation rates and overall extents of condensation are often controlled by r and the extent of complexation with multidentate chelating or bridging ligands such as acetylacetonate. The evaporation rate along with the sample dimensions establish the time period during which condensation can occur.

For bulk gels, many studies have established that aging under conditions that promote condensation (with no evaporation) strengthen the gel and reduce the drying shrinkage, allowing the formation of highly porous xerogels [1]. By comparison, Figures 4 a [12] and b show that a reduction of the condensation rate (as judged by a reduction in the reciprocal gel time) has the effect of reducing the xerogel pore volume and pore size and narrowing the pore size distribution as evidenced by a sharper Type I N_2 sorption isotherm. Corresponding studies of thin film membranes prepared from compositions identified as A2 and A2** in Figure 4 showed molecular sieving effects only for the A2** membrane consistent with a reduction in pore size and/or pore size distribution due to the reduced condensation rate [12].

As pointed out above, the overlap of gelation and drying combined with the thin dimensions of films cause the time scale of film deposition to be very short, limiting the overall extent of condensation. Thus films and samples prepared in thin sheets normally have lower volume fraction porosities and smaller pore sizes than their bulk counterparts. As an example of this behavior, we find that films and sheets prepared from the A2** composition exhibit Type II N_2 sorption isotherms

Table 2. Summary of film porosities as a function of sol aging times. Aging serves to grow polymers prior to their deposition on the substrate.

Sample Aging Times[a]	Refractive Index	Porosity[b] %	Median Pore Radius (nm)	Surface Area[b] (m^2/g)
Unaged	1.45	0	d	1.2-1.9
3 days	1.31	16	1.5	146
1 week	1.25	24	1.6	220
2 weeks	1.21	33	1.9	263
3 weeks[d]	1.18	52	3.0	245

a Aging of dilute sol at 50°C and pH ~ 3 prior to film deposition.

b Determined from N_2 adsorption isotherm.

c The 3 week sample gelled. It was re-liquified at high shear rates and diluted with ethanol prior to film deposition.

d N_2 adsorption isotherms are of type II

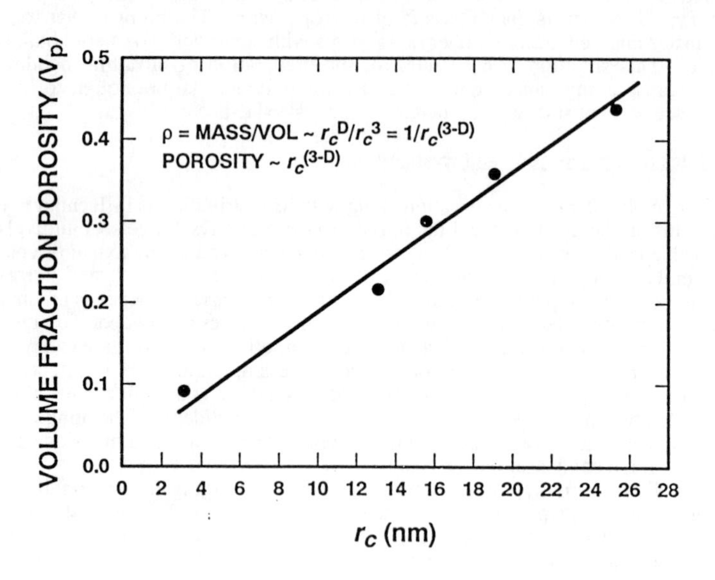

Figure 2. Volume fraction film porosity versus average sol cluster size r_c measured prior to film deposition.

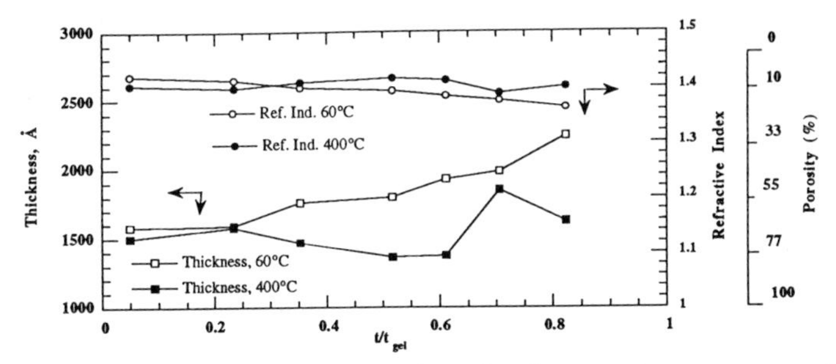

Figure 3. Film thickness, refractive index and volume fraction porosity versus normalized sol aging time t/t_{gel} prior to film deposition.

(a)

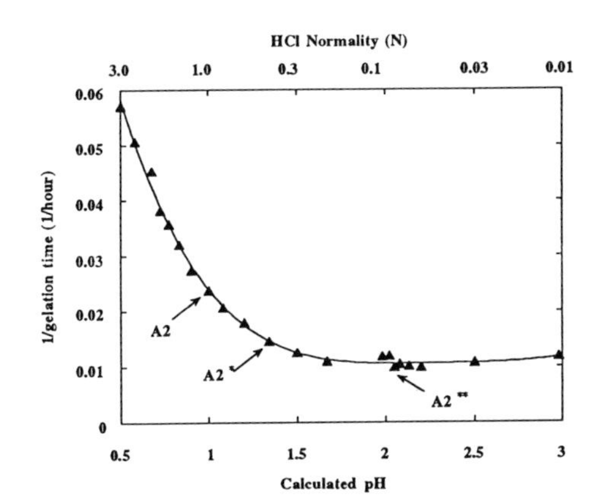

(b)

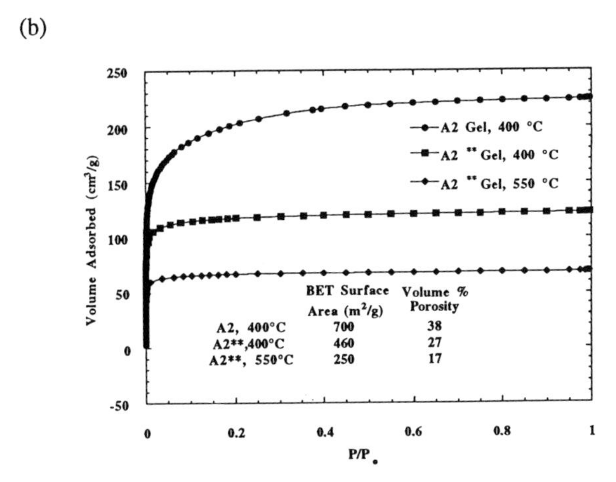

Figure 4. (a). Reciprocal gel times (proportional to average condensation rate) versus calculated pH for a series of silica sols where HCl normality refers to the normality of the acid used in the hydrolysis steps [12]. (b). N_2 sorption isotherms of silica gels (referred to as A2 or A2** in Figure 4a) annealed at 400 or 550°C.

(-196°C) and Type I CO_2 sorption isotherms (0°C), after outgassing at 180°C, indicative of yet a further reduction in pore size [13].

Ambient Pressure Aerogels

A special circumstance related to control of the relative rates of condensation and evaporation occurs when chemical condensation is largely precluded by reaction of the sol or gel surfaces with organic ligands, for example organosilanes, prior to drying. In this situation drying shrinkage is shown to be reversible for both bulk and thin film specimens. Figure 5 [14] compares the pore volumes and bulk densities of bulk silica xerogels and aerogels prepared from identical wets gels by: 1) evaporation of the original pore fluid (xerogel), 2) ethanol washing and supercritical ethanol extraction (high temp. aerogel), 3) solvent exchange and supercritical CO_2 extraction (low temp. aerogel), or 4) surface modification followed by evaporation (ambient pressure aerogel). We observe that the surface modification procedure results in "ambient pressure" aerogels with porosities and densities comparable to classical aerogels prepared under supercritical conditions. Figure 6 shows the film thickness profile, determined *in situ* during film deposition, for an organically-modified silica sol[15]. We see that the film shrinks to a minimum thickness and then "springs back" to a volume more than twice that of the fully compacted state. The ambient pressure aerogel process, demonstrated thus far for silica, should be generally applicable to the wide range of single and multi-component oxides of interest for catalytic applications where high porosities and surface areas are of concern.

Management of Capillary Pressure

As discussed in conjunction with Eq. 3, the drying shrinkage and, hence, pore volume and average pore size depend on a balance between the capillary pressure that serves to compact the gel and the network modulus or viscosity that resists compaction. Elimination of capillary pressure through supercritical processing is the classical means of avoiding drying shrinkage in the preparation of aerogels. Recently, Deshpande and co-workers [16] have shown that, for aprotic pore fluids that do not react chemically with the gel network, increasing the pore fluid surface tension causes a general trend of reduction in pore volume and pore size. These studies have also elucidated the importance of the pore surface chemistry in dictating the extent of drying shrinkage. Figure 7 shows the average pore radius versus the pore fluid surface tension for two sets of silica gels prepared with either hydroxylated or ethoxylated surfaces. The general trend of decreasing pore size with increasing surface tension is observed, but, in addition, we observe that samples prepared with hydroxylated surfaces exhibit consistently larger pore sizes for the same value of surface tension (a similar trend is observed for pore volume). Since Wilhemy plate studies [17] indicated complete wetting of this series of pore fluids on both surfaces, this trend is attributed to a greater extent of condensation accompanying drying for the case of hydroxylated surfaces, which serves to increase the network modulus and reduce drying shrinkage.

Organic Template Approaches

In the approaches described above, the xerogel microstructure is established by the extent of drying shrinkage. A serious deficiency of these approaches is the inherent correlation of pore volume and average pore size. In addition these approaches result in a distribution of pore sizes. To overcome these deficiencies, a number of research groups are exploring organic template approaches to prepare meso- and microporous materials. In these approaches fugitive organic templates (pendant or bridging ligands, amphiphiles, or polymers) are incorporated in a dense inorganic matrix. Removal of the template, normally by pyrolysis, creates pores that mimic the size and shape of the template. An advantage of this approach is that the pore volume is controlled by the volume fraction of the template constituent(s), so it should now be possible to prepare materials with large volume fractions of ultramicropores for membrane applications where high flux is desirable. A second advantage is that it should be possible to precisely control pore size, shape, orientation, and conceivably chirality through the choice and organization of the template constituent(s). A third advantage is the excellent potential for creating complex hierarchical microstructures comprising a variety of pore sizes, shapes, orientations and connectivities.

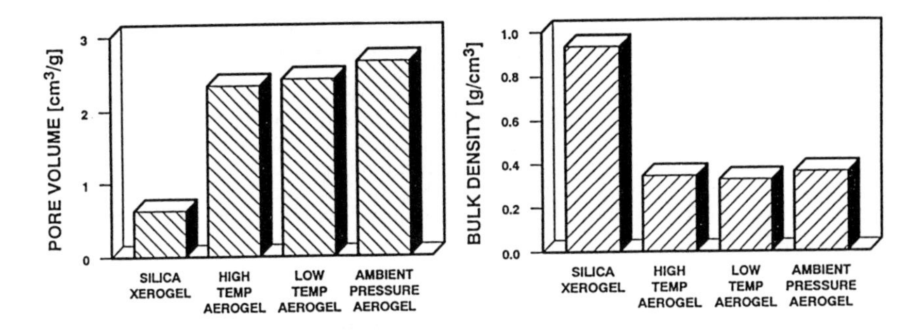

Figure 5. Comparison of porosities of aerogels prepared by classical routes to those of the corresponding ambient pressure aerogel and silica xerogel [12].

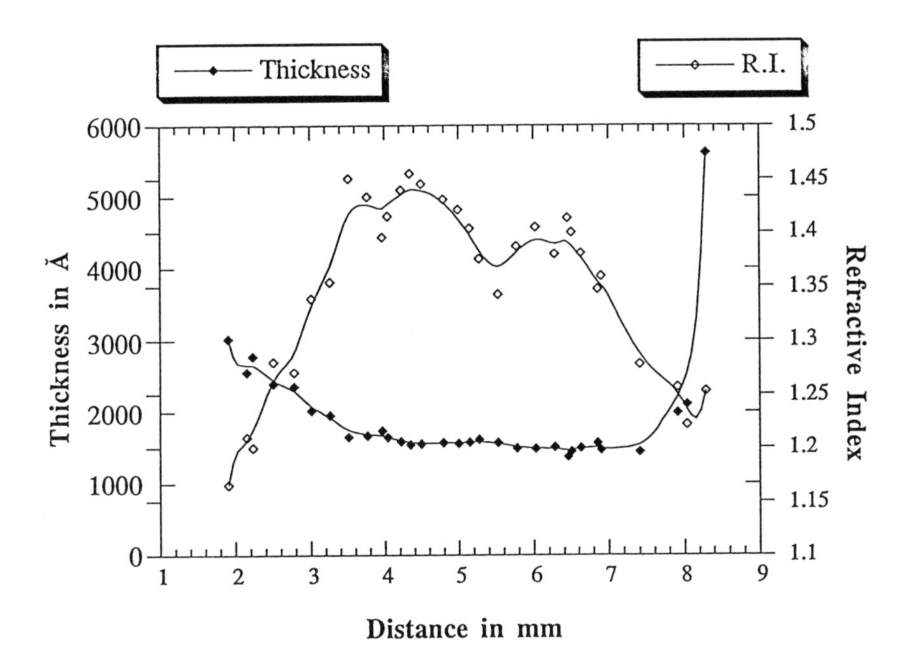

Figure 6. Thickness and refractive index of ambient pressure aerogel films measured *in situ* during film deposition by imaging ellipsometry.

Excellent examples of the success of the template approach in "sol-gel-related" areas are zeolites [18] and the Mobil family of mesoporous materials referred to as MCM-41 [19]. These materials are generally produced by organization of anionic silicates, aluminosilicates, etc. around cationic or amphiphilic templates, producing crystalline or pseudo-crystalline powders. Sol-gel approaches rely on co-polymerization of inorganic precursors, e.g. tetraethoxysilane (TEOS) 1 with organically-modified metal alkoxides such as methyltriethoxysilane (MTES) 2 or methacryloxypropylsilane (MPS) 3. Alternatively it is possible to synthesize homopolymers or co-polymers that incorporate more than one organic ligand. The major challenges of template approaches are to obtain a dense matrix prior to template removal and (for membranes) to achieve pore connectivity, for example by exceeding the percolation threshold of the template constituent, while avoiding phase separation that upon template removal would create pores larger than the characteristic template size.

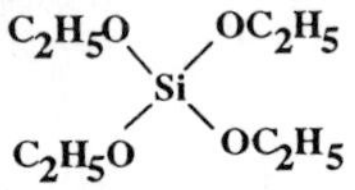

1 Tetraethylorthosilicate (TEOS)

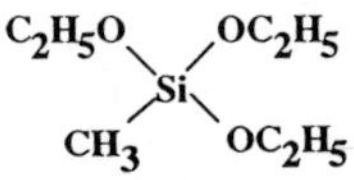

2 Methyltriethoxysilane (MTES) 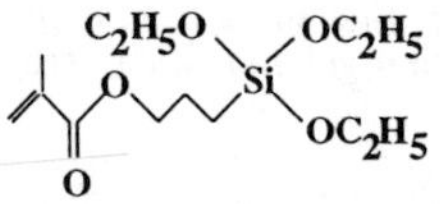 **3 Methacryloxypropyltrimethoxysilane (MPS)**

As an example of a successful sol-gel approach, Figure 8 shows N_2 sorption isotherms for a 4:1 TEOS:MPS co-polymer after heating to 150°C or 500°C in air at 1°C/min. [20]. There is no detectable N_2 adsorption for the 150°C sample, consistent with a relatively dense matrix. After heating to 500°C, where TGA data show pyrolysis of organic constituents to be complete, the isotherm is of Type I, clearly showing the creation of microporosity by template removal. The inset compares N_2 sorption isotherms of the 500°C sample with that of zeolite ZSM-5 ($r_p \approx 0.3$ nm) over the relative pressure range 10^{-6} to 1. This comparison suggests that the xerogel has a similar average pore size but a slightly broader pore size distribution than ZSM-5.

A second example is that of 10:1 TEOS:MTES co-polymers prepared as rapidly dried powders or supported membranes [21]. Figure 9 shows N_2 sorption isotherms of rapidly dried xerogels after heating to 150, 400 or 550°C at 1°C/min. and holding isothermally for 30 min. or 4.0 hours (550°C sample only). The isotherms of the 150 and 400°C samples are of Type I, and there is practically no detectable nitrogen sorption for the 550°C sample. Related dilatometry and TGA experiments indicate that these results are attributable to a progressive consolidation of the microporous network promoted in some way by the inclusion (and ensuing removal) of the methyl ligands. For powders (unconstrained by attachment to a substrate), the creation of porosity by methyl ligand pyrolysis at 500°C is completely masked by the enhanced consolidation of the matrix. Samples prepared as supported membranes show somewhat different behavior. The CO_2 permeance increases from 2.29 x 10^{-3} to 1.81 x 10^{-2} (cm^3/cm^2-s-cm Hg) after heating to 400°C and then decreases to 2.57 x 10^{-3} cm^3/cm^2-s-cm Hg after heating to 550°C. Corresponding CO_2/CH_4 separation factors vary from 1.5 to 1.2 to 12.2. Comparing the 150 and 550°C results, the 550°C membranes exhibit higher flux and higher separation factors. This is attributed to the creation of microporosity by template pyrolysis over the range 450-500°C. Apparently the constraint of the underlying support stabilizes the membrane porosity to higher temperatures.

Sintering/Surface Derivatization

Sintering and surface derivatization are two strategies that can be used in combination with any of those previously discussed to further reduce pore size and alter the pore surface chemistry.

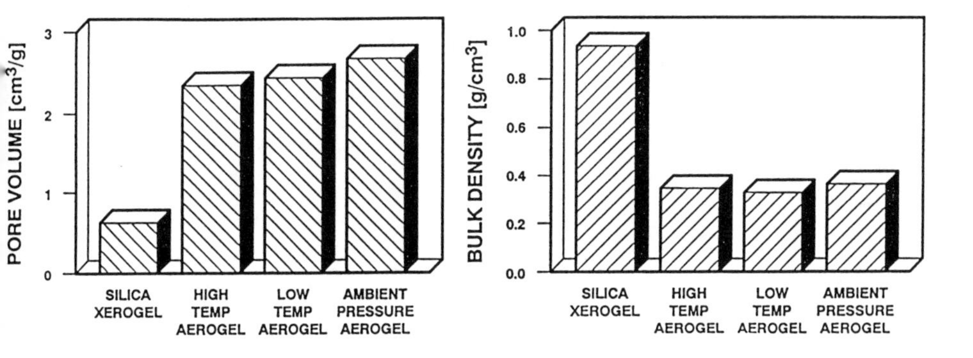

Figure 7. Average pore radius determined by N_2 sorption versus surface tension of pore fluid used during drying. Prior to drying the pore surfaces were treated to create primarily hydroxylated surfaces or ethoxylated surfaces [16].

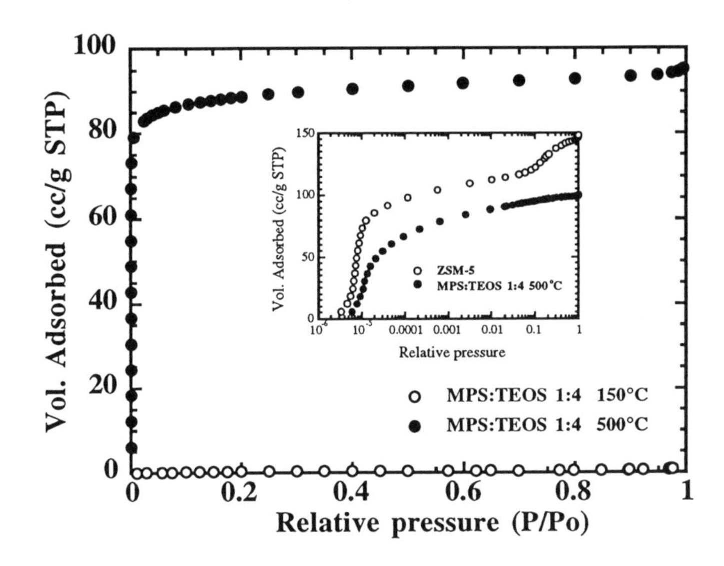

Figure 8. N_2 sorption isotherms measured at -196°C for TEOS:MPS (4:1) co-polymers after heating to 150 or 500°C. Inset compares the isotherm over the pressure range 10^{-6} - 1 to that of ZSM-5 zeolite ($r_p = 0.3$ nm) [20].

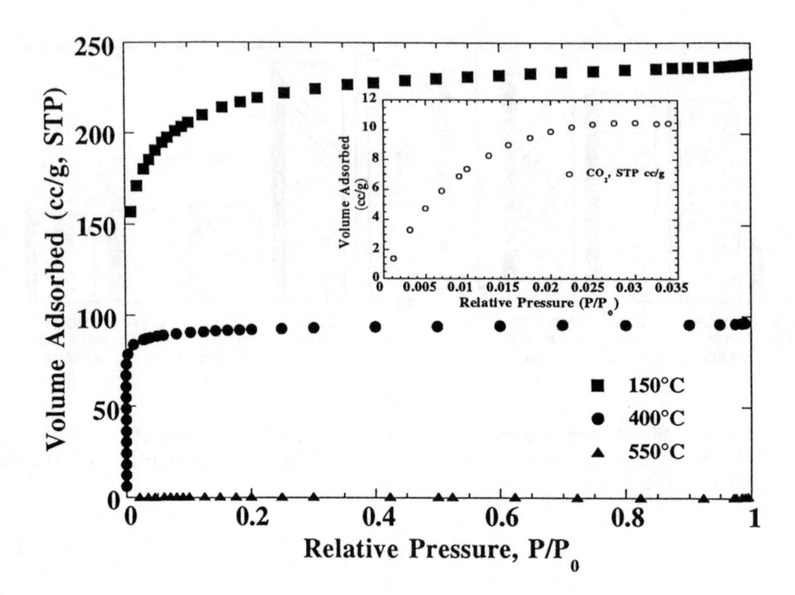

Figure 9. N_2 sorption isotherms measured at -196C for TEOS:MTES (9:1) after heat treatments to 150, 400 or 550°C. The inset shows a type-I microporous CO2 adsorption isotherm (@ 0°C) for the 550°C sample.

Sintering is a process of consolidation driven by a reduction in the solid-vapor interfacial energy [1]. The rate of viscous sintering is proportional to the total surface energy divided by the product of viscosity and pore size. For silica xerogels with exceptionally small pore sizes, complete densification has been observed at temperatures below 550°C. Figure 4b compares N_2 sorption isotherms for microporous silica xerogels after heating to 400 or 550°C. We see that the 550°C heat treatment has the effect of reducing the pore volume and further narrowing the pore size distribution as is evident from the sharper Type I isotherm. For high temperature catalytic membrane reactors, it will be necessary to "fine tune" the initial membrane pore size to achieve stable pores of the correct size at the operating temperature.

Typical xerogel pore surfaces are hydroxylated or, in some cases, partially alkoxylated. A fully hydroxylated silica surface has, for example, a hydroxyl coverage of 4.9 OH/nm^2. These hydroxyl groups can serve as functional sites for surface derivatization reactions of the general type:

$$M\text{-}OH_{(surface)} + X_xM'R_{N-x} \Rightarrow M_{(surface)}\text{-}O\text{-}M'R_{N-x}X_{x-1} + HX \qquad (10)$$

carried out in either the liquid or vapor phase, where M and M' are metals, N is the coordination number of M', X is typically a halide or alkoxide ligand, and R is an organic ligand. R may be chosen to reduce the pore size according to the steric bulk of the organic ligand, provide surface hydrophobicity, or provide functional surface sites. An example of the latter case is the reaction of *N*-(2-aminoethyl)-3-aminopropyltrimethoxysilane with the silica surface [22]:

$$\equiv Si\text{-}OH + (MeO)_3Si\text{—}(CH_2)_3NH(CH_2)_2NH_2 \longrightarrow \equiv Si\text{-}O\text{—}Si(MeO)_2\text{—}(CH_2)_3NH(CH_2)_2NH_2 + MeOH \quad (11)$$

Membrane Surface

The bi-dentate diamine ligand can then be used to complex metals as shown by the hypothesized mechanism for complexation of $[(1,5\text{-COD})RhCl]_2$ where (1,5-COD) is 1,5-cyclooctadiene [22]:

(12)

Reactions 11 and 12 performed on the surface of a microporous silica membrane followed by reduction in flowing hydrogen at *ca* 225°C resulted in a uniform deposition of 6 nm rhodium particles on the exterior membrane surface [22]. The steric bulk of the diamine functionalized silylating reagent prevented derivatization of the membrane interior. In general surface derivatization schemes may be designed to control the location of catalyst as well as tailor pore size and pore surface chemistry.

When $M \neq M'$ in Reaction 10, we modify the composition of the oxide framework. Compositional modification may be performed to alter the surface acidity or chemical stability and hinder sintering or phase transitions in addition to modifying the pore size. When the surface derivatizing agent is a metal alkoxide or metal halide, sequential reaction/hydrolysis steps should result in monolayer-by-monolayer reduction in pore size. For example, derivatization of a microporous silica membrane with titanium iso-propoxide followed by calcination at 400°C resulted in a membrane having He or H_2/CH_4 separation factors exceeding 1000 over the temperature range 40-225°C [23].

CONCLUSIONS

The preceding discussion has illustrated that sol-gel processing is a versatile means of preparing ceramics with controlled microstructures. Using strategies such as aggregation of fractals, controlled (and reversible) drying shrinkage, pyrolysis of organic templates, partial sintering and surface derivatization, we have shown that volume fraction porosity may be varied from about 5 to over 95% and that pore size may be varied over the range 3 nm $\geq r_p \geq$ 0.2 nm. Although these microstructural properties mimic those obtainable in classical aerogels and zeolites, we can now achieve these properties using facile *ambient pressure* routes amenable to the low cost production of catalysts and membranes.

ACKNOWLEDGEMENTS

We gratefully acknowledge support from DOE-Basic Energy Sciences, the National Science Foundation (CTS9101658), the Gas Research Institute, the Electric Power Research Institute, and Morgantown Energy Technology Center. Sandia National Laboratories is a U.S. Department of Energy facility supported by DOE Contract Number DE-AC04-76-DP00789.

REFERENCES

1. C.J. Brinker and G.W. Scherer, Sol-Gel Science: The Physics and chemistry of Sol-Gel Processing (Academic Press, San Diego, 1990).

2. F.A.L. Dullien, Porous Media, Fluid Transport and Pore Structure, (Academic Press, New York, 1979).

3. G.W. Scherer, J. Non-Cryst. solids **109**, 183 (1989).

4. G.W. Scherer, J. Non-Cryst. Solids **155**, 1 (1993).

5. G.W. Scherer, in Better Ceramics Through Chemistry VI, eds. A. Cheetham, C.J. Brinker, M.L. Mecartney, C. Sanchez (Materials Research Society **346**, Pittsburgh, 1994) p. 209.

6. G.W. Scherer, J. Non-Cryst. Solids **109**, 171 (1989).

7. A.F.M. Leenars, K. Kreizer, and A.J. Burggraaf, J. Mat. Sci. **10**, 1077 (1984).

8. C. Sanchez, M. In, J. Non-Cryst. Solids **147/148**, 1 (1992).

9. S.G. Kroll, J. Appl. Polym. Sci. **23**, 847 (1979).

10. T.J. Garino, Ph.D. Dissertation, MIT, Cambridge, 1986.

11. D.L. Logan, C.S. Ashley, C.J. Brinker in Better Ceramics Through Chemistry V, eds., M.J. Hampden-Smith, W.G. Klemperer, C.J. Brinker (Materials Research Society **271**, Pittsburgh, 1992) p. 541.

12. C.J. Brinker, R. Sehgal, S.L. Hietala, R. Deshpande, D.M. Smith, D. Loy, C.S. Ashley, J. Membrane Sci. **94**, 85 (1994).

13. C.J. Brinker, T.L. Ward, R. Sehgal, N.K. Raman, S.L. Hietala, D.M. Smith, D.-W. Hua, and T.J. Headley, J. Membrane Sci. **77**, 165 (1993).

14. D.M. Smith, R. Deshpande, C.J. Brinker, Better Ceramics Through Chemistry V, eds., M.J. Hampden-Smith, W.G. Klemperer, C.J. Brinker (Materials Research Society **271**, Pittsburgh, 1992) p. 567.

15. S.S. Prakash and C.J. Brinker, J. Non-Cryst. Solids, submitted.

16. R. Deshpande, D.-W Hua, D.M. Smith, and C.J. Brinker, J. Non-Cryst. Solids **144**, 32 (1992).

17. D. Stein and D.M. Smith, J. Non-Cryst. Solids, submitted.

18. D.W. Breck, Zeolite Molecular Sieves, (R.E. Kreiger, Malabar, FL, 1984).

19. C.T. Kresge, M.E. Leonowicz, W.J. Roth, J.C. Vartuli, J.S. Beck, Nature **359**, 710 (1992).

20. C.J. Brinker, N.K. Raman, L. Delattre, S.S. Prakash, Proceedings of The Third International Conference on Inorganic Membranes (Worcester, MA, 1994) ed. Y. Ma, to be published.

21. N.K. Raman and C.J. Brinker, J. Membrane Sci, submitted.

22. N.K. Raman, T.L. Ward, C.J. Brinker, R. Sehgal, D.M. Smith, Z. Duan, M.J. Hampden-Smith, J.K. Bailey, and T.J. Headley, Applied Catalysis A: General **69**, 65 (1993).

23. R. Sehgal and C.J. Brinker, unpublished results.

EFFECT OF PREHYDROLYSIS ON THE TEXTURAL AND CATALYTIC PROPERTIES OF TITANIA-SILICA AEROGELS* James B. Miller, Scott T. Johnston, and Edmond I. Ko, Department of Chemical Engineering, Carnegie Mellon University, Pittsburgh, PA 15213-3890.

A sol-gel based preparation of a mixed oxide allows considerable control over the extent of component mixing. When the reactivities of the alkoxide precursors are evenly matched, a homogeneously mixed sample can be produced; a large reactivity difference results in a segregated mixed oxide product. We have prepared and characterized two sets of mixed titania-silica aerogels, each covering the entire composition range. One set was made using silicon precursor prehydrolysis to minimize the reactivity difference and promote homogeneous component mixing; the second set was made without prehydrolysis.

As shown in Figure 1, the maximum catalytic activity for 1-butene isomerization, a reaction requiring a weak Brønsted acid, occurs at 67 mol% titania within both sample sets. This observation, supported by Brønsted/Lewis acid ratio estimates, suggests that Brønsted acid sites result from formation of 'hetero linkages' (Ti-O-Si) containing titanium atoms in six-fold coordination. Six-fold coordination is expected to dominate at high titania content (1). Furthermore, according to the model of Tanabe et al. (2), a 'six-fold coordinated hetero-linkage' provides the 'charge imbalance' necessary to form an acid site in the mixed oxide. At higher silica contents, however, hetero linkages with tetrahedrally coordinated titanium atoms appear in significant numbers (2). Consistent with Tanabe's criterion, which predicts an absence of charge imbalance in this case (1), our data suggest that the 'tetrahedrally coordinated hetero linkage' does not contribute to either Brønsted acidity or 1-butene isomerization activity.

Samples prepared without prehydrolysis of the silicon precursor exhibit relatively low isomerization activities (see Figure 1). This result is indicative of a smaller population of hetero linkages and, therefore, partial segregation of titania and silica. Low activities coupled with low fractional Brønsted populations and low total acid site densities suggest a specific form of component segregation in the non-prehydrolyzed

---

* The work described in this extended abstract has been published elsewhere (3). We acknowledge the support of the Division of Chemical Sciences, Office of Basic Energy Sciences, U.S. Department of Energy (Grant DE-FG02-93ER14345).

samples, one in which inactive silica-rich clusters obscure much of the surface of a titania-enriched 'core.'

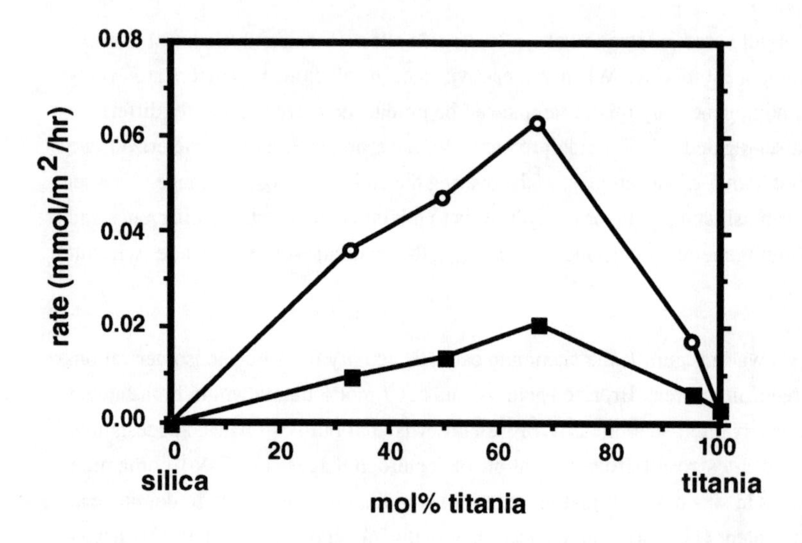

Figure 1. 1-butene isomerization activity of titania-silica aerogels as functions of preparation method and composition. Open symbols represent prehydrolyzed samples; closed symbols, non-prehydrolyzed. Isomerization activity is reported as reaction rate (cis-2-butene and trans-2-butene products) at 95 min time on stream. Reaction conditions: 423K, 1 atm, 5 sccm 1-butene, 95 sccm He, ~0.2 g catalyst. All samples were calcined at 773K in oxygen for 2 hr before activity testing.

REFERENCES

1. Liu, Z., Tabora, J., and Davis, R.J., *J. Catal.*, **149**, 117(1994).

2. Tanabe, K., Sumiyoshi, T., Shibata, K., Kiyoura, T., and Kitigawa, J., *Bull. Chem. Soc., Jpn.* **47**(5), 1064 (1974).

3. Miller, J.B., Johnston, S.T., and Ko, E.I., *J. Catal.* **150**, 311 (1994).

ADVANCES IN THE DESIGN OF PILLARED CLAY CATALYSTS BY SURFACTANT AND POLYMER MODIFICATION

T. J. PINNAVAIA, JEAN-RÉMI BUTRUILLE, LAURENT J. MICHOT AND JINGIE GUAN
Department of Chemistry and Center for Fundamental Materials Research, Michigan State University, East Lansing, Michigan 48824

ABSTRACT

The catalytic properties of pillared clays for organic chemical conversions, especially those occurring in liquid media, can be substantially improved by surfactant modification of the pore structure during the pillaring reaction. For instance, the incorporation of a non-ionic surfactant, such as the alkylated polyethylene oxide $C_{12-14}H_{25-29}O(CH_2CH_2O)_5H$, in the synthesis of alumina pillared montmorillonite results in a dramatic increase in the interparticle (textural) mesoporosity of the final calcined products. These surfactant-modified mesoporous pillared clays are exceptionally active as catalysts for the diffusion controlled liquid phase alkylation of biphenyl. The enhanced mesoporosity facilitates access of the reactants to the active acid sites in the interlayer nanopores of the pillared clay tactoids. Organic polymeric molecules also can be used to mediate the pore structure of pillared clay materials. The reaction of aluminum chlorohydrate oligomers with Na^+ rectorite in the presence of polyvinyl alcohol as a pillaring precursor affords a supergallery alumina pillared rectorite with a basal spacings of 52 Å and a corresponding gallery height of 33 Å under air- dried conditions. A stable gallery height of 23 Å is observed even after treatment with 100% steam at 800°C for 17 hour. The new supergallery intercalate is characterized by a surface area, pore volume, and catalytic cracking activity superior to conventional alumina pillared rectorite and related smectites with 9 Å gallery heights.

INTRODUCTION

Zeolites typically exhibit outstanding shape selective catalytic properties [1-3]. However, conventional zeolites often are limited to catalytic applications involving relatively small molecules. Moreover, the diffusion limitations in these materials can be severe [2]. Thus, there is considerable interest in the design of mesoporous materials to overcome these limitations [3-6]. Alumina pillared clays, a versatile class of microporous acid catalysts, have been shown to be attractive materials for the alkylation of liquid-phase substrates [6]. By appropriate choice of clay and pillar, it is possible to prepare pillared clays with tailored gallery height [7,8] and interpillar distances [9]. In order to modify the porous properties of alumina pillared clays, pillared clay syntheses have been carried out in the presence of space-filling polymers, i.e., polyvinyl alcohols [10] or non-ionic surfactants [11], such as the alkylated polyethylene oxide $C_{12-14}H_{25-29}O(CH_2CH_2O)_5H$ [12, 13]. Since these modified pillared clays exhibit enhanced textural

Mat. Res. Soc. Symp. Proc. Vol. 368

mesoporosity they are potentially capable of facilitating access to catalytic sites, especially for reaction carried out in condensed media.

The present work relates in part to our efforts to improve pillared clays as catalysts for condensed phase chemical conversions of large organic substrates and potential environmental applications [14]. The properties of a surfactant-modified alumina pillared montmorillonite are reported for propylene alkylation of liquid phase biphenyl. We find that the presence of a non-ionic surfactant during the pillaring reaction results in a product with increased textural mesoporosity and dramatically enhanced catalytic activity relative to pillared analogs prepared without surfactant modification.

Modification of the pore structure of pillared clays can also be achieved by incorporating organic polymers into the pillaring reaction. Smectite clays pillared by inorganic metal oxide aggregates normally have basal spacings of 18.0-22.0 Å and gallery heights of 8.4 - 12.4 Å[15-17]. The agents that serve to prop open the interlayer space oftentimes are displaced from their original interlayer positions at elevated temperatures, causing the galleries to collapse, especially in the presence of water vapor. For instance, alumina pillared smectites treated at 730 °C for 4 hours in steam experience a 90% loss in surface area and a 80% decrease in microporous volume[18]. We have recently found that the incorporation of polyvinyl alcohol as a precursor in the pillaring reaction of Al_{13} oligomers of rectorite clay affords a supergallery pillared derivative [19]. The term "supergallery" has been used previously in the literature to describe microporous 2:1 layered derivatives in which the gallery height is substantially larger than the thickness of the clay layers [20]. In order to further illustrate the utility of surfactant modification in pillared clay synthesis, the present work also includes a summary of some of the principal properties of this unusual alumina pillared clay relative to a conventional alumina pillared rectorite with 9 Å galleries.

EXPERIMENTAL SECTION

The pillaring reactions of Na^+ montmorillonite (cation exchange capacity, 115 meq/ 100 g) with base hydrolyzed $Al_{13}O_4(OH)_{24}(H_2O)_{12}^{7+}$ cations (OH^- / Al^{3+} = 2.4.) were carried out in the presence of the alkylated polyethylene oxide surfactant $C_{12-14}H_{25-29}O(CH_2CH_2O)_5H$, henceforth abbreviated $C_{12-14}E_5$. The molar ratio of surfactant : clay was varied over the range 0 - 1.4 molecules per O_{20} unit cell, but the ratio of Al^{3+} : clay was held constant at 15 mmol Al^{3+} per meq clay. The reaction mixtures were allowed to age overnight, and the products were collected by centrifugation, washed, and air-dried at room temperature. The final alumina pillared products were calcined in a programmable oven at 500 °C for 12 hours.

A supergalley pillared rectorite clay was prepared by the reaction of aluminum chlorohydate oligomers with the sodium - exchanged form of the mineral in the presence of polyvinyl alcohol (PVA) as a pillar precursor according to previously described methods [19]. The pillaring agent was a commercially available aluminum chlorohydrate solution, $Al_2(OH)_5Cl$ (Reheiss Chemical Co.). The raw clay for the preparation of the stable supergallery pillared derivatives was naturally occurring rectorite (China) with a cation exchange capacity (CEC) of 60 meq / 100 g. The intercalated PVA was removed by heating to 650 °C in nitrogen for 2 h, exposing the hot sample to air, and allowing it to cool slowly. Alternatively, the PVA could be desorbed by exposing the pillared product to 100% steam at 800°C for 17 hours and then allowing the sample to cool slowly in air. For comparison purposes a conventional alumina pillared rectorite with a 9 Å gallery height (d_{001} = 29 Å) was prepared by the reaction of Na^+ rectorite with base hydrolyzed aluminum chloride at a pH of 4.5 , as reported in the literature [21]. Nitrogen adsorption and desorption isotherms were obtained using an Omnisorb 360 CX sorptometer. Surface areas were determined by the BET method. Pore volumes were calculated by the t - plot method.

Catalytic activity for the cracking of light diesel oil (bp = 205 - 330 °C) was carried out at 460 °C by the microactivity test (MAT) using a weight hourly space velocity (WHSV) 16 h^{-1} and catalyst to oil ratio (w/w) 4.0.

RESULTS AND DISCUSSION

Surfactant Modification in the Pillaring of Montmorillonite

Figure 1 presents a plot of $C_{12\text{-}14}E_5$ surfactant loading versus the initial surfactant content of the pillaring solution. The uptake of surfactant by the $Al_{13}O_4(OH)_{24}(H2O)_{12}^{7+}$ clay is linear over the entire concentration range. Significantly, the Al_{13} content of the clay remains essentially constant near a value of 0.091 ± 0.005 moles per $O_{20}(OH)_4$ formula unit of clay, regardless of surfactant loading. That is, the pillar population density inside the galleries is *not* significantly affected by the co-adsorption of the surfactant.

Nitrogen BET surface area (S.A.) and porosity studies provide useful insights into the nature of $C_{12\text{-}14}E_5$ binding. Figure 2 illustrates the relationship between surface area and surfactant loading. In the absence of adsorbed surfactant Al_{13} montmorillonite outgassed at 150 °C exhibits a BET S.A. of 305 m^2 / g and a microporous liquid volume of 0.10 cm^3 / g. Most of the S.A. arises from the presence of the micropores, the non-microporous value being only 32 m^2 / g. The addition of surfactant causes dramatic changes in both surface area and pore structure. An almost linear decrease in S.A. occurs with $C_{12\text{-}14}E_5$ loading up to 0.3 molecules / cell, whereupon the S.A. value drops to a *non-microporous* value of 32 m^2 / g. The dramatic reduction in microporous surface area indicates that the surfactant binds preferentially to the gallery surfaces between the pillaring Al_{13} cations. Increasing the loading to 0.67 molecules per cell causes an increases in microporous S.A., most likely due to further intercalation and swelling of the galleries by some of the surfactant. However, increasing the loading beyond 0.67 molecules / cell again eliminates all microporosity, signaling the filling of the galleries and the further binding of surfactant to the external surfaces of the particles. Significantly, most of the non-microporous surface area observed at high surfactant loading arises from the presence of mesopores. This is illustrated by the pore size distributions shown in Figure 3 for loading of 0, 0.19, and 0.67 molecules per cell.

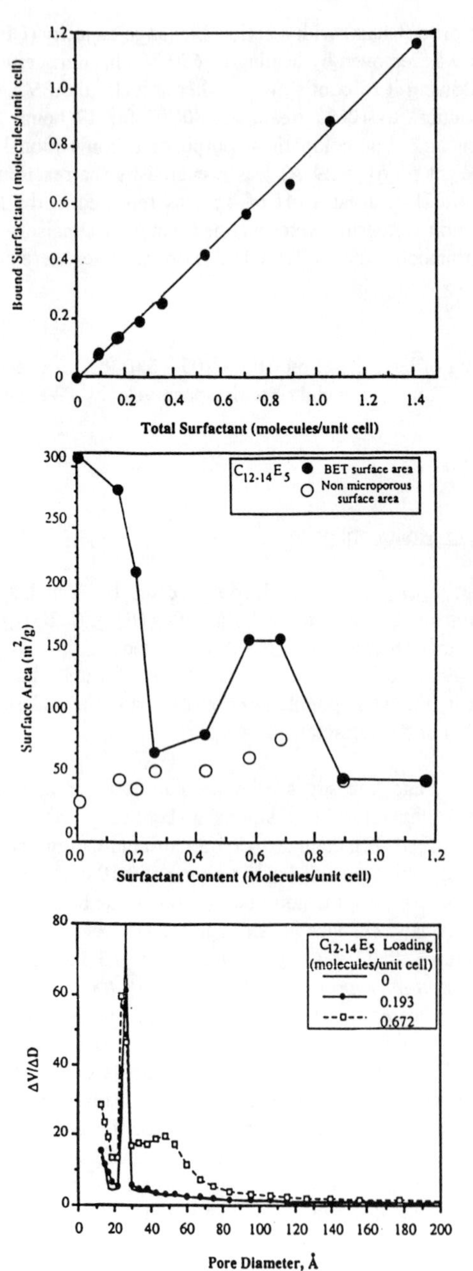

Figure 1 Final $C_{12\text{-}14}E_5$ surfactant loading in Al_{13} montmorillonite *vs.* surfactant content of the pillaring solution.

Figure 2 Dependence of BET and non-microporous surface areas on surfactant loading in C_{12}-$14E_5$-Al_{13} montmorillonite.

Figure 3 Mesopore size distributions for $C_{12\text{-}14}E_5$ - Al_{13} montmorillonites

In all of the above experiments the surface area values are for the surfactant-loaded clays. Removing the intercalated surfactant by calcination restores the intragallery micropore volume. Moreover, removal of the surfactant from the external surfaces of the clay tactoids makes available the textural (interparticle) mesoporosity. Nitrogen adsorption - desorption measurements on surfactant modified alumina pillared Arizona montmorillonites calcined at 500 °C verifies that the micropore volume (0.10 ± 0.01 cm^3 / g) is independent of the total surfactant loading. Thus, the surfactant does not substatically alter the lateral pillar separation. However, the textural mesoporosity increased from 0.08 cm^3 / g for the pillared clay prepared in the absence of surfactant to 0.63 cm^3 / g for the calcined product prepared at optimal surfactant loading. We can now consider the affect of textural mesoporosity on the catalytic properties of the pillared clay.

Alumina pillared montmorillonites have been previously shown to be effective solid acid catalysts for the propylene alkylation of liquid phase biphenyl[6, 22, 23]. For reaction at 250 °C and 140 psi the conversion of substrate is under diffusion control. Thus, access to catalytically active sites should be greatly affected by interparticle mesoporosity. As can be seen from the results presented in Table I increasing the mesoporosity of the catalyst results in a dramatic increase in activity. The conversion improves from 4% for a conventional pillared clay with 0.08 cm^3 / g to 100% for a surfactant-modified derivative with a mesoporosity of 0.63 cm^3 / g. The exception activity of this latter catalyst also is reflected in the higher yields of tri- and tetra-alkylated products. No other catalyst that we have tested for this reaction [6] was as active as the modified pillared clay with 0.63 cm^3 / g mesoporosity. The number of acid sites on this latter catalyst is not exceptional, and the Brønsted acidity is not especially strong. Moreover, for all the surfactant-modified pillared clays, the acidity is very similar.

Under our experimental conditions, propene alkylation of biphenyl is mainly diffusion controlled. Decreasing the particle size of the catalyst from 75-150 μm to < 45μm more than doubles the biphenyl conversion. The enhanced mesoporous volume of a surfactant modified pillared clay is almost certainly a decisive factor in increasing the diffusion rate within the catalyst particles. It is of interest to compare the catalytic activities of the surfactant-modified pillared clays and the alumina delaminated Laponite (ADL), which also is a mesoporous solid acid, but lacking microporosity. ADL affords 89% conversion with 0.7 wt% catalyst, whereas only 0.25 wt% of the high mesoporosity pillared clay affords 100% conversion. Thus, it appears that *mesoporosity in combination with microporosity* is important in optimizing the activity of a pillared clay for condense phase reactions. The mesopores act as transport pores to facilitate access of acid sites that are located mainly within the gallery micropores of the structure.

Polymer Modification of Pillared Rectorite

Rectorite is a mixed layer 2:1 silicate containing regularly alternating galleries of non - swellable, high -charge density, mica - type galleries and swellable, low - charge density, smectite - type galleries. Only the smectite galleries are readily available for intercalation and ion exchange. The presence of the mica-type galleries causes the layer thickness of rectorite (19.6 Å) to be twice that of a normal smectite.

TABLE I CATALYTIC ACTIVITY OF SURFACTANT MODIFIED PILLARED MONTMORILLONITE FOR PROPENE ALKYLATION OF BIPHENYL[a]

Mesoporosity, Cm^3/g	0.08	0.34	0.63
Catalyst Weight, g	0.08	0.08	0.08
Conversion, %	4	46	100

[a] Reaction conditions: 30g (0.194 mol) biphenyl; 9 atm. propene; 250°C; 20 hours.

TABLE II PROPERTIES OF SUPERGALLERY PILLARED RECTORITE

Treatment	Spacing Å	Area $m2_{/g}$	Vol. $ml3_{/g}$	MAT 5
Air dried. 25°C	52	27.3	0.11	17
Calcined, 650°C/2h	40	218	0.25	62
Steamed, 800°C/17h	42	200	0.20	64

Alumina pillared clays normally are prepared by the reaction of a hydrophilic exchange form of a smectite clay, typically a Na^{+}- or Ca^{2+}- exchanged derivative, with Al_{13} polycations. Previous studies of related clay systems interlayered by non-ionic surfactants have shown that reducing the hydrophilic properties of the gallery surfaces greatly improves the hydrolytic stability of intercalated aluminum polycations [12-14]. The analogous reaction of Na^{+} rectorite with aluminum chlorohydrate polycations in the presence of PVA at pH values of 7.0 affords intercalated products with unusually large basal spacings. Figure 4 illustrates the XRD patterns obtained for a representative reaction product. The fresh, air-dried sample exhibits a first order reflection at 51.1 Å and a second order peak at 25.4 Å, indicating the presence of an ordered supergallery structure. A somewhat smaller gallery height of 40.2 Å is obtained upon thermolysis in nitrogen and then in air at 650 °C. Exposing the sample to 100% steam at 800°C for up to 40 h and then to air to remove intercalated PVA results in a single supergallery reflection at 42-43 Å. This spacing corresponds to a gallery height of about 23. Å

The alumina pillared rectorite prepared by surfactant modification not only has a supergallery structure, but it also represents a new type of *highly stable* porous structure. In typical structural models for alumina pillared smectites the 2:1 mica-type layers are presumed to be separated by single Al_{13} pillaring units [15-17]. Also, alumina pillared rectorites prepared by conventional methods contain "doubly - thick" 2:1 layered silicate host layers separated by single Al_{13} pillars [2]. Thus, the gallery heights of conventional alumina pillared smectites and rectorites are no more than 10Å. In contrast, the basal spacings of the pillared rectorite prepared by PVA modification corresponds to a gallery height that is twice as large as conventional pillared clays. That is, our new supergallery rectorite has an unprecedented doubly - thick layer / doubly - thick pillar structure relative to conventional pillared smectites.

The surface areas and pore volumes of our supergallery pillared rectorite after calcination at 650°C for 2h and after steam treatment at 800°C for 17 h are given in Table II. . The sample formed by thermal activation exhibits a surface area (218 m^2/g) comparable to that for the material activated in steam (200 m^2/g). Also, the supergallery pillared rectorite subjected to hydrothermal treatment at 800°C for 17 h possesses both Bronsted and Lewis acidity. The FTIR ring stretching frequencies of chemisorbed pyridine indicate the presence of both Lewis sites (1453-1457 cm^{-1}) and Bronsted sites (1545-1547 cm^{-1}). The absorptions characteristic of both Bronsted and Lewis acid sites are more intense for the supergallery clay than for conventional pillared rectorite. Thus, on the basis of the improvement in surface area and acidity, supergallery pillared rectorite is expected to be a more active acid catalysts than a conventional pillared analog.

In order to assess the catalytic properties of supergallery pillared rectorite, we have carried out microactivity tests (MAT) of gas oil cracking. Table II reports the diesel oil conversions for thermally and hydrothermally activated supergallery pillared rectorites. Both materials exhibit conversions in the range 62-64% under standard MAT reaction conditions (see experimental section). In comparison, a conventional alumina pillared rectorite with a gallery height of 9 Å has a MAT value of 54%. It is noteworthy that air-dried supergallery rectorite gives a relatively low MAT conversion (17%). Thus, the complete removal of the intercalated PVA from the supergallery structure by pre-calcination or steaming is essential for realizing high catalytic conversions.

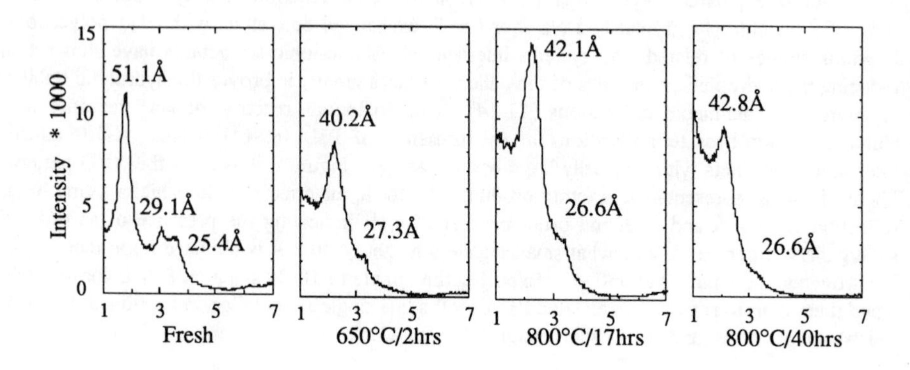

FIGURE 4. XRD (Cu-K∝) of Calcined Supergalley Pillared Rectorite.

CONCLUSIONS

Our results suggest that surfactant and polymer modification of the pillaring reactions of smectite clays can be very promising approaches to improving access to the pore structure of pillared clay catalysts. The incorporation of a non-ionic surfactant such as $C_{12\text{-}14}H_{25\text{-}29}O(CH_2CH_2O)_5H$ in the synthesis of an alumina pillared clay substantially increases the interparticle (textural) mesoporosity of the aggregated tactoids without altering the microporosity. One of the advantages of enhanced mesoporosity is the greatly improved activity of the catalyst for liquid phase reactions, such as the propylene alkylation of liquid phase biphenyl. The textural mesoporosity most likely facilitate diffusion-limited mass transport to the gallery micropores where the majority of acid sites in the pillared clay are located.

Complexing polymers, such as polyvinyl alcohol (PVA), also are potentially effective agents for improving access to the pore structure of pillared clay catalysts. The incorporation of PVA into the synthesis of alumina pillared rectorite results in the formation of supergallery derivatives. The role of the polymer is to stabilize intragallery polycations that are much larger than the Al_{13} oligomers normally intercalated by the clay during a conventional pillaring reaction. Consequently, the gallery height of the PVA-modified pillared rectorite (23Å) is more than twice that of a pillared rectorite prepared in the absence of polymer. Thus catalytically more accessible pillared clay structures might be anticipated through the use of polymer modification.

REFERENCES

1. P.B. Weisz, Chemtech., 498 (1973).

2. I.E. Maxwell, J. Incl. Phenom. **4,** 1(1986).

3. G.S. Lee, J.J. Mag, S.C. Rocke and J.M. Garcés, Catal. Lett. **2**, 197 (1989).

4. W.A. Wachter, US Patent No.5 051 385 (1991).

5. C.T. Kresge, M.E. Leonowicz, W.J. Roth and J.C. Vartuli, US Patent No. 5 098 684 (1992).

6. J.-R. Butruille and T.J. Pinnavaia, Catal. Today **14**, 141 (1992).

7. E. Kikuchi, T. Matsuda , H. Fujiki and Y. Morita, Appl. Catal. **11**, 331 (1984).

8. J. Sterte, Clays Clay Miner. **39,** 167 (1991).

9. T. Mori and K. Suzuki, Chem. Lett. 2165 (1989).

10. K. Suzuki, H. Masakazu, H. Masuda and T. Mori, J. Chem. Soc., Chem. Commun. 873 (1991).

11. D.R. Fahey, K.A. Williams, R.J. Harris and P.R. Stapp, US Patent 4,845,066 (1989).

12. L.J. Michot and T.J. Pinnavaia, Chem. Mater., **4**, 1433 (1992).

13. L.J. Michot, O. Barres, E. L. Hegg, and T. J. Pinnavaia, Langmuir **9**, 1794 (1993).

14. L.J. Michot and T.J. Pinnavaia, Clays Clay Miner., **39,** 634 (1991).

15. R. Burch, Catal. Today, **2,** 2,3 (1988)

16. F. Figueras, Catal. Rev., **30,** 457 (1988).

17. R. A. Schoonheydt, Stud Surf. Sci. Catal., **58,** 201 (1991).

18. M.L. Occelli, Ind. Eng. Chem. Prod. Res. Dev., **22,** 553 (1983).

19. J. Guan, T. J. Pinnavaia, Proc. Inter. Conf. on Soft Chemistry Routes to New Materials, Nante, September 6-10, 1993, J. Rouxel, M. Tournoux and R. Brec, Eds., Trans Tech Publ., pp. 109-114 (1994).

20.. A. Moini, T. J. Pinnavaia,:Solid State Ionics, **26,** 119 (1988).

21. J. Guan, E. Min, Z. Yu, Proc. 9th Int. Congr. Catal., Ottawa, Canada, 104, (1988).

22. J.R. Butruille and T. J. Pinnavaia, Catal. Lett. **12**, 187 (1992).

23. J. R. Butruille and T. J. Pinnavaia in Multifunctional Mesoporous Inorgnic Inorganic solids, edited by C. A. C Sequeira and M. J. Hudson **259,** 272 Kluwer, Amsterdam (1993)

COMPUTER SIMULATION OF DIFFUSION AND ADSORPTION IN PILLARED CLAYS

XIAOHUA YI, MUHAMMAD SAHIMI AND KATHERINE S. SHING
Department of Chemical Engineering, University of Southern California, Los Angeles,CA 90089-1211

ABSTRACT

We developed a model to describe the morphology and energetics of pillared clays. Grand-Canonical Ensemble Monte Carlo and Molecular Dynamics simulations are used to study diffusion and adsorption of finite-size molecules in such systems, and the effect of various factors on these processes is investigated.

INTRODUCTION

Diffusion and reaction in porous catalysts have been the subject of considerable research activity in the last few years [1]. These systems, in addition to their great industrial importance, also represent ideal model porous systems well suited for theoretical and experimental studies of hindered (restricted) diffusion, adsorption and reaction phenomena [2,3]. Such phenomena, which involve the transport and reaction of large molecules in small pores, occur also in many processes of current scientific and industrial interest, such as separation processes, solvent swelling rubbers, polyelectrolyte gels, enzyme immobilization in porous solids, and size exclusion chromatography.

Among all catalytic systems zeolites have received the greatest attention [4]. But considerably less attention has been focussed on studying another class of catalytic materials, namely, pillared clays [5-7]. The original idea for producing pillared clays, due to Barrer and MacLeod [8], was to insert molecules into clay minerals to prop apart the aluminosilicate sheets, thereby producing larger pores than in native clays, or even in zeolites. However, such materials did not have the thermal stability that zeolites usually possess. But pillars of hydroxyaluminum and other cations, which are capable of being dehydrated to oxide pillars and to support temperatures of up to 500°C without structural collapse under catalytic cracking conditions, are relatively new and have been reported by several groups [9-11].

In general, pillared montmorillomites are 2:1 dioctahedral clay minerals consisting of layers of silica in tetrahedral coordination, holding in between them a layer of alumina in octahedral coordination. Substituting Si^{4+} with Al^{3+}, or Al^{3+} with Mg^{2+} gives the silicate layer a negative net charge, which is normally compensated by Na^{+}, Ca^{2+} and Mg^{2+} ions [12]. By exchanging the charge compensating cations with large cationic oxyaluminum polymers, one can synthesize molecular sieve-type materials [10,11]. These inorganic polymers, when heated, form pillars which prop open the clay layer structure and form permanent pillared clays. Such catalytic materials have shown high catalytic activities for gas oil cracking, and large initial activities towards methanol conversion to olefins and toluene ethylation. They have also been suggested as a new class of sorbents for gas separations [13]. Moreover, their pore sizes can be made larger than those of faujastic zeolites, and since access to the interior pore volume of pillared clays is controlled by the distance between the silicate layers and also the distance between the pillars, one or both distances may be adjusted to

Mat. Res. Soc. Symp. Proc. Vol. 368 © 1995 Materials Research Society

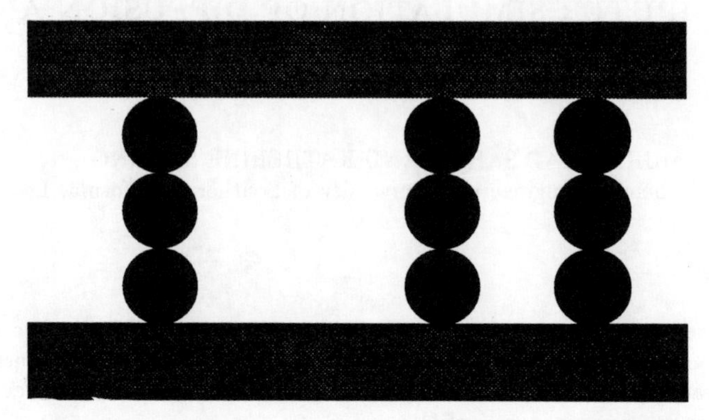

Figure 1: Schematic representation of the pillared clays.

suit a particular application. However, despite their industrial importance and the wealth of experimental information currently available, only recently has fundamental theoretical effort been undertaken [14-16] to investigate diffusion and adsorption in pillared clays. In this paper we use Molecular Dynamics (MD) and Grand-Canonical Ensemble Monte Carlo (GCEMC) simulations to study diffusion and adsorption of finite-size molecules in model pillared clays.

THE MODEL OF PILLARED CLAYS

Our model of pillared clays is shown schematically in Fig. 1. The surface of the solid walls is assumed to be the (100) face of a face-centered cubic solid. The pillars consist of a given number of atoms or groups, and are assumed to be distributed either randomly or uniformly between the solid walls. Two parameters, the separation between the solid walls h and the porosity of the system φ are used to characterize the morphology of the clays.

MOLECULAR DYNAMICS SIMULATION

The following potential function

$$\phi(r) = \begin{cases} \phi_{LJ}(r) - \phi_{LJ}(r_c) & \text{if } r \leq r_c \\ 0 & \text{if } r > r_c \end{cases} \tag{1}$$

is used to describe the interactions between sorbate particles as well as between sorbate particles and the pillars, where r_c is the truncation distance of the potential, and ϕ_{LJ} is the standard Lennard-Jones $6-12$ potential. The advantage of this potential is that it avoids the long-range corrections, which are difficult to evaluate accurately for our system due to the presence of the solid walls [17]. Assuming pairwise additivity and a rigid solid, the instantaneous potential energy between all sorbate particles and the pillars, and that associated with the sorbate-sorbate interactions are easily obtained. The potential function to describe the interaction between the N_s sorbate particles and the solid walls is given by [18]

$$\psi_{sw} = 2\pi\epsilon_{sw}\sum_{i=1}^{N_s}\left[\frac{2}{5}\left(\frac{\sigma_{sw}}{z_i}\right)^{10} - \left(\frac{\sigma_{sw}}{z_i}\right)^4 - \frac{0.4714}{(z_i/\sigma_{sw}+0.4314)^3}\right], \tag{2}$$

where ϵ_{sw} and σ_{sw} are the potential parameters that describe the interaction between the sorbate particles and the solid walls, and z_i is the vertical distance between the center of the ith sorbate particle and the solid wall.

At the outset of the simulation, several variables are specified. These are the geometric parameters h and φ, the interaction parameters of the potential functions, the temperature T of the system and the density of the sorbate particles, and the simulation parameters such as the magnitude of the time step, number of simulation steps, and the cut-off distance r_c. The characteristic diameter of a pillar unit is chosen the same as the particle size parameter σ, and thus all lengths are expressed in units of σ. In pillared clays h is about $12-20\mathring{A}$, and we use $\sigma = 3.405\mathring{A}$, the value for argon, so that $h = 4-6$ (in units of σ). Periodic boundary conditions are used in the open directions of the system, and the size of the system is selected such that the number of sorbate particles it contains can statistically represent the desired density. The porosity φ was varied between 1 (a slit pore) and the percolation threshold φ_c of the system, the critical porosity below which no macroscopic diffusion is possible, and the density ρ_s of the sorbate particles was varied from to 0.6 to study the density dependence of the diffusivity. The spatial distribution of the pillars is not completely understood yet [19]. Therefore, we studied two different spatial distributions of the pillars, namely, a random distribution and a uniform one. In the uniform distribution, the pillars are distributed on the site of a square lattice.

A standard fifth-order Gear predictor-corrector algorithm [20] was used to integrate Newton's equation of motion. The simulation begins with a random assignment of momenta and positions to the N_s sorbate particles, subject to the constraint of zero total momentum. The first several thousands time steps of the simulations are spent for system equilibration, in which the velocities of particles are scaled and adjusted periodically according to the specified temperature of the system. These adjustments are terminated once the difference between the simulated and the specified temperatures was less then 0.005. Then the trajectories of the molecules are determined for several tens of thousands of simulation time steps, and the effective diffusivity D of the particle are evaluated from their mean square displacements (MSD), or from the Green-Kubo equation that relates D to the velocity autocorrelation function. Although both methods were used in this study, we report here the results that were obtained using the MSD method, because the results obtained with both methods are comparable for simulations of limited durations, and the MSD method is generally considered more reliable for describing the long time diffusive behavior of the molecules than the Green-Kubo equation.

GRAND-CANONICAL ENSEMBLE MONTE CARLO SIMULATION

To study the adsorption behavior and to facilitate comparison with experimental studies where the sorbate loading is controlled externally by varying its chemical potential (via pressure monitoring), we made use of the GCEMC method. In this method we specify the sorbate chemical potential μ, the temperature T and the volume V of the system. The standard GCEMC algorithm of Adams [21] was used, where sorbate molecules are added or removed randomly with the acceptance probability

$$p^+ = \min\left\{1, \frac{V}{N_s+1}\exp[(\mu-\psi)/kT]\right\} , \tag{3}$$

$$p^- = \min\left\{1, \frac{N_s}{V}\exp[(\psi-\mu)/kT]\right\} , \tag{4}$$

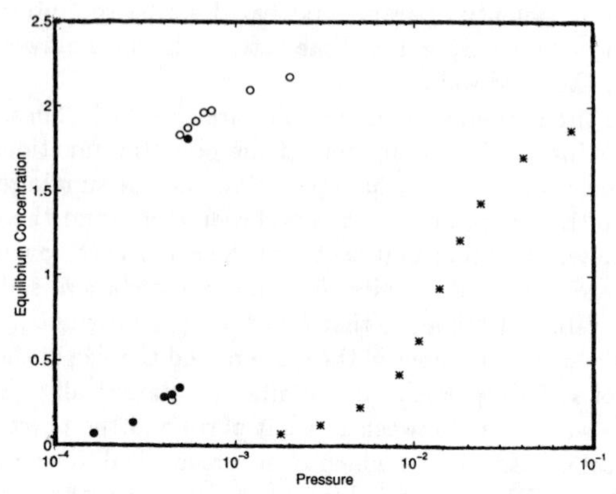

Figure 2: Adsorption isotherms at $T = 1.2$ (*) and $T = 0.8$ (circles).

where ψ is the interaction potential energy experienced by the sorbate molecule to be added or removed. The equilibrium sorbate loading is then evaluated as the ensemble average $\langle N_s \rangle / V$. Adsorption isotherms are obtained by systematically varying μ and evaluating the corresponding sorbate loading. Each GCEMC simulation consists of a minimum of 4×10^6 MC configurations after about 10^6 configurations for equilibrium.

ADSORPTION ISOTHERMS

Figure 2 shows the adsorption isotherms at two different (dimensionless) temperatures and the porosity $\varphi = 0.97$. At the higher temperature we have a continuous adsorption isotherm in which the equilibrium concentration of the adsorbed molecules rises with the external pressure P, as expected. The relationship between P and μ (set in the simulation) is given by an equation of state (the Lennard-Jones fluid in our case). However, at the lower temperature there is a sharp discontinuous (first-order) transition as the pressure rises, which can be interpreted as a sort of condensation. At this temperature and at lower pressures one has two monolayers of the molecules, one on each solid wall. However, as the pressure rises an additional monolayer is formed in between the two solid walls. Figure 2 also shows the hysteresis phenomenon that is routinely observed in adsorption phenomena, namely, the two branches of the isotherms are not idential if one starts from a low pressure and increases it and vice versa.

DIFFUSION COEFFICIENT

Figure 3 shows the corresponding diffusivities as a function of the pressure at the two temperatures. As expected, D decreases continuously with the pressure at the higher T. However, at the lower temperature there is a sharp discontinuity in the curve, corresponding to the one seen in Fig. 2. The lower branch of this curve corresponds to the upper branch of the corresponding adsorption isotherm shown in Fig. 2, where three monolayers of the molecules have been formed, while the upper branch with the higher diffusivities corresponds to the lower branch of the adsorption isotherm with two monolayers.

Elsewhere [16] we have determined the dependence of the diffusivity on the porosity of the system, the molecular density, and the spatial distribution of the pillars. We have

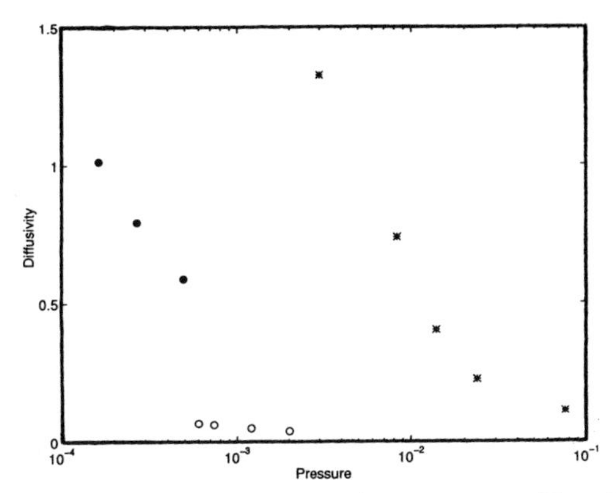

Figure 3: The dependence of the diffusivity on the pressure at $T = 1.2$ ($*$) and $T = 0.8$ (circles).

shown that while at high porosities (larger than 0.9) uniform and random distributions of the pillars may yield comparable estimates of various quantities of interest, the difference between the two systems increases as the porosity of the system decreases, and may become quite large at low porosities near the percolation threshold. We have also determined [16] the dependence of D on the separation distance h between the solid walls of the system, and have shown that for small values of h the diffusivity increases with h, but for $h > 5$ it attains an essentially constant value, implying that for $h > 5$ the system is effectively three dimensional, which is why D becomes independent of h. Note that with $\sigma = 3.405\AA$ that we used in our simulations, $h = 6$ is equivalent to $h \simeq 20\AA$, which is about the upper limit of the distance between the silicate layers in most pillared clays. Another important characteristic property of pillared clays is their percolation threshold φ_c [22]. Our MD simulations yielded [18] $\varphi_c \simeq 0.8250$. It can be shown [16] that for our model of pillared clays the critical porosity is $\varphi_c = 1 - \pi/18 \simeq 0.8255$, in perfect agreement with our estimate, which is also indicative of the accuracy of our simulations. From their experimental data, Chen *et al.* [23] estimated that $\varphi_c \simeq 0.79$, also in agreement with our result.

We have studied many other properties of diffusion and adsorption in pillared clays, including the dependence of D, the adsorption isotherms, the molecular density profile and the solvation force on the porosity, the temperature, the molecular density, and the spatial distribution of the pillars. The results are given elsewhere [18,26].

ACKNOWLEDGMENTS

This work was supported by the National Science Foundation under Grant No. CTS-9122529. The work of M. S. was also supported in part by the Department of Energy. We are grateful to Professor Dietrich Wolf for allowing us to have continued access to the computer facilities at the HLRZ Center in KFA Jülich, Germany.

REFERENCES

1. M. Sahimi, G. R. Gavalas and T. T. Tsotsis, *Chem. Eng. Sci.* **45**, 1443 (1990).

2. W. M. Deen, *AIChE J.* **33**, 1409 (1987).

3. M. Sahimi, *J. Chem. Phys.* **96**, 4718 (1992).

4. G. T. Kerr, *Sci. Am.* **261**, 100 (1989).

5. T. J. Pinnavaia, *Science* **220**, 365 (1983); P. Laszlo, *ibid.* **235**, 1473 (1987).

6. Y. W. Lee, R. H. Raythatha and B. J. Tatarchuk, *J. Catal.* **115**, 159 (1989).

7. M. L. Occelli, R. A. Innes, F. S. S. Hwu and J. W. Hightower, *Appl. Catal.* **14**, 69 (1985).

8. R. M. Barrer and D. M. MacLeod, *Trans. Faraday Soc.* **51**, 1290 (1955).

9. G. W. Brindley and R. E. Sempels, *Clays Clay Miner.* **12**, 229 (1977).

10. H. Lahav, V. Shani and J. Shabtai, *Clays Clay Miner.* **26**, 107 (1978).

11. D. E. W. Vaughan and R. J. Lussier, in *Proceedings of the 5th International Conference on Zeolites*, Naples (1980).

12. R. E. Grim, *Clay Mineralogy* (McGraw-Hill, New York, 1986).

13. R. T. Yang and M. S. Baksh, *AIChE J.* **37**, 679 (1991).

14. M. Sahimi, *J. Chem. Phys.* **92**, 5107 (1990).

15. R. Cracknell, C. A. Koh, S. M. Thompson and K. E. Gubbins, *MRS Proc.* **290**, 135 (1993).

16. X. Yi, K. S. Shing and M. Sahimi, *AIChE J.* **41** (February 1995).

17. I. K. Snook and W. van Megan, *J. Chem. Phys.* **72**, 2907 (1980).

18. W. A. Steele, *Surf. Sci.* **36**, 317 (1973).

19. M. S. Baksh and R. T. Yang, *AIChE J.* **38**, 1357 (1992).

20. C. W. Gear, *Numerical Initial Value Problems in Ordinary Differential Equations* (Prentice-Hall, Englewood Cliffs, New Jersey, 1971).

21. D. J. Adams, *Mol. Phys.* **29**, 307 (1975).

22. M. Sahimi, *Applications of Percolation Theory* (Taylor and Francis, London, 1994).

23. B. Y. Chen, H. Kim, S. D. Mahanti, T. J. Pinnavaia and Z. X. Cai, *J. Chem. Phys.* **100**, 3872 (1994).

24. X. Yi, K. S. Shing and M. Sahimi, *Chem. Eng. Sci.* (to be published).

REACTIVITY OF INTERSTITIAL Cl^- ANIONS IN SYNTHETIC HYDROTALCITE-LIKE MATERIALS IN THE HALIDE EXCHANGE REACTION WITH ALKYL HALIDES

Esteban López-Salinas*, Yoshio Ono** and Eiichi Suzuki**
*Instituto Mexicano del Petróleo, Gerencia de Catálisis y Materiales, Eje Central Lázaro Cárdenas 152, CP 07730 México, D.F.
**Tokyo Institute of Technology, Department of Chemical Engineering, Ookayama, Meguro-ku, 152 Tokyo, Japan

ABSTRACT

Two types of synthetic hydrotalcite-like $[Mg_6Al_2(OH)_{16}][A^{n-}]_{2/n}$ (HT) were prepared where A = Cl^- anions (Cl-HT), and $NiCl_4^{2-}$ complex anions (NiCl-HT), in order to examine: (1) the reactivity of Cl^- anion (or ligand) towards alkyl halides and (2) their catalytic behavior in the Finkelstein reaction. For instance, 0.4 g of NiCl-HT (Cl^-: 1.41 mmol) suspended in 20 cm^3 of toluene with 14.1 mmol of butyl bromide at 373 K yielded 42 % butyl chloride after 15 min of reaction. In comparison, $NiCl_4(Et_4N)_2$ (precursor to prepare NiCl-HT) yielded at the same conditions only 2.4 % of butyl chloride. This indicates a great mobility of Cl^- in the interlayer region of HTs. The reaction rate is greatly accelerated using DMF instead of toluene and butanol totally suppressed it. This is consistent with a S_N2 mechanism. The halide-exchange reaction between benzyl chloride and butyl bromide is catalyzed by both Cl-HT and NiCl-HT.

INTRODUCTION

Hydrotalcite (HT), $Mg_6Al_2(OH)_{16}CO_3.4H_2O$, a naturally occurring anionic clay comprises positively charged layers of Mg-Al hydroxides separated by compensating CO_3^{2-} anions and H_2O molecules which lie in the interlayer region. The use of hydrotalcite-like compounds as catalysts of several reactions such as aldol condensation[1] , and polymerization of β-propiolactone[2] or of propylene oxide[3] has been reported. However, the real catalyst in those studies is a mixture of paracrystalline metal oxides since the hydrotalcite-like material was calcined at ca. 723 K prior to its use as a catalyst. This procedure destroys the laminar structure[4]

The use of pristine HTs, that is with the layers intact,is very scarce. It has been reported that interlayer I^- anions in a hydrotalcite-like material, $Zn_2Cr(OH)_6I.2H_2O$, function as a nucleophile which reacts with butyl bromide to give butyl iodide[5]. Suzuki et al.[6] have reported that several synthetic hydrotalcites catalyze the halide-exchange between alkyl halides. In this study two types of synthetic hydrotalcite-like compounds, $[Mg_6Al_2(OH)_{16}][A^{n-}]_{2/n}$ where A = Cl^- (Cl-HT), and $NiCl_4^{2-}$ (NiCl-HT), were synthesized with a view to examine the reactivity of interstitial Cl^- anions (or ligands) towards alkyl bromides in the liquid-phase and as catalysts in the halide-exchange reaction between alkyl halides (Finkelstein reaction).

Mat. Res. Soc. Symp. Proc. Vol. 368 © 1995 Materials Research Society

EXPERIMENTAL

Synthesis of the hydrotalcite-like materials

Hydrotalcite-like materials were synthesized following the method of Miyata[7]. Cl-HT was obtained by adding aqueous NaOH dropwise into an aqueous solution containing $MgCl_2.6H_2O$ and $AlCl_3.6H_2O$ with stirring at 313 K. The slurry was transferred to an autoclave and aged at 448 K for 18 h. After repeated washing with decarbonated water, the Cl-HT was dried in vacuo at 393 K for 4 h. NiCl-HT was prepared by anion-exchanging $[NiCl_4]^{2-}$ (in an ethanol solution) with NO_3^- in a NO_3-HT as reported elsewhere[8].

Halide exchange reactions

The halide exchange reactions between Cl-HT (or NiCl-HT) and the alkyl bromides were carried out in a 50 cm^3 flask equipped with a condenser. The reactions were conducted under a nitrogen atmosphere. A portion of the liquid-phase was withdrawn periodically and analyzed by gas chromatography on 2 m long SE-30 column and a flame ionization detector.
A 0.4 g of dried NiCl-HT (Cl^-: 1.41 mmol) was suspended in 20 cm^3 of a solvent and then 14.1 mmol of alkyl halide were added. In the case of Cl-HT, 1 g (Cl^-: 3.25 mmol) of dried Cl-HT and 32.5 mmol of alkyl halide in 30 cm^3 of solvent were used.

For the halide exchange reaction between benzyl chloride and butyl bromide, 14.1 mmol of each of the reactants were added to 0.4 g of NiCl-HT in 20 cm^3 of solvent. When using Cl-HT as the catalyst, 1 g of Cl-HT and 32.5 mmol of each of the reactants in 30 cm^3 of solvent were used.

UV-vis Diffuse Reflectance spectroscopy

The UV-vis Diffuse Reflectance (DR) spectra were recorded at room temperature on a Shimadzu MPS-2000 spectrophotometer using $BaSO_4$ as a baseline standard. The quartz cell used for the measurements could be connected to a conventional high vacuum system. The intensity of the bands is reported as the Schuster-Kubelka-Munk function.

RESULTS AND DISCUSSION

Halide exchange between butyl bromide and NiCl-HT (or Cl-HT)

Table I shows the results of reacting various Cl-containing compounds with butyl bromide at 373 K using toluene as the solvent. NiCl-HT yields 42.1 and 84.4 % mol of butyl chloride after 15 and 150 min of reaction, respectively. In contrast, the salt used to intercalate $[NiCl_4]^{2-}$ anions into HT, $NiCl_4[Et_4N]_2$, yielded only 2.4 and 6.9 % mol of butyl chloride at 15 and 150 min of reaction, respectively. $NiCl_2$ showed no reactivity. Cl-HT seems to exchange its Cl^- faster than NiCl-HT. These results suggest that the hydrotalcite layers may be exherting an activation effect on interstitial Cl^- anions (or ligands).

Table I. Reactivity of various Cl-containing compounds in the halide exchange with butyl bromide.

Cl-compounds	C_4H_9Cl yield / % mol	
	15 min	150 min
NiCl-HT (a)	42.1	84.4
$NiCl_4[Et_4N]_2$ (b)	2.4	6.9
$NiCl_2$ (c)	0	0
Cl-HT (d)	45.0	95.0

Reaction conditions: Butyl bromide (14.1 mmol) in toluene (20 cm^3) at 373 K with (a) 0.4 g of NiCl-HT (Cl: 1.41 mmol), (b) 0.16 g of $NiCl_4[Et_4N]_2$, (c) 0.09 g of $NiCl_2$ and (d) 0.38 g of Cl-HT.

Alkyl group and solvent effects

In Table II the effect of increasing the chain of the alkyl group in the alkyl bromide or using a bromide bonded to a secondary carbon (isopropyl bromide) results in a considerable decrease in the halide exchange rate, suggesting an S_N2 mechanism. The same trend was observed when using indistinctively Cl-HT or NiCl-HT. This can be more straightforwardly demonstrated by changing the nature of the reaction solvent. In Fig. 1 the time course of butyl chloride formation using various solvents in the halide exchange between butyl bromide and NiCl-HT at 373 K is shown. The reaction proceeds with non-polar n-decane or toluene solvents. Using polar aprotic DMF or DMSO solvents accelerates the reaction enormously, exchanging all interstitial Cl^- in less than 10 min. However, when polar hydroxilic n-butanol or ethanol are the reaction solvents, the halide exchange is practically suppressed. The effect of these solvents is typical when a S_N2 mechanism operates[9]. An examination by XRD of the NiCl-HT (or Cl-HT), after being used as catalysts with the above solvents, indicates that the layered structure and the interlayer distance remain unaltered (ca. 3.3 Å for NiCl-HT).Therefore, it seems unlikely that alkyl bromides or solvent molecules diffuse into the interlayer region, pointing out that the halide exchange may be taking place in the edges of the interlamelar region. A great mobility of Cl^- (or Br^-) in the interlayer is suggested.

Table II. Effect of type of alkyl group on the halide exchange between NiCl-HT and alkyl bromides.

Alkyl bromide	Alkyl Chloride yield/ %mol	
	15 min	150 min
Propyl bromide	27.0	64.3
Butyl bromide	24.3	60.3
Isopropyl bromide	12.1	41.9

Reaction conditions: Alkyl bromide (14.1 mmol) and 0.4 g of NiCl-HT (Cl: 1.41 mmol) in toluene (20 cm^3) at 353 K

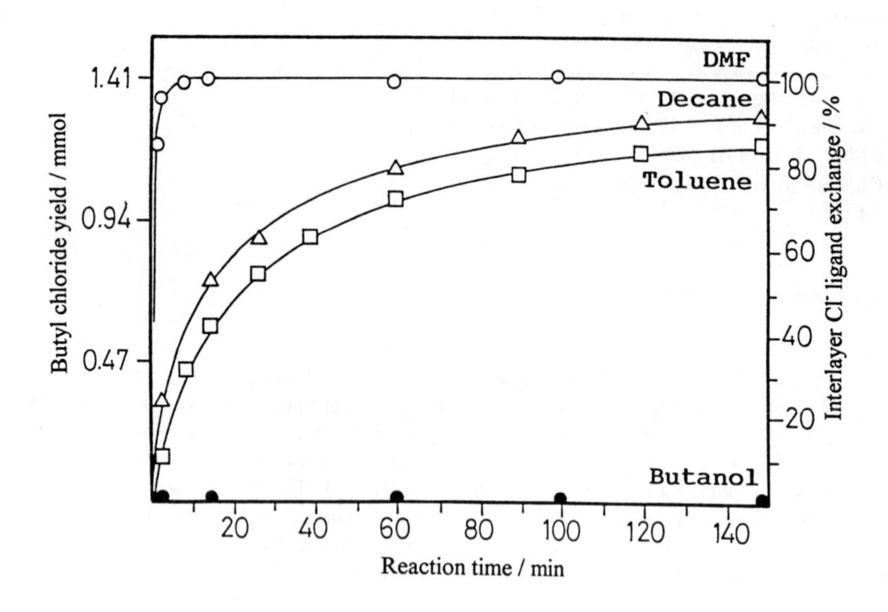

Fig. 1 Effect of the reaction solvent in the halide exchange between NiCl-HT and butyl bromide. Reaction conditions: Butyl bromide (14.1 mmol) and 0.4 g of NiCl-HT (Cl : 1.41 mmol) in 20 cm^3 of solvent at 373 K.

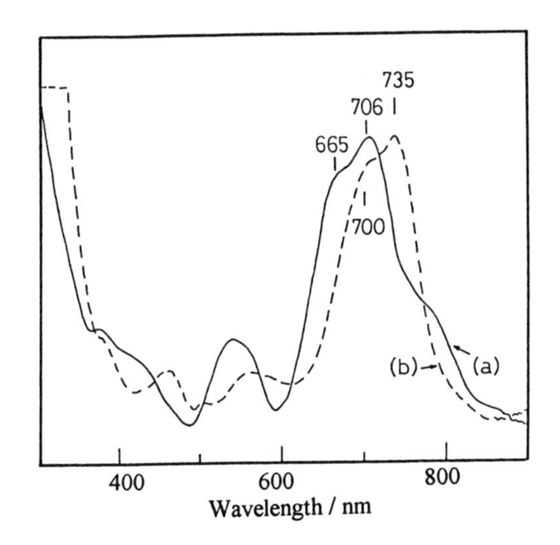

Fig. 2 UV-vis Diffuse Reflectance spectra of NiCl-HT. (a) After heating under vacuum at 473 K (before the halide exchange reaction). (b) After 150 min of reaction with butyl bromide in DMF, washing with toluene and heating under vacuum at 473 K.

Coordination environment of interstitial Ni(II)

As described above, Cl^- in NiCl-HT can be replaced by Br^- from alkyl bromides. The problem here is whether the incoming Br^- exists as Br^- ion or as Ni-coordinated ligands in the interlayers. To throw light on this problem, the coordination environment of Ni(II), before and after the halide exchange reaction, was examined by UV-vis DR spectroscopy. In Fig. 2(b) the spectrum of a NiCl-HT after reaction in DMF with butyl bromide (see Fig. 1), washed with anhydrous toluene and dried in vacuo at 473 K, shows a clearly shifted band compared to that of the initial NiCl-HT before reaction (spectrum (a)). The doublet band at 700, 735 nm agrees very well with that of tetrahedral $[NiBr_4]^{2-}$, which exhibits a doublet at 707, 756 nm from the $^3T_1(F)$-$^3T_1(P)$ term[10]. These results indicate that Br^- from butyl bromide indeed diffuses through the interlayer and exchanges with Cl^- in $[NiCl_4]^{2-}$ (ligand exchange reaction) preserving the initial tetrahedral environment around Ni(II).

Catalysis by NiCl-HT

In Table III the halide exchange between benzyl chloride and butyl bromide with various catalysts is shown. The reaction is nule when using toluene and no catalyst. However, when using DMF as the solvent (without a catalyst) the reaction proceeds a little. On the other hand, when NiCl-HT is added, whether in toluene or DMF, 40 and 61 % mol of benzyl bromide, respectively, formed after 60 min of reaction. These results indicate that NiCl-HT acts like a catalyst. Noteworthy, NiCl-HT (heterogeneous catalysts) showed very close activity to that of $NiCl_4[Et_4N]_2$ (homogeneous catalyst, because it dissolves in DMF)), both in DMF, suggesting that hydrotalcite layers could be exploited to heterogenize catalytically active anions.

Table III. Catalysis of halide exchange between benzyl chloride and butyl bromide at 373 K.

Catalyst	Solvent	Benzyl bromide yield / % mol at reaction time (min)	
		15 min	60 min
None	Toluene	0	0
	DMF	4	10
NiCl-HT	Toluene	22	40
	DMF	45	61
$NiCl_4[Et_4N]_2$	Toluene*	0	0
	DMF*	58	60

Reaction conditions: $C_6H_5CH_2Cl$=C_4H_9Br (14.1 mmol), NiCl-HT (0.4 g), 373 K, solvent (20 cm^3). * $NiCl_4[Et_4N]_2$ is insoluble in toluene and soluble in DMF.

CONCLUSIONS

Alkyl bromides undergo nucleophilic substitution by Cl^- anions (or ligands) intercalated in the hydrotalcite layers in the liquid phase. The selection of the reaction solvent being crucial. Thus, in DMF, n-decane or toluene butyl bromide is converted to butyl chloride, but n-butanol and ethanol inhibit the reaction. Interstitial tetrahedral $[NiCl_4]^{2-}$ in NiCl-HT are isomorphically converted into $[NiBr_4]^{2-}$ after halide exchange of NiCl-HT with butyl bromide. The halide exchange between benzyl chloride and butyl bromide is catalyzed by both NiCl-HT and Cl-HT.

REFERENCES

1. E. Suzuki and Y. Ono, Bull. Chem. Soc. Jpn. **61**, 1008 (1988).
2. T. Nakatsuka, H. Kawasaki, S. Yamashita and S. Kohjiya, Bull. Chem. Soc. Jpn. **52**, 2449 (1979).
3. S. Kohjiya, T. Sato, T. Nakamura and S. Yamashita, Makromol. Chem., Rapid Commun. **2**, 231 (1981).
4. W. T. Reichle, J. Catal. **94**, 547 (1985).
5. K. J. Martin and T. J.Pinnavaia, J. Am. Chem. Soc.**108**, 541 (1986).
6. E. Suzuki, M. Okamoto and Y. Ono, J. Mol. Catal. **61**, 283 (1990).
7. S. Miyata, Clays and Clay Miner.**23**, 369 (1975).
8. E. López-Salinas and Y. Ono, Microporous Mats. **1**, 33 (1992).
9. P. Sykes, A guidebook to Mechanism in Organic Chemistry, 5th edn. (Longman, New York, 1982), p. 80.
10. A.B.P. Lever, Inorganic Electronic Spectroscopy (Studies in Physical and Theoretical Chemistry, vol. 33), 2nd edn. (Elsevier, 1984), p. 530.

MOLECULAR SIEVE SYNTHESIS USING METALLOCENES AS STRUCTURE DIRECTING AGENTS.

KENNETH J. BALKUS, JR.*, ALEXEI G. GABRIELOV AND NETANYA SANDLER
University of Texas at Dallas, Department of Chemistry,
Richardson, TX 75083-0688

ABSTRACT

We have been exploring molecular sieve synthesis using metal complexes as templates. Our objectives were to affect the crystallization of new structures as well as prepare bifunctional catalysts via the encapsulation of metal complexes during synthesis. In this paper the results for the synthesis of SiO_2, $AlPO_4$ and SAPO molecular sieves in the presence of metallocenes will be presented. In particular, Cobalticinium ion, Cp_2Co^+, was found to be a template for Nonasil, $AlPO_4$-5, $AlPO_4$-16, SAPO-16 and CoAPO-16 which are known phases but require templates. The larger decamethyl derivative, $Cp^*_2Co^+$ has produced a novel high silica zeolite molecular sieve, UTD-1 as well as a new SAPO phase UTD-2 which appear to be large pore materials. In all cases, the metal complex is occluded intact and cannot be extracted. Aspects of the synthesis and characterization of these microporous materials are presented below.

INTRODUCTION

The synthesis of crystalline molecular sieves often requires an organic additive to act as a template or structure directing agent [1]. Organic molecules such as aliphatic amines are frequently employed to alter the gel chemistry and/or act as a void filler. In contrast, there are relatively few examples where metal complexes have been employed as structure directing agents. We have developed a method for synthesizing a zeolite around well defined metal complexes in an effort to prepare *ship-in-a-bottle* type complexes [2]. Such host-guest materials may function as hybrid catalysts by coupling the reactivity of the metal complex, which is not necessarily bound to the surface, with the shape selectivity of the molecular sieve [3]. In particular, zeolite NaX or Y entrapped metal phthalocyanines have shown promise as oxidation catalysts. Recently, polymer embedded zeolite NaY containing iron phthalocyanines were found to be catalysts for the oxidation of alkanes with activity comparable to enzymes [4]. The addition of metal complexes during molecular sieve crystallization is clearly a viable strategy for the preparation of these *ship-in-a-bottle* type catalysts, however, certain metal complexes may also affect the synthesis of novel materials. Such a metal complex with templating properties has been realized with bis(cyclopentadienyl)cobalt(III) ion, Cp_2Co^+. The cage type zeolites ZSM-51 [5] and ZSM-45 [6] were the first molecular sieves prepared using Cp_2CoPF_6. Subsequently, we

Mat. Res. Soc. Symp. Proc. Vol. 368

found Cp_2CoOH was a template for the all silica clathrate type molecular sieve Nonasil [7,8] using both hydroxide and fluoride mineralizing agents. More recently, Cp_2CoF was used to prepare Octadecasil and Dodecasil-1H [9]. We have also found that Cp_2CoOH produces the aluminum phosphate molecular sieves $AlPO_4$-5 and $AlPO_4$-16 [10]. All of these molecular sieves normally require an organo-cation to form which further supports the idea that cobalticinium ion acts as a structure directing agent. In this paper we report the preparation and characterization of the silicoaluminum phosphate SAPO-16 and the cobalt aluminum phosphate analog, CoAPO-16.

It is becoming apparent that Cp_2Co^+ ion has a propensity to generate small pore, cage type molecular sieves. If we could alter the size, shape and symmetry of this complex by functionalizing the cylopentadienyl rings, then one might anticipate the structure directing properties would be proportionately modified. Therefore, we have employed the larger bis(pentamethylcyclopentadienyl)cobalt(III) ion, $Cp^*_2Co^+$, and report herein the preliminary templating properties observed for this complex. We have crystallized a large pore, high silica zeolite, UTD-1, as well as a novel SAPO phase, UTD-2. Details of the synthesis and preliminary characterization will be presented.

EXPERIMENTAL

The metal complex Cp_2CoOH was prepared as previously described [7]. $Cp^*_2CoPF_6$ was prepared according to the published procedure [11] and then transformed to the hydroxide salt in the same manner as Cp_2CoOH. All other reagents were used as received.

SAPO-16 was prepared by adding 0.512 grams of fumed silica to 7.86 grams of a 31% aqueous Cp_2CoOH solution. The resulting gel was stirred at room temperature for 2 hours. Then 4.0 grams of aluminum hydroxide, 4.4 grams of 85% H_3PO_4 and 4.5 grams of deionized water were combined and stirred for one hour. The silicate and aluminum phosphate solutions were then mixed, resulting in a gel with a molar ratio of Al_2O_3:P_2O_5:SiO_2:$[Cp_2Co^+]$:H_2O = 1:0.9:0.4:0.56:40. The gel was aged with stirring for 15 hours and then transferred to a 25 mL Teflon-lined pressure reactor (Parr) .The reactor was heated at 200°C under static conditions for 20 hours. The resulting yellow crystalline product was isolated by centrifugation, washed with deionized water, suction filtered and then dried at room temperature for 24 hours.

The CoAPO-16 can be prepared by first combining 3.4 mL of deionized water, 3.2 g 85% H_3PO_4 and 1.5 grams $CoSO_4 \cdot 7H_2O$. After stirring for thirty minutes, 2.0 grams $Al(OH)_3$ were slowly added to the solution and mixed for an additional hour. This was followed by the addition of 10.6 grams of a 26.7% Cp_2CoOH solution, resulting in a gel with a molar ratio of Al_2O_3:P_2O_5:CoO:$[Cp_2Co]^+$:H_2O = 0.8:1:0.4:1:55. The gel was aged for 4 hours, transferred into an autoclave and then heated at 175°C under static conditions for 40 hours. The blue-green CoAPO-16 was isolated by centrifugation, washed with deionized water and dried at room temperature.

All silica UTD-1 was prepared by combining 0.0275 grams of NaOH, 6.7 mL of deionized water, 0.73 grams of 31.6% Cp^*_2CoOH aqueous solution and 0.4 grams of fumed silica (Aldrich). The resulting gel having a molar ratio of SiO_2:Na_2O:$[Cp^*_2Co]^+$:H_2O = 1:0.05:0.1:60 was aged for 1 hour at room temperature. The gel was then transferred to a 25 cm^3 teflon-lined pressure reactor and heated at 175°C for 2 days. The crystallization mixture was then cooled to RT, diluted with 250 mL of deionized water, suction filtered, washed with deionized water and dried at 90°C for 2 hours.

SAPO UTD-2 was prepared by first slowly mixing 2 grams of aluminum hydroxide to 2.22 g of 85 wt % H_3PO_4 and 2.0 mL of deionized water. This gel was combined with 0.58 grams of a 30 wt % $Cp^*{}_2CoOH$ aqueous solution, 0.256 grams of fumed silica and 9.28 grams of 40 wt % tetrapropylammonium hydroxide (TPAOH). The resulting gel having a molar ratio of $Al_2O_3:P_2O_5:SiO_2:[TPA]^+:[Cp^*{}_2Co]^+:H_2O$ = 1:0.9:0.4:1.7:0.05:50 was aged with stirring for 6 hours. The gel was then transferred to a 25 mL Teflon-lined pressure reactor and heated at 200°C for 24 hours. The crystallization mixture was then cooled to RT, diluted with 100 mL of deionized water, centrifuged, washed with deionized water, isolated by suction filtration and dried at RT for 24 hours.

X-ray powder diffraction patterns were recorded on a Scintag XDS 2000 diffractometer using CuKα radiation with a chopper increment of 0.01 and a scan rate of 1 deg min^{-1}. CaF_2 was used as an internal standard. Scanning electron micrographs were obtained using a Phillips XL SEM.

RESULTS AND DISCUSSION

The metal complex Cp_2CoOH as been employed in the synthesis of SAPO-16 and CoAPO-16. The hydroxide salt of the Cp_2Co^+ ion is preferred because of the higher solubility and ability to buffer the pH. Both SAPO-16 and CoAPO-16 are isostructural with $AlPO_4$-16 and Octadecasil which were previously prepared using Cp_2Co^+ as the template. The AST topology consists of double 4 rings connected by single tetrahedral atoms. This results in 18 hedron cages consisting of 4 and 6 membered rings which is large enough to accommodate the metal complexes (~5.2 x 4.86 Å). However, the Cp_2Co^+ ions are trapped and cannot be removed without destroying the lattice. We find ~4 complexes per unit cell which is what would be expected for entrapped template molecules.

The AST structure is apparently not easy to form since only quinuclidene was previously known as a template for $AlPO_4$-16, CoAPO-16 [12] or SAPO-16 [13]. The formation of $AlPO_4$-16 [10] using Cp_2CoOH is complicated by the formation of $AlPO_4$-5. In contrast, SAPO-16 and CoAPO-16 can be prepared as a highly crystalline single phase materials. There is no evidence of SAPO-5 or CoAPO-5 even at different temperatures. Figure 1 shows the XRD pattern for as synthesized SAPO-16 containing the Cp_2Co^+ ion which is essentially the same for CoAPO-16. The unit cells are slightly larger than those for the same phases prepared with quinuclidene [12,13] but otherwise no significant changes. The XRD data and unit cell parameters will be reported elsewhere [14]. The Scanning electron micrograph of as synthesized CoAPO-16 containing Cp_2Co^+ ion is shown in figure 2. These tetrahedral shaped crystals ~3 - 5 microns in diameter, appear to represent a characteristic morphology for molecular sieves having the AST topology. We observe the same crystal shape for $AlPO_4$-16, SAPO-16 and CoAPO-16. Additionally, the organo-cation templated $AlPO_4$-16 is also reported to crystallize as tetrahedra.

So far at least five different topologies (AST, NON, LEV, DOH, AFI) have been prepared using Cp_2Co^+ as a template. In most cases with the exception of $AlPO_4$-5, the resulting molecular sieves have clathrate or cage type structures. This could be useful in designing a synthesis such as SAPO-37 [13] where two templates are employed presumably to form the small cages and then connect them.

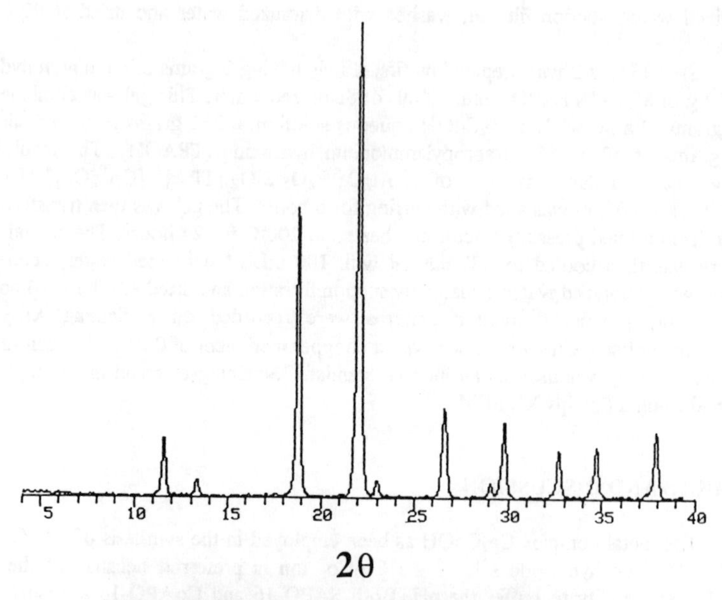

Figure 1. XRD pattern for as synthesized SAPO-16

Figure 2. Scanning electron micrograph of as synthesized CoAPO-16

We now turn our attentions to the structure directing properties of the larger bis(pentamethylcylopentadienyl)cobalt (III) ion, $Cp^*{}_2Co^+$. The addition of methyl groups to the Cp ring increase the effective diameter of the molecule by ~2Å. Therefore, we anticipated larger pore molecular sieves might result with this complex. In a first attempt we tried to approximate the conditions used for Cp_2CoOH in the synthesis of Nonasil [7] but the resulting material was clearly not a small pore clathrate. Instead we discovered a novel large pore molecular sieve we refer to as UTD-1. The XRD pattern for as synthesized UTD-1 containing the $Cp^*{}_2Co^+$ ion is shown below in Figure 3. The scanning electron micrograph for UTD-1 is shown in Figure 4 which reveals bundles of two dimensional planks. The width of these crystals is ~2 microns across. UTD-1 is thermally stable up to

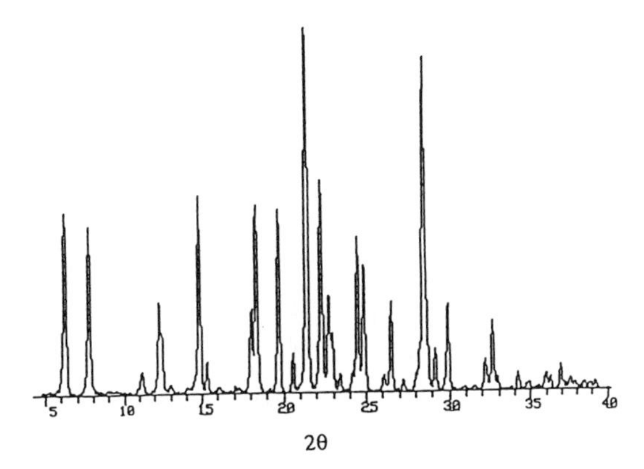

Figure 3. XRD pattern for as synthesized UTD-1

Figure 3. Scanning electron micrograph of high silica zeolite UTD-1

to 500°C with decomposition of the occluded metal complex at ~375°C. Preliminary, adsorption studies are consistent with a large pore channel type molecular sieve [15]. The calcined all silica UTD-1 has Bronsted acidity as evidenced by the formation of hydrocarbons from methanol. The presence of hexamethylbenzene in the products is also indicative of a large pore molecular sieve. A small amount of aluminum (Si/Al >350) can be introduced into the framework which enhances the catalytic activity.

These results encouraged us to examine the synthesis of SAPO phases and we were pleased to discover another new phase, UTD-2. This molecular sieve can be prepared using only Cp^*_2CoOH as the template. However, we obtained better crystallinity if we also include TPA in the gel mixture. The XRD pattern for as synthesized UTD-2 containing the $Cp^*_2Co^+$ ion is shown in Figure 4. The scanning electron micrograph in Figure 5 reveals hexagonal plate like crystals ~3 microns in diameter. UTD-2 is stable to

250°C if heated slowly in a vacuum. Adsorption and catalytic activity studies is currently in progress.

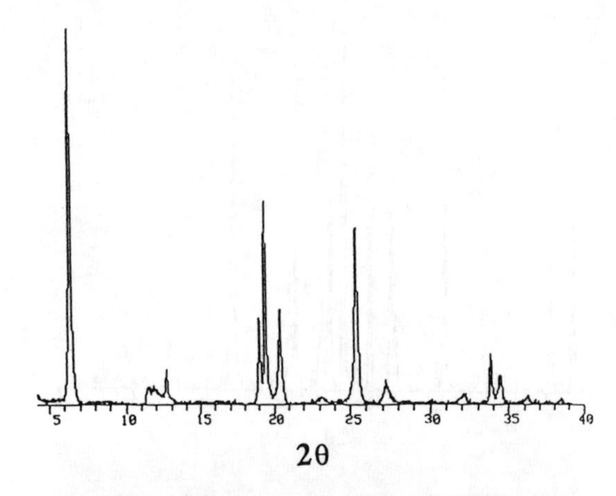

Figure 4. XRD pattern for as synthesized UTD-2

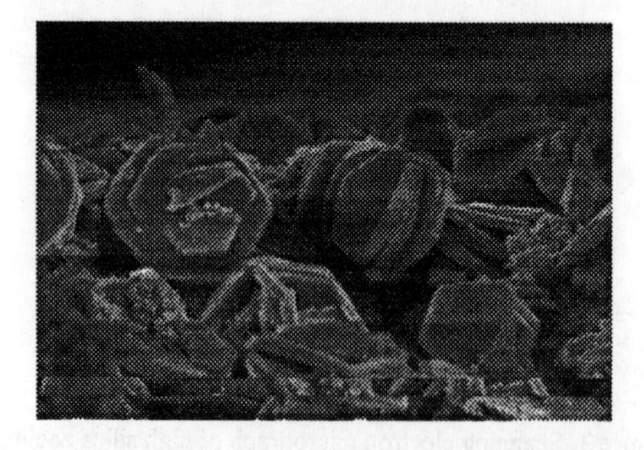

Figure 5. Scanning electron micrograph of SAPO molecular sieve UTD-2.

In conclusion, we prepared SAPO-16 and CoAPO-16 using Cp_2CoOH as the template. The application of Cp^*_2CoOH as a structure directing agent has resulted in two novel molecular sieves that appear to be large pore materials with catalytic potential.

ACKNOWLEDGMENTS

We wish to thank the National Science Foundation (CHE-9157014) and the Robert A. Welch Foundation for financial support of this work. We also thank the Clark Foundation for a summer student fellowship (NS).

REFERENCES

1. R. Szostak *Molecular Sieves,* (van Nostrand, New York, 1989).
2. K. J. Balkus, Jr. and S. Kowalak, U.S. Patent No. 5,167,942 (1992).
3. Balkus, Jr., K. J.; and Gabrielov, A. G. in *Inclusion Chemistry with Zeolites, Nanoscale Materials by Design*, edited by N. Herron and D. Corbin, (Kluwer, In Press).
4.. R. F. Parton, I. F. J. Vankelecom, M. Casselman, C. P. Bezoukhanova, J. B. Uytterhoeven and P. A. Jacobs, Nature, **370**, 541 (1994).
5. E. W. Valyocsik, U. S. Patent No. 4,568,654 (1986).
6. E. W. Valyocsik, U. S. Patent No. 4,556,549 (1985).
7. K. J. Balkus, Jr. and S. Shepelev, Micropor. Mater., **1**, 383 (1993).
8. K. J. Balkus, Jr. and S. Shepelev, Petrol. Preprints, **38,** 512 (1993).
9. P. Behrens and G. van der Goor, Poster at the 8th International Symposium on Molecular Recognition and Inclusion, Ottawa, (1994).
10. K. J. Balkus, Jr., A. G. Gabrielov and S. Shepelev, Micropor. Mater., In Press.
11. Kölle, U.; Khouzami, F. Chem. Ber., *114*, 2929 (1981).
12. B. M. Lok, C. A. Messina, R. L. Patton, R. T. Gajek, R. T. Cannan and E. M. Flanigen, U. S. Patent No. 4,440,871 (1984).
13. S. T. Wilson and E. M. Flanigen, U. S. Patent No. 4,567,029 (1986).
14. K. J. Balkus, Jr., A. G. Gabrielov and N. Sandler, J. Porous Mater., In Press.
15. K. J. Balkus, Jr., A. G. Gabrielov and S. I. Zones, Manuscript in Preparation.

Green Technology for the 21st Century: Ceramic Membranes

Water Chemistry Program
University of Wisconsin-Madison

Anderson, M.A., Zeltner, W.A. and Merritt, C.M.

Professor Marc A. Anderson
Water Chemistry Program
University of Wisconsin-Madison
660 N. Park Street
Madison, WI 53706

GREEN TECHNOLOGY FOR THE 21st CENTURY: CERAMIC MEMBRANES

Background

The nineties have brought us to an era of environmental crisis. We are faced with problems of air pollution, both indoor and outdoor, acid rain, water pollution, hazardous wastes, toxic landfills and leaking storage tanks in our soil, to name a few. Our rapid advances into the industrial and technological age have contributed to these problems. However, our advancing technology in the areas of remediation and such "green" engineering tools as ceramic membranes can certainly make a significant contribution in terms of environmental clean-up and a healthier world.

The market potential and applications of ceramic membrane technology are enormous. Clean water is one of the main issues of the nineties. For example, certain towns and villages in the U.S. are unable to drink municipal water due to chemical contamination. Three quarters of the U.S. population live in areas where the ground water that supplies their drinking water, is 50% contaminated. There are approximately 400 manufacturers in the home water treatment business and it is estimated that sales of these products are expected to reach over $1 billion a year in 1995.

The Water Chemistry Program at the University of Wisconsin-Madison is the only one of its kind in the world. Our group is at the cutting edge of research leading to innovative materials that can be used commercially to produce clean air and clean water. Ceramic membranes represent a relatively new class of materials. They can be produced from a variety of starting materials and processed in different ways to yield products with a broad range of physical-chemical characteristics and an equally large range of applications.

Mat. Res. Soc. Symp. Proc. Vol. 368 © 1995 Materials Research Society

These inorganic membranes possess many advantages over organic membranes, which frequently suffer catastrophic degradation in the presence of elevated levels of chlorine, a common cleaning agent. Ceramic membranes are impervious to attack by either aromatic compounds or microbial agents. They are stable over relatively broad ranges of pH and temperature. In fact, their robust character enables ceramic membranes to withstand high temperatures, elevated pressures, organic solvents and chemical and heat sterilization. These characteristics indicate that ceramic membranes could replace their organic polymeric counterparts in many applications where these conditions would otherwise preclude using an organic polymer membrane. Pore size can be controlled in these materials and is typically from 5-100A in diameter. Coupled with the large variety of surface chemistries which can be expressed within these nano-cavities, one can design these materials to perform a variety of tasks. These applications include separations, catalysis, photocatalysis and energy storage systems for solar cells and micromotor devices.

At this point in time, however, we are at a turning point. We do have 23 licensed patents to our name and a third of our money comes from industry. However, we are actively seeking additional industrial partners who are interested in commercializing this technology. Although there are several companies in Europe and Asia who are interested in working with us, we prefer to keep this technology and its development in the United States, if at all possible. Already, we lag behind Germany with its advanced environmental program. Germany, in a ten year time frame, has jumped into environmental leadership, beating out the U.S., Canada, Japan and Sweden. This lead also provides a premier position in world markets with associated economic benefits for that country. Germany has developed a close relationship between the areas of environmental protection, government regulation and economic development. This is currently making a positive impact fiscally and will certainly provide a strong lead into the 21st century. It is possible for us to compete on an even keel with Germany providing this window of opportunity is taken up.

The objective of this white paper is to present a case for American investment in ceramic membrane research. This cutting edge technology could well provide economic growth for existing industries, in addition to facilitating new growth industries in environmental remediation. A market survey published in 1992 on the economic benefits from alternate applications of inorganic membrane technology indicated that active implementation of such technology could well result in a $2 billion dollar per year sales market, a $16.6 billion increase in the national GDP, a $2 billion dollar improvement in the balance of trade and a decrease in energy use of 6 quads per year. No such study has occurred on the impact of such technology on environmental restoration and waste management, but it is projected that the above estimates could probably be doubled.

The advantages of ceramic membranes over their organic counterparts in separation systems will be discussed. We will also provide an overview of "sol-gel" chemical techniques used to prepare these novel membranes and examine the broad array of applications that we envisage for these materials as our society enters the 21st century. In particular, we focus attention on the colloid chemistry of particle preparation and particle-particle interactions as we attempt to control the physical-chemical properties of the membranes. We are able to produce a wide array of membrane devices for multiple applications by controlling the membrane and through selection of appropriate supporting materials. Properties of the applications will be discussed, as well as their relevance to American industries.

Ceramic Membrane Synthesis

We produce inorganic membranes through "sol-gel" chemical techniques. In this process, either inorganic polymers or particles are created through the controlled hydrolysis and condensation of inorganic or organometallic precursors. A suspension of these polymers or particles (a sol) is coated onto a support. As the solvent evaporates, the sol forms a "gel" on the support surface. This gel is then dried and fired into a supported membrane, or porous thin film. To date these sol-gel techniques for fabricating inorganic membranes are not only being applied in our own laboratories at the University of Wisconsin but also in many research and industrial settings throughout the world.

The successful synthesis of ceramic membranes requires control over five major processing steps:

1. **sol preparation**
2. **choice and pre-treatment of the support**
3. **coating the sol on the support**
4. **gelling and drying the supported membrane**
5. **firing the supported membrane.**

Within each of these processing steps, there are a subset of variables, some more important than others. We will examine some of the most important variables and our attempts to design the membrane for a particular application.

1. Sol Preparation

The most critical step in membrane fabrication is the preparation of a stable colloidal suspension or sol which then leads to membranes whose properties are both reproducible and controllable. Many variables must be considered in synthesizing such sols. Of great importance is the choice of starting materials or chemical precursors. This choice is largely dictated by the physical-chemical properties desired in the final membrane. Single, double and even triple component materials can be fabricated (e.g., a superconductive membrane such as a composite yttrium, barium and copper oxide).

It is possible to vary the nature of the inorganic membranes produced by sol-gel processes from polymeric to particulate by changing the molar ratio of water to the metal alkoxide, as well as by changing the solvent from water to alcohol. However, our interest has been in producing membranes with extremely small pores with a narrow range of pore size. These materials should be more easily prepared from particulate systems.

When synthesizing particulate sols, we prefer to produce particles with a uniform size that do not grow. Thus, we try to prevent the particles from contacting each other. For oxide particles such as TiO_2, Al_2O_3, Fe_2O_3, SiO_2, ZrO_2, etc., we can stabilize these particulates by adjusting the pH of the suspension. Many of the oxides are amphoteric solids whose surface charge is a function of the pH of the solution.

Interestingly, the isoelectric concept can also be used to determine whether a two component system is completely mixed or whether one oxide has coated the surface of the other. By monitoring the electrophoretic mobilities of the particles as a function of pH, we can determine conditions that produce charge stabilized sols and the acid-base behavior of these particles. Steric methods can also be used to produce stable colloidal systems. We have even produced metal membranes using surfactants to sterically stabilize these particles.

2. Choosing and Preparing the Support

Ceramic membranes by themselves are quite fragile materials which are only used to determine their physico-chemical characteristics. However, most applications require a robust system, so the stable sol must be coated onto a suitable supporting material. For instance, the support for applications involving separations would be a porous material. Other applications such as filtration catalysts or photocatalysts could employ non-porous materials as supports sensors.

A major application for ceramic membranes at present is for separation processes, such as ultrafiltration and gas phase separations; processes which require a porous support. The support may be metal (porous stainless steel), a ceramic (porous aluminae and zirconia) or even porous carbon. Porous supports are generally available as tubes, flat plates, hollow fibres, honeycomb monoliths, porous spheres, although other configurations are possible. The supporting material must be physically and chemically compatible with the membrane and meet the demands of the application. In the case of high temperature membrane reactors, for example, the membranes need to survive temperatures higher than 600°C.

There are two major problems in commercializing ceramic membranes for separation processes. One is the need to coat the sol on the porous support and then treat the system in such a way that the resulting coating does not crack. The other problem is the need to optimize the geometry of the system so that the final module has a high surface area to volume ratio. This property affects the efficiency of the separation process, but tubes and flat plates have low surface area to volume ratios. It is expected that the next generation of ceramic membranes will employ geometric configurations that tend to optimize separations, through-put, energy requirements, etc.

These problems are of less concern when using non-porous supports. In our applications we utilize glass supports, such as tubes, plates or spheres. These are usually acid and base cleaned, fired and sometimes chemically etched before coating with the membrane. We have also used metallic and semi-conductor supports for particular applications.

3. Coating the Sol

Slipcasting, dipping, spinning and spraying the support with the particulate suspension have all been utilized to coat the supports. In our experience, all of these processes have proven to be quite useful according to the nature of the support and desired product.

4. Gellation and Drying

Particulate gels can be defined as an array, or matrix, of particles which encapsulate the solvent and have a semi-solid character. In the case of oxide particles, these gels can be produced by three mechanisms: adding electrolyte to the suspension to lower the repulsive forces between particles, changing the pH of the suspension so as to reduce the charge on the surface of the particles and increasing the concentration of particles through evaporation or centrifugation in order to increase the number of particle-particle collisions.

Control of the drying of the supported membrane is crucial to fabricating a crack-free membrane. There are three main methods for drying: slow evaporation of the solvent, addition of drying control aids and critical point extraction. The first method is common in the areas of ceramic films and membranes, while the second and third methods are more commonly employed to fabricate ceramic and glass monoliths.

5. Firing

Ceramic membranes do not have sufficient structural strength when dried to be used in practical applications. Thus, these membranes are sintered to enable physical interactions to occur between the particles in the membrane and to allow the support and the membrane to become a coherent mass. As these materials are heated to higher temperatures there is a concomitant drop in surface area and a loss of membrane porosity. One of the major problems facing ceramic membrane research is how to preserve membrane porosity at higher temperatures.

In particular, we are attempting to preserve microstructural porosity at high firing temperatures (up to 1000°C) for membranes fabricated from nano-sized particles (ca. 10Å diameter) in order to develop membranes for gas-phase separations and reverse osmosis. This approach is in contrast to that of other ceramics researchers who are using nano-phase particles in order to sinter materials to theoretical densities at relatively low temperatures. We wish to do the reverse and preserve porosity in nano-sized systems at elevated temperatures. Our approach has been to molecularly dope our oxides with other elements.

In the final analysis, the controlled hydrolysis and condensation of organic alkoxides yields a final membrane which is thermally stable to temperatures in excess of 400°C, with pore diameters as low as 10Å. In addition, many of these membranes have specific surface areas in excess of 300 m^2/g and extremely narrow pore size distributions. These attributes have largely been achieved by producing very small particles of uniform size that are packed into a random close-packed array and sintered onto a support through firing.

Applications

The word *"membrane"* stems from the Latin root *"membrana"*, which means skin or parchment. Traditionally, ceramic membranes have been used as ultrafilters to remove solutes from organic and aqueous process streams. In the near future, they will be employed in high pressure reverse osmosis processes. Electrodialysis could also be performed by applying a potential to these membranes. As these membranes are refined, they will see more applications in gas separations and in ceramic membrane reactors which couple gas separation with catalytic reactors. The area of catalytic reactors is one of the most hotly pursued applications for these materials. Other areas of application include catalysts, photocatalysts, sensors, wave guides, batteries, fuel cells, optical windows, new lasing materials and protective coatings, etc.

"Proof of concept" tests for many of these technological applications have already been performed in our laboratory. We will discuss the more traditional separations and move forward to more advanced concepts, which require a longer technological lead time before commercial scale applications can be brought to fruition.

1. Ultrafiltration

The first membrane prepared in this laboratory was an Al_2O_3 membrane, which was supported on a slip-cast porous clay support. This membrane was coated on the solvent impregnated support by withdrawing the support from a suspension of Al_2O_3 particles. When fired at 400°C, this membrane had a typical pore diameter of 40Å, although the individual alumina particles had diameters on the order of 10-20nm.

Currently, particles with diameters of 2-5nm are being produced out of a variety of oxides and mixed oxides. Membranes produced from these particles have theoretical pore diameters (assuming random close packing) as low as 5Å. To date, we have produced membranes with pore diameters as low as 8Å.

2. Gas Separation by Ceramic Membranes

In order to penetrate the field of gas separations, ceramic membranes must overcome such established technologies as:

Polymeric membranes for separating gases and

Other separation techniques such as pressure swing adsorption or cryogenics.

Although other separation methods have long been used and improved, all of them have their deficiencies. Some are inadequate for large-scale processes and some cannot provide products of high quality, while advanced membranes can make the separation process continual, simple, more effective and selective.

Membranes can be divided into two groups: organic and inorganic membranes. Although development of both types began approximately forty years ago, polymeric membranes have enjoyed much greater commercial success. In the past most investigations of gas separation processes have utilized polymeric membranes and the first polymeric membrane for gas separation was commercialized ten years ago. This field is still rapidly expanding and separation studies of many gas systems such as CO_2/CH_4, O_2/N_2, H_2/CO_2, SO_2/air and He/other gases have been reported with high selectivity and permeability.

Until recently, the application of ceramic membranes for gas separation has been in the laboratory stage, except for the success in uranium isotope enrichment. However, a growing interest in developing ceramic membranes for gas separation has occurred for two reasons.

1. Some high temperature reactors in chemical industry demand both thermal and chemical stability of the membrane while exposed to organic solvents. These factors limit the use of polymeric membranes.
2. Ceramic membranes have been greatly improved as a result of recent development efforts.

The transport mechanisms of gases through porous ceramic membranes have been extensively studied. These include mainly Knudsen diffusion, molecular sieving, surface diffusion and capillary condensation. However, there are still problems with the application of ceramic membranes for gas separation. Commercial utilization of ceramic membranes will depend upon how their separation efficiencies, selectivities, permeabilities and costs compare with those of polymeric membranes and other separation techniques.

The characteristics of ceramic membranes that affect their ability to perform gas phase separations are their pore size, porosity, thickness and surface properties.

Pore Size

In order to achieve high separation factors pore sizes of ceramic membranes must be as small as possible. Sol-gel techniques have been shown to produce ceramic membranes with small pore sizes and narrow pore size distributions by controlling the chemistry during the process of particle formation and packing. Both $\rho\gamma$-Al_2O_3 and silica membranes manufactured by sol-gel techniques at Twente University of Technology in the Netherlands were characterized by pore diameters of 25Å. We have produced membranes with pore diameters of less than 15Å of SiO_2, TiO_2, ZrO_2 and TiO_2/ZrO_2, but some of these membranes still crack. Newly developed membranes composed of even smaller particles (<30Å) have pore radii less than 5Å.

Porosity

In addition to small pore sizes, ceramic membranes need high porosity if they are to be highly permeable. Sol-gel processing enables us to produce materials that possess good microstructures by allowing us to control the particle size. In the worst case scenario, particulate membranes have a porosity of approximately 30% if they are not fired at too high a temperature. Membranes of TiO_2 and ZrO_2 with pore radii of less than 15Å and porosities up to 55% have been manufactured.

Thickness

Thickness is an important factor because the gas flow through the porous material is inversely proportional to the thickness of the membrane. We usually fabricate the membrane by dip coating or slipcasting a sol onto a porous substrate followed by thermal treatment. The factors which affect the thickness of such a coating are the rate of withdrawal of the substrate from the sol, the viscosity of the suspension, the morphology of the substrate and for slipcasting, the time of contact between the sol and the support. Organic binders (eg. polyvinyl, alcohol or hydroxl cellulose) can be added to the suspension in order to enhance strength and avoid cracking. Typically, the thickness of the fired coating is between 0.1 and this behaviour. In addition, separation membranes composed of the catalytic material itself and characterized by the presence of nanopores could also be expected to have different properties than both conventional porous materials and catalysts supported on pellets. This area of research has been quite promising.

Catalysis

The recent development in our laboratories of alumino-silicate membranes, stable at high temperatures, makes possible the use of ceramic membrane technology in the catalytic processing of petroleum-based feedstocks. For example, silica-alumina catalysts prepared using techniques developed for the preparation of ceramic membranes can have small pores and narrow pore size distributions. Hence these catalysts can potentially have properties that combine the essential features of zeolites and amorphous alumino-silicates. Such catalysts, when used in catalytic cracking units, can reduce the production of aromatics and enhance the production of olefins that can be used not only in the petrochemical industry, but also for the production of the alkylates and oxygenates used in reformulated gasolines that will be required by recently enacted environmental regulations (Clean Air Act of 1990).

Efforts have been made to produce new silica-alumina catalysts that not only have the desired selectivity for olefin production, but also reduce coke formation and possess increased activity. Controlling the pore size and pore size distribution of the catalyst may be important in attaining this goal. Several efforts that achieved similiar results were made in the 1970's. The major problem of these efforts was their failure to produce catalysts with narrow pore size distributions. However, increased activity, modified selectivity and even shape selectivity, compared to conventional amorphous silica-alumina catalysts, were reported. Other materials, e.g.. pillared clays, have also attracted interest, particularly for cracking bulky hydrocarbon molecules that can fit in their pores, which are larger than those of the zeolites. Unfortunately pore hydrothermal stability and excessive coking problems severely constrain the usefulness of these materials for catalytic cracking operations.

An important feature of ceramic membrane technology involves the catalytic properties of the novel materials that can be produced. For example, we have produced silica-alumina membranes composed of nano-particles. Thus, they have a narrow pore size distribution, with a mean value in the same range as the characteristic dimensions of the openings in zeolites. Our nano-particulate membranes present the possibility of producing a catalyst which would combine the essential features of zeolites with those of conventional silica-aluminas. Such a catalyst would combine the limited hydrogen transfer ability of amorphous silica-alumina with the limited over-cracking and low coking properties of zeolites. It would appear to be possible to utilize these materials to produce an olefin-rich product while utilizing current FCC units. In such applications, our catalyst most likely would be supported on conventional silica-alumina, but other supports will also be considered. Because these catalysts can be produced with different pore sizes, another advantage of our approach would be a capability of preparing catalysts which have pore sizes suitable for processing heavier hydrocarbon feedstocks. Such feedstocks cannot be easily processed by zeolites, because these feedstocks contain molecules too large to fit in the zeolitic cavities.

1.0 microns.

Surface Properties

Surface diffusion has been studied for several gas systems. In some systems, such as O_2/N_2 separations, Knudsen flow only gives a small separation factor (1.07 for O_2/N_2) due to the similarity in molecular weights of the two gases being separated. Surface diffusion of zeolite deposited on an alumina membrane, however, can enhance this separation factor to 3.3. The adsorption and migration of oxygen molecules on the membrane wall makes the separation of these two gases possible. In general, surface diffusion occurs when one of the gas components has a special affinity with the membrane surface. That is why the properties of the membrane surface affects the efficiency of the separation.

Various thin-film deposition techniques have been used to minimize the pore size and to modify the surface properties of the substrate. Heavy metals are frequently used as surface components because their adsorption and catalysis properties have long been studied. Another common method of modifying the ceramic membrane surface is to adsorb the metal on the surface of the porous membrane.

Ceramic Membrane Reactors for Catalysis

Three kinds of membranes have attracted most interest for use in catalytic processes: noble metal membranes, ceramic membranes and ion conductive membranes. Perm selective ceramic membranes have only recently been studied for high temperature catalytic reactions because only recently have technological advances made them a reasonable and reliable alternative. Noble metal membranes, however, have been studied since the sixties as the phenomenon on which they are based, hydrogen diffusion through the metal, has been known for almost a century.

The properties a membrane must have in order to be useful for catalytic reactions are:

1. stability at high temperatures
2. reasonable permeability and selectivity
3. high rate of heat transfer from or to the reaction zone.

Given those properties, a membrane catalytic reactor can allow integration of reaction and separation processes, increased selectivity and enhanced production of thermodynamically limited reactions. All of these factors eventually lead to lower costs and lower energy consumption.

To date, research efforts have focused on constructing new reactors to exploit the advantages offered by catalytic membranes, choosing the most appropriate catalytic membranes and optimizing other design and operational parameters to achieve the best possible performance. Little work has been done to explore the unique catalytic properties that these ceramic membrane materials may possess. In most cases, the membranes used in these reactors consist of commercially available separation membranes (usually alumina membranes with pore diameters of ca. 40Å) doped with a metal in order to induce catalytic activity. Many times, the reactor itself is a packed bed reactor in which the membrane acts only to separate some of the products from the reaction mixture. Although it is reasonable to expect that porous supports with pore sizes on the order of a micron will have similiar surface properties as conventional catalyst supports, membranes with much smaller pores should deviate from

Photocatalysis

Many inorganic membranes are also semiconductors (e.g., TiO_2,ZnO, CdS). When these particulate membranes are irradiated with light energy, an electron-hole pair is produced if the energy of irradiation is greater than the band gap energy of the semi-conductor. If these electrons and holes can be captured by surface adsorbed species before they have a chance to recombine, they can be utilized to perform reductive and oxidative chemical reactions respectively. Suspended particles of TiO_2 have been shown to photocatalyze the decomposition of many organic compounds such as 3,4-dichlorobiphenyl. Several researchers have shown that many of these organic compounds can be mineralized or degraded to carbon dioxide, water and inorganic compounds on particulate semiconductor surfaces with no other byproducts formed.

We have utilized a photocatalytic reactor which incorporates thin films of TiO_2 to study the rate of photodegradation of 3-chlorosalicyclic acid when illuminated with a UV light source. Our research has also demonstrated that this same system is capable of reducing Cr(VI) to Cr(III), thus showing that either the electron or the hole can be utilized in this process. We have shown these same materials to be even more effective in the destruction of volatile organic compounds such as trichloroethylene, our nations number one groundwater contaminant.

The prime advantage of ceramic membrane photocatalysts over particulate systems is that the membranes do not suffer from a particulate attrition problem and do not require a separation system. These membranes may also be electrically biased, which slows down the recombination reaction and provides a higher efficiency. It is possible with this system that ions such as silver might be salvaged from waste streams by photoplating onto a counterelectrode while organic materials are degraded by the TiO_2 membrane.

Sensors

Ceramic membranes could provide new materials for chemical and biochemical sensing devices because some of these membranes combine high surface areas with the ability to transmit optical and/or electrical signals, these materials could be used either alone or in host configurations to detect various analytes. Another approach is to use these materials as hosts for enzymes that are porous specific for given analytes. Optical signals resulting from these enzyme-analyte interactions could be observedby various techniques (e.g., fluorescence spectroscopy). Another approach would utilize inorganic materials that fluoresce directly to act as gas sensors. We have examined the fluorescence of ZnO membranes which was shown to be related to the amount of water bound to the surface.

Batteries

The most obvious use for nanoporous ceramic membranes in battery devices is as a separator which is placed between the cathode and anode to prevent electrical short-circuits between the electrodes, while offering minimal resistance to the transport of ionic charge carriers. In addition, the separator must, in some cases selectively inhibit the transport of certain chemical species and also allow for gas transport. Currently, organic membranes such as polypropylene or cellophane serve this function. A porous inorganic membrane such as those discussed here can function in the same manner. These membranes are mechanically robust and will withstand high operating temperatures. They can also be impregnated with dissolved ionic charge carriers to establish electrical conductivity between the electrodes. Ceramic membranes are ideal candidates for this application as they are typically less than 1 micron in thickness.

An additional bonus in battery research is that both electrodes, as well as the separator, can be fabricated using the sol-gel membrane technology described herein. To illustrate this principle, we are constructing a MnO_2-ZrO_2-Zn alkaline battery. In this battery, an MnO_2 coating is spun onto a thin supporting layer of Zn foil. This membrane covered foil is fired in order to sinter the MnO_2 to the supporting metal. A membrane layer of ZrO_2 is next spun onto the MnO_2 and the unit is fired again. Finally, a colloidal Zn layer is spun on the surface of the ZrO_2 and fired under reducing conditions. This process is repeated many times until the final layer of Zn is repeated. The final step is to dip the device into a solution of concentrated KOH which serves as the electrolyte. The capillarity of this membrane causes the KOH to be pulled into each of the ZrO_2 membrane layers. Thus, this membrane device is composed of many cells that possess a total thickness of approximately 2-3 microns and operates at a thermodynamic efficiency of 1.4 volts per cell. This type of technology makes the production of more powerful, smaller and lighter batteries feasible. Costs can be lowered for certain types of batteries and in addition, new forms of batteries can be produced that are incorporated as integral parts of the structures or devices that they power.

Although the results obtained to date are promising, further funding of this research is necessary to continue its development. Such funding is particularly important as, beginning in 1998, the California Air Resources Board is requiring the sale of "zero-emission vehicles" in the state of California. This law has been welcomed by the business and political establishment as they view it as an excellent economic development opportunity for the state. A UCLA study has predicted that the electric car industry could provide an additional 24,000 manufacturing jobs, a 5% increase for that region. Although these numbers are not huge, this philosophy has also been adopted by New York and Massachussetts where electric car mandates have already been passed. Twelve northeastern states, plus the District of Columbia, have petitioned the U.S. EPA to impose those same emission standards on their entire region. If this were to occur then approximately half the population of the U.S. would be brought into this market. Interestingly, as of June 1994, an initiative passed in Sacramento County, California that requires homebuilders to install electric vehicle charging units in all new garages (at a cost of less than $40). Sacramento County is the first metropolitan area in the U.S. to encourage the use of zero emmission vehicles through a building ordinance. Thus, the need for advanced battery concepts is great, as is the potential market.

Wave Guides

Many of our membranes contain particles with diameters less than 50Å that do not scatter visible light. When placed on a non-porous smooth surface, these membranes act like dense thin films in that they could precisely and perfectly guide a laser beam. This capacity enables ceramic membranes to be used in thin film sensing devices, such as those described above, where a guided wave in a porous medium may be required. Ceramic membranes may well serve in integrated systems, also, where they can be incorporated in IC devices much as photoresists are spun onto silicon wafers.

Summary

We believe that inorganic membranes are materials that will find a large array of applications in the near future, including some that we have not considered. In some cases they will replace organic membranes in classical separation processes. In other cases. they will bring membrane separations to areas not currently being served by organic membranes due to particular environmental conditions. Finally, ceramic membranes will, in the future, be utilized in a wide variety of systems which are non-classical and are only now just beginning to be developed in research laboratories.

Author Index

Subject Index